NEURAL NETWORKS
AND
FUZZY SYSTEMS

NEURAL NETWORKS
AND
FUZZY SYSTEMS

A DYNAMICAL SYSTEMS APPROACH
TO MACHINE INTELLIGENCE

Bart Kosko
University of Southern California

PRENTICE HALL, Englewood Cliffs, NJ 07632

Library of Congress Cataloging-in-Publication Data

Kosko, Bart.
  Neural networks and fuzzy systems : a dynamical systems approach
to machine intelligence / Bart Kosko.
    p.   cm.
  Includes bibliographical references and index.
  Contents: v. 1. Neural network theory
  ISBN 0-13-611435-0
  1. Expert systems (Computer science) 2. Artificial intelligence.
3. Neural networks (Computer science) 4. Fuzzy systems. I. Title.
QA76.76.E95K67 1992
006.3--dc20   90-28790
                                                  CIP

Editorial/production supervision
    and interior design: Joe Scordato
Jacket design: Wanda Lubelska
Manufacturing buyers: Linda Behrens and Dave Dickey
Acquisitions editor: Pete Janzow

© 1992 by Prentice-Hall, Inc.
A Simon & Schuster Company
Englewood Cliffs, New Jersey 07632

IBM is a registered trademark of International Business Machines Corporation.

Printed in the United States of America

10 9 8 7

ISBN 0-13-611435-0

Prentice-Hall International (UK) Limited, *London*
Prentice-Hall of Australia Pty. Limited, *Sydney*
Prentice-Hall Canada Inc., *Toronto*
Prentice-Hall Hispanoamericana, S.A., *Mexico*
Prentice-Hall of India Private Limited, *New Delhi*
Prentice-Hall of Japan, Inc., *Tokyo*
Simon & Schuster Asia Pte. Ltd., *Singapore*
Editora Prentice-Hall do Brasil, Ltda., *Rio de Janiero*

ISBN 0-13-611435-0

9 780136 114352

90000>

For my daughter, Victoria Kosko

CONTENTS

FOREWORD *by Lotfi A. Zadeh*   *xvii*

FOREWORD *by James A. Anderson*   *xix*

PREFACE   *xxv*

1   NEURAL NETWORKS AND FUZZY SYSTEMS   *1*

Neural and Fuzzy Machine Intelligence   2

     Neural Pre-Attentive and Attentive Processing, 2

Fuzziness as Multivalence   3

     Bivalent Paradoxes as Fuzzy Midpoints, 4
     Fuzziness in the Twentieth Century, 5
     Sets as Points in Cubes, 7
     Subsethood and Probability, 9

The Dynamical-Systems Approach to Machine Intelligence:
  The Brain as a Dynamical System   12

  Neural and Fuzzy Systems as Function Estimators, 13
  Neural Networks as Trainable Dynamical Systems, 14
  Fuzzy Systems and Applications, 18

Intelligent Behavior as Adaptive Model-Free Estimation   19

  Generalization and Creativity, 20
  Learning as Change, 22
  Symbols vs. Numbers: Rules vs. Principles, 24
  Expert-System Knowledge as Rule Trees, 24
  Symbolic vs. Numeric Processing, 25
  Fuzzy Systems as Structured Numerical Estimators, 26
  Generating Fuzzy Rules with Product-Space Clustering, 28
  Fuzzy Systems as Parallel Associators, 29
  Fuzzy Systems as Principle-Based Systems, 32

References   34

Problems   36

Part 1   Neural Network Theory   38

  2   *NEURONAL DYNAMICS I: ACTIVATIONS AND SIGNALS*   39

Neurons as Functions   39

Signal Monotonicity   40

  Signal and Activation Velocities, 41

Biological Activations and Signals   41

  Competitive Neuronal Signals, 43

Neuron Fields   44

Neuronal Dynamical Systems   44

  Neuronal State Spaces, 45
  Signal State Spaces as Hypercubes, 46
  Neuronal Activations as Short-Term Memory, 47

Common Signal Functions   48

Pulse-Coded Signal Functions   50

  Velocity-Difference Property of Pulse-Coded Signals, 51

References   52

Problems   53

3   NEURONAL DYNAMICS II: ACTIVATION MODELS   **55**

Neuronal Dynamical Systems   55

Additive Neuronal Dynamics   56

   Passive Membrane Decay, 56
   Membrane Time Constants, 57
   Membrane Resting Potentials, 57
   Additive External Input, 58

Additive Neuronal Feedback   59

   Synaptic Connection Matrices, 59
   Bidirectional and Unidirectional Connection Topologies,
      60

Additive Activation Models   61

Additive Bivalent Models   63

   Bivalent Additive BAM, 63
   Bidirectional Stability, 68
   Lyapunov Functions, 69
   Bivalent BAM Theorem, 73

BAM Connection Matrices   79

   Optimal Linear Associative Memory Matrices, 81
   Autoassociative OLAM Filtering, 83
   BAM Correlation Encoding Example, 85
   Memory Capacity: Dimensionality Limits Capacity, 91
   The Hopfield Model, 92

Additive Dynamics and the Noise-Saturation Dilemma   94

   Grossberg's Saturation Theorem, 95

General Neuronal Activations: Cohen-Grossberg and
      Multiplicative Models   99

References   103

Problems   106

Software Problems   108

   Part I: Discrete Additive Bidirectional Associative
      Memory (BAM), 108
   Part II, 109

4 SYNAPTIC DYNAMICS I: UNSUPERVISED LEARNING   111

Learning as Encoding, Change, and Quantization   111

> Supervised and Unsupervised Learning in Neural
>   Networks, 113

Four Unsupervised Learning Laws   115

> Four Deterministic Unsupervised Learning Laws, 116
> Brownian Motion and White Noise, 118

Probability Spaces and Random Processes   119

> Measurability and Sigma-Algebras, 119
> Probability Measures and Density Functions, 122
> Gaussian White Noise as a Brownian Pseudoderivative
>   Process, 127

Stochastic Unsupervised Learning and Stochastic
    Equilibrium   131

> Stochastic Equilibrium, 133

Signal Hebbian Learning   138

> Recency Effects and Forgetting, 138
> Asymptotic Correlation Encoding, 138
> Hebbian Correlation Decoding, 140

Competitive Learning   145

> Competition as Indication, 146
> Competition as Correlation Detection, 147
> Asymptotic Centroid Estimation, 148
> Competitive Covariance Estimation, 149

Differential Hebbian Learning   152

> Fuzzy Cognitive Maps, 152
> Adaptive Causal Inference, 158
> Klopf's Drive Reinforcement Model, 159
> Concomitant Variation as Statistical Covariance, 161
> Pulse-Coded Differential Hebbian Learning, 163

Differential Competitive Learning   166

> Differential Competitive Learning as Delta Modulation,
>   168

References   170

Problems   173

Software Problems 175

Part I: Competitive Learning, 175
Part II: Differential Competitive Learning, 176

5 SYNAPTIC DYNAMICS II: SUPERVISED LEARNING 179

Supervised Function Estimation 180

Supervised Learning as Operant Conditioning 181

Supervised Learning as Stochastic Pattern Learning with
Known Class Memberships 183

Supervised Learning as Stochastic Approximation 185

The Perceptron: Learn Only If Misclassify, 187
The LMS Algorithm: Linear Stochastic Approximation,
190

The Backpropagation Algorithm 196

History of the Backpropagation Algorithm, 196
Feedforward Sigmoidal Representation Theorems, 199
Multilayer Feedforward Network Architectures, 201
Backpropagation Algorithm and Derivation, 203
Backpropagation as Stochastic Approximation, 210
Robust Backpropagation, 211
Other Supervised Learning Algorithms, 212

References 213

Problems 215

Software Problems 218

Part I: Exclusive-OR (XOR), 218
Part II: Sine Function, 219
Part III: Training Set versus Test Set, 220

6 ARCHITECTURES AND EQUILIBRIA 221

Neural Networks as Stochastic Gradient Systems 221

Global Equilibria: Convergence and Stability 223

Synaptic Convergence to Centroids: AVQ Algorithms 225

Competitive AVQ Stochastic Differential Equations, 225
Competitive AVQ Algorithms, 227
Unsupervised Competitive Learning (UCL), 227
Supervised Competitive Learning (SCL), 228

Differential Competitive Learning (DCL), 228
Stochastic Equilibrium and Convergence, 228

Global Stability of Feedback Neural Networks   232

ABAMs and the Stability-Convergence Dilemma, 233
Stability-Convergence Dilemma, 235
The ABAM Theorem, 236
Higher-Order ABAMs, 239
Adaptive Resonance ABAMs, 240
Differential Hebbian ABAMS, 241

Structural Stability of Unsupervised Learning   242

Random Adaptive Bidirectional Associative Memories   243

Noise-Saturation Dilemma and the RABAM
Noise-Suppression Theorem, 247
RABAM Noise-Suppression Theorem, 248
RABAM Annealing, 253

References   255

Problems   257

Software Problems   258

Part I: Random Adaptive Bidirectional Associative
Memory (RABAM), 258
Part II: Binary Adaptive Resonance Theory (ART-1), 259

Part 2   Adaptive Fuzzy Systems   **262**

7   *FUZZINESS VERSUS PROBABILITY*   **263**

Fuzzy Sets and Systems   263

Fuzziness in a Probabilistic World   264

Randomness vs. Ambiguity: Whether vs. How Much   265

The Universe as a Fuzzy Set   268

The Geometry of Fuzzy Sets: Sets as Points   269

Paradox at the Midpoint, 273
Counting with Fuzzy Sets, 274

The Fuzzy Entropy Theorem   275

The Subsethood Theorem   278

Bayesian Polemics, 289

The Entropy-Subsethood Theorem   293

References   294

Problems   296

8   FUZZY ASSOCIATIVE MEMORIES   **299**

Fuzzy Systems as Between-Cube Mappings   299

Fuzzy and Neural Function Estimators   302

  Neural vs. Fuzzy Representation of Structured Knowledge,
    304
  FAMs as Mappings, 306
  Fuzzy Vector-Matrix Multiplication: Max-Min
    Composition, 307

Fuzzy Hebb FAMs   308

  The Bidirectional FAM Theorem for Correlation-Minimum
    Encoding, 310
  Correlation-Product Encoding, 311
  Superimposing FAM Rules, 313
  Recalled Outputs and "Defuzzification", 314
  FAM System Architecture, 316
  Binary Input-Output FAMs: Inverted-Pendulum Example,
    317
  Multiantecedent FAM Rules: Decompositional Inference,
    322
  Adaptive Decompositional Inference, 326

Adaptive FAMs: Product-Space Clustering in FAM Cells
    327

  Adaptive FAM-Rule Generation, 328
  Adaptive BIOFAM Clustering, 329
  Adaptive BIOFAM Example: Inverted Pendulum, 333

References   335

Problems   336

Software Problems   337

**9   COMPARISON OF FUZZY AND NEURAL TRUCK
    BACKER-UPPER CONTROL SYSTEMS**   **339**

Fuzzy and Neural Control Systems   339

Backing up a Truck   340

  Fuzzy Truck Backer-Upper System, 340

Neural Truck Backer-Upper System, 345
Comparison of Fuzzy and Neural Systems, 346
Sensitivity Analysis, 347
Adaptive Fuzzy Truck Backer-Upper, 348
Fuzzy Truck-and-Trailer Controller, 352
BP Truck-and-Trailer Control Systems, 356
AFAM Truck-and-Trailer Control Systems, 356
Conclusion, 360

References   361

10   FUZZY IMAGE TRANSFORM CODING   **363**

Transform Image Coding with Adaptive Fuzzy Systems   363

Adaptive Cosine Transform Coding of Images, 365

Adaptive FAM systems for Transform Coding   366

Selection of Quantizing Fuzzy-Set Values, 367
Product-Space Clustering to Estimate FAM Rules, 368
Differential Competitive Learning, 370
Simulation, 373
Conclusion, 374

References   377

Problems   378

**11   COMPARISON OF FUZZY AND KALMAN-FILTER
       TARGET-TRACKING CONTROL SYSTEMS**   **379**

Fuzzy and Math-Model Controllers   379

Real-Time Target Tracking   381

Fuzzy Controller   382

Fuzzy-Centroid Computation, 386
Fuzzy-Controller Implementation, 390

Kalman-Filter Controller   392

Fuzzy and Kalman-Filter Control Surfaces, 394

Simulation Results   396

Sensitivity Analysis, 399
Adaptive FAM (AFAM), 402

Conclusion   406

References   406

APPENDIX: NEURAL AND FUZZY SOFTWARE INSTRUCTIONS

407

Neural Network Software Instructions: Using the OWL
  Demonstration Programs (IBM-PC/AT) 408

  General, 408
  ART, 409
  BAM, 411
  BKP, 413
  CL, 414
  RABAM, 416

Fuzzy-Associative-Memory Software Instructions 418

  Fuzzy Truck Backer-Upper Control System, 418
  Fuzzy Target-Tracking Demonstration, 419
  Adaptive Fuzzy Control of Inverted Pendulum, 421

INDEX

425

FOREWORD

Lotfi A. Zadeh

It is hard to overestimate the importance of this textbook. Bart Kosko has broken new ground with an outstanding work on a subject, adaptive fuzzy systems, certain to play an increasingly central role in our understanding of human cognition and our ability to build machines that simulate human decision making in uncertain and imprecise environments.

This is what artificial intelligence (AI) was supposed to do when it was conceived in the mid-1950s. Since then, traditional AI, based almost entirely on symbol manipulation and first-order logic, has attracted a great deal of attention, a large following, and massive financial support. The AI community can point with pride to its accomplishments in expert systems, game-playing systems, and, to a lesser extent, natural language processing. Yet many of us believe that traditional AI has not lived up to its expectations. AI has not come to grips with common sense reasoning. It has not contributed significantly to the solution of real-world problems in robotics, computer vision, speech recognition, and machine translation. And AI arguably has not led to a significantly better understanding of thought processes, concept formation, and pattern recognition.

I believe AI would have made much more progress toward its goals if it had not committed itself so exclusively to symbol manipulation and first-order logic. This commitment has made AI somewhat inhospitable to methods that involve numerical computations, including neural and fuzzy methods, and has severely limited its ability to deal with problems where we cannot benignly neglect uncertainty and imprecision. Most real-world problems fall into this category.

With this in view, we can better understand the growing popularity of numerical methods that deal with a wide range of real-world problems, problems AI has failed to solve if even address. Prominent among these numerical techniques are neural network theory and fuzzy theory. Separately and in combination, neural networks and fuzzy systems have helped solve a wide variety of problems ranging from process control and signal processing to fault diagnosis and system optimization. Professor Kosko's *Neural Networks and Fuzzy Systems*, along with its companion applications volume *Neural Networks for Signal Processing*, is the first book to present a comprehensive account of neural-network theory and fuzzy logic and how they combine to address these problems. Having contributed so importantly to both fields, Professor Kosko is uniquely qualified to write a book that presents a unified view of neural networks and fuzzy systems. This unified view is a direction certain to grow in importance in the years ahead.

Interpolation plays a central role in both neural network theory and fuzzy logic. Interpolation and learning from examples involve the construction of a model of a system from the knowledge of a collection of input-output pairs. In neural networks, researchers often assume a feedforward multilayer network as an approximation framework and modify it with, say, the backpropagation gradient-descent algorithm. In the case of fuzzy systems, we usually assume the input-output pairs have the structure of fuzzy if-then rules that relate linguistic or fuzzy variables whose values are words (fuzzy sets) instead of numbers. Linguistic variables facilitate interpolation by allowing an approximate match between the input and the antecedents of the rules. Generally, fuzzy systems work well when we can use experience or introspection to articulate the fuzzy if-then rules. When we cannot do this, we may need neural-network techniques to generate the rules. Here arise adaptive fuzzy systems.

One cannot be but greatly impressed by Professor Kosko's accomplishment as author of *Neural Networks and Fuzzy Systems*. This seminal work is a landmark contribution that will shape the development of neural networks and fuzzy systems for years to come.

Lotfi A. Zadeh
Department of Electrical Engineering
   and Computer Science
Computer Science Division
University of California at Berkeley

FOREWORD

James A. Anderson

We live in a world of marvelous complexity and variety, a world where events never repeat exactly. Heraclitus commented two and a half millennia ago that "We never step twice into the same river." But even though events are never exactly the same, they are also not completely different. There is a thread of continuity, similarity, and predictability that allows us to generalize, often correctly, from past experience to future events.

This textbook joins together two techniques—neural networks and fuzzy systems—that seem at first quite different but that share the common ability to work well in this natural environment. Although there are other important reasons for interest in them, from an engineering point of view much of the interest in neural networks and fuzzy systems has been for dealing with difficulties arising from uncertainty, imprecision, and noise. The more a problem resembles those encountered in the real world—and most interesting problems are these—the better the system must cope with these difficulties.

Neural networks, neurocomputing, or 'brainlike' computation is based on the wistful hope that we can reproduce at least some of the flexibility and power of the human brain by artificial means. Neural networks consist of many simple computing elements—generally simple nonlinear summing junctions—connected together by connections of varying strength, a gross abstraction of the brain, which consists of very large numbers of far more complex neurons connected together with far more complex and far more structured couplings.

Neural-network architectures cover a wide range. In one sense, every computer is a neural net, because we can view traditional digital logic as constructed from interconnected McCullough-Pitts 'neurons'. McCullough-Pitts neurons were proposed in 1943 as models of biological neurons and arranged in networks for the specific purpose of computing logic functions. The architectures of current neural networks are massively parallel and concerned with approximating input-output relationships, with very many units arranged in large parallel arrays and computing simultaneously. Massive parallelism is of great importance now that the speed of light begins to constrain computers of standard serial design. Large-scale parallelism provides a way, perhaps the only way, to significantly increase computer speed. Even limited insights and crude approximations into how the brain combines slow and noisy computing devices into powerful systems can offer considerable practical value.

As this textbook shows, even the simple networks we now work with can perform interesting and useful computations if we properly choose the problem. Indeed, the problems where artificial neural networks have the most promise are those with a real-world flavor: signal processing, speech recognition, visual perception, control and robotics.

Neural networks help solve these problems with natural mechanisms of generalization. To oversimplify, suppose we represent an object in a network as a pattern of activation of several units. If a unit or two responds incorrectly, the overall pattern stays pretty much the same, and the network still responds correctly to stimuli. Or, if an object, once seen, reappears, but with slight differences, then the pattern of activation representing the object closely resembles its previous appearance, and the network still tends to respond almost as it did before. When neural networks operate, similar inputs naturally produce similar outputs. Most real-world perceptual problems have this structure of input-output continuity.

If neural networks, supposedly brain-like, show intrinsic generalization, we might wonder if we observe such effects in human psychology. Consider the psychological problem of categorization. Why do we call a complex manufactured object found in a house, an object we have not seen before, a "chair" because it has a more-or-less flat part a couple of feet off the floor, has four legs, consists of wood, and so on? One approach to categorization—popular with computer scientists—makes a list of properties and matches the new object with the property list. If the new object matches a listed property, then we conclude that the object is an example of the category; otherwise, we conclude that it is not. One quickly discovers with this approach that it does not work in practice. Natural categories tend to be messy: Most birds fly, but some do not. Chairs can consist of wood, plastic, or metal and can have almost any number of legs, depending on the whims of the designer. It seems practically impossible to come up with a property list for any natural category that excludes all examples that are not in the category and includes all examples that are in the category.

The "prototype" model provides a model for human categorization with a good deal of psychological support. Instead of forming a property list for a category, we store a "best example" of the category (or possibly a few best examples). The system computes the similarity between a new example and the prototype and classifies the new example as an example of the category in the nearest-neighbor sense—if the new example is "close enough" to the prototype.

This computational strategy leads to some curious human psychology. For example, it seems that most people in the United States imagine a prototype bird that looks somewhat like a robin or a sparrow. (Of course, the prototype will depend, sometimes in predictable ways, on individual experience.) So Americans tend to judge ostriches or penguins as "bad" birds because these birds do not resemble the prototype bird, even though they are birds. "Badness" shows up in a number of ways: when people are asked to give a list of examples of "birds," prototypical birds tend to head the list; the response times to verify sentences such as "Penguins are birds" tend to be longer then to "Robins are birds;" and they put the prototypes into sentences as defaults in comprehension—the bird in "I saw a bird on the lawn" is probably not a turkey.

Neural networks naturally develop this kind of category structure. In fact, we can hardly stop neural networks from doing it, which points out a serious potential weakness of neural networks. Classification by similarity causes neural networks great distress in situations where we cannot trust similarity.

A famous example is "parity"—whether there are an even or an odd number of ones in a bit vector of ones and zeros. If we change only one element, then the parity changes. So nearest neighbors always have opposite parity. Parity causes no difficulties for digital logic. But it is so difficult for simple neural networks to compute the parity function that it, and related problems, caused the engineering community to lose interest in neural networks in the 1960's when computer scientists first pointed out this limitation. Yet such a pattern of computational strengths and weaknesses was exactly what excited the interest of psychologists and cognitive scientists at the same time. The problems that neural networks solved well and solved poorly were those where humans showed comparable strengths and weaknesses in their "cognitive computations." For this reason until quite recently most of the study of neural networks has been carried out by psychologists and cognitive scientists who sought models of human cognitive function.

Engineering techniques for dealing with uncertainty are sometimes as much statements about human psychology as they are about engineering. Neural networks deal with uncertainty as humans do, not by deliberate design, but as a byproduct of their parallel-distributed structure. It would be equally possible, and perhaps desirable, for us to directly build these insights about categorization into an artificial system. Fuzzy systems take this approach.

Fuzzy or multivalued set theory develops the basic insight that categories are not absolutely clear cut: a particular example of a category can "belong" to lesser or greater degree to that category. This assumption captures quite nicely the psychological observation that a particular object can be a better or worse example of chair, depending on other members of the category. For example, an appropriately shaped rock in the woods can be a "chair" even though it is a very bad example of the chair category. So we can consider the rock as weakly belonging to the chair category. When such insights are properly (and elegantly) quantified, as in this textbook, "fuzzy" systems can be just as well defined and useful as the more traditional formulations of statistics and probability.

The theory of probability arose historically from an attempt to quantify odds in gambling, particularly the statistics of thrown dice, and in the more respectable type of gambling called insurance. Assumptions that color the basic structure of probability theory and statistics may arise from trying to explain a system that was specifically designed by humans for a particular purpose. For example, the rules for wining and losing in a game should be clear and precise because money may change hands. Every possible game outcome must fall into a predetermined category. These categories are noise free: the sum of the dice equals six and not seven; a coin comes up heads or tails. Individuals in the vital statistics of a population are alive or dead, baptised or unbaptised.

Because games are precise by design, traditional probability theory assumes an accuracy and precision of categorization of the world that may not represent many important problems. We must wonder what would have happened if, instead of being concerned with gambling or insurance, probability theory had been initially developed to predict the weather, where there are continuous gradations between overlapping linguistic categories: dense fog, drizzle, light rain, heavy rain, and downpour. Perhaps fuzzy systems would have become the mainstream of uncertainty-reasoning formalisms and 'traditional' probability an extreme approximation useful in certain special cases. The reaction of most people when they first hear about fuzzy logic is the subjective feeling "Yes, this formulation makes sense psychologically."

Because general statements about both human psychology and the structure of the world embed so deeply in both neural networks and fuzzy systems, the introductory parts of this book contain several examples drawn from philosophy, biology, cognitive science, and even art and law. These examples and references are not there to show the author's erudition, but to illuminate and make explicit the basic assumptions made when building the models. Unfortunately, neuroscientists and engineers often lie in unconscious bondage to the ideas of dead philosophers and psychologists when they assume that the initial formulations and basic assumptions of their abstract systems are "obvious." Like social customs, these assumptions are obvious only if you grew up with them.

There are also significant differences between neural networks and fuzzy systems. There are formal similarities between them, as Professor Kosko points out, but they are also very different in detail. The noise and generalization abilities of neural networks grow organically out of the structure of the networks, their dynamics, and their data representation. Fuzzy systems start from highly formalized insights about the psychology of categorization and the structure of categories found in the real world. Therefore, the "theory of fuzziness" as developed is an abstract system that makes no further claims about biological or psychological plausibility. This abstract system may sometimes be easier to use and simpler to apply to a particular problem than neural networks may be. The reverse may also hold. Whether to use one or another technology depends on the particular application and on good engineering judgement.

Both neural networks and fuzzy systems break with the historical tradition, prominent in Western thought, that we can precisely and unambiguously characterize the world, divide it into categories, and then manipulate these descriptions according to precise and formal rules. Other traditions have a less positive approach to explicit, discrete categorization, one more in harmony with the ideas presented here. Huang Po, a Buddhist teacher of the ninth century, observed that "To make use of your minds to think conceptually is to leave the substance and attach yourself to form," and "from discrimination between this and that a host of demons blazes forth!"

James A. Anderson
Department of Cognitive and Linguistic Sciences
Brown University

PREFACE

Neural networks and fuzzy theory have been underground technologies for many years. They have had far more critics than supporters for most of their brief histories. Until recently most neural and fuzzy researchers published and presented papers in non-neural and non-fuzzy journals and conferences. This has prevented standardization and suggested that the two fields represent pantheons of ad hoc models and techniques.

This textbook presents neural networks and fuzzy theory from a unified engineering perspective. The basic theory uses only elementary calculus, linear algebra, and probability, as found in upper-division undergraduate curricula in engineering and science. Some applications use more advanced techniques from digital signal processing, random processes, and estimation and control theory. The text and homework problems introduce and develop these techniques.

Neural networks and fuzzy systems estimate functions from sample data. Statistical and artificial intelligence (AI) approaches also estimate functions. For each problem, statistical approaches require that we guess how outputs functionally depend on inputs. Neural and fuzzy systems do not require that we articulate such a mathematical model. They are model-free estimators.

We can also view AI expert systems as model-free estimators. They map conditions to actions. Experts do not articulate a mathematical transfer function from the condition space to the action space. But the AI framework is symbolic. Symbolic processing favors a propositional and predicate-calculus approach to machine intelligence. It does not favor numerical mathematical analysis or hardware implementation. In particular symbols do not have derivatives. Only sufficiently smooth functions have derivatives. Symbolic systems may change with time, but they are not properly dynamical systems, not systems of first-order difference or differential equations.

Neural and fuzzy systems are numerical model-free estimators, and dynamical systems. Numerical algorithms convert numerical inputs to numerical outputs. Neural theory embeds in the mathematical fields of dynamical systems, adaptive control, and statistics. Fuzzy theory overlaps with these fields and with probability, mathematical logic, and measure theory. Researchers and commercial firms have developed numerous neural and fuzzy integrated-circuit chips. High-speed modems, long-distance telephone calls, and some airport bomb detectors depend on adaptive neural algorithms. Fuzzy systems run subways, tune televisions and computer disc heads, focus and stabilize camcorders, adjust air conditioners and washing machines and vacuum sweepers, defrost refrigerators, schedule elevators and traffic lights, and control automobile motors, suspensions, and emergency braking systems. In these cases, and in general, we use neural networks and fuzzy systems to increase *machine IQ*.

The book contains two large sections. The first section develops neural network theory. It presents neural dynamical systems as stochastic gradient systems. Chapter 1 previews the main neural and fuzzy themes developed throughout the text, focusing on hybrid neural-fuzzy systems for estimation and control. Chapter 2 presents neurons as signal functions and establishes notation. Chapter 3 develops both simple and general models of how the neuron's membrane evolves in time. Chapter 4 focuses on unsupervised synaptic learning and reviews probability theory and random processes. Chapter 5 presents supervised synaptic learning as stochastic approximation. Chapter 6 combines the material in previous chapters, allows neurons and synapses to change simultaneously, and proves global equilibrium and stability theorems for feedforward and feedback neural networks.

The second section examines fuzziness and adaptive fuzzy systems. Chapter 7 presents the new geometric theory of fuzzy sets as points in unit hypercubes. Chapter 7, as its "Fuzziness versus Probability" title suggests, deals in part with the apparent conflict between probability theory and fuzzy theory. This unavoidably evokes a polemical flavor and may jolt the unsuspecting reader. The pungency of the underlying journal article, of the same title, has helped secure it a much wider audience than I had thought possible. After teaching and lecturing on fuzzy theory for several years, I have found that audiences always ask the questions Chapter 7 anticipates. For these reasons I have included it in this textbook in its fuzzy-versus-probability form. I develop formal probability theory in the first four chapters of the textbook. I view its relation to fuzzy theory as taxonomical not adversarial. Chapter 8 develops a new theory of fuzzy systems and shows how to combine neural and fuzzy systems to produce adaptive fuzzy systems. Chapter 9 compares neural and adaptive fuzzy systems for backing up a truck, and truck-and-trailer, to a loading dock in a parking lot. Chapter 10 applies the adaptive fuzzy methodology to signal processing and compares a fuzzy image transform-coding system with a popular algorithmic approach. Chapter 11 benchmarks an adaptive fuzzy system against a Kalman-filter controller for real-time target tracking. The Appendix discusses how to use the accompanying neural-network and fuzzy-system software, developed by HyperLogic Corporation, and Olmsted & Watkins, and Togai Infralogic.

For this textbook I have also edited for Prentice Hall a companion volume of neural applications, *Neural Networks for Signal Processing*. The edited volume extends and applies several of the neural-network models developed in this textbook. All chapters in the edited volume include detailed homework problems to assist in classroom instruction.

ACKNOWLEDGMENTS

This book took over a year to write and rewrite. Many colleagues assisted me in the process. I would like to thank my colleagues who patiently read and commented on drafts of chapters. In particular I would like to thank Seong-Gon Kong, Jerry Mendel, Peter Pacini, Rod Taber, Fred Watkins, and Lotfi Zadeh. Of course I assume all responsibility for chapter content. I would like to thank the staffs of HyperLogic Corporation, Olmsted & Watkins, and Togai InfraLogic for their tireless development of the fuzzy and neural software, software instructions, and software homework problems. I would like to thank Tim Bozik, Executive Editor at Prentice Hall, for his courage in instigating and supporting the evolving manuscript. Most of all I want to thank Delsa Tan for her persistence and equanimity as she processed every word and equation of the text in countably many iterations.

Bart Kosko
Los Angeles, California

NEURAL NETWORKS AND FUZZY SYSTEMS

From causes which appear similar, we expect similar effects. This is the sum total of all our experimental conclusions.

David Hume
An Inquiry Concerning Human Understanding

A learning machine is any device whose actions are influenced by past experiences.

Nils Nilsson
Learning Machines

Man is a species that invents its own responses. It is out of this unique ability to invent, to improvise, his responses that cultures are born.

Ashley Montagu
Culture and the Evolution of Man

NEURAL AND FUZZY MACHINE INTELLIGENCE

This book examines how adaptive systems respond to stimuli. Systems map inputs to outputs, stimuli to responses. Adaptation or learning describes how data changes the system, how sample data changes system parameters, how training changes behavior.

Neural Pre-Attentive and Attentive Processing

The human visual system behaves as an adaptive system. Consider how it responds to this stimulus pattern:

What do we see when we look at the Kanizsa [1976] square? We see a square with bright interior. We see illusory boundaries. Or do we? We *recognize* a bright square. Technically we do not see it, because it is not there.

The Kanizsa square exists in our brain, not "out there" in physical reality on the page. Out there only four symmetric ink patterns stain the page.

In the terminology of eighteenth-century philosopher Immanuel Kant [1783, 1787], the four ink stains are *noumena*, "things in themselves." Light photons bounce off the noumena and stimulate our surface receptors, retinal neurons in this case. The noumena-induced sensation produces the Kanizsa-square *phenomenon* or perception in our brain. There would be no Kanizsa squares in the spacetime continuum without brains or brainlike systems to perceive them.

Today we understand many of the neural mechanisms of perception that Kant could only guess at. The real-time interaction of millions of competing and cooperating neurons produces the Kanizsa-square illusion [Grossberg, 1987] and everything we "see."

We take for granted our high-speed, distributed, nonlinear, massively parallel **pre-attentive processing**. In our visual processing we pay no attention to how we segment images, enhance contrasts, or discount background luminosity. When we process sound, we pay no attention to how our cochleas filter out high-frequency signals [Mead, 1989] or how our auditory cortex breaks continuous speech into syllables and words, compensates for rhythmic changes in speech duration, and detects and often corrects errors in pronunciation, grammar, and meaning. We likewise ignore our real-time pre-attentive processing in the other sense modalities, in smell, taste, touch, and balance.

We experience these pre-attentive phenomena, but we ignore them and cannot control or completely explain them. Natural selection has ensured only that we perform them, ceaselessly and fast.

Attention precedes recognition. We recognize segmented image pieces and parsed speech units. An emergent "searchlight," perhaps grounded in thalamic neurons [Crick, 1984], seems to selectively focus attention in as few as 70 to 100 milliseconds. We look, see, pay attention, then recognize.

Neural network theory studies both pre-attentive and attentive processing of stimuli. This leaves unaddressed the higher cognitive functions involved in reasoning, decision making, planning, and control. The asynchronous, nonlinear neurons and synapses in our brain perform these functions under uncertainty. We reason with scant evidence, vague concepts, heuristic syllogisms, tentative facts, rules of thumb, principles shot through with exceptions, and an inarticulable pantheon of inexact intuitions, hunches, suspicions, beliefs, estimates, guesses, and the like.

Natural selection evolved this uncertain cognitive calculus. Our cultural conditioning helps refine it. A fashionable trend in the West has been to denigrate this uncertainty calculus as illogical, unscientific, and nonrigorous. We even call it "fuzzy reasoning" or "fuzzy thinking." Modern philosophers [Churchland, 1981] often denigrate the entire cognitive framework as "folk psychology."

Yet we continue to use our fuzzy calculus. With it we run our lives, families, careers, industries, hospitals, courts, armies, and governments. In all these fields we employ the products of exact science, but as tools and decision aids. The final control remains fuzzy.

FUZZINESS AS MULTIVALENCE

Fuzzy theory holds that all things are matters of degree. It mechanizes much of our "folk psychology." Fuzzy theory also reduces black-white logic and mathematics to special limiting cases of gray relationships. Along the way it violates black-white "laws of logic," in particular the law of noncontradiction *not-(A and not-A)* and the law of excluded middle *either A or not-A*, and yet resolves the paradoxes or antinomies [Kline, 1980] that these laws generate. Does the speaker tell the truth when he says he lies? Is set *A* a member of itself if *A* equals the set of all sets that are not members of themselves? Fuzziness also provides a fresh, and deterministic, interpretation of probability and randomness.

Mathematically fuzziness means multivaluedness or multivalence [Rosser, 1952; Rescher, 1969] and stems from the Heisenberg position-momentum uncertainty principle in quantum mechanics [Birkhoff, 1936]. Three-valued fuzziness corresponds to truth, falsehood, and indeterminacy, or to presence, absence, and ambiguity. Multivalued fuzziness corresponds to degrees of indeterminancy or ambiguity, partial occurrence of events or relations.

Bivalent Paradoxes as Fuzzy Midpoints

Consider the bivalent paradoxes again. A California bumpersticker reads TRUST ME. Suppose instead a bumpersticker reads DON'T TRUST ME. Should we trust the driver? If we do, then, as the bumpersticker instructs, we do not. But if we don't trust the driver, then, again in accord with the bumpersticker, we do trust the driver. The classical liar paradox has the same form. Does the liar from Crete lie when he says that all Cretans are liars? If he lies, he tells the truth. If he tells the truth, he lies. Russell's barber is a man in a town whose advertises his services with the logo "I shave all, and only, those men who don't shave themselves." Who shaves the barber? If he shaves himself, then according to his logo he does not. If he does not, then according to his logo he does. Consider the card that says on one side "The sentence on the other side is true," and says on the other side "The sentence on the other side is false."

The "paradoxes" have the same form. A statement S and its negation not-S have the same *truth-value* $t(S)$:

$$t(S) = t(\text{not-}S) \tag{1-1}$$

The two statements are both TRUE (1) or both FALSE (0). This violates the laws of noncontradiction and excluded middle. For bivalent truth tables remind us that negation reverses truth value:

$$t(\text{not-}S) = 1 - t(S) \tag{1-2}$$

So (1-1) reduces to

$$t(S) = 1 - t(S) \tag{1-3}$$

If S is true, if $t(S) = 1$, then $1 = 0$. $t(S) = 0$ also implies the contradiction $1 = 0$.

The fuzzy or multivalued interpretation accepts the logical relation (1-3) and, instead of insisting that $t(S) = 0$ or $t(S) = 1$, simply solves for $t(S)$ in (1-3):

$$2t(S) = 1 \tag{1-4}$$

or

$$t(S) = \frac{1}{2} \tag{1-5}$$

So the "paradoxes" reduce to literal half-truths. They represent in the extreme the uncertainty inherent in every empirical statement and in many mathematical statements. Geometrically, the fuzzy approach places the paradoxes at the midpoint of the one-dimensional unit hypercube $[0, 1]$. More general paradoxes reside at the midpoint of n-dimensional hypercubes, the unique point equidistant to all 2^n vertices.

Multivaluedness also resolves the classical *sorites* paradoxes. Consider a heap of sand. Is it still a heap if we remove one grain of sand? How about two grains? Three? If we argue bivalently by induction, we eventually remove all grains and still conclude that a heap remains, or that it has suddenly vanished. No single grain takes us from heap to nonheap. The same holds if we pluck out hairs from a nonbald scalp or remove 5%, 10%, or more of the molecules from a table or brain. We transition gradually, not abruptly, from a thing to its opposite. Physically we experience degrees of occurrence. In terms of statements about the physical processes, we arrive again at degrees of truth.

Suppose there are n grains of sand in the heap. Removing one grain leaves $n-1$ grains and a truth value $t(S_{n-1})$ of the statement S_{n-1} that the $n-1$ sand grains are a heap. The truth value $t(S_{n-1})$ obeys $t(S_{n-1}) < 1$ in general. $t(S_{n-1})$ may be close to unity, but we have some nonzero doubt d_{n-1} about the truth of the matter. (The argument still holds if there exist no doubting creatures in the universe.) For instance [Gaines, 1983],

$$t(S_n) \;=\; 1 - d_n \tag{1-6}$$

where $0 \leq d_n \leq d_{n-1} \leq \ldots \leq d_{n-m} \leq \ldots \leq 1$. So $t(S_{n-m})$ approaches zero as m increases to n. If we argue inductively, we can interpret the overall inference as the forward chain "(If S_n, then S_{n-1}) and (If S_{n-1}, then S_{n-2}) and ... and (If S_1, then S_0)." If we multiplicatively interpret the conjunction operator, then

$$t(S_n \longrightarrow S_{n-m}) \;=\; \prod_{k=0}^{m}(1 - d_{n-k}) \tag{1-7}$$

If we interpret the conjunction operator as the minimum operator, as discussed in the homework problems at the end of the chapter, then

$$t(S_n \longrightarrow S_{n-m}) \;=\; \min(1 - d_n, \ldots, 1 - d_{n-m}) \tag{1-8}$$

$$=\; 1 - \max(d_n, \ldots, d_{n-m}) \tag{1-9}$$

In both cases the implication truth value $t(S_n \longrightarrow S_0)$ equals zero (or some small number). We pay a truth-value fee for each application of *modus ponens*, of concluding B from A and $A \longrightarrow B$. The overall inference is vacuous. This reflects the everyday epistemological precept that the longer an explanation, the less we tend to trust it.

Fuzziness in the Twentieth Century

Logical paradoxes and the Heisenberg uncertainty principle led to the development of multivalued or "fuzzy" logic in the 1920s and 1930s. Quantum theorists allowed for indeterminacy by including a third or middle truth value in the bivalent logical framework. The next step allowed degrees of indeterminacy, viewing TRUE and FALSE as the two limiting cases of the spectrum of indeterminacy.

Polish logician Jan Lukasiewicz [Rescher, 1969] first formally developed a three-valued logical system in the early 1930s. Lukasiewicz extended the range of truth values from $\{0, 1/2, 1\}$ to all rational numbers in $[0, 1]$, and finally to all numbers in $[0, 1]$ itself. Logics that use the general truth function t: {Statements} $\longrightarrow$ $[0, 1]$ define continuous or "fuzzy" logics. Logicians refer to this system as $\mathbf{L}_1$. The exercises at the end of the chapter develop Lukasiewicz's fuzzy logic.

In the 1930s quantum philosopher Max Black [1937] applied continuous logic componentwise to sets or lists of elements or symbols. Historically, Black drew the first fuzzy-set membership functions. Black called the uncertainty of these structures *vagueness*. Anticipating Zadeh's fuzzy-set theory, each element in Black's multivalued sets and lists behaved as a statement in a continuous logic.

In 1965 systems scientist Lotfi Zadeh [1965] published the paper "Fuzzy Sets" that formally developed multivalued set theory, introduced the term *fuzzy* into the technical literature, and inaugurated a second wave of interest in multivalued mathematical structures, from systems to topologies. The recent emergence of fuzzy commercial products, as well as new theory, has generated a third wave of interest in multivalued systems.

Zadeh extended the bivalent **indicator function** I_A of nonfuzzy subset A of X,

$$I_A(x) \;=\; \begin{cases} 1 & \text{if } x \in A \\ 0 & \text{if } x \notin A \end{cases} \qquad (1\text{-}10)$$

to a multivalued indicator or **membership function** $m_A \colon X \longrightarrow [0, 1]$. This allows us to combine such multivalued or fuzzy sets with the pointwise operators of indicator functions:

$$I_{A \cap B}(x) \;=\; \min(I_A(x),\, I_B(x)) \qquad (1\text{-}11)$$

$$I_{A \cup B}(x) \;=\; \max(I_A(x),\, I_B(x)) \qquad (1\text{-}12)$$

$$I_{A^c}(x) \;=\; 1 - I_A(x) \qquad (1\text{-}13)$$

$$A \subset B \text{ iff } I_A(x) \leq I_B(x) \text{ for all } x \text{ in } X \qquad (1\text{-}14)$$

The membership value $m_A(x)$ measures the elementhood or degree to which element x belongs to set A:

$$m_A(x) \;=\; \text{Degree}(x \in A) \qquad (1\text{-}15)$$

Just as the individual indicator values $I_A(x)$ behave as statements in bivalent propositional calculus, membership values $m_A(x)$ correspond to statements in a continuous logic. If A defines a fuzzy subset of the real line, as in Figure 1.1 below, then in principle we can graph $m_A \colon R \longrightarrow [0, 1]$ in two dimensions. In practice indicator functions I_A graph as step functions or rectangular pulses on the real line.

Sets as Points in Cubes

Fuzziness prevents logical certainty at the level of black-white axioms. This seems unsettling to some [Quine, 1981] and liberating to others.

At the system level fuzziness allows us to build computer chips and systems that "intelligently" control subways, automobile systems, and numerous consumer electronic and other devices. At this level fuzzy processing may resemble neural processing.

Neural networks and fuzzy systems process inexact information and process it inexactly. Neural networks recognize ill-defined patterns without an explicit set of rules. Fuzzy systems estimate functions and control systems with partial descriptions of system behavior. Experts may provide this heuristic knowledge, or, as we illustrate in Chapters 8 through 11, neural networks may adaptively infer it from sample data.

Neural and fuzzy systems share a more formal mathematical property. They share the same state space. A set of n neurons defines a family of n-dimensional continuous or "fuzzy" sets. The neurons emit bounded signals.

The neuronal signals range from some minimum value to some maximum value, say from 0 to 1. At each instant the n-vector of neuronal outputs defines a fuzzy unit or fit vector. Each fit value indicates the degree to which the neuron or element belongs to the n-dimensional fuzzy set.

The neuronal state space, the set of all possible neural outputs, equals the set of all n-dimensional fit vectors, the fuzzy power set. Both equal the unit hypercube $I^n = [0, 1]^n = [0, 1] \times \cdots \times [0, 1]$, the set of all vectors of length n and with coordinates in the unit interval $[0, 1]$. Chapter 8 discusses fuzzy systems and associative memories, which map unit cubes to unit cubes, fuzzy sets to fuzzy sets. We shall use this recent geometric view of *sets as points* [Kosko, 1987, 1990] throughout this book.

The 2^n vertices of I^n represent extremized neuronal-output combinations, as we often find in networks of competitive or laterally inhibitive neurons. Many feedback neural networks [Hopfield, 1984] drive initial states inside the unit cube to nearest vertices. These systems dynamically disambiguate fuzzy input descriptions by minimizing their fuzzy entropy. The midpoint of the cube, where a fuzzy set A equals its own opposite A^c, has maximal fuzzy entropy, as we discuss in Chapter 7. The black-white vertices have minimal fuzzy entropy.

Proper fuzzy sets, nonvertex points, A violate the "laws" of noncontradiction and excluded middle: $A \cap A^c \neq \emptyset$ and $A \cup A^c \neq X$. In Chapter 7 we show that fuzzy entropy, the measure of fuzziness, balances the fuzzy count of the *overlap* $A \cap A^c$ and *underlap* $A \cup A^c$ in a simple ratio:

$$E(A) = \frac{M(A \cap A^c)}{M(A \cup A^c)}$$

There are 2^n bit vectors of length n. They define the vertices of I^n. So the vertices also represent the nonfuzzy power set of the n elements $x_1, \ldots, x_n$, the set

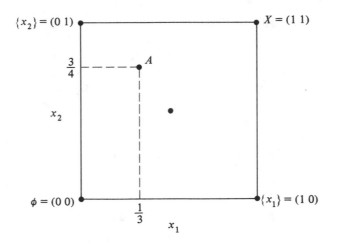

FIGURE 1.1   Fuzzy power set $F(2^X)$ of X corresponds to the unit square when $X = \{x_1, x_2\}$. The four nonfuzzy subsets in the nonfuzzy power set 2^X correspond to the four corners of the 2-cube. The fuzzy subset A corresponds to the fit vector $(1/3, 3/4)$ and to a point inside the 2-cube if $m_A(x_1) = 1/3$ and $m_A(x_2) = 3/4$. The midpoint **M** of the unit square corresponds to the maximally fuzzy set. Long diagonals connect nonfuzzy set complements. *Source:* Kosko, B., "Fuzziness vs. Probability," International Journal of General Systems, Volume 17, Numbers 2–3, Gordon and Breach Science Publishers, 1990.

of all nonfuzzy subsets of the n elements. The bit value 0 in the ith slot of a bit vector indicates the absence of element x_i in that subset. The bit value 1 indicates the presence of x_i in the subset. The bit vector $(1\ 0\ 1\ 0\ 0)$ indicates the subset $\{x_1, x_3\}$ of set $\{x_1, x_2, x_3, x_4, x_5\}$.

Fit values equal the membership values $m_A(x_i)$ discussed above. Fit values measure partial set membership or degrees of *elementhood*. The fit value 1/5 indicates that element x_i belongs only slightly to the fuzzy subset A. The fit value 1/2 indicates that x_i belongs to fuzzy set A as much as it does not—as much as it belongs to the complement fuzzy set A^c.

Consider the set X of two elements x_1 and x_2. The *power set* of X, denoted 2^X, contains the four subsets of X: $2^X = \{\emptyset, \{x_1\}, \{x_2\}, X\}$. These four nonfuzzy sets correspond to four bit vectors:

$$\emptyset \;=\; (0\ 0)$$

$$\{x_1\} \;=\; (1\ 0)$$

$$\{x_2\} \;=\; (0\ 1)$$

$$X \;=\; (1\ 1)$$

The fuzzy power set $F(2^X)$, which contains all continuum-many fuzzy subsets of X, corresponds to the unit square. Figure 1.1 displays the fuzzy power set $F(2^X)$.

Figure 1.1 represents the fuzzy subset A as a point inside the 2-dimensional unit hypercube. If A has membership degrees or fit values $m_A(x_1) = 1/3$ and $m_A(x_2) = 3/4$—so x_1 belongs to A less than x_2 does—then A corresponds to the fit vector $(1/3, 3/4)$.

The cube midpoint corresponds to the maximally fuzzy set $\mathbf{M}$. The midpoint set $\mathbf{M}$ uniquely obeys the peculiar relation $\mathbf{M} = \mathbf{M} \cap \mathbf{M}^c = \mathbf{M} \cup \mathbf{M}^c = \mathbf{M}^c$, and so maximally violates the bivalent laws of noncontradiction and excluded middle. The classical paradoxes of logic and set theory correspond to midpoint phenomena. Note that the cube midpoint in Figure 1.1 is uniquely equidistant to all 2^2 vertices. The cube midpoint behaves as the black hole of set theory.

Subsethood and Probability

Elementhood represents a special case of *subsethood*. Subsethood measures the degree to which set A belongs to set B, the degree to which A is a subset of B. We denote this subsethood measure as $S(A, B)$:

$$S(A, B) \;=\; \text{Degree}(A \subset B) \tag{1-16}$$

Subsethood provides a unified set-theoretic framework for fuzziness and probability. For instance, in the simplest case A equals the singleton set $\{x_i\}$. Then the subsethood of $\{x_i\}$ in B equals the membership or elementhood value $m_B(x_i)$:

$$S(\{x_i\}, B) \;=\; m_B(x_i) \tag{1-17}$$

Equation (1-17) follows directly from the subsethood theorem (1-22) below when we interpret $\{x_i\}$ as a bit vector with a 1 in the ith slot and 0s elsewhere.

Subsethood reveals the connection between fuzziness and randomness. Subsethood reduces probability to set theory. Randomness does not depend on the fuzziness or ambiguity of an event. It depends on the uncertainty between certain events. Randomness equals the uncertainty that arises when a nonfuzzy set B is partially contained in one of its own nonfuzzy subsets A. $S(A, B) = 1$, since A is a subset of B. But in general multivaluedness holds. The converse subsethood $S(B, A)$ is less than one but greater than zero:

$$0 \;<\; S(B, A) \;<\; 1 \tag{1-18}$$

Classical set theory implicitly forbids the strict inequalities in (1-18). The law of excluded middle dictates that every set either is or is not a subset of every other set. As a result, for centuries theorists have had to arbitrarily define probability as a frequency ratio or stipulate that it obeyed certain axioms. They could not *derive* probability from more fundamental concepts.

Fuzzy theory derives the axioms of the conditional probability measure $P(B \mid A)$,

$$P(B \mid A) \;=\; \frac{P(A \cap B)}{P(A)} \tag{1-19}$$

FUZZY SUBSETHOOD

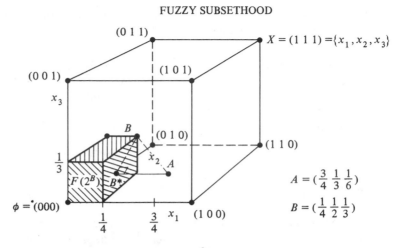

FIGURE 1.2 Subsethood theorem in R^3. X contains 3 elements, x_1, x_2, and x_3, and 8 nonfuzzy subsets. Fuzzy subset $B = (1/4, 1/2, 1/3)$ contains infinitely many fuzzy subsets B' such that $S(B', B) = 1$. They define the shaded hyper-rectangle. $S(A, B) < 1$, since A lies outside the hyperrectangle. The closer A is to the hyperrectangle, the larger the subsethood $S(A, B)$. B^* denotes the sub-set of B closest to A. B^* equals $A \cap B$ and uniquely defines an orthogonal or Pythagorean relationship between A and B.

the probability that B occurs given that A occurs, from the properties of the subset-hood measure $S(A, B)$. If X defines the "sample space" of all elementary outcomes of an experiment, then X is a "sure event," since $P(X) = 1$. Then (1-19) implies that every probability $P(A)$ equals the conditional probability $P(A \,|\, X)$:

$$P(A) \;=\; P(A \,|\, X) \tag{1-20}$$

This identity reflects the general subsethood relationship

$$P(A) \;=\; S(X, A) \tag{1-21}$$

On the surface the subsethood relation (1-21) seems absurd. How can superset X belong to one of its own subsets? How can the whole be part of one of its own parts? X cannot *totally* belong to A unless $X = A$. But X can *partially* belong to A. The subsethood theorem in Chapter 7 proves that this partial containment depends directly on the overlap between X and A, the intersection $X \cap A$. Figure 1.2 illustrates the Pythagorean geometry of the subsethood theorem in three dimensions. The shaded hyperrectangle defines $F(2^B)$, the fuzzy power set of B.

The subsethood theorem relates $S(A, B)$ to the magnitudes of A, B, and $A \cap B$:

$$S(A, B) \;=\; \frac{M(A \cap B)}{M(A)} \tag{1-22}$$

The ratio in (1-22) resembles, behaves as, and generalizes the defining ratio (1-19) of conditional probability. $M(A)$ denotes the fuzzy count of fit vector A:

$$M(A) \quad = \quad a_1 + \cdots + a_n \tag{1-23}$$

$M(A)$ generalizes the classical cardinality count, which sums only 1s and 0s. In the infinite case appropriate integrals replace summations. Equation (1-22) implies that the fuzzy entropy $E(A)$ of A equals the degree to which $A \cap A^c$ contains its own superset $A \cup A^c$: $E(A) = S(A \cup A^c, A \cap A^c)$.

In Figure 1.2, $A = (3/4, 1/3, 1/6)$ and $B = (1/4, 1/2, 1/3)$. Then the closest subset B^* to A that satisfies the total-subsethood condition

$$b_1^* \leq b_1, \ldots, b_n^* \leq b_n \tag{1-24}$$

corresponds to $B^* = (\frac{1}{4} \ \frac{1}{3} \ \frac{1}{6})$, which also equals the pairwise minimum of A and B. Equation (1-24) generalizes (1-14) above. As discussed in Chapter 7, the subsethood theorem ensures this in general:

$$B^* \quad = \quad A \cap B \tag{1-25}$$

Equation (1-23) implies that $M(A) = 15/12 = 5/4$, and $M(A \cap B) = 3/4$. Then the subsethood theorem gives $S(A, B) = (3/4)/(5/4) = 3/5 = 60\%$.

Relative frequency provides the clearest example of between-set fuzziness. Suppose we flip a coin, draw balls from urns, or shoot at a target. The elementary events in X are trials. Each trial is successful or unsuccessful. So X does not possess fuzzy subsets in its event space (its sigma-algebra). Each coin flip results in a head or a tail, not something in between. Suppose A defines the subset of successful trials. If X contains n trials, then A corresponds to a vertex of I^n and equals a bit vector of 1s and 0s. Suppose there are n_A successes out of n trials, where 1s indicate successes and 0s indicate failures. The event X equals total success, the bit vector of all 1s. X contains n successes. Then, since $A \cap X = A$, the subsethood theorem (1-22) gives

$$S(X, A) \quad = \quad \frac{n_A}{n} \tag{1-26}$$

Historically probability theorists have called the subsethood ratio in (1-26), or its limit, the "probability of success" or $P(A)$. This adds only a cultural tag. The success ratio n_A/n behaves no differently in its deterministic subsethood framework than it did in its "random" framework. The relative-frequency ratio still provides a stable estimate for probability values in our physical, engineering, economic, and gambling models. It still implies all the theorems it has always implied.

But we cannot derive the relative-frequency ratio from between-set relationships if we deny the strict inequality (1-18) and insist that subsethood is two-valued. Bivalence forces us to assume the ratio as a theoretical primitive.

Whether by design or by accident we have historically followed the bivalent path in mathematics for almost three thousand years. Bivalence has simplified our formal frameworks but at a cost. It has led to logical paradoxes (bivalent contradictions), unexplained primitives, and "randomness" in a universe that seems to obey physical laws and where events have causes.

THE DYNAMICAL-SYSTEMS APPROACH TO MACHINE INTELLIGENCE: THE BRAIN AS A DYNAMICAL SYSTEM

Several engineering and scientific disciplines study how adaptive systems respond to stimuli. Electrical engineers study the topic as signal processing, nonlinear filtering, coding theory, circuit design, and adaptive control. Computer scientists study it as algorithm and automata theory, computer design, robotics, and artificial intelligence. Mathematicians study it as function approximation, statistical estimation, combinatorial optimization, and dynamical systems. Philosophers study it as epistemology, causality, and action. Biologists study it as neuroscience, biophysics, ecology, evolution, and population biology. Psychologists study it as reinforcement learning, psychometrics, and cognitive science. Economists study it as utility maximization, game theory, econometrics, and market equilibrium theory. Cultural anthropologists study it as culture.

We shall emphasize electrical engineering as we seek general principles of how adaptive systems process information. We call these principles *machine-intelligence* principles. We shall draw freely from the related fields of engineering and science.

The term *artificial intelligence* usually refers to the computer-scientific approach to machine intelligence. This approach emphasizes symbolic processing and tree search. AI has become the emblem for a popular computer-age view of the brain: *brain = computer*. This view ranges from classical science-fiction speculation (the computer HAL in *2001: A Space Odyssey*) to proposed space-based weapons systems.

We shall explore machine intelligence from a *dynamical-systems* viewpoint: *brain = dynamical system*. On this view a maple leaf falling to a potential-energy minimum on the ground better describes brain activity than does a computer executing instructions. The dynamical models we shall study are cast as large systems of differential or difference equations. The principles describe local or global interactions of nonlinear parallel processes.

Some of these machine-intelligence principles and mechanisms may explain natural phenomena and processes. Some already extend our theoretical and mathematical knowledge. But ultimately they should help us build smarter machines. They should give rise to new computational devices–electrical, optical, molecular, plasma, fluid, or other devices.

In this sense machine intelligence becomes an engineering discipline. Nearly a half century ago Norbert Wiener [1948] outlined the first incarnation of such a machine-intelligence engineering. Wiener called it *cybernetics*.

We shall focus our analysis on artificial neural networks and fuzzy systems. These new, related systems represent broad classes of "machine-intelligent" adaptive systems. Chapters 2 through 6 describe neural network theory. Chapters 7 through 11 present a geometric theory of fuzzy sets and systems and its neural extension to adaptive fuzzy systems. The companion volume [Kosko, 1991] describes engineering applications of neural networks.

Neural and Fuzzy Systems as Function Estimators

Neural networks and fuzzy systems estimate input-output functions. Both are trainable dynamical systems. Sample data shapes and "programs" their time evolution. Unlike statistical estimators, they estimate a function without a mathematical model of how outputs depend on inputs. They are *model-free* estimators. They "learn from experience" with numerical and, sometimes, linguistic sample data.

Neural and fuzzy systems encode sampled information in a parallel-distributed framework. Both frameworks are numerical. We can prove theorems to describe their behavior and limitations. We can implement neural and fuzzy systems in digital or analog VLSI circuitry or in optical-computing media, in spatial-light modulators and holograms.

Artificial neural networks consist of numerous, simple processing units or "neurons" that we can globally program for computation. We can program or train neural networks to store, recognize, and associatively retrieve patterns or database entries; to solve combinatorial optimization problems; to filter noise from measurement data; to control ill-defined problems; in summary, to estimate sampled functions when we do not know the form of the functions.

The human brain contains roughly 10^{11} or 100 billion neurons [Thompson, 1985]. That number approximates the number of stars in the Milky Way Galaxy, and the number of galaxies in the known universe. As many as 10^4 synaptic junctions may abut a single neuron. That gives roughly 10^{15} or 1 quadrillion synapses in the human brain. The brain represents an asynchronous, nonlinear, massively parallel, feedback dynamical system of cosmological proportions.

Artificial neural systems may contain millions of nonlinear neurons and interconnecting synapses. Future artificial neural systems may contain billions of real or virtual model neurons. In general no teacher supervises, stabilizes, or synchronizes these large-scale nonlinear systems.

Many feedback neural networks can learn new patterns and recall old patterns simultaneously, and ceaselessly. Supervised neural networks can learn far more input-output pairs, or stimulus-response associations, than the number of neurons and synapses in the network architecture. Since neural networks do not use a mathematical model of how a system's output depends on its input—since they behave as model-free estimators—we can apply the same neural-network architecture, and dynamics, to a wide variety of problems.

Like brains, neural networks recognize patterns we cannot even define. We call this property *recognition without definition*. Who can define a tree, a pillow, or their own face to the satisfaction of a computer pattern-recognition system? These and most concepts we learn *ostensively*, by pointing out examples. We do not learn them as we learn the definition of a circle. We abstract these concepts from sample data, just as a child abstracts the color red from observed red apples, red wagons, and other red things, or as Plato abstracted triangularity from considered sample triangles.

Recognition without definition characterizes much intelligent behavior. It enables systems to generalize. Dogs, lizards, and slugs recognize multitudes of unforeseen, complex patterns without, of course, any ability to define them. Descriptive natural languages developed only yesterday in human evolution. Yet a great deal of modern philosophy, influenced by formal logic and behaviorist psychology, has insisted on concept definition preceding recognition or even discussion. Below we discuss how this insistence has helped shape the field of artificial intelligence and its emblem, the expert system.

Neural networks store pattern or function information with distributed encoding. They superimpose pattern information on the same associative-memory medium—on the many synaptic connections between neurons. Distributed encoding enables neural networks to complete partial patterns and "clean up" noisy patterns. So it helps neural networks estimate continuous functions.

Distributed encoding endows neural networks with fault tolerance and "graceful degradation." If we successively rip out handfuls of synaptic connections from a neural network, the network tends to smoothly degrade in performance, not abruptly fail. Computers and digital VLSI chips do not gracefully degrade when their components fail. Natural selection seems to have favored distributed encoding in brains, at least in sections of brains.

Neural networks, and brains, pay a price for distributed encoding: crosstalk. Distributed encoding produces crosstalk or interference between stored patterns. Similar patterns may clump together. New patterns may crowd out older learned patterns. Older patterns may distort newer patterns.

Crosstalk limits the neural network's storage capacity. Different learning schemes provide different storage capacities. The number of neurons bounds the number of patterns a neural network can store reliably with the simplest unsupervised learning schemes. Even for more sophisticated supervised learning schemes, storage capacity ultimately depends on the number of network neurons and synapses, as well as on their function. *Dimensionality limits capacity.*

Biological neurons and synapses motivate the neural network's topology and dynamics. We interpret neurons as simple input-output functions, threshold switches for two-state neurons and asymptotic threshold switches for continuous neurons. We interpret synapses as adjustable weights. In neural analog VLSI chips [Mead, 1989], operational amplifiers model nonlinear neurons, and resistors model synapses.

The overall network behaves as an adaptive function estimator. Indeed, commercial adaptive estimators are simple, usually linear, neural networks. These include antennae beam formers, high-speed modems, and echo-cancellers for long-distance telephone calls.

Neural Networks as Trainable Dynamical Systems

Neural networks *geometrize* computation. Network activity burrows a trajectory in a state space of large dimension, say R^n. Each point in the state space defines a snapshot of a possible neural-network configuration.

The trajectory begins with a computational problem and ends with a computational solution. The user or the environment specifies the system's initial conditions, which define where the trajectory begins in the state space. In pattern learning, the pattern to be learned defines the initial conditions. In pattern recognition or recall, the pattern to be recognized defines the initial conditions.

Most of the trajectory corresponds to *transient* behavior or computations. Synaptic values gradually change to learn new pattern information. Neuronal outputs fluctuate.

The trajectory ends where the system reaches equilibrium, if it ever reaches equilibrium. In the simplest and rarest case, the equilibrium attractor is a **fixed point** of the dynamical system. Most popular neural networks converge to fixed points. In more complicated cases the equilibrium attractor is a **limit cycle** or **limit torus**. In Chapter 4 we discuss a crude method for storing discrete time-varying patterns as limit cycles in feedback networks. The equilibrium attractors are *robust* or *structurally stable* if small perturbations do not distort or destroy them.

In general, and in most dynamical systems, the equilibrium attractor is *aperiodic* or **chaotic**. Once the network enters this region of the state space, it wanders forever without apparent structure or order. Yao and Freeman [1990] have used dynamical neural models and time-series data to argue that rabbit olfactory bulbs process odor information with chaotic attractors. As discussed in the homework problems, the function $x_{k+1} = cx_k(1 - x_k)$ behaves as a chaotic dynamical system for values of c near 4 and x values in the open unit interval $(0, 1)$.

In Chapter 3 we discuss global Lyapunov functions for proving that certain feedback neural networks converge to fixed points from any initial conditions. Geometrically we can view the Lyapunov function as a surface sculpted by learned pattern information, as in Figure 1.3.

Figure 1.3 illustrates the geometry of fixed-point stability in feedback neural networks. Patterns behave as rocks on the rubber sheet of learning. The patterns, as well as "spurious" or unlearned patterns, dig out attractor basins in the state space and tend to rest at the local Lyapunov minimum of the attractor. The Lyapunov sheet changes shape as the system learns new patterns. Input patterns Q rapidly classify to nearest stored neighbors as if they were ball bearings rolling into local depressions in a gravity field. In fixed-point attractor basins the state-trajectory balls stop at the local minima (or hover arbitrarily close to it). In limit-cycle attractors, the ball Q would rotate in an elliptical orbit inside the attractor basin. In limit-tori attractors, Q would cycle toroidally in the attractor basin, as if, in R^3, winding around the surface of a bagel. In chaotic attractors, Q would wander aperiodically within the attractor region.

In all these cases, the *number* of attractor basins does not affect the speed of convergence, the rate at which Q falls into the attractor basin. The dimensionality of the state space also does not in principle affect the convergence rate. In practice, Q converges exponentially quickly. This suggests that global stability may underlie our biological neural networks' ability to rapidly recognize patterns, generate

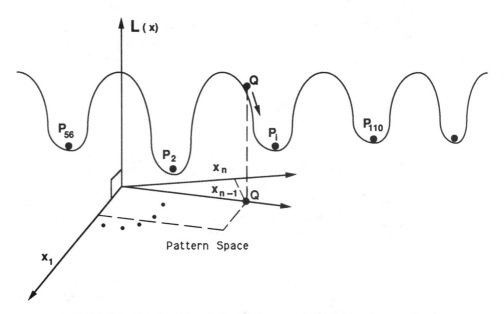

FIGURE 1.3 Global stability of a feedback neural network. Learning encodes the vector patterns P_1, P_2, ... by gradually sculpting a Lyapunov or "energy" surface in the augmented state space R^{n+1}. Input vector pattern Q rapidly "rolls" into the nearest attractor basin, where the system classifies Q as a learned pattern P or misclassifies Q as a spurious pattern. Q's descent rate does not depend on the number of stored patterns.

answers, and exhibit appropriate muscle reflexes independent of the amount of pattern information in our brains. Computer-type storage devices tend to slow as the number and complexity of patterns stored in them increases.

Mathematically we can describe the time evolution of the neural network by the (autonomous) dynamical system equation

$$\dot{\mathbf{x}}(t) = \mathbf{f}(\mathbf{x}) \tag{1-27}$$

where the overdot denotes time differentiation. The state vector $\mathbf{x}(t)$ describes all neuronal and synaptic values of the neural network at time t. The neural network reaches *steady state* when

$$\dot{\mathbf{x}} = \mathbf{O} \tag{1-28}$$

holds indefinitely or until new stimuli perturb the system out of equilibrium. Neural computation seeks to identify the steady-state condition (1-28) with the solution of a computational problem, whether in pattern recognition, image segmentation, optimization, or numerical analysis.

We can locally linearize $\mathbf{f}$ by replacing $\mathbf{f}$ with its Jacobian matrix of partial derivatives $\mathbf{J}$. The eigenvalues of $\mathbf{J}$ describe the system's local behavior about an equilibrium point. For instance, if all eigenvalues have negative real parts, then the local equilibrium is a fixed point and the system converges to it exponentially

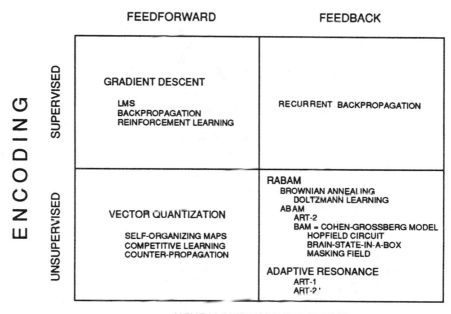

FIGURE 1.4 Taxonomy of neural-network models. *Source:* "Unsupervised Learning in Noise," *IEEE Trans. on Neural Networks*, Volume 1, Number 1, March 1990. (© 1990 IEEE)

quickly. More abstractly, generalized eigenvalues or *Lyapunov exponents* describe the underlying dynamical contraction and expansion that may produce chaos.

We can classify neural-network models depending on whether they learn with supervision (pattern-class information) and whether they contain closed synaptic loops or feedback. Figure 1.4 provides a rough taxonomy of several popular neural-network models.

Supervised feedforward models provide the most tractable, most applied neural models. We discuss these stochastic gradient systems in Chapter 5 and mention recent attempts to extend these supervised systems into the feedback domain. Unsupervised feedback models are biologically the most plausible, but mathematically the most complicated. These networks simultaneously learn and recall patterns. Both neurons and synapses change state when these systems learn and when they recall, recognize, or reconstruct pattern information. Chapter 6 proves global stability for many of these adaptive dynamical systems in the RABAM theorem. Unsupervised feedforward neural networks tend to converge to locally sampled pattern-class centroids, as discussed in Chapters 4 and 6, and in Chapter 1 of the companion volume [Kosko, 1991].

Fuzzy Systems and Applications

Fuzzy systems store banks of fuzzy associations or common-sense "rules." A fuzzy traffic controller might contain the fuzzy association "If traffic is heavy in this direction, then keep the light green longer." Fuzzy phenomena admit degrees. Some traffic configurations are *heavier* than others. Some green-light durations are *longer* than others. The single fuzzy association (HEAVY, LONGER) encodes all these combinations.

Fuzzy systems are even newer than neural systems. Yet already engineers have successfully applied fuzzy systems in many commercial areas. Fuzzy systems "intelligently" automate subways; focus cameras and camcorders; tune color televisions; control automobile transmissions, cruise controllers, and emergency braking systems; defrost refrigerators; control air conditioners; automate washing machines and vacuum sweepers; guide robot-arm manipulators; invest in securities; control traffic lights, elevators, and cement mixers; recognize Kanji characters; select golf clubs; even arrange flowers.

Most of these applications originated in Japan, though fuzzy products are sold and applied throughout the world. Until very recently, Western scientists, engineers, and mathematicians have overlooked, discounted, or even attacked early versions of fuzzy theory, usually in favor of probability theory. Below, and especially in Chapter 7, we examine this philosophical resistance in more detail and present a new geometrical theory of multivalued or "fuzzy" sets and systems.

Fuzzy systems "reason" with parallel associative inference. When asked a question or given an input, a fuzzy system fires each fuzzy rule in parallel, but to different degree, to infer a conclusion or output. Thus fuzzy systems reason with sets, "fuzzy" or multivalued sets, instead of bivalent propositions. This generalizes the Aristotelian logical framework that still dominates science and engineering. In one second a digital fuzzy VLSI chip may execute thousands, perhaps millions, of these parallel-associative set inferences. We measure such chip performance in FLIPS, fuzzy logical inferences per second.

Fuzzy systems estimate sampled functions from input to output. They may use linguistic (symbolic) or numeric samples. An expert may articulate linguistic associations such as (HEAVY, LONGER). Or a fuzzy system may adaptively infer and modify its fuzzy associations from representative numerical samples.

In the latter case, neural and fuzzy systems naturally combine. The combination resembles an adaptive system with sensory and cognitive components. Neural parameter estimators embed directly in an overall fuzzy architecture. Neural networks "blindly" generate and refine fuzzy rules from training data. Chapters 8 through 11 describe and illustrate these adaptive fuzzy systems.

Adaptive fuzzy systems learn to control complex processes very much as we do. They begin with a few crude rules of thumb that describe the process. Experts may give them the rules. Or they may abstract the rules from observed expert behavior. Successive experience refines the rules and, usually, improves performance.

Chapter 9 applies this adaptive cognitive process to backing up a truck-and-trailer rig to a loading dock. (A supervised neural system can also solve this problem, though at much greater computational cost. So far the truck-and-trailer dynamical system has eluded mathematical characterization.) The fuzzy system quickly learns a set of governing fuzzy rules as it samples actual truck-and-trailer trajectories. Additional training samples improve only marginally the fuzzy system's performance. This property is better experienced than explained. As an exercise, you might try backing your car into the same parking space five times from five different starting positions.

INTELLIGENT BEHAVIOR AS ADAPTIVE MODEL-FREE ESTIMATION

Below we discuss neural and fuzzy systems in more detail. First we examine the properties neural and fuzzy systems share with us and, more broadly, with all intelligent systems. These properties reduce to the single abstract property of adaptive model-free function estimation: *Intelligent systems adaptively estimate continuous functions from data without specifying mathematically how outputs depend on inputs.* We now elaborate this thesis.

A **function** f, denoted $f: X \to Y$, maps an input domain X to an output range Y. For every element x in the input domain X, the function f uniquely assigns the element y in the output range Y. We denote this unique assignment as $y = f(x)$. $f(x) = x^3$ defines a cubic function. $f(x_1, x_2, x_3) = (x_1, x_2, x_1^2 - x_2^2)$ defines a "saddle" or hyperbolic-paraboloid vector function in physical or three-dimensional space R^3. Pressure is a function of temperature, mass of energy ($e = mc^2$), gravity of mass, erosion of gravity, consumption of income. Functions define causal hypotheses. Science and engineering paint our pictures of the universe with functions.

Humans, animals, reptiles, amphibians, and others also estimate functions. We all respond to stimuli. We associate responses with stimuli. We associate actions with scenarios, class labels with patterns, effects with causes. Equivalently, we map stimuli to responses.

Mathematically, all these systems transform inputs to outputs. The transformation defines the input-output function $f: X \to Y$. Indeed the transformation defines the system. We can operatively characterize any system—atomic, molecular, biological, ecological, economic or legal, geological, galactic—by how it transforms input quantities into output quantities.

We call system behavior "intelligent" if the system emits appropriate, problem-solving responses when faced with problem stimuli. The system may use an associative memory embedded in the resistive network of an analog VLSI chip or embedded in the synaptic webs of its brain. Or the system may use a mathematical algorithm to search a decision tree, as in computer chess programs.

Generalization and Creativity

Intelligent systems also *generalize*. Their behavioral repertoires exceed their experience. Eighteenth-century philosopher David Hume saw why: Intelligent systems associate similar responses with similar stimuli. Small input changes produce small output changes. Hence they estimate *continuous* functions. The pilot lands the airplane at night the same way if only a few of the runway lights are out or if the new runway differs only slightly from more familiar runways. The leopard stalks like prey in like ways in like circumstances. Each minnow in a school smoothly adjusts its swimming behavior to the position of its smoothly moving neighbors.

Function continuity accounts for much novel or creative behavior, if not all of it. We call system behavior "novel" if the system emits appropriate responses when faced with new or unexpected stimuli. "Novel ideas," says behaviorist psychologist B. F. Skinner [1953], are "responses never made before under the same circumstances.... Novel contingencies generate novel forms of behavior." Usually these new stimuli resemble known or learned stimuli, and our responses usually resemble known responses.

Geometrically, when systems generalize or "create," they map stimulus balls to response balls. Consider a known stimulus-response pair $(\mathbf{x}, \mathbf{y})$. Stimulus $\mathbf{x}$ defines a point in the stimulus space $\mathbf{S}$, the set of all possible stimuli for the problem at hand. In practice $\mathbf{S}$ often corresponds to the real Euclidean vector space R^n. Response $\mathbf{y}$ defines a point in the response space $\mathbf{R}$, which may correspond to R^p.

Now imagine a stimulus ball $\mathbf{B_x}$ centered about stimulus $\mathbf{x}$ and a response ball $\mathbf{B_y}$ centered about response $\mathbf{y}$. All the stimuli $\mathbf{x}'$ in $\mathbf{B_x}$ resemble stimulus $\mathbf{x}$. The closer stimulus $\mathbf{x}'$ is to stimulus $\mathbf{x}$, and hence the smaller the distance $d(\mathbf{x}', \mathbf{x})$, the more $\mathbf{x}'$ resembles $\mathbf{x}$. The responses $\mathbf{y}'$ in $\mathbf{B_y}$ behave similarly.

Suppose $\mathbf{y} = f(\mathbf{x})$ for some unknown continuous function $f \colon R^n \to R^p$. The function f defines the sampled system. Suppose further that f generates the response ball from the stimulus ball: $\mathbf{B_y} = f(\mathbf{B_x})$. So for every similar response $\mathbf{y}'$ in $\mathbf{B_y}$, we can find some similar stimulus $\mathbf{x}'$ in $\mathbf{B_x}$ such that $\mathbf{y}' = f(\mathbf{x}')$. Formally f maps the stimulus ball *onto* the response ball.

(We use the term "ball" loosely. Technically, $f(\mathbf{B_x})$ need not define an open ball in R^p. Thus we measure $\mathbf{B_y}$ with a volume measure below in (1-29). The *open mapping theorem* in real analysis [Rudin, 1974] implies that all bounded onto linear transformations f map the open ball $\mathbf{B_x}$ to some set in R^p that contains the open ball $\mathbf{B_y}$, where $\mathbf{y} = f(\mathbf{x})$. At best we can only locally approximate most system transformations f as linear transformations.)

Then we can measure the **creativity** $C_{B_r}(f)$ of system f, given the stimulus ball $\mathbf{B_x}$, by the volume ratio

$$C_{B_r}(f) \;=\; \frac{V(\mathbf{B_y})}{V(\mathbf{B_x})} \tag{1-29}$$

where the V operator (Lebesgue measure) measures ball volume in R^n or R^p. C_{B_r}, crude as it is, captures many intuitions. It also resembles a spectral transfer function.

Consider the extreme cases of infinite and zero creativity. For a fixed non-degenerate response ball $\mathbf{B_y}$, as the stimulus ball $\mathbf{B_x}$ contracts to $\mathbf{x}$, the creativity measure $C_{B_x}(f)$ increases to infinity. (The point $\mathbf{x}$ has zero volume.) $C_{B_x}(f)$ also increases to infinity if the stimulus ball is constant and nondegenerate but the response ball $\mathbf{B_y}$ expands without bound as its radius approaches infinity. In both cases an infinitely creative system emits infinitely many responses when presented with, in the first case, a vanishingly small number of stimuli or, in the second case, a fixed set of stimuli.

Infinite creativity need not represent infinite problem solving. The reinforcing environment selects "solutions" from our varied or creative responses. Most creative solutions are impractical. We can emit creative responses without solving problems or contributing to our genetic fitness. Sometimes we call these responses "art" or "play."

At the other extreme, zero creativity occurs when the response ball $\mathbf{B_y}$ vanishes or when the stimulus ball expands without bound as its radius grows to infinity. In the first case the system f is a constant function. It maps all stimuli in $\mathbf{B_x}$ to a single value $\mathbf{y}$ in R^p. Such an f is "dumb" or "dull." In the second case, for an infinite-radius stimulus ball $\mathbf{B_x}$, the stimuli overwhelm the system's response repertoire. Such systems resemble classical pattern-recognition devices that are sensitive only to well-defined, well-centered patterns (faces, zip codes, bar codes).

Small variations in input provide the simplest novel stimuli. The physical or cultural environment may produce these variations. Or we may systematically produce them as grist for our analytical mill. We may vary stimuli to solve a crossword puzzle, to fit physical variables to astronomical data, or to formulate and resolve a mathematical conjecture.

We are all forward-looking creatures. We tend not to see the gradual causal chains that precede our every action, idea, and innovation. Even Beethoven's Fifth Symphony appears less a discontinuity when we examine Beethoven's note-books and a variety of preceding musical compositions by him and by other composers.

Variation and selection drive biological and cultural evolution. Physical and cultural environments drive the selection process. Function continuity, and other factors, drive variation.

Nature and man experiment with local variations of input parameters. This generates local variations of output parameters. Then selection processes filter the new outputs. More accurately, they filter the corresponding new systems. We call the new systems "winners" or "fit" if they pass through the selection filters, "losers" or "unfit" if they do not pass through.

Variation and selection *rates* may vary, especially over long stretches of geological or cultural time. Different perturbed processes unfold at different speeds. So some evolutionary stretches appear more "punctuated" than others [Gould, 1980]. This means some measures of change—ultimately time derivatives—are nonlinear. It does not mean that the underlying input-output functions are discontinuous.

Learning as Change

Intelligent systems also *learn* or *adapt*. They learn new associations, new patterns, new functional dependencies. They sample the flux of experience and encode new information. They compress or quantize the sampled flux into a small, but statistically representative, set of prototypes or exemplars. Sample data changes system parameters.

"Learning" and "adaptation" are linguistic gifts from antiquity. They simply mean *parameter change*. The parameters may be numerical weights in an inner-product sum, average neurotransmitter release rates at synaptic junctions, or gene (allelle) frequencies at chromosomal loci in populations.

"Learning" usually applies to synaptic changes in brains or nervous systems, coefficient changes in estimation or control algorithms or devices, or resistor changes in analog VLSI circuitry. Sometimes we synonymously apply "adaptation" to the same changing parameters. In evolutionary theory "adaptation" applies to positive changes in gene frequencies [Wilson, 1975].

In all cases learning means change. Formally, a system learns if and only if the system parameter vector or matrix has a nonzero time derivative. In neural networks we usually represent the synaptic web by an adjacency or connection matrix M of numerical synaptic values. Then learning is *any change in any synapse*:

$$\dot{M} \neq O \qquad (1\text{-}30)$$

We can learn well or learn badly. But we cannot learn without changing, and we cannot change without learning.

Learning laws describe the synaptic dynamical system, how the system encodes information. They determine how the synaptic-web process unfolds in time as the system samples new information. This shows one way that neural networks compute with dynamical systems. Neural networks also identify neural activity with dynamical systems. This allows the systems to decode information.

In principle we can harness any dynamical system to encode and decode some information. We can view a kinetic swirl of molecules, a joint population of lynxes and rabbits, and a solar system as systems that transform input states to output states. Initial conditions and perturbations encode questions. Transient behavior computes answers. Equilibrium behavior provides answers. In the extreme case we can even view the universe as a dynamical-system "computer." A godlike entity may choose Big-Bang initial conditions, and there are infinitely many, to encode certain information or to ask certain questions. The dynamical system computes as the universe expands transiently. Universal equilibrium behavior could represent the computational output: a heat-death pattern or perhaps a periodic or chaotic oscillation of expansion and contraction.

Consider mowing a lawn of green grass. The mower "teaches" the lawn the short-grass pattern. The lawn consists of a parallel field of grass blades. Grass blades learn what they are cut. The lawn behaves as a semipermanent, yet plastic, information-storage medium. It tolerates faults and distributes cut patterns over large numbers of parallel units. We can mow our name in the lawn, and read or decode it from a rooftop. In principle we can encode all known information in a sufficiently big lawn. Eventually the lawn will forget this information if we do not resample comparable data, if we do not re-mow the lawn to a similar shape.

Ultimately learning provides only a means to some computational end. Neural networks learn patterns or functions or probability distributions to recognize future patterns, filter future input streams of data, or solve future combinatorial optimization problems. Fuzzy systems learn associative rules to estimate functions or control systems. We climb the ladder of learning and kick it away when we reach the roof of computation. We care how the learned parameter performs in some computational system, not how it was learned, just as we applaud the piano recital and not the practice sessions.

Neural and fuzzy systems ultimately learn some unknown probability (subsethood) function $p(\mathbf{x})$. The probability density function $p(\mathbf{x})$ describes a distribution of vector patterns or signals $\mathbf{x}$, a few of which the neural or fuzzy system samples. When a neural or fuzzy system estimates a function $f: X \rightarrow Y$, it in effect estimates the joint probability density $p(\mathbf{x}, \mathbf{y})$. Then solution points $(\mathbf{x}, f(\mathbf{x}))$ should reside in high-probability regions of the input-output product space $X \times Y$.

We do not need to learn if we know $p(\mathbf{x})$. We could proceed directly to our computational task with techniques from numerical analysis, combinatorial optimization, calculus of variations, or any other mathematical discipline. The need to learn varies inversely with the quantity of information or knowledge.

Sometimes the patterns cluster into exhaustive decision classes $D_1, \ldots, D_k$. The decision classes may correspond to high-probability regions or "mountains." (If the pattern vectors are two-dimensional, then $p(\mathbf{x})$ defines a hilly surface in three-dimensional space R^3.) Then class boundaries correspond to low-probability regions or "valleys" on the probability surface.

Supervised learning uses class-membership information. *Unsupervised* learning does not. An unsupervised learning system processes each sample $\mathbf{x}$ but does not "know" that $\mathbf{x}$ belongs to class D_i and not to class D_j. Unsupervised learning uses unlabelled samples. Neither supervised nor unsupervised learning systems assume knowledge of the underlying probability density function $p(\mathbf{x})$.

Suppose we want to train a speech-recognition system at an international airport. We want the German lightbulb to light up when someone speaks German to the speech-recognition system, the Hindi lightbulb to light up when someone speaks Hindi, and so on. The system learns as we feed it training waveforms or spectrograms.

We supervise the learning if we label each training sample as German, Hindi, Japanese, etc. We may do this to compute an error. If the English lightbulb lights up for a German sample, we may algorithmically punish the system for this misclassification.

An unsupervised system learns only from the raw training samples. We do not indicate language class labels. Unsupervised systems adaptively cluster like patterns with like patterns. The speech-recognition system gradually clumps German speech patterns together. In competitive learning, for instance, the system learns class centroids, centers of pattern mass.

Unsupervised learning may seem difficult and unreliable. But most learning is unsupervised, since we do not know accurately the labels of most sample data, especially in real-time processing. Every second our biological synapses learn without supervision on a single pass of noisy data.

Symbols vs. Numbers: Rules vs. Principles

We all share another property: We cannot articulate the mathematical rules that describe, if not govern, our behavior. We can ask a violinist how she plays, and she can tell us. But her answer will not be a mathematical function. In general her answer will not enable us to reproduce her behavior.

All lifeforms recognize vast numbers of patterns. The most primitive patterns relate to how an organism forages, avoids predators, and reproduces [Wilson, 1975].

On this planet only man articulates rules, and he articulates very few. We articulate some rules in grammar, common law, and science ("physical laws"). Yet all our natural languages, living and dead, and all our systems of law have culturally evolved without conscious design and not in accord with articulated principles [Hayek, 1973]. To some extent this also holds for our accumulated knowledge of medical, biological, and social science.

There have been exceptions, and the exceptions have helped create the field of artificial intelligence. Last century linguists developed the articulated language *Esperanto*. Mathematician Giuseppe Peano similarly devised the language *Interlingua*. A few fans still learn and speak *Esperanto* and *Interlingua*, but far fewer speak them than speak Latin. This century computer scientists have consciously created the many computer programming languages. Today programmers frequently use *C*, Pascal, and even Fortran, and infrequently use Algol and Jovial.

Computer scientists developed artificial intelligence in large part around the computer language Lisp, or *list processing*, and more recently around Prolog, or *logic programming*. Lisp and Prolog process symbols and lists of symbols. Symbolic logic, the bivalent propositional and predicate calculi, underlies their processing structure.

Expert-System Knowledge as Rule Trees

AI systems store and process propositional *rules*. The rules are logical implications: IF A, THEN B. They associate actions B with conditions A. The rule antecedents and consequents correspond to step functions defined on their universes of discourse. One part of the input space activates or "fires" A as true, and the other part does not activate A.

Collections of rules define "knowledge bases" or "rulebases." The rule $A \rightarrow B$ locally structures the knowledge of A and B as a logical implication. The knowledge base globally structures the rules as an acyclic tree (or forest). The logical-implication paths $A \rightarrow B \rightarrow C \rightarrow D \rightarrow \ldots$ flow from the tree's root nodes or antecedents to its leaf nodes or consequents. The term *knowledge base* stems from the computer-scientific term *database*. Because of the tree structure of knowledge bases, we might more accurately call them *knowledge trees*. Chapter 4 discusses fuzzy cognitive maps, which use feedback and vector-matrix operations to convert knowledge trees to knowledge networks.

Knowledge engineers search the knowledge tree to enumerate logical paths. Path enumeration defines the *inference* process. Forward-chaining inference proceeds from knowledge-tree antecedents to consequents. Backward-chaining inference proceeds from consequents or observations to plausible antecedents or hypotheses. Forward-chaining inference answers what-if questions. It derives effects from causes. Backward-chaining inference answers why or how-come questions. It suggests causes for observed effects. Path-enumeration complexity increases nonlinearly with the number of rules stored. Real-time path enumeration in large knowledge trees may be combinatorially prohibitive, requiring heuristic or approximate search strategies [Pearl, 1984].

Knowledge engineers acquire, store, and process the bivalent rules as symbols, not as numerical entities. This often allows knowledge engineers to rapidly acquire structured knowledge from experts and to efficiently process it. But it forces experts to articulate the propositional rules that approximate their expert behavior, and this they can rarely do.

Symbolic vs. Numeric Processing

Symbolic processing fits naturally in the brain-as-computer framework. Language strings model thoughts or short-term memory. Rules and relations between language strings model long-term memory. Programming replaces learning. Logical inference replaces time evolution and nonlinear dynamics. Feedforward flow through knowledge trees replaces feedback equilibria.

But we cannot take the derivative of a symbol. We require a sufficiently continuous function. Symbol processing precludes mathematical analysis in the traditional senses of engineering and the physical sciences. The symbolic framework allows us to quickly represent structured knowledge as rules, but prevents us from directly applying the tools of numerical mathematics and from directly implementing AI systems in large-scale integrated circuits.

Figure 1.5 provides a taxonomy of model-free estimators. The taxonomy divides the knowledge type into structured (rule-like) and unstructured types and divides the framework into symbolic and numeric. All entries define model-free estimators, because users need not state how outputs depend mathematically on inputs.

Figure 1.5 outlines the advantages and disadvantages of machine-intelligent systems. AI expert systems exploit structured knowledge, when knowledge engi-

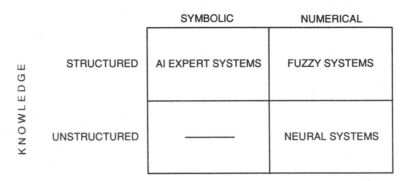

FIGURE 1.5 Taxonomy of model-free estimators. User need not state how system outputs explicitly depend on inputs.

neers can acquire it, but store and process it outside the analytical and computational numerical framework.

Neural networks exploit their numerical framework with theorems, efficient numerical algorithms, and analog and digital VLSI implementations. But neural networks cannot directly encode structured knowledge. They superimpose several input-output samples $(\mathbf{x}_1, \mathbf{y}_1)$, $(\mathbf{x}_2, \mathbf{y}_2)$, ..., $(\mathbf{x}_m, \mathbf{y}_m)$ on a black-box web of synapses. Unless we check all input-output cases, we do not know what the neural system has learned, and in general we do not know what it will forget when it superimposes new samples $(\mathbf{x}_k, \mathbf{y}_k)$ atop the old. We cannot directly encode the common-sense traffic-light rule "If traffic is heavy in one direction, keep the light green longer in that direction." Instead we must present the system with a sufficiently large set of input-output pairs, combinations of numerical traffic-density measurements and green-light duration measurements.

Fuzzy Systems as Structured Numerical Estimators

Fuzzy systems directly encode structured knowledge but in a numerical framework. We enter the fuzzy association (HEAVY, LONGER) as a single entry in a FAM-rule matrix. Each entry defines a fuzzy associative memory (FAM) "rule" or input-output transformation. In Chapter 8 we discuss the fuzzy control of an inverted pendulum. Figure 1.6 shows a bank of FAM rules sufficient to control an inverted pendulum.

θ, $\Delta\theta$, and v define fuzzy variables. Fuzzy variables θ and $\Delta\theta$ define the system's state variables. The angle fuzzy variable θ measures the angle the pendulum shaft makes with the vertical and ranges from -90 to 90. The angular-velocity fuzzy $\Delta\theta$ variable measures the instantaneous rate of change of angle values. In practice it measures the difference between successive angle values. Output fuzzy

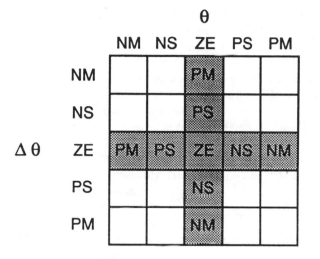

FIGURE 1.6 Bank of FAM rules to control an inverted pendulum. Each entry in the FAM matrix defines a fuzzy association between output fuzzy sets and paired input fuzzy sets.

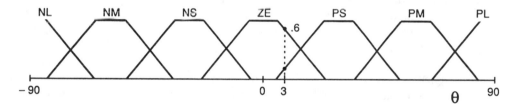

FIGURE 1.7 Seven trapezoidal fuzzy-set values assumed by fuzzy variable θ. Each value of θ belongs to each fuzzy set to some, but usually zero, degree. The exact value 3 belongs to the zero fuzzy number ZE to degree 0.6, to the positive small fuzzy number PS to degree 0.2, and to positive medium PM to degree 0.

variable v measures the current to a motor controller that adjusts the pendulum shaft.

Each fuzzy variable can assume five fuzzy-set values: Negative Medium (NM), Negative Small (NS), Zero (ZE), Positive Small (PS), and Positive Medium (PM). The entry at the center of the FAM matrix defines the steady-state FAM rule: "IF θ = ZE AND $\Delta\theta$ = ZE, THEN v = ZE."

We usually define the fuzzy-set values NM, ..., PM as trapezoids or triangles over regions of the real line. For the fuzzy angle variable θ, we can define ZE as a narrow triangle centered at the zero value in the interval $[-90, 90]$. Then the angle value 0 belongs to the fuzzy set ZE to degree 1. The angle values 3 and -3 may belong to ZE only to degree 0.6. Figure 1.7 shows seven trapezoidal fuzzy-set values assumed by fuzzy variable θ.

Fuzzy systems allow users to articulate linguistic FAM rules by entering values in a FAM matrix. Once a fuzzy engineer defines variables and fuzzy sets, the engineer can design a prototype fuzzy system in minutes.

Chapter 8 shows that a large neural-type matrix encodes each FAM rule. When fuzzy variables assume fuzzy subsets of the real line, as when we define ZE as a triangle centered about 0, then these associative matrices have uncountably infinite dimension. This endows each FAM rule with rich structure and "memory capacity." FAM systems do not add these matrices together, which avoids neural-type crosstalk.

A virtual representation scheme allows us to exploit the coding and capacity properties of these infinite matrices without actually writing them down. This holds for binary input-output FAMs (BIOFAMs), which includes all fuzzy systems used in commercial applications. BIOFAMs accept nonfuzzy scalar inputs, such as $\theta = 15$ and $\Delta\theta = -10$, and generate nonfuzzy scalar outputs, such as $v = -3$.

Generating Fuzzy Rules with Product-Space Clustering

Neural networks can adaptively generate the FAM rules in a fuzzy system. We illustrate this in Chapters 8 through 11 with the new technique of unsupervised *product-space clustering*. Synaptic vectors quantize the input-output space. Clustered synaptic vectors track how experts associate appropriate responses with input stimuli. Each synaptic cluster estimates a FAM rule. The experts who generate the input-output data need not articulate the FAM rules. They need only behave as experts. The key geometric idea is *cluster equals rule*.

Consider the input-output product space of the inverted-pendulum system. There are two input variables and one output variable, so the input-output product space equals R^3 (in practice a three-dimensional subcube within R^3). Each input-output triple $(\theta, \Delta\theta, v)$ defines a point in R^3. The time evolution of the inverted-pendulum system defines a smooth curve or trajectory in R^3. As the fuzzy system stabilizes the inverted pendulum to its vertical position, the trajectory may spiral into the origin of R^3, where the above steady-state FAM rule keeps the system in equilibrium until perturbed.

Each fuzzy variable can assume five fuzzy subsets of the x, y, or z coordinate axes of R^3. The Cartesian product of these fuzzy subsets defines 125 ($5 \times 5 \times 5$) *FAM cells* in the input-output product space R^3. Most system trajectories pass through only a few FAM cells. We show in Chapter 8 that these FAM cells equal FAM rules because the FAM cells equal fuzzy Cartesian products, and the uncountably infinite entries in the associative matrices correspond to these Cartesian products. So a FAM rule equals an associative (fuzzy Hebb) matrix, which equals a fuzzy Cartesian product, which equals a FAM cell.

Unsupervised neural clustering algorithms efficiently track the density of input-output samples in FAM cells. We need only count the number of synaptic vectors in each FAM cell at any instant to estimate, and to weight, the underlying FAM rules used by the expert or physical process that generates the input-output data. This produces an *adaptive histogram* of FAM-cell occupation. Chapters 8 through 11 apply the adaptive product-space clustering methodology to inverted-pendulum control, backing up a truck-and-trailer in a parking lot, and real-time target tracking.

Suppose a system contains n fuzzy variables, and each fuzzy variable can assume m fuzzy-set values. This defines m^n FAM cells in the input-output product space R^n. Different fuzzy variables can assume different types and different numbers of fuzzy-set variables. So in general there are $m_1 \times \cdots \times m_n$ FAM cells. Suppose $n = m = 3$. Suppose the fuzzy sets are low, medium, and high and have bounded extent. Then a Rubik's cube represents the input-output product space partitioned into 27 FAM cells if the fuzzy sets do not overlap. In general FAM cells have nonempty but fuzzy intersection.

If we define n fuzzy variables, each with m fuzzy-set values, then there are 2^{m^n} possible fuzzy systems. Expert articulation, fuzzy engineering, and adaptive estimation produce only a small fraction of the total number 2^{m^n} of possible fuzzy systems. Different fuzzy-set definitions and different encoding or decoding strategies ("inferencing" techniques) produce different classes of 2^{m^n} possible fuzzy systems.

Fuzzy Systems as Parallel Associators

Fuzzy systems store and process FAM rules in parallel. Mathematically a fuzzy system maps points in an input product hypercube (possibly of infinite dimension) to points in an output hypercube. Fuzzy systems associate output fuzzy sets with input fuzzy sets, and so behave as associative memories. Unlike neural associative memories, fuzzy systems do not sum the associative matrices that represent FAM rules. *Neural networks sum throughputs. Fuzzy systems sum outputs.*

Summing outputs avoids crosstalk and achieves modularity. We can meaningfully look inside the black box of a fuzzy model-free estimator. Figure 1.8 displays the generic fuzzy system architecture for a single-input, single-output FAM system.

Fuzzy inference computes the output fuzzy sets B'_j, weights them with the scalar weights w_j, and sums them to produce the output fuzzy set B:

$$B = \sum_j w_j B'_j \qquad (1\text{-}31)$$

In principle in (1-31) we sum over all m^n possible FAM rules, since most rules have weight $w_i = 0$. Chapter 8 discusses the mechanism of the two types of fuzzy inference, correlation-product and correlation-minimum inference.

Adaptive fuzzy systems use sample data and neural or statistical algorithms to choose the coefficients w_j and thus to define the fuzzy system at each time instant. Adaptation changes the system structure. Geometrically, a time-varying between-cube mapping defines an adaptive fuzzy system. In the simplest case, if the input fuzzy sets define points in the unit hypercube I^n, and the output fuzzy sets define points in the unit hypercube I^p, then transformation $\mathbf{S}$ defines a fuzzy system if $\mathbf{S}$ maps I^n to I^p, $\mathbf{S}: I^n \longrightarrow I^p$. Then $\mathbf{S}$ associates fuzzy subsets of the output space

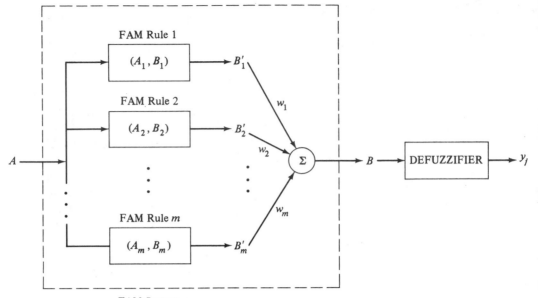

FAM System

FIGURE 1.8   Fuzzy-system architecture. The system maps input fuzzy sets A to output fuzzy sets B. The system stores separate FAM rules and in parallel fires each FAM rule to some degree for each input. Experts or adaptive algorithms determine the FAM-rule weights w_j. Experts may use only $w_j = 1$ (articulates rule) or $w_j = 0$ (omits rule). Centroidal output converts fuzzy-set vector B to a scalar. In BIOFAM systems A defines a unit binary vector or delta pulse.

Y with fuzzy subsets of the input space X. So $\mathbf{S}(A) = B$. $\mathbf{S}$ defines an adaptive fuzzy system if $\mathbf{S}$ changes with time:

$$\frac{d\mathbf{S}}{dt} \neq \mathbf{O} \qquad (1\text{-}32)$$

BIOFAM systems convert the vector B into a single scalar output value $y \in Y$. We call this process *defuzzification*, although to defuzzify a fuzzy set formally means to round it off from some point in a unit hypercube to the nearest bit-vector vertex. Fuzzy engineers sometimes compute y as the mode $y_{\max}$ of the B distribution,

$$m_B(y_{\max}) = \sup\{m_B(y): y \in Y\} \qquad (1\text{-}33)$$

Here m_B denotes the fuzzy membership function $m_B: Y \longrightarrow [0, 1]$ that assigns fit values or occurrence degrees to the elements of Y. If the output space Y equals a finite set of values $\{y_1, \ldots, y_p\}$, as in some computer discretizations, then we can replace the supremum in (1-33) with a maximum:

$$m_B(y_{\max}) = \max_j \; m_B(y_j) \qquad (1\text{-}34)$$

The more popular centroidal defuzzification technique uses all, and only, the information in the fuzzy distribution B to compute y as the centroid $\bar{y}$ or center of mass of B:

$$\bar{y} \;=\; \frac{\displaystyle\int_{-\infty}^{\infty} y m_B(y)\, dy}{\displaystyle\int_{-\infty}^{\infty} m_B(y)\, dy} \tag{1-35}$$

provided the integrals exist. In practice we restrict fuzzy subsets to finite stretches of the real line. In Chapter 11 we prove that if the fuzzy variables assume only symmetric trapezoidlike fuzzy-set values, then (1-35) reduces to a simple discrete ratio. The numerator and denominator contain only m products. This discrete centroid trivializes the computational burden of defuzzification and admits direct VLSI implementation.

Figure 1.8 and Equation (1-31) additively combine the weighted fuzzy sets B_j'. Earlier fuzzy systems [Mamdani, 1977] combined output fuzzy sets with pairwise maxima. Unfortunately, the maximum combination technique,

$$B \;=\; \max_j \; \min(w_j,\, B_j') \tag{1-36}$$

based upon the so-called "extension principle" of classical fuzzy theory [Klir, 1988], tends to produce a uniform distribution for B as the number of combined fuzzy sets increases [Kosko, 1987]. A uniform distribution always has the same mode and centroid. So, ironically, as the number of FAM rules increases, system sensitivity decreases.

The additive combination technique (1-31) tends to invoke the fuzzy version of the central limit theorem. The added fuzzy waveforms pile up to approximate a symmetric unimodal, or bell-shaped, membership function. Different fuzzy waveforms produce similarly *shaped* output distributions B but centered about different places on the real line. We consistently observe this tendency toward a Gaussian membership function after summing only a few fuzzy waveforms. (Technically the CLT requires normalization by the square root of the number of summed waveforms. Equation (1-31) does not normalize B because, for defuzzification, we care only about the relative values in B, the relative degrees of occurrence of output values.)

The maximum combination technique (1-36) forms the envelope of the weighted fuzzy sets B_j'. Then B resembles the silhouette of a desert full of sand dunes. As the number of sand dunes increases, the silhouette becomes flatter. The additive combination technique (1-31) piles the sand dunes atop one other to form a sand mountain.

Fuzzy inference allows us to reason with sets as if they were propositions. The virtual-representation scheme for FAM rules greatly simplifies the fuzzy inference process if we use exact numerical inputs. Figure 1.9 illustrates the FAM (correlation-minimum) inference procedure derived in Chapter 8. We can apply this inference procedure in parallel to any number of FAM rules with any number of antecedent fuzzy-variable conditions.

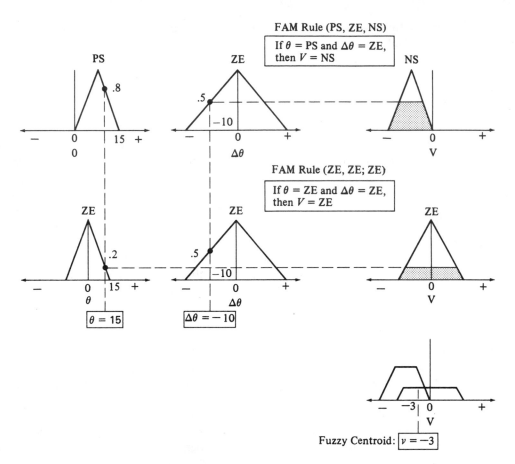

FIGURE 1.9 FAM inference procedure. The fuzzy system converts the numerical inputs, $\theta = 15$ and $\Delta\theta = -10$, into the numerical output $v = -3$. Since the FAM rules combine the antecedent terms with AND, the smaller of the two fit values scales the output fuzzy set. If the FAM rules combined antecedents disjunctively with OR, the larger of the fit values would scale the output fuzzy set.

Fuzzy Systems as Principle-Based Systems

AI expert systems chain through rules. Inference proceeds down, or up, branches of a decision tree. Except for chess trees or other game trees, in practice these search trees are wider than they are deep. Shallow trees (or forests) can exaggerate the all-or-none effect of bivalent propositional rules. Relative to deeper trees, shallow trees use a smaller proportion of their stored knowledge when they inference. They are noninteractive.

Fuzzy systems are shallow but fully interactive. Every inference fires every FAM rule, itself a fuzzy expert system, to some degree. A similar property holds for the feedback fuzzy cognitive maps discussed in Chapter 4.

Consider an AI judge and a fuzzy judge. Opposing counsel present the same evidence and testimony to both judges. The AI judge rounds off the truth value of every key statement or alleged fact to TRUE or FALSE (1 or 0), opens a rule book, uses the true statements to activate or choose the antecedents of some of the rules, then logically chains through the rule tree to reach a decision. A more sophisticated AI judge may chain through the rule tree with uncertainty-factor algorithms or heuristic search algorithms.

The fuzzy judge weights the evidence to different degrees, say with fractional values in the unit interval [0, 1]. The fuzzy judge does not use a rule book. Instead the fuzzy judge determines to what degree the fuzzy evidence invokes a large set of vague legal *principles*. The fuzzy judge may cite case precedents to enunciate these principles or to illustrate their relative importance. The fuzzy judge reaches a decision by combining these fuzzy facts and fuzzy principles in an unseen act of intuition or judgment. If pressed, the fuzzy judge may defend or explain the decision by citing the salient facts and relevant legal principles, precedents, and perhaps rules. In general the fuzzy judge cannot articulate an exact legal audit trail of the decision process.

The distinction between the AI judge and the fuzzy judge reduces to the distinction between rules and principles. Recently legal theorists [Dworkin, 1968, 1977; Hayek, 1973] have focused on this distinction and challenged the earlier "positivist" legal theories of law as articulated rules [Kelsen, 1954; Hart, 1961].

Rules, as Dworkin [1977] says, apply "in an all-or-none fashion." Principles "have a dimension that rules do not—the dimension of weight or importance," and the court "cites principles as its justification for adopting and applying a new rule." Rules greatly outnumber principles. Principles guide while rules specify:

> Only rules dictate results, come what may. When a contrary result has been reached, the rule has been abandoned or changed. Principles do not work that way; they incline a decision one way, though not conclusively, and they survive intact when they do not prevail.

Rules tend to be black or white. They abruptly come into and out of existence. We post rules on signs, vote on them as propositions, and send them in memos: must be 18 to vote, open from 8 a.m. to 5 p.m., $500 fine for littering, office term lasts four years, can take only five sick days a year, and so on. Rules come and go as culture evolves.

Principles evolve as culture evolves. Most legal principles in the United States grew out of medieval British common law. Each year their character changes slightly, adaptively, as we apply them to novel circumstances. These principles range from very abstract principles, such as presumption of innocence or freedom of contract, to more behavioral principles, such as that no one can profit from a crime or you cannot challenge a contract if you acquiesce to it and act on it.

Each principle admits a spectrum of exceptions. In each case a principle holds only to some, often slight, degree. Judges cite case precedents in effect to estimate the current weight of principles. All the principles "hang together" to some degree in each decision, just as all the fuzzy rules (principles) in Figure 1.8 contribute to some degree to the final inference or decision.

We often call AI expert systems rule-based systems because they consist of a bank or forest of propositional rules and an "inference engine" for chaining through the rules. The rule in rule-based emphasizes the articulated, expertly precise nature of the stored knowledge.

The AI precedent and modern legal theory suggest that we should call fuzzy systems **principle-based systems**. The fuzzy rules or principles indicate how entire clumps of output spaces associate with clumps of input spaces. Indeed FAM rules often behave as partial derivatives. Many applications require only a few FAM rules for smooth system control or estimation. In general AI rule-based systems would require vastly more precise rules to approximate the same system performance.

Adaptive fuzzy systems use neural (or statistical) techniques to abstract fuzzy principles from sampled cases and to gradually refine those principles as the system samples new cases. The process resembles our everyday acquisition and refinement of common-sense knowledge. Future machine-intelligent systems may match, then someday exceed, our ability to learn and apply the fuzzy common-sense knowledge—knowledge we can articulate only rarely and inexactly—that we use to run our lives and run our world.

REFERENCES

Birkhoff, G., and von Neumann, J., "The Logic of Quantum Mechanics," *Annals of Mathematics*, vol. 37, no. 4, 823-843, October 1936.

Black, M., "Vagueness: An Exercise in Logical Analysis," *Philosophy of Science*, vol. 4, 427-455, 1937.

Churchland, P. M., "Eliminative Materialism and the Propositional Attitudes," *Journal of Philosophy*, vol. 78, no. 2, 67-90, February 1981.

Crick, F., "Function of the Thalamic Reticular Complex: The Searchlight Hypothesis," *Proceedings of the National Academy of Sciences*, vol. 81, 4586-4590, 1984.

Dworkin, R. M., "Is Law a System of Rules?" in *Essays in Legal Philosophy*, R. S. Summers (ed.), Oxford University Press, New York, 1968.

Dworkin, R. M., *Taking Rights Seriously*, Harvard University Press, Cambridge, MA, 1977.

Grossberg, S., "Cortical Dynamics of Three-Dimensional Form, Color, and Brightness Perception: I. Monocular Theory," *Perception and Psychophysics*, vol. 41, no. 2, 87-116, 1987.

Gould, S. J., "Is a New and General Theory of Evolution Emerging?" *Paleobiology*, vol. 6, no. 1, 119-130, 1980.

Hart, H. L. A., *The Concept of Law*, Oxford University Press, New York, 1961.

Hayek, F. A., *Law, Legislation, and Liberty*, vol. I: *Rules and Order*, University of Chicago Press, Chicago, 1973.

Hopfield, J. J., "Neurons with Graded Response Have Collective Computational Properties like Those of Two-State Neurons," *Proceedings of the National Academy of Sciences*, vol. 81, 3088-3092, 1984.

Kanizsa, G., "Subjective Contours," *Scientific American*, vol. 234, 48-52, 1976.

Kant, I., *Prolegomena to any Future Metaphysics*, 1783.

Kant, I., *Critique of Pure Reason*, 2nd ed., 1787.

Kelsen, H., *General Theory of Law and State*, Harvard University Press, Cambridge, MA, 1954.

Kline, M., *Mathematics: The Loss of Certainty*, Oxford University Press, New York, 1980.

Klir, G. J., and Folger, T. A., *Fuzzy Sets, Uncertainty, and Information*, Prentice Hall, Englewood Cliffs, NJ, 1988.

Kosko, B., "Fuzzy Entropy and Conditioning," *Information Sciences*, vol. 40, 165-174, 1986.

Kosko, B., *Foundations of Fuzzy Estimation Theory*, Ph.D. dissertation, Department of Electrical Engineering, University of California at Irvine, June 1987; Order Number 8801936, University Microfilms International, 300 N. Zeeb Road, Ann Arbor, MI 48106.

Kosko, B., "Fuzziness vs. Probability," *International Journal of General Systems*, vol. 17, no. 2, 211-240, 1990.

Kosko, B., *Neural Networks for Signal Processing*, Prentice Hall, Englewood Cliffs, NJ, 1991.

Mamdani, E. H., "Application of Fuzzy Logic to Approximate Reasoning Using Linguistic Synthesis," *IEEE Transactions on Computers*, vol. C-26, no. 12, 1182-1191, December 1977.

Mead, C., *Analog VLSI and Neural Systems*, Addison-Wesley, Reading, MA, 1989.

Pearl, J., *Heuristics: Intelligent Search Strategies for Computer Problem Solving*, Addison-Wesley, Reading, MA, 1984.

Quine, W. V. O., "What Price Bivalence?" *Journal of Philosophy*, vol. 78, no. 2, 90-95, February 1981.

Rescher, N., *Many-Valued Logic*, McGraw-Hill, New York, 1969.

Rosser, J. B., and Turquette, A. R., *Many-Valued Logics*, North-Holland, New York, 1952.

Rudin, W., *Real and Complex Analysis*, McGraw Hill, New York, 1974.

Skinner, B. F., *Science and Human Behavior*, Macmillan, New York, 1953.

Thompson, R. F., *The Brain: An Introduction to Neuroscience*, W. H. Freeman & Company, New York, 1985.

Wiener, N., *Cybernetics: Control and Communication in the Animal and the Machine*, M.I.T. Press, Cambridge, MA, 1948.

Wilson, E. O., *Sociobiology: The New Synthesis*, Harvard University Press, Cambridge, MA, 1975.

Yao, Y., and Freeman, W. J., "Model of Biological Pattern Recognition with Spatially Chaotic Dynamics," *Neural Networks*, 153-170, vol. 3, no. 2, 1990.

Zadeh, L. A., "Fuzzy Sets," *Information and Control*, vol. 8, 338-353, 1965.

PROBLEMS

1.1. Lukasiewicz's multivalued or "fuzzy" logic ($\mathbf{L}_1$ logic) uses a continuous-valued truth function $t: \mathbf{S} \longrightarrow [0, 1]$ defined on the set $\mathbf{S}$ of statements. Lukasiewicz defined the generalized conjunction (AND), disjunction (OR), negation (NOT) operators respectively as

$$t(A \text{ AND } B) = \min(t(A), t(B))$$
$$t(A \text{ OR } B) = \max(t(A), t(B))$$
$$t(\text{NOT-}A) = 1 - t(A)$$

for statements A and B. Prove the generalized noncontradiction-excluded-middle law:

$$t(A \text{ AND } \sim A) + t(A \text{ OR } \sim A) = 1$$

The equality above implies that the classical bivalent law of noncontradiction, $t(A \text{ AND } \sim A) = 0$, holds if and only if the classical bivalent law of excluded middle, $t(A \text{ OR } \sim A) = 1$, holds. Note that in the case of bivalent "paradox," when $t(A) = t(\text{NOT-}A)$, the equality reduces to the equality $1/2 + 1/2 = 1$.

1.2. Let $t: \mathbf{S} \longrightarrow [0, 1]$ be a continuous or "fuzzy" truth function on the set $\mathbf{S}$ of statements. Define the Lukasiewicz implication operator as the truth function $t_L(A \longrightarrow B) = \min(1, 1 - t(A) + t(B))$ for statements A and B. Then prove the following generalized fuzzy *modus ponens* inference rule:

$$t_L(A \longrightarrow B) = c$$
$$t(A) \geq a$$

Therefore $\qquad \overline{\qquad t(B) \quad \geq \quad \max(0, a + c - 1)}$

Hence if $t(A) = t_L(A \longrightarrow B) = 1$, then $t(B) = 1$, which generalizes classical bivalent *modus ponens*.

1.3. Use the multivalued logic operations in Problem 1.2 to prove the following generalized fuzzy *modus tollens* inference rule:

$$t_L(A \longrightarrow B) = c$$
$$t(B) \leq b$$

Therefore $\qquad \overline{\qquad t(A) \quad \leq \quad \min(1, 1 - c + b)}$

Hence if $t_L(A \longrightarrow B) = 1$ and $t(B) = 0$, then $t(A) = 0$, which generalizes classical bivalent *modus tollens*.

1.4. Define the Gaines implication operator as

$$t_G(A \longrightarrow B) = \begin{cases} \min(1, t(B)/t(A)) & \text{if } t(A) > 0 \\ 1 & \text{if } t(A) = 0 \end{cases}$$

Use the Gaines implication operator $t_G(A \longrightarrow B)$ to derive generalized fuzzy *modus ponens* and *modus tollens* inference rules. The conclusions of the inference rules should differ from the conclusions of the inference rules in Problems 1.2 and 1.3.

1.5. Exclusive-or (XOR) equals negated logical equivalence:

$$t(A \text{ XOR } B) \quad = \quad 1 - t(A = B)$$

Equivalence equals biconditionality. Bivalent statements are equivalent if and only if the two statements have the same truth values. So the exclusive-or relation holds between two bivalent statements if and only if the two statements have opposite truth values.

Fuzzy statements can be equivalent to different degrees. But equivalence still equals biconditionality:

$$t(A = B) \quad = \quad t((A \longrightarrow B) \text{ AND } (B \longrightarrow A))$$

Prove that if we use the Lukasiewicz implication operator, then exclusive-or equals the absolute difference (or l^1 or fuzzy Hamming distance) of the truth values $t(A)$ and $t(B)$:

$$t_L(A \text{ XOR } B) \quad = \quad |t(A) - t(B)|$$

1.6. Set X contains n elements $x_1, \ldots, x_n$. So X contains 2^n nonfuzzy subsets A. Define the bivalent indicator function I_A of nonfuzzy set A as

$$I_A(x_i) \quad = \quad \begin{cases} 1 & \text{if} \quad x_i \in A \\ 0 & \text{if} \quad x_i \notin A \end{cases}$$

So I_A defines the mapping $I_A: X \longrightarrow \{0, 1\}$.

Suppose we extend I_A to a multivalued mapping by augmenting its range from $\{0, 1\}$ to $\{y_1, \ldots, y_m\}$, where $y_1 = 0$, $y_m = 1$, and $0 < y_j < 1$ if $1 < j < m$. Then I_A defines the mapping $I_A: X \longrightarrow \{y_1, \ldots, y_m\}$. How many multivalued subsets does X have? In the two-dimensional case, $X = \{x_1, x_2\}$, draw the planar lattice that describes the multidimensional power set of X, all its multidimensional subsets, when $m = 3$, and when $m = 5$.

1.7. Consider the discrete dynamical system

$$\begin{aligned} x_{k+1} \quad &= \quad f(x_k) \\ &= \quad c x_k(1 - x_k) \end{aligned}$$

for x values in $(0, 1)$ and $0 < c \leq 4$. Many dynamical systems transition into chaos as we increase a control or gain parameter, such as c. Select $c = 3.5$ and use the two choices of initial conditions, $x_0 = 0.5$ and $x_0 = 0.51$, to generate $x_1, \ldots, x_{20}$. Plot the two trajectories on graph paper. Are they aperiodic (chaotic) or periodic? Does a difference of 0.01 in initial condition significantly affect the overall shape of the discrete trajectory?

Now repeat the above experiment but use the gain parameter $c = 3.9$ (or $c = 4$). No matter how close two initial conditions, in a chaotic dynamical system they always produce divergent trajectories. Does $c = 3.9$ produce chaos?

PART ONE

NEURAL NETWORK
THEORY

2

NEURONAL DYNAMICS I: ACTIVATIONS AND SIGNALS

NEURONS AS FUNCTIONS

Neurons behave as functions. Neurons transduce an unbounded input **activation** $x(t)$ at time t into a bounded output **signal** $S(x(t))$. Usually a sigmoidal or S-shaped curve, as in Figure 2.1, describes the transduction. A sigmoidal curve also describes the input-output behavior of many operational amplifiers.

For instance, the *logistic* signal function

$$S(x) \;=\; \frac{1}{1 + e^{-cx}}$$

is sigmoidal and strictly increases for positive scaling constant $c > 0$. Strict monotonicity implies that the activation derivative of S is positive:

$$S' \;=\; \frac{dS}{dx} \;=\; cS(1 - S) \;>\; 0$$

The threshold signal function (dashed line) in Figure 2.1 illustrates a nondifferentiable signal function. In general, signal functions are piecewise differentiable. The family of logistic signal functions, indexed by c, approaches asymptotically the

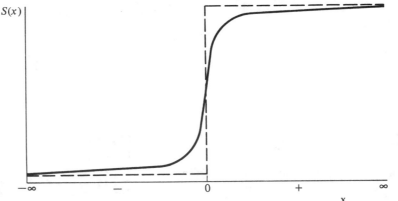

FIGURE 2.1 Signal $S(x)$ as a bounded monotone-nondecreasing function of activation x. Dashed curve defines a threshold signal function.

threshold function as c increases to positive infinity. Then S transduces positive activations x to unity signals, negative activations to zero signals. S would transduce the four-neuron vector of activations $(-6\ \ 350\ \ 49\ \ -689)$ to the four-dimensional bit vector of signals $(0\ \ 1\ \ 1\ \ 0)$. A discontinuity occurs at the zero activation value, which equals the signal function's "threshold." We can arbitrarily transduce zero activations to unity, zero, or the previous signal value. Zero activation values occur less frequently in networks with many neurons.

SIGNAL MONOTONICITY

In general, signal functions are **monotone nondecreasing**: $S' \geq 0$. Increasing activation values can only increase the output signal or leave it unchanged. They can never decrease the signal. In practice this means signal functions have an upper bound or saturation value. One admissible possibility, often found in simulation discretizations, is an increasing *staircase* signal function. The staircase signal function is a piecewise-differentiable monotone-nondecreasing signal function.

Gaussian signal functions represent an important exception to signal monotonicity. Chapter 7 in the companion volume [Kosko, 1991] applies generalized Gaussian signal functions or potential functions to supervised estimation of sampled functions. We can use Gaussian or bell-shaped signal functions in feedforward neural networks. Gaussian signal functions take the form $S(x) = e^{-cx^2}$ for $c > 0$. Then $S' = -2cxe^{-cx^2}$. So the sign of the signal-activation derivative S' is opposite the sign of the activation x. We shall assume signal functions are monotone nondecreasing unless stated otherwise.

Generalized Gaussian signal functions define potential or **radial basis functions** $S_i(\mathbf{x})$:

$$S_i(\mathbf{x}) \;=\; \exp\left[-\frac{1}{2\sigma_i^2} \sum_j^n (x_j - \mu_j^i)^2\right]$$

for input activation vector $\mathbf{x} = (x_1, \ldots, x_n) \in R^n$, variance σ_i^2, and mean vector $\boldsymbol{\mu}_i = (\mu_1^i, \ldots, \mu_n^i)$. Each radial basis function S_i defines a spherical *receptive field* in R^n. The ith neuron emits unity, or near-unity, signals for sample activation vectors $\mathbf{x}$ that fall in its receptive field. The mean vector $\boldsymbol{\mu}$ centers the receptive field in R^n. The variance σ_i^2 localizes it. The radius of the Gaussian spherical receptive field shrinks as the variance σ_i^2 decreases. The receptive field approaches R^n as σ_i^2 approaches ∞. Radial basis functions are not proper signal functions, because they depend on all neuronal activations not just the ith activation x_i. We shall consider only scalar-input signal functions $S_i(x_i)$.

Signal monotonicity means neurons are nonlinear but not too much so. The neural-network literature sometimes refers to this property as "semilinearity," especially when exemplified by logistic signal functions. Linear signal functions make computation, and analysis, comparatively easy. But linear signal functions do not suppress noise. So linear networks are not robust. Nonlinearity increases a network's computational richness and facilitates noise suppression. But nonlinearity risks computational, and analytical, intractability. Nonlinearity also favors dynamical instability. Signal monotonicity seems to be nature's evolved balance, since almost all biological neurons have sigmoidal signal characteristics.

Signal and Activation Velocities

The **signal velocity** dS/dt, denoted $\dot{S}$, measures a signal's instantaneous time change. The chain rule gives $\dot{S}$ as the product of the change in the signal due to the activation and the change in the activation with time:

$$\begin{aligned} \dot{S} &= \frac{dS}{dx}\frac{dx}{dt} \\ &= S'\dot{x} \end{aligned}$$

So signal velocities depend explicitly on activation velocities. This dependence, largely overlooked in the neural literature, increases the number of unsupervised learning laws that adapt with **locally available** information. In some of the Hebbian-type synapses discussed in Chapter 4, a synapse has access only to presynaptic signals and the postsynaptic neuron it abuts. So the synapse has access only to the postsynaptic neuron's membrane potential or activation. Then the chain rule implies that changes in the postsynaptic neuron's output signal, perhaps due to competition with neighboring neurons, may causally (molecularly) affect abutting synapses through changes in the postsynaptic neuron's activation.

BIOLOGICAL ACTIVATIONS AND SIGNALS

The neurophysiological interpretation of the activation $x(t)$ and the signal $S(x(t))$ involves electrical pulses of potential difference and their temporal sum-

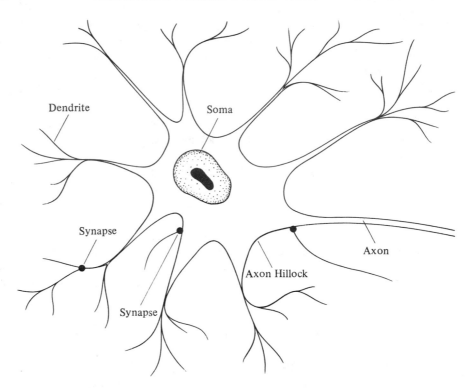

FIGURE 2.2 Neuron anatomy. Dendritic tree extends from the cell nucleus or soma. Longer branches are axons. Synapses bridge axons from other neurons to the cell membrane or its dendrites. Cell capacitance time-integrates arriving synaptic signals.

mation. Activations involve small membrane pulses. Signals involve large axonal pulses or action potentials.

A potential difference in electrical charge occurs between the inside and outside of a neuron's surface membrane. Impinging neurotransmitter chemical signals, sent from other neurons across abutting synapses, modulate the neuron's membrane potential. Synapses transduce pulse-coded electrical neuronal signals into neurotransmitter signals. Figure 2.2 shows the key functional units of a biological neuron.

Membrane potential differences or pulses accumulate at the membrane of the **axon hillock**, where the neuron connects to one of its **axons** or long branches [Shepherd, 1979]. The axon-hillock membrane generates a **signal pulse** or **action potential** if the small potential-difference pulses arriving at the neuron's axon hillock exceed a time-varying threshold of activity. The large signal pulse propagates down the axon and its branches, where axonal insulators restore and amplify the signal as it propagates, until it arrives at a **synaptic junction**. Synaptic processes transduce the electrical signal into a chemical neurotransmitter signal, which in turn affects the potential difference across the membrane of the abutting postsynaptic neuron. The threshold potential takes values near -40 millivolts.

At the molecular level, ions drive changes in potential difference. **Ionic flow** changes the axon-hillock membrane potential by modifying the membrane's conductance. Conductance change involves selectively opening and closing **molecular channels**. Sodium (Na^+) and potassium (K^+) ions flow across the membrane through these molecular channels. The sodium and potassium ions change their conductance, their ability to flow, as they flow across the membrane by altering these molecular channels in complex ways. Sodium and potassium are an **antagonistic** ionic pair. Sodium ions are **excitatory**. They increase the membrane potential. Potassium ions are **inhibitory**. They decrease the membrane potential. The signal threshold summarizes the molecular channels through which the ions flow. The threshold, or more accurately the threshold effect, varies inversely with the number of open molecular channels. The more open molecular channels available for current flow, the lower the threshold that temporally summed potential-difference pulses must exceed to emit a signal pulse. Calcium, chloride, and other ions also affect neuronal signal transmission and reception.

Mathematically, the real-valued activation $x(t)$ represents the membrane potential or voltage difference across the neuron's surface membrane at time t, at least in the sense of temporal summation of small pulses. The activation can be positive or negative. In theory the activation can even be infinite.

The signal $S(x)$ induced by the activation x represents the neuron's **firing frequency** of action potentials, or pulses, in a sampling interval. In biological neurons the sampling interval may consist of the past 10 to 30 milliseconds. The firing frequency equals the average number of pulses emitted in a sampling interval. This lumped-frequency interpretation suggests signal values $S(x)$ should be nonnegative. This need not be the case. We shall allow negative frequencies, or **signed signals**, to occur.

Competitive Neuronal Signals

Nonnegative signals often describe the **competitive** status of neurons "competing" in a laterally inhibitory neuron field. Here each neuron excites itself, and possibly near neighbors, and inhibits distant neighbors according to the sign of the synaptic connections emanating from the neuron to its neighbors. The neurons "compete" for activation and thus for signal strength. In the competitive setting a signal function, taking values in the unit interval $[0, 1]$, keeps score of how the neuron fares in the competition. The neuron "wins" at time t if $S(x(t)) = 1$, "loses" if $S(x(t)) = 0$, and otherwise possesses a fuzzy win-loss status between 0 and 1. Competitive neurons usually possess steep signal functions that approximate threshold signal functions.

In practice signal values are usually binary or bipolar. **Binary** signal functions take values in the unit interval $[0, 1]$. **Bipolar** signals are signed. They take values in the bipolar interval $[-1, 1]$. Binary and bipolar signals transform into each other by simple scaling and translation. For instance, the bipolar logistic signal function takes the form

$$\frac{2}{1 + e^{-cx}} - 1$$

Neurons with bipolar threshold signal functions are sometimes called **McCulloch-Pitts neurons**.

NEURON FIELDS

In neural networks we deal with **fields** of neurons. A field of neurons is a topological grouping. In general, neural networks contain many fields of neurons. Neurons within a field are topologically ordered, often by proximity. Planar hexagonal packing of neurons provides a different topological ordering. Three-dimensional or volume proximity packing, often found in mammalian brains, provides another topological ordering.

In the simplest case, which we assume thoughout this book, neurons are not topologically ordered. They are related only by the synaptic connections between them. Kohonen [1988] calls this lack of topological structure in a field of neurons the **zeroth-order topology**. This lack of topological structure suggests that neural-network models are indeed abstractions, not descriptions, of mammalian-brain neural networks. In the brain, order matters.

We denote by F_X the default field of neurons. We will usually denote a second field of neurons by F_Y, and a third by F_Z. We shall develop many neural network results valid for any number of neuron fields. In this case we usually use the minimal neural hierarchies $\{F_X, F_Y\}$ or $\{F_X, F_Y, F_Z\}$ instead of the more accurate, more cumbersome indexed-hierarchy notation $\{F_1, \ldots, F_k\}$. For instance, we denote a three-layer feedforward neural network as $F_X \rightarrow F_Y \rightarrow F_Z$. Here F_X and F_Z represent input and output fields.

We use Kohonen's notation for dimensionality. Input field F_X contains n neurons. Output field F_Y contains p neurons. The neural system samples or "experiences" m vector associations $(\mathbf{x}_i, \mathbf{y}_i)$. The notation reminds us that the neural system samples a function $f : R^n \longrightarrow R^p$ m times to generate the associated pairs $(\mathbf{x}_1, \mathbf{y}_1), \ldots, (\mathbf{x}_m, \mathbf{y}_m)$. The overall neural network behaves as an **adaptive filter**. F_X neurons and synapses transduce or filter input data streams into data streams at F_Y, and often vice versa (simultaneously). Sample data changes network parameters.

NEURONAL DYNAMICAL SYSTEMS

We describe the **neuronal dynamical system** by a system of first-order differential or difference equations that govern the time evolution of the neuronal activations or membrane potentials. Different differential equations govern the synaptic dynamical system, as we discuss in Chapters 4 through 6. For the fields F_X and F_Y, we denote the activation differential equations as

$$\dot{x}_1 \;=\; g_1(F_X, F_Y, \ldots) \tag{2-1}$$

$$\vdots$$

$$\dot{x}_n \;=\; g_n(F_X, F_Y, \ldots) \tag{2-2}$$

$$\dot{y}_1 \;=\; h_1(F_X, F_Y, \ldots) \tag{2-3}$$

$$\vdots$$

$$\dot{y}_p \;=\; h_p(F_X, F_Y, \ldots) \tag{2-4}$$

or, in vector notation,

$$\dot{\mathbf{x}} \;=\; \mathbf{g}(F_X, F_Y, \ldots) \tag{2-5}$$

$$\dot{\mathbf{y}} \;=\; \mathbf{h}(F_X, F_Y, \ldots) \tag{2-6}$$

where x_i and y_j denote respectively the activation time functions of the ith neuron in F_X and the jth neuron in F_Y. The arguments of g_i and h_j functions also include synaptic and input information.

We do not include time as an independent variable. As a result, in dynamical systems theory, neural-network models classify as **autonomous** dynamical systems. Nonautonomous dynamical systems might allow the change in activation x_i to depend, say, additively on t^2. Autonomous systems are usually easier to analyze than nonautonomous systems.

Time does play a special role in neuronal dynamics: time is "fast" at the neuronal level. In mammalian neural systems, membrane fluctuations occur at the millisecond level. In hardware or computer implementations of neural networks, neuronal fluctuations can in principle occur at the nanosecond level. In contrast, time is "slow" at the synaptic level. In mammalian neural systems, synaptic fluctuations occur at the second or minute level. We think faster than we learn.

Note the absence of second-order time derivatives (accelerations) and partial derivatives in (2-1) through (2-6). This represents one distinction between neural-network models and the models of classical "neural modeling," where often the differential equations give a detailed, multivariable description of how individual neurons or synapses behave.

Neuronal State Spaces

We define the **state** of the neuronal dynamical system at time t with the instantaneous vectors of activations

$$X(t) \;=\; (x_1(t), \ldots, x_n(t))$$

$$Y(t) \;=\; (y_1(t), \ldots, y_p(t))$$

For convenience, sometimes we shall identify the topological groupings or fields F_X and F_Y with their respective activation state vectors X and Y.

The **state space** of field F_X is the extended real vector space R^n (with positive and negative infinities appended). Similarly R^p is the state space of F_Y. The state space of the entire or joint neuronal dynamical system is the product space $R^n \times R^p$. A point in the state space specifies a snapshot of all neuronal behavior.

A smooth time-indexed curve defines a **trajectory** in the state space. A trajectory describes the time evolution of the network activations.

We can concatenate the fields F_X and F_Y into a single larger field $F_Z = [F_X \mid F_Y]$, the **augmented field**. Then the network state space corresponds to the higher-dimension vector space R^{n+p}. In Kohonen's [1988] terminology, this augmentation converts a **heteroassociative** network into an **autoassociative** network. We must use this formal equivalence with care. Just as the numbers 4 and $1+1+1+1$ are formally equivalent but have different computational properties, concatenated fields can have different computational, metrical, or other properties. For example, metrical or nearest-neighbor classification often degrades in higher dimensions. (Example of dimension-dependent effect: the volume of unit Euclidean hyperspheres increases up to the fifth dimension, then decreases for higher dimensions.) When the fields F_X and F_Y possess topological structure other than connectivity, their concatenation becomes even less plausible.

Signal State Spaces as Hypercubes

We observe the **signal state** $S(X)$ of field F_X at time t. $S(X)$ denotes the n-vector of signals emitted by the neurons in F_X:

$$S(X(t)) \quad = \quad (S_1^X(x_1(t)), \ldots, S_n^X(x_n(t)))$$

S_i^X denotes the signal function of the ith neuron in field F_X. Different neurons can have different nonlinear signal characteristics. For notational simplicity we omit the superscripted field identification. So $S_i(x_i)$ and $S_j(y_j)$ denote respectively the signal function of the ith neuron in field F_X and the signal function of the jth neuron in field F_Y.

The **signal state space** of F_X consists of all possible signal states. The boundedness of signal functions implies that the signal state space is an n-dimensional hypercube. Anderson [1983] refers to this as "**brain states in a box.**"

In the special but common case where signal functions take values in the unit interval $[0, 1]$, the signal state space equals the unit hypercube I^n, or $[0, 1]^n$. In general we can scale and translate signal functions so that the unit hypercube equals the signal state space of the network. Then the product cube $I^n \times I^p$ defines the signal state space of the two-field network $\{F_X, F_Y\}$, and I^{n+p} defines the signal state space of the concatenated one-field network $[F_X \mid F_Y]$.

In stable feedback neural networks, equilibrium points tend to occur at or near the 2^n vertices of the cube. Signal-function monotonicity produces an extremizing effect that feedback tends to amplify. Hopfield and Tank [1985] exploit this property when constructing networks for solving optimization problems, in particular the traveling-salesman problem. Chapters 2 and 3 in the companion volume [Kosko, 1991] apply this optimization technique to image processing.

The unit hypercube I^n, and hence the binary signal state space, also defines the **fuzzy power set** $F(2^X)$ on n elements, the set of all fuzzy subsets of the set $X = \{x_1, \ldots, x_n\}$ of n elements x_i. As discussed in Chapter 7, a multivalued or **fuzzy set** A corresponds to a point in I^n. The ith coordinate of A indicates the degree to which element x_i belongs to set A, the **fit** (fuzzy unit) value of x_i. So the usual 2^n subsets of X correspond to the 2^n vertices of the unit cube I^n. Hence geometrically many neural and fuzzy computations coincide. A common application of this isomorphism in feature detection interprets the signal value $S(x_i(t))$ as the degree of membership of x_i in feature set A at time t, or the degree to which a sampled input pattern belongs to the ith pattern class.

Infinite sets, such as R^n, also contain fuzzy subsets. Radial basis functions define spherical receptive-field fuzzy subsets of R^n.

Neuronal Activations as Short-Term Memory

We model **short-term memory** (STM) in neural networks as the activation state vectors X and Y. Patterns or waveforms that persist, however briefly, among the activations represent STM events, purely psychological phenomena. We usually forget in a matter of seconds a new name or telephone number when we meet several people at a busy convention or party. This provides informal evidence that activation patterns compete for storage in STM.

Synapses encode **long-term memory** (LTM) pattern information. Another neural-network conjecture concerns sleep. Sleep may allow "winning" STM activation patterns to be downloaded into LTM. High arousal (hormone) levels may favor concurrent STM activation patterns in the competitive encoding process: *Arousal up, learning up*. The emotional highs of our day invade our dreams of night.

Physiology and psychology, flesh and spirit, connect when we identify neuronal activation patterns with STM. Membrane potentials provide a modern version of Descartes' pineal glands, which the seventeenth-century naturalist believed connected mind and body.

We can also identify neuronal activations with sense modalities. Activation patterns may represent auditory, olfactory, tactile, or visual patterns. Or they may represent rates of muscle contraction in motor systems. This representational generality unifies and strengthens neural-network theory. Fields of neuronal activations, of membrane potentials, compose the common language of the brain, a language that adaptively integrates sensory, motor, and cognitive activity. Fluctuating neuronal fields may provide dynamic phonemes of thought.

COMMON SIGNAL FUNCTIONS

There are at least as many bounded monotone-nondecreasing signal functions as there are real numbers. This vast function space allows model neurons to possess widely varying signal characteristics. It also bestows great generality on the global stability theorems derived in Chapter 6. We now review some of the more popular signal functions.

The **logistic** signal function remains the most popular binary signal function. We define it as

$$S(x) = \frac{1}{1 + e^{-cx}}$$

where here and throughout c denotes a positive scaling constant. The activation derivative obeys $S' = cS(1 - S) > 0$. So S is monotone increasing.

The logistic signal function equals an exponential divided by a sum of two exponentials, or a Gibbs probability density function. So the logistic signal function is the *maximum-entropy* [Jaynes, 1983] signal function, which, along with its computational simplicity, may account for its primacy in neural-network models.

An infinitely steep logistic signal function gives rise to a **threshold** signal function

$$S(x^{k+1}) = \begin{cases} 1 & \text{if } x^{k+1} > T \\ S(x^k) & \text{if } x^{k+1} = T \\ 0 & \text{if } x^{k+1} < T \end{cases}$$

for an arbitrary real-valued threshold T. The index k indicates the discrete time step. The index notation implies that threshold signal functions instantaneously transduce discrete activations to signals. When the activation equals the threshold at time $k+1$, the signal maintains the same value it had at time k. The neuron does not make a state-update "decision" at time $k + 1$. Hopfield [1982] observed that we can model asynchronous update policies if each neuron in a network randomly makes update decisions, perhaps with a fixed mean update rate. We discuss this interpretation in Chapter 3.

A naturally occurring bipolar signal function is the **hyperbolic-tangent** signal function

$$S(x) = \tanh(cx)$$

with activation derivative $S' = c(1 - S^2) > 0$. Similar to the logistic signal function, the hyperbolic tangent equals a ratio of summed exponentials:

$$\tanh(cx) = \frac{e^{cx} - e^{-cx}}{e^{cx} + e^{-cx}}$$

The exponentials give rise to the simple form of the derivative and its positivity.

The **threshold linear** signal function is a binary signal function often used to approximate neuronal firing behavior:

$$S(x) = \begin{cases} 1 & \text{if} & cx \geq 1 \\ 0 & \text{if} & cx < 0 \\ cx & \text{else} \end{cases}$$

which we can rewrite as

$$S(x) = \min(1, \max(0, cx))$$

Between its upper and lower bounds the threshold linear signal function is trivially monotone increasing, since $S' = c > 0$.

The threshold linear function stays as linear as a signal function can stay and yet remain bounded. For the **linear** signal function, $S(x) = cx$, is unbounded and not a proper signal function. Suitably scaled logistic and hyperbolic-tangent signal functions behave linearly in their midregions and thus can approximate threshold linear signal functions.

A related signal function is the **threshold exponential** signal function:

$$S(x) = \min(1, e^{cx})$$

When $e^{cx} < 1$, $S' = ce^{cx} > 0$. The second derivative is also positive, $S'' = c^2 e^{cx} > 0$, as indeed are all higher-order activation derivatives, $S^n = c^n e^{cx} > 0$. The threshold exponential rapidly *contrast enhances* signal patterns. This speeds network convergence.

Many signal functions arise if negative activations emit zero signals. Only positive activations generate *supra*threshold signals. So we can use familiar probability *distribution* functions as binary signal functions. For instance, though the Gaussian probability *density* function does not produce a monotone-nondecreasing signal function, its integral, the Gaussian probability distribution function, does. Unfortunately this integral is difficult to compute.

A more tractable example is the **exponential-distribution** signal function, which provides another type of thresholded exponential signal function:

$$S(x) = \max(0, 1 - e^{-cx})$$

For $x > 0$, suprathreshold signals are monotone increasing since $S' = ce^{-cx} > 0$. Note, though, that the exponential-distribution signal function is strictly convex, since $S'' = -c^2 e^{-cx} < 0$. Strict convexity produces a "diminishing returns" effect as the signal approaches saturation.

Another example is the family of **ratio-polynomial** signal functions:

$$S(x) = \max\left(0, \frac{x^n}{c + x^n}\right)$$

for $n > 1$. For positive activations, suprathreshold signals are monotone increasing, since

$$S' = \frac{cnx^{n-1}}{(c + x^n)^2} > 0$$

PULSE-CODED SIGNAL FUNCTIONS

Electrical pulse trains propagate down the axons emanating from a neuron. Biological neuronal systems use complex electrochemical mechanisms to amplify these pulses on their journey to synaptic junctions. Individual pulses seem to carry little information, though below we suggest they indicate the sign of signal velocities. Pulse frequencies—pulse trains arriving in a sampling interval—seem to be the bearers of neuronal signal information.

We can reliably decode **pulse-coded** signal information. We can more easily decode the occurrence or absence of a pulse in a sampling interval than we can decode a specific graded signal value. Arriving individual pulses can be somewhat corrupted in shape and still accurately decoded as present or absent. The sampling interval in biological neurons may be only a few milliseconds. During this moving window the neuron counts, sums, or integrates the most recent pulses. We can naturally represent this windowing effect with a "fading memory" weighted time integral or sum. In the simplest case we use an exponentially weighted time average of sampled binary pulses:

$$S_i(t) = \int_{-\infty}^{t} x_i(s)e^{s-t}ds \qquad (2\text{-}7)$$

$$S_j(t) = \int_{-\infty}^{t} y_j(s)e^{s-t}ds \qquad (2\text{-}8)$$

In the pulse-coded formulation x_i and y_j denote binary pulse functions that summarize the excitation of the membrane potentials of the ith and jth neurons in the respective fields F_X and F_Y. The function x_i equals one if a pulse arrives at time t, and zero if no pulse arrives:

$$x_i(t) = \begin{cases} 1 & \text{if a pulse occurs at } t \\ 0 & \text{if no pulse at } t \end{cases} \qquad (2\text{-}9)$$

and similarly for $y_j(t)$.

Pulse-coded signal functions are binary signal functions. Pulse-coded signal functions are bipolar if we use bipolar pulses instead of binary pulses—if $x_i(t) = -1$ when no pulse occurs at t. We shall assume binary pulses throughout.

Direct integration proves that pulse-coded signals take values in the unit interval $[0, 1]$. For, consider the two extreme cases of (2-7). At the first extreme, $x_i(t) = 0$ occurs for the entire time history. Then the integral trivially equals zero.

At the other extreme, $x_i(t) = 1$ occurs for the entire time history. All other pulse trains produce integrals bounded between these two extremes. But when $x_i = 1$ everywhere,

$$
\begin{aligned}
S_i(t) &= \int_{-\infty}^{t} e^{s-t} ds \\
&= e^{t-t} - \lim_{s \to -\infty} e^{s-t} \\
&= 1
\end{aligned}
$$

When x_i increases from 0 to 1, the pulse count S_i can only increase. So S_i is monotone nondecreasing. The exponential weighting factor further ensures strict monotonicity. (Strictly speaking, the zero and unity instantaneous pulses have measure zero and make no contribution to the integral. Instead small epsilon-intervals of pulses ensure monotonicity. For discrete sums of exponentially weighted pulses, though, the argument remains valid as stated.)

Velocity-Difference Property of Pulse-Coded Signals

The exponentially weighted integrals in (2-7) and (2-8) should look familiar. Consider the first-order linear inhomogenous differential equation:

$$\dot{x} + p(t)x = q(t) \tag{2-10}$$

Suppose $p(t) = 1$ for all t, and for simplicity suppose time increases from zero to positive infinity. Then the solution to this differential equation involves an exponentially weighted integral:

$$x(t) = x(0)e^{-t} + \int_{0}^{t} q(s)e^{s-t} ds \tag{2-11}$$

The pulse-coded signal functions (2-7) and (2-8) have the same form as (2-11) when minus infinity is the lower bound on the time integration. For then the exponential weight on the initial condition $x(0)$ in (2-11) equals zero. Now interpret x in (2-10) and (2-11) as the signal function S, remember that $p(t) = 1$ everywhere, and interpret $q(t)$ as a binary pulse function. Then equation (2-10) gives a surprisingly simple form for the signal velocity:

$$\dot{S}_i(t) = x_i(t) - S_i(t) \tag{2-12}$$

$$\dot{S}_j(t) = y_j(t) - S_j(t) \tag{2-13}$$

So pulse-coded signal functions are once but not twice time differentiable. We can integrate pulse functions, but we cannot formally differentiate them.

We have just derived the central result of pulse-coded signal functions: *The instantaneous signal-velocity equals the current pulse minus the current expected pulse frequency.* Gluck and Parker [1988, 1989] first noted this velocity representation of pulse-coded signal functions in their study of unsupervised signal-velocity learning laws, in particular a version of the differential Hebbian law discussed in Chapter 4. This important relationship warrants a name. We shall call it the **velocity-difference property** of pulse-coded signal functions, because it identifies the signal velocity with a pulse-signal difference.

The velocity-difference property implies that signal-velocity computation is *easy* and *local*. Neurons and synapses can compute the time derivative without complicated, approximate, and perhaps unstable differencing techniques. We cannot plausibly propose that biological or electrical synapses can compute with such techniques in real time. Equations (2-12) and (2-13) give an exact form for the instantaneous signal velocity. Moreover, the current pulse and the current signal, or expected pulse frequency, are both **locally available** quantities, a necessary property for biological synapses and a desirable property for artificial synapses. Synapses can compute signal velocities on the spot in real time.

Pulse-coded signal functions offer another computational advantage. A synapse may need only the sign of the signal derivative at any instant. The synapse can determine the sign simply by examining the current pulse value, since pulse-coded signal functions take values in $[0, 1]$. If a pulse arrives at time t, $1 - S(t) > 0$ unless, in the extreme case, a sustained pulse, or a high-density pulse train, has recently arrived. If a pulse is absent, $-S(t) < 0$ unless, in the other extreme, no pulses, or very low-density pulse trains, have recently arrived. In both extremes the velocity equals zero. The signal has stayed constant for some relatively long length of time. Other mechanisms can check for this condition.

In Chapter 4 we will see how pulse-coded signal functions suggest reinterpreting classical Hebbian and competitive learning. There we obtain methods of local unsupervised learning as special cases of the corresponding differential Hebbian and differential competitive learning laws when pulse-coded signal functions replace the more popular, less realistic logistic or other non-pulse-coded signal functions.

We have examined pulse-coded signal functions in detail because of their coding-decoding importance to biological and artificial neurons and synapses, and because of their near absence from the neural-network literature. Pulse-coded neural systems, like signal-velocity learning schemes, remain comparatively unexplored. They offer a practical reformulation of neural-network models and promise fresh research and implementation insights.

REFERENCES

Anderson, J. A., "Cognitive and Psychological Computation with Neural Models," *IEEE Transactions on Systems, Man, and Cybernetics*, vol. SMC-13, 799-815, September-October 1983.

Gluck, M. A., Parker, D. B., and Reifsnider, E., "Some Biological Implications of a Differential-Hebbian Learning Rule," *Psychobiology*, vol. 16, no. 3, 298-302, 1988.

Gluck, M. A., Parker, D. B., and Reifsnider, E. S., "Learning with Temporal Derivatives in Pulse-Coded Neuronal Systems," *Proceedings 1988 IEEE Neural Information Processing Systems (NIPS) Conference*, Denver, CO, Morgan Kaufman, 1989.

Hopfield, J. J., "Neural Networks and Physical Systems with Emergent Collective Computational Abilities," *Proceedings of the National Academy of Science*, vol. 79, 2554-2558, 1982.

Hopfield, J. J., and Tank, D. W., "Neural Computation of Decisions in Optimization Problems," *Biological Cybernetics*, vol. 52, 141-152, 1985.

Jaynes, E. T., *Papers on Probability, Statistics, and Statistical Physics,* Rosenkrantz, R. D. (ed.), Reidel, Boston, 1983.

Kosko, B., *Neural Networks for Signal Processing*, Prentice Hall, Englewood Cliffs, NJ, 1991.

Kohonen, T., *Self-Organization and Associative Memory*, 2nd ed., Springer-Verlag, New York, 1988.

Shepherd, G. M , *The Synaptic Organization of the Brain*, 2nd ed., Oxford University Press, New York, 1979.

PROBLEMS

2.1. Show that the following signal functions S have activation derivatives dS/dx, denoted S', of the stated form:

   (a) If $S(x) = \dfrac{1}{1 + e^{-cx}}$, then $S'(x) = cS(1 - S)$.

   (b) If $S = e^{-cx^2}$, then $S' = -2cxe^{-cx^2}$.

   (c) If $S = \tanh(cx)$, then $S' = c(1 - S^2)$.

   (d) If $S = e^{cx}$, then $d^n S/dx^n = c^n e^{cx}$.

   (e) If $S = 1 - e^{-cx}$, then $S'' = -c^2 e^{-cx}$.

   (f) If $S = \dfrac{x^n}{c + x^n}$, then $S' = \dfrac{cnx^{n-1}}{(c + x^n)^2}$.

2.2. Consider the logistic signal function:

$$S(x) \;\;=\;\; \frac{1}{1 + e^{-cx}}, \quad c > 0$$

   (a) Solve the logistic signal function $S(x)$ for the activation x.

   (b) Show that x strictly increases with logistic S by showing that $dx/dS > 0$.

   (c) Explain why in general the "inverse function" x increases with S if $S' > 0$.

2.3. Consider the general first-order linear inhomogeneous differential equation:

$$\dot{x} + p(t)x \;\;=\;\; q(t)$$

for $t \geq 0$. Show that the solution is

$$x(t) \;\;=\;\; e^{-\int_0^t p(s)ds}\left[x(0) + \int_0^t q(s)e^{\int_0^s p(w)dw}\, ds\right]$$

2.4. Consider the pulse-coded signal function:

$$S(t) \;=\; \int_{-\infty}^{t} x(s) e^{s-t} \, ds$$

Use the general solution in Problem 2.3 to derive the velocity-difference property:

$$\dot{S} \;=\; x - S$$

3

NEURONAL DYNAMICS II: ACTIVATION MODELS

NEURONAL DYNAMICAL SYSTEMS

Neuronal activations change with time. How they change depends on the dynamical equations (2-1) through (2-6) in Chapter 2. These equations specify how the activations x_i and y_j, the membrane potentials of the respective ith and jth neurons in fields F_X and F_Y, change as a function of network parameters:

$$\dot{x}_i = g_i(X, Y, \ldots) \qquad (3\text{-}1)$$

$$\dot{y}_j = h_j(X, Y, \ldots) \qquad (3\text{-}2)$$

Below we develop a general family of activation dynamical systems, study some of their steady-state behavior, and show how we can derive some from others. We begin with the simplest activation model, the passive-decay model, and gradually build up to the most complex models. We postpone the general global analysis of these models until Chapter 6. The global analysis requires incorporating synaptic dynamics as well. In this chapter we assume the neural systems do not learn when they operate. Synapses either do not change with time or change so slowly we can assume them constant. Chapters 4 and 5 study synaptic dynamics.

ADDITIVE NEURONAL DYNAMICS

What affects a neuron? External inputs and other neurons may affect it. Or nothing may affect a neuron—then the activation should decay to its resting value. So in the absence of external or neuronal stimuli, first-order **passive decay** describes the simplest model of activation dynamics:

$$\dot{x}_i = -x_i \tag{3-3}$$

$$\dot{y}_j = -y_j \tag{3-4}$$

Then the membrane potential x_i decays exponentially quickly to its zero resting potential, since $x_i(t) = x_i(0)e^{-t}$ for any finite initial condition $x_i(0)$.

In the simplest case we add to (3-3) and (3-4) external inputs and information from other neurons. First, though, we examine and interpret simple additive and multiplicative scaling of the passive-decay model.

Passive Membrane Decay

We can add or multiply constants as desired in the neural network models in this book. In practice, for instance, the **passive-decay rate** $A_i > 0$ scales the rate of passive decay to the membrane's resting potential:

$$\dot{x}_i = -A_i x_i \tag{3-5}$$

with solution $x_i(t) = x_i(0)e^{-A_i t}$. The default passive-decay rate is $A_i = 1$. A relation similar to (3-5) holds for y_j as well, but for brevity we omit it.

Electrically the passive-decay rate A_i measures the cell membrane's resistance, or "friction," to current flow. The larger the decay rate A_i, the faster the decay—the less the resistance to current flow. So A_i varies inversely with the **resistance** R_i of the cell membrane:

$$A_i = \frac{1}{R_i} \tag{3-6}$$

Ohm's law states that the **voltage drop** V_i in volts across a resistor equals the **current** I_i in amperes scaled by the resistance R_i in ohms

$$V_i = I_i R_i \tag{3-7}$$

If we interpret the activation variable x_i as the voltage V_i, (3-6) allows us to interpret the passive-decay term $A_i x_i$ as the current I_i flowing through the resistive membrane.

The inverse resistance $1/R_i$—the passive-decay rate A_i—summarizes the **conductance** g_i of the cell membrane. We measure g_i in amperes per volt. The conductance measures the *permeability* of membrane, axonal, or synaptic ions. The most common neuronal ions are calcium (Ca^{2+}), chloride (Cl^-), potassium (K^+), and sodium (Na^+) ions. Each ion type has its own conductance characteristic g_i. The notation g_i is standard and usually indexed by ion type. This index and context should successfully distinguish it from the nonlinear function g_i described above in Equation (3-1).

Membrane Time Constants

The **membrane time constant** $C_i > 0$ scales the time variable of the activation dynamical system. The notation C_i suggests that the *membrane capacitance* affects the time scale of activation fluctuation. The smaller the capacitance, the faster things change. As the membrane capacitance increases toward positive infinity, membrane fluctuation slows to a stop. So the multiplicative constant C_i scales the entire function g_i in (3-1) or h_i in (3-2) above, in this case:

$$C_i \dot{x}_i \;=\; -A_i x_i \tag{3-8}$$

with solution $x_i(t) = x_i(0)e^{-(A_i/C_i)t}$. The default neuronal time constant is $C_i = 1$.

The voltage drop V_i across the ith capacitor with **capacitance** C_i in farads and stored **charge** Q_i in coulombs equals the ratio

$$V_i \;=\; \frac{Q_i}{C_i} \tag{3-9}$$

Since C_i is constant and the current is the time derivative of the charge, $I_i = dQ_i/dt$, time differentiation of (3-9) gives the membrane current as the left-hand side of (3-8):

$$I_i \;=\; C_i \dot{V}_i \;=\; C_i \dot{x}_i \tag{3-10}$$

Kirchoff's conservation laws underlie the electrical interpretation of (3-8). **Kirchoff's voltage law** states that *the sum of the voltages in a closed electrical loop equals zero.* **Kirchoff's current law** states that *the sum of currents directed toward a circuit junction equals the sum of currents directed away from the junction.* Equations (3-6), (3-7), and (3-10) show that Equation (3-8) obeys Kirchoff's current law. Equation (3-8) equivalently states that the sum of the voltage drops across the membrane resistor and membrane capacitor equals the zero applied voltage.

Membrane Resting Potentials

The **resting potential** P_i is an additive constant. It need not be positive. The default resting potential equals zero: $P_i = 0$. We define the resting potential as the activation value to which the membrane potential equilibrates in the absence of external or neuronal inputs:

$$C_i \dot{x}_i \;=\; -A_i x_i + P_i \tag{3-11}$$

with solution

$$x_i(t) \;=\; x_i(0)e^{-(A_i/C_i)t} + \frac{P_i}{A_i}(1 - e^{-(A_i/C_i)t}) \tag{3-12}$$

The time-scaling capacitance C_i enters the solution only transiently by scaling the exponential. The capacitance does not affect the asymptotic or *steady-state* solution P_i/A_i, which also does not depend on the choice of the finite initial condition $x_i(0)$.

Steady state means zero system time derivative in a differentiable model. So we can often directly find the steady-state solution by setting a dynamical equation, in particular Equation (3-11), equal to zero and solving for the variable in question. Of course this shortcut gives no insight into the transient behavior that precedes steady state—if *fixed-point* steady state ever occurs. Most dynamical systems have oscillatory or aperiodic (chaotic) equilibrium behavior. In many biological or implementation instances, transient behavior may be more important than steady-state behavior, especially if convergence is slow or if one needs to estimate or manipulate the rate of convergence, as in the "simulated annealing" discussed in Chapter 6.

If $A_i = 1$ in (3-11), the steady-state activation value equals the resting potential P_i. If the resting potential P_i is large enough (*supra*threshold), the neuron will spontaneously generate signals or pulse trains in the absence of external or neuronal stimuli, an observed biological effect. If $P_i = 0$ and if the logistic signal function, $S_i(x_i) = (1 + e^{-cx})^{-1}$, transduces membrane activations into signals, then at steady state $S_i = \frac{1}{2}$, which may correspond to a moderately dense pulse train in the pulse-coded formulation.

In general the parameters A_i, C_i, and P_i vary slowly with time. They are functionals, or mathematical summaries, of complex biophysical processes. They change so slowly relative to activation or membrane fluctuations that we can assume them constant during a neuronal epoch of equilibration. So (3-12) still describes transient and steady-state behavior. This property increases significantly the biological and implementation generality of neuronal models.

The resting potential also accounts for the frequent use of the term *potential difference* to describe the membrane activation. For the **potential difference** x equals the difference between the *measured voltage* V_{measured} of the cell membrane and its *resting voltage* or potential V_{rest}:

$$x \quad = \quad V_{\text{measured}} - V_{\text{rest}}$$

So the activation x can equal a positive or negative real number even if the voltages are nonnegative.

Additive External Input

Suppose we apply a relatively constant numerical **input** I_i to a neuron. The notation is standard but ambiguous, since we have so far used I_i to denote the membrane current instead of an applied input current. In general I_i will represent input except where stated otherwise.

The simplest way the input I_i can affect the membrane potential or activation value is additively:

$$\dot{x}_i \quad = \quad -x_i + I_i \tag{3-13}$$

where we have set scaling parameters to their default values: unity passive-decay rate $A_i = 1$, unity time-scale capacitance $C_i = 1$, and zero resting potential $P_i = 0$. Then exponentially quickly the membrane potential adapts its behavior to the applied external stimulus:

$$x_i(t) = x_i(0)e^{-t} + I_i(1 - e^{-t}) \longrightarrow I_i \qquad (3\text{-}14)$$

which describes a special case of (3-12) above.

This asymptotic property underlies the practical expedient of entering data into a network by identifying some neuronal output values, usually those in the first or input field F_X, with the input data. Strictly speaking, this practice is sound only if such input neurons have linear, and hence unbounded, signal functions. In feedforward networks we can assume this with impunity.

The input I_i can represent the magnitude of directly experienced sensory information or directly applied control information. Biologically I_i often represents the steady-state output from a neuron in an adjoining neural network. Then the input only indirectly represents sensory, control, or other information. Mathematically we require only that the input be numerical—nonnegative if it represents a signal intensity—and that it *change slowly* in time relative to activation changes. Slow change allows us to assume the input I_i constant.

ADDITIVE NEURONAL FEEDBACK

Neurons do not compute alone. Neurons modify their state activations with external input and with feedback from one another.

This feedback takes the form of *path-weighted signals* from synaptically connected neurons. Signal functions instantly transduce the ever-changing activations into bounded signals. The signals propagate down axons and through synapses, after which the signals perturb postsynaptic neuron membranes and hence activations. The process is asynchronous, nonlinear, massively fed back, fast, and ceaseless. The global stability theorems in Chapter 6 summarize how computational order emerges from this high-dimensional dynamical anarchy. We now develop the basic concepts and notations of synaptic webs represented by connection matrices.

Synaptic Connection Matrices

Suppose the n neurons in field F_X synaptically connect to the p neurons in field F_Y. Imagine an axon from the ith neuron in F_X that terminates in a synapse m_{ij} that abuts the jth neuron in F_Y. We assume that the real number m_{ij} summarizes the synapse, and that m_{ij} changes so slowly relative to activation fluctuations that it is constant. Thus we assume no learning: $\dot{m}_{ij} = 0$ for all t. The synaptic value m_{ij} might represent the average *rate* of release of a neurotransmitter such as norepinephrine. So, as a rate, m_{ij} can be positive, negative, or zero.

The **synaptic matrix** or **connection matrix** M is an n-by-p matrix of real numbers whose entries are the **synaptic efficacies** m_{ij}. The ijth synapse is **excitatory** if $m_{ij} > 0$, **inhibitory** if $m_{ij} < 0$.

The matrix M describes the **forward projections** from neuron field F_X to neuron field F_Y. Similarly, the p-by-n synaptic matrix N describes the **backward projections** from F_Y to F_X. We can specify the *neural network* by the 4-tuple (F_X, F_Y, M, N), where F_X represents not only the collection of topological neurons, but also their activation and signal computational characteristics. Kohonen [1988] calls two-layer networks **heteroassociative**, one-layer networks **autoassociative**. We can generalize this notation to arbitrarily many fields with arbitrary *inter*field connection topologies.

In general M and N differ in structure. For instance, in the first processing stages of vision, the feedforward projections M greatly outnumber the feedback projections N between neuron fields, allowing us to approximate such networks as pure feedforward networks. This tends to even out deeper into cortex, but significant variations still occur. Mathematically, when M and N differ, fixed-point equilibrium behavior tends not to occur. Instead equilibrium behavior may be oscillatory or aperiodic. The characterization of this equilibrium behavior remains an open research problem in dynamical systems theory.

Bidirectional and Unidirectional Connection Topologies

An important special case occurs when the forward and backward synaptic projections, M and N, have the same, or approximately the same, structure. Then $M = N^T$ and $N = M^T$, where M^T and N^T denote respectively the matrix transposes of M and N. This defines the minimal two-layer feedback network in the sense that all other feedback projections N use more information than is available in the feedforward projections M. Such networks are **bidirectional networks**. When the activation dynamics of F_X and F_Y lead to overall stable behavior, these networks are *bidirectional associative memories* or BAMs. The ART-2 model of adaptive resonance, discussed in Chapter 6, provides an important example.

Unidirectional networks occur when a neuron field synaptically *intra*connects to itself. So unidirectional connection topologies are special cases of bidirectional topologies when the two fields F_X and F_Y coincide: $F_X = F_Y$. Then matrix M is n-by-n, or square.

If M is also symmetric, $M = M^T$, then the unidirectional network defines a BAM. The non-BAM interpretation of the synaptic coefficient m_{ij} views the corresponding axonal-synaptic fiber as the same fiber that corresponds to the coefficient m_{ji}, and moreover the coefficients are numerically equal: $m_{ij} = m_{ji}$. The BAM interpretation assumes that two different axonal-synaptic fibers connect the ith and jth neurons, and the synaptic coefficients m_{ij} and m_{ji} are numerically equal, or approximately so. In a two-field network, the matrices M and N represent different synaptic webs even though $M = N^T$. In implementations, it is often sufficient to use or store only M. Just as we can concatenate two neuron fields F_X and F_Y into the augmented field $F_Z = [F_X \,|\, F_Y]$, so too can we extend heteroassociators to autoassociators. In particular, we can extend bidirectional networks to unidirectional networks. If M connects F_X to F_Y, and N connects F_Y to F_X, then the augmented field F_Z intraconnects to itself by the square block matrix B

$$ B \;=\; \begin{pmatrix} \mathbf{O} & M \\ N & \mathbf{O} \end{pmatrix} $$

In the BAM case, when $N = M^T$, then $B = B^T$. Hence a BAM *symmetrizes* an arbitrary rectangular matrix M.

In general, matrices M and N *inter*connect fields F_X and F_Y. Matrices P and Q *intra*connect F_X and F_Y. P denotes an n-by-n matrix. Q denotes a p-by-p matrix. Then the block matrix C intraconnects the augmented field $F_Z = [F_X \,|\, F_Y]$:

$$ C \;=\; \begin{pmatrix} P & M \\ N & Q \end{pmatrix} $$

Then the neurons in F_Z are symmetrically intraconnected, $C = C^T$, if and only if $N = M^T$, as in the BAM case, and in addition $P = P^T$ and $Q = Q^T$.

In biological networks P and Q are often symmetric. The symmetry reflects a **lateral inhibition** or **competitive** connection topology. In the simplest case this means P and Q have positive main diagonal entries, $p_{ii} > 0$ and $q_{jj} > 0$, and negative or zero off-diagonal elements, $p_{ij} \le 0$ and $q_{jk} \le 0$. The strength of inhibitory connections is often **distance dependent**, typically decreasing with physical separation. Then the symmetry of the distance relation ensures the symmetry of the intraconnection matrices P and Q.

ADDITIVE ACTIVATION MODELS

A system of $n+p$ coupled first-order differential equations defines the **additive activation model** that interconnects fields F_X and F_Y through constant synaptic matrices M and N:

$$ \dot{x}_i \;=\; -A_i x_i + \sum_{j=1}^{p} S_j(y_j) n_{ji} + I_i \tag{3-15} $$

$$ \dot{y}_j \;=\; -A_j y_j + \sum_{i=1}^{n} S_i(x_i) m_{ij} + J_j \tag{3-16} $$

The additive autoassociative ($F_X = F_Y$) model corresponds to a system of n coupled first-order differential equations:

$$ \dot{x}_i \;=\; -A_i x_i + \sum_{j=1}^{n} S_j(x_j) m_{ji} + I_i \tag{3-17} $$

In the general heteroassociative case M and N^T differ in (3-15) and (3-16). In the general autoassociative case M and M^T differ in (3-17). In all cases we assume we can add and multiply constants at will.

The classical neural circuit, reviewed by Perkel [1981], describes a special case of the additive autoassociative model (3-17):

$$C_i \dot{x}_i = -\frac{x_i}{R_i} + \sum_j \frac{x_j - x_i}{r_{ij}} + I_i \qquad (3\text{-}18)$$

$$= -\frac{x_i}{R_i'} + \sum_j S_j(x_j)m_{ij} + I_i \qquad (3\text{-}19)$$

where r_{ij} measures the cytoplasmic resistance between neurons i and j. Equation (3-19) follows from (3-18) if we expand the summation in (3-18), define the lumped resistance R_i' as

$$\frac{1}{R_i'} = \frac{1}{R_i} + \sum_j^n \frac{1}{r_{ij}} \qquad (3\text{-}20)$$

put $m_{ij} = 1/r_{ij}$, and assume that the neuron uses a linear (thus unbounded) signal function: $S_i(x_i) = x_i$.

The **Hopfield circuit** [1984] arises from (3-19) if each neuron has a strictly increasing bounded signal function ($S' > 0$) and if the synaptic connection matrix is symmetric ($M = M^T$):

$$C_i \dot{x}_i = -\frac{x_i}{R_i} + \sum_j S_j(x_j)m_{ij} + I_i \qquad (3\text{-}21)$$

The Hopfield circuit belongs to the important class of feedback neural-network models that are *globally stable*. They converge exponentially quickly to fixed points for all inputs, as we prove in Chapter 6. This convergence depends on the boundedness of signal functions. In many cases bounded nonlinear signals distinguish modern neural-network models from the models of classical biological neural modeling, which, as in the Perkel model (3-18), often use linear signal functions.

Continuous additive bidirectional associative memories define the heteroassociative analogues of Hopfield circuits:

$$\dot{x}_i = -A_i x_i + \sum_j^p S_j(y_j)m_{ij} + I_i \qquad (3\text{-}22)$$

$$\dot{y}_j = -A_j y_j + \sum_i^n S_i(x_i)m_{ij} + J_j \qquad (3\text{-}23)$$

where again we can add or multiply constants as desired. As discussed in Kosko [1987–88], (3-22) and (3-23) describe globally stable networks. This remains true even if the synaptic connections change due to many types of unsupervised learning, as summarized by the ABAM Theorems in Chapter 6.

ADDITIVE BIVALENT MODELS

Discrete additive activation models correspond to neurons with threshold signal functions. The neurons can assume only two values: ON and OFF. ON represents the signal value $+1$. OFF represents 0 or -1. These two-valued, or **bivalent**, neural-network models stem from the classical neural model of McCulloch and Pitts [1943].

Bivalent models can represent asynchronous and stochastic behavior. These properties offer little practical value in small-scale simulations. Indeed they are seldom simulated. Instead neurons process signals deterministically and synchronously. This is unrealistic when modeling arbitrarily large collections of biological neurons. Then we find no global mechanisms of synchronization. (Some biological neurons may synchronize the firing of comparatively small local groups of neurons.) Different neurons emit action potentials at different times and often on different time scales. Apparent random behavior may represent the net effect of perhaps thousands of neural, hormonal, glial, and other inputs that ceaselessly perturb the neuron. Additive bivalent models describe asynchronous and stochastic behavior with the same mathematical mechanism: at each moment, each neuron can randomly "decide" whether to change state, whether to emit a new signal given its current activation.

We present two additive bivalent neural networks: the nonadaptive, additive, bivalent bidirectional associative memory (BAM) and the autoassociative Hopfield model. We present the BAM model first, though the Hopfield model historically precedes it. The BAM model reduces to the Hopfield model in the autoassociative case when we allow only one neuron to change state at a time.

Bivalent Additive BAM

The neural literature often refers to the discrete version of (3-22)–(3-23) as a **bidirectional associative memory**, or **BAM**. In fact this network represents the simplest type of BAM network: additive, two-valued, and nonadaptive. As such its most immediate "application" remains as an expository tool for illustrating feedback nonlinear neural networks, though researchers [Mathai, 1989; Wang 1990] have studied such BAMs in several research contexts. We define a discrete additive BAM with threshold signal functions, arbitrary thresholds and inputs, an arbitrary but constant synaptic connection matrix M, and discrete time steps k:

$$x_i^{k+1} = \sum_j^p S_j(y_j^k)m_{ij} + I_i \qquad (3\text{-}24)$$

$$y_j^{k+1} = \sum_i^n S_i(x_i^k)m_{ij} + J_j \qquad (3\text{-}25)$$

The signal functions S_i and S_j in (3-24) and (3-25) represent binary or bipolar threshold functions. For instance, threshold binary signal functions correspond to

$$S_i(x_i^k) = \begin{cases} 1 & \text{if} \quad x_i^k > U_i \\ S_i(x_i^{k-1}) & \text{if} \quad x_i^k = U_i \\ 0 & \text{if} \quad x_i^k < U_i \end{cases} \qquad (3\text{-}26)$$

$$S_j(y_j^k) = \begin{cases} 1 & \text{if} \quad y_j^k > V_j \\ S_j(y_j^{k-1}) & \text{if} \quad y_j^k = V_j \\ 0 & \text{if} \quad y_j^k < V_j \end{cases} \qquad (3\text{-}27)$$

for arbitrary real-valued thresholds $U = (U_1, \ldots, U_n)$ for F_X neurons and $V = (V_1, \ldots, V_p)$ for F_Y neurons. The bipolar versions of (3-26) and (3-27) yield the signal value -1 when $x_i < U_i$, or when $y_j < V_j$.

The stay-the-same value that occurs when $x_i = U_i$ or $y_j = V_j$ occurs infrequently in large networks. Activations can take on increasingly more values as we add more neurons to the network. The stay-the-same value conveniently breaks ties and simplifies the discrete stability proof below.

The bivalent signal functions (3-26)–(3-27) allow us to model complex *asynchronous* state-change patterns. At any moment different neurons can "decide" whether to compare their current activation to their threshold. They need not do so all at once. At each moment any of the 2^n subsets of F_X neurons, or the 2^p subsets of F_Y neurons, can decide to change state. The choice of subset may be probabilistic. Each neuron may randomly decide whether to check the threshold conditions in (3-26) or (3-27). We may describe the "randomness" with slowly varying means and variances of state-change frequency. Then the network behaves as a *vector stochastic process*. Each neuron in turn behaves as a scalar stochastic process. At each moment each neuron defines a random variable that can assume the value ON (+1) or OFF (0 or -1). The neuron's observed time history represents only one of many possible sample functions from the stochastic process. If the neurons are spatially ordered—as, for example, two-dimensional neural picture elements (pixels) or three-dimensional neural volume elements (voxels)—then the network behaves as a *random field*.

In practice we often assume that the network is deterministic, and state changes are **synchronous**: we update an entire field (vector) of neurons at a time. The other extreme is **simple asynchrony**: only one neuron makes a state-change decision at a time. This serial-processing case corresponds to the discrete Hopfield model [1982]. In general state-change decisions are **subset asynchronous**: one subset of neurons per field makes state-change decisions at a time. When the subsets represent singleton sets, simple asynchronous state change results. When the subsets represent the entire fields F_X and F_Y, synchronous state change results.

We can illustrate the BAM model with *any* real-valued matrix. The BAM network converges to fixed points for every matrix with arbitrary subset asynchronous state-change policies, as proved below. Chapters 4 and 6 discuss Hebbian encoding of pattern information in real matrices. For now we assume learning is so slow as to leave the synaptic connection matrix unchanged during neuronal operation. We have already encoded pattern information in the matrix M. These patterns consist of pairs of binary or bipolar signal state vectors.

Consider an additive bivalent BAM with 4 neurons in F_X and 3 neurons in F_Y. A 4-by-3 matrix M represents the forward synaptic projections from F_X to F_Y. The 3-by-4 matrix transpose M^T represents the backward projections from F_Y to F_X. In particular, suppose M and M^T correspond to

$$
M = \begin{pmatrix} -3 & 0 & 2 \\ 1 & -2 & 0 \\ 0 & 3 & 2 \\ -2 & 1 & -1 \end{pmatrix}, \qquad M^T = \begin{pmatrix} -3 & 1 & 0 & -2 \\ 0 & -2 & 3 & 1 \\ 2 & 0 & 2 & -1 \end{pmatrix}
$$

Suppose at initial time k all the neurons in F_Y are ON. So the signal state vector $S(Y_k)$ at time k corresponds to

$$
S(Y_k) = (1 \ \ 1 \ \ 1)
$$

Suppose also that the joint effects of feedback from F_Y and prior initial inputs produce the activation-state vector X_k at F_X, where

$$
X_k = (x_1^k, x_2^k, x_3^k, x_4^k) = (5 \ -2 \ 3 \ 1)
$$

Suppose for simplicity that all thresholds equal zero—$U_i = V_j = 0$ for all i and j—and that all signals are binary.

Now consider how the BAM network behaves if all neuronal state-change decisions are synchronous. First, at time $k+1$ the F_X neurons transduce their real-valued activations into a binary signal state vector $S(X_k)$. Synchronous operation means that each F_X neuron thresholds its activation in parallel according to (3-26) with zero thresholds. The result gives the binary signal vector

$$
S(X_k) = (1 \ \ 0 \ \ 1 \ \ 1)
$$

The index notation in $S(X_k)$ implicitly assumes that neurons instantaneously transduce activations to signals. If this is unrealistic, we can introduce an extra time step to model the time lag. For simplicity we shall assume instantaneous activation transduction.

Next, at time $k+1$ these F_X signals pass "forward" through the filter M to affect the activations of the F_Y neurons. The three F_Y neurons compute three dot products, or correlations. The signal state vector $S(X_k)$ multiplies each of the three columns of M. Since $S(X_k)$ denotes a row vector, we can write this parallel dot-product computation as the vector matrix multiplication

$$S(X_k)M = (\sum_{i=1}^{4} S_i(x_i^k)m_{i1}, \sum_{i=1}^{4} S_i(x_i^k)m_{i2}, \sum_{i=1}^{4} S_i(x_i^k)m_{i3})$$

$$= (-5 \ \ 4 \ \ 3)$$

$$= (y_1^{k+1}, y_2^{k+1}, y_3^{k+1})$$

$$= Y_{k+1}$$

The third equality follows from the activation equation (3-25), since we assume all external inputs J_j equal zero.

We synchronously compute the new signal state vector $S(Y_{k+1})$ by applying in parallel the threshold law (3-27), with zero thresholds, to each F_Y neuron. The result gives the new F_Y signal state vector:

$$S(Y_{k+1}) = (0 \ \ 1 \ \ 1)$$

Thus the first F_Y neuron changes state from ON to OFF: $S_1(y_1^{k+1}) - S_1(y_1^k) = 0 - 1 < 0$. An asynchronous state-change policy may not produce this state change. For instance, if we update only the third F_Y neuron—or only the second and third F_Y neurons—at time $k + 1$, then $S(Y_{k+1}) = (1 \ \ 1 \ \ 1) = S(Y_k)$.

The signal state vector $S(Y_{k+1})$ then passes "backward" through the synaptic filter M^T at time $k + 2$:

$$S(Y_{k+1})M^T = (2 \ \ -2 \ \ 5 \ \ 0)$$

$$= (x_1^{k+2}, x_2^{k+2}, x_3^{k+2}, x_4^{k+2})$$

$$= X_{k+2}$$

according to the activation equation (3-24) with all inputs I_i equal to zero.

Synchronous thresholding at F_X at time $k + 2$ now reveals a BAM fixed-point equilibrium:

$$S(X_{k+2}) = (1 \ \ 0 \ \ 1 \ \ 1)$$

$$= S(X_k)$$

Note that, according to (3-26), the fourth F_X neuron maintains its previous value— ON $(+1)$ in this case—because its activation equals the zero threshold when the neuron makes the state-change decision.

Since $S(X_{k+2}) = S(X_k)$, passing $S(X_{k+2})$ forward through M will produce $S(Y_{k+1})$ at F_Y at time $k + 3$. Passing $S(Y_{k+3}) = S(Y_{k+1})$ backward through M^T will produce $S(X_{k+2})$ again at F_X. These same two signal state vectors will pass back and forth in *bidirectional equilibrium* forever—or until new inputs perturb the system out of equilibrium.

Asynchronous state changes may lead to different bidirectional equilibria. The discrete BAM theorem below, though, guarantees that the system will reach such equilibria, and reasonably quickly. Suppose in the above example we keep the first F_Y neuron ON. We update only the second and third F_Y neurons. At time k, all the F_Y neurons are ON. The $k + 1$ vector of F_X signals $S(X_k)$ is the same as before, and so leads to the same F_Y activation vector Y_{k+1}:

$$Y_{k+1} = S(X_k)M$$
$$= (-5 \ 4 \ 3)$$

This time, however, the first F_Y neuron does not threshold its activation to zero in accord with (3-27). It does not threshold its activation at all, since the first F_Y neuron does not make a state-change decision at the moment. Instead the neuron maintains its previous signal value, which in this case is ON $(+1)$. The second and third F_Y neurons do threshold their activations in accord with (3-27). Other subset asynchronous state-change policies may update only the second neuron, or only the third neuron, or neither. In the present case the new F_Y signal state vector at time $k + 1$ equals

$$S(Y_{k+1}) = (1 \ 1 \ 1)$$

Thus no F_Y neuron changes signal state, even though the input activations change. The new F_X activation state vector equals

$$X_{k+2} = S(Y_{k+1})M^T$$
$$= (-1 \ -1 \ 5 \ -2)$$

which synchronously thresholds to

$$S(X_{k+2}) = (0 \ 0 \ 1 \ 0)$$

This binary vector differs from the first $S(X_{k+2})$ binary vector above by 2 bits or 50% (Hamming distance of 2) and reflects the inhibitory influence of the first F_Y neuron. Passing this vector forward to F_Y gives

$$Y_{k+3} = S(X_{k+2})M$$
$$= (0 \ 3 \ 2)$$

which thresholds to

$$S(Y_{k+3}) = (1 \ 1 \ 1)$$
$$= S(Y_{k+1})$$

for any asynchronous (or synchronous) state-change policy. Similarly, $S(X_{k+4}) = S(X_{k+2}) = (0 \ 0 \ 1 \ 0)$ for any asynchronous state-change policy we apply to the F_X neurons. The system has reached a new equilibrium. The binary pair $\{(0 \ 0 \ 1 \ 0), (1 \ 1 \ 1)\}$ represents a fixed point of the BAM dynamical system.

This example illustrates that different subset asynchronous state-change policies applied to the same data need not produce the same fixed-point equilibrium. They *tend* to produce the same equilibria. "Nearby" initial conditions also tend to converge toward the same fixed-point equilibrium, though the tendency falls off with distance. This tendency underlies the *content-addressable memory* (continuous function estimation) property characteristic of neural networks.

The example suggests that all BAM state changes lead to fixed-point stability. They do. This property holds for synchronous as well as asynchronous state changes. We can immediately exploit it with vector-matrix multiplication devices, optical [Guest, 1987] or otherwise. More important, this property holds for all subset asynchronous state-change policies and suggests a global mechanism of neuronal organization.

Bidirectional Stability

A BAM system (F_X, F_Y, M) is **bidirectionally stable** if all inputs converge to fixed-point equilibria. Bidirectional stability provides one example of *global* or *absolute* stability.

Bidirectional stability is a dynamic equilibrium. The same signal information flows back and forth in a bidirectional fixed point. Suppose for example that A denotes a binary n-vector in $\{0, 1\}^n$, and B denotes a binary p-vector in $\{0, 1\}^p$. Suppose A enters the BAM system as input. Then we can represent how a BAM equilibrates to the bidirectional fixed point (A_f, B_f) as

$$
\begin{array}{ccccc}
A & \longrightarrow & M & \longrightarrow & B \\
A' & \longleftarrow & M^T & \longleftarrow & B \\
A' & \longrightarrow & M & \longrightarrow & B' \\
A'' & \longleftarrow & M^T & \longleftarrow & B' \\
& & \vdots & & \\
A_f & \longrightarrow & M & \longrightarrow & B_f \\
A_f & \longleftarrow & M^T & \longleftarrow & B_f \\
& & \vdots & &
\end{array}
$$

where A', A'', ... and B', B'', ... represent intermediate or transient signal state vectors between, respectively, A and A_f, and B and B_f.

We can also view a bidirectional equilibrium as a *resonant* state. When the synaptic topology M (and thus M^T) changes—when learning occurs—the bidirectional stability of the F_X and F_Y neurons may erode. Persistent oscillations or, worse, chaotic wandering may occur. When both the changing neurons (STM) and the changing synapse (LTM) equilibrate, an adaptive BAM fixed point results. Grossberg [1982] alternatively refers to this joint equilibration of neurons and synapses as *adaptive resonance*.

We now prove the bivalent BAM theorem that *every matrix is bidirectionally stable*. In Chapter 6 we shall extend this theorem to, ultimately, the RABAM theorem for nonlinear stochastic dynamical systems. To motivate the proof we briefly review Lyapunov functions.

Lyapunov Functions

How do we prove stability for a system defined with arbitrarily many inter-locked differential or difference equations? The first, or direct, approach "simply" solves the equations and then studies how the system evolves with time. This is seldom feasible in the high-dimensional nonlinear case.

The second approach finds a Lyapunov function. The Lyapunov approach offers a shortcut to proving global stability of a dynamical system. If we cannot find a Lyapunov function, nothing follows. The dynamical system may or may not be stable. But if we can find a Lyapunov function, stability holds. Often, though, we cannot establish anything else. In general the Lyapunov approach reveals only the existence of stable points, not their number or nature.

This limits stable neural networks used as content-addressable memories. Sta-bility ensures only that retrieval occurs when we present an input stimulus to the memory. It does not indicate which item in memory the network recalls. This re-sembles not knowing at which airport your airplane will land, though being assured that wherever you land it will be an airport. The same problem limits stable feedback networks used for combinatorial optimization. Stability ensures only that the net-work will equilibrate to local minima of the payoff function. To know more we must know how the system encoded pattern information into the network's synaptic web.

The bivalent BAM theorem illustrates a global stability theorem proved with a weak type of Lyapunov function. We borrow the essential proof technique from con-trol theory, where a quadratic form usually forms the basis of the Lyapunov function. Amari [1977] seems to have been the first neural theorist to apply the Lyapunov technique to symmetric autoassociative networks of McCulloch-Pitts neurons. Hop-field [1982] extended Amari's proof technique to the case of simple asynchronous bivalent neurons.

A **Lyapunov function** L maps system state variables to real numbers and decreases with time. In the BAM case, L maps the bivalent product space to real numbers. So $L: B^n \times B^p \longrightarrow R$, where B^n denotes either the Boolean n-cube $\{0, 1\}^n$ or the bipolar n-cube $\{-1, 1\}^n$.

In control theory a Lyapunov function usually involves a quadratic form. To see why, suppose L is sufficiently differentiable to apply the chain rule:

$$\dot{L} = \sum_i^n \frac{\partial L}{\partial x_i} \frac{dx_i}{dt}$$

$$= \sum_i \frac{\partial L}{\partial x_i} \dot{x}_i \qquad (3\text{-}28)$$

If I denotes the n-by-n identity matrix, and $\mathbf{x} = (x_1, \ldots, x_n)$ defines a vector of state variables, consider the quadratic choice of L:

$$L = \frac{1}{2} \mathbf{x} I \mathbf{x}^T$$

$$= \frac{1}{2} \sum_i x_i^2 \tag{3-29}$$

Now suppose the dynamical system describes the passive decay system

$$\dot{x}_i = -x_i \tag{3-30}$$

We can solve these linear equations simply enough:

$$x_i(t) = x_i(0)e^{-t} \tag{3-31}$$

The solution decays exponentially quickly to the origin. So we know in advance how the system behaves globally. For comparison, take the partial derivative of the quadratic L in (3-29):

$$\frac{\partial L}{\partial x_i} = x_i \tag{3-32}$$

and substitute it and (3-30) into (3-28):

$$\dot{L} = -\sum_i x_i^2 \tag{3-33}$$

or, equivalently,

$$\dot{L} = -\sum_i \dot{x}_i^2 \tag{3-34}$$

In either case,

$$\dot{L} < 0 \tag{3-35}$$

along system trajectories—so long as at least one state variable changes or differs from zero. At equilibrium,

$$\dot{L} = 0 \tag{3-36}$$

and this occurs if and only if all velocities equal zero:

$$\dot{x}_i = 0 \tag{3-37}$$

An instructive, and easy, exercise shows that gradient systems—$\dot{x}_i = -\partial L/\partial x_i$ for some scalar-valued "potential" function L—are stable. Most neural-network models are gradient systems.

   The above example illustrates the basic properties of a classical Lyapunov function. The origin represents an equilibrium, indeed the only equilibrium in this case. The quadratic Lyapunov function L in (3-29) defines a positive-definite quadratic form, since, trivially, the identity matrix is symmetric with all eigenvalues equal to 1. (Recall that a real symmetric matrix A is **positive definite** if and only if, for all nonnull state vectors $\mathbf{x}$, $\mathbf{x}A\mathbf{x}^T > 0$. This holds if and only if all the eigenvalues of A are real and positive.) So $\dot{L} < 0$ except at the origin, where $\dot{L} = 0$. By (3-35), L strictly decreases in time along trajectories. L stops changing if and only if the state vector stops changing, stops moving in the state space.

A dynamical system is **stable** [Elbert, 1984] *if* some Lyapunov function L decreases along trajectories: $\dot{L} \leq 0$. Strict equality need not hold in (3-35). A dynamical system is **asymptotically stable** *if* it strictly decreases along trajectories: $\dot{L} < 0$. In a stable equilibrium the trajectory may hover arbitrarily close to the equilibrium point without reaching it. In an asymptotically stable system the state trajectory reaches the equilibrium, and in general reaches it exponentially fast [Anderson, 1979].

Monotonicity of a Lyapunov function provides a *sufficient not necessary* condition for stability and asymptotic stability. Inability to produce a Lyapunov function proves nothing. The system may or may not be stable. Demonstration of a Lyapunov function, any Lyapunov function, proves stability. Unfortunately, other than a quadratic guess, in general we have no constructive procedure for generating Lyapunov functions.

The linear system (3-30) is asymptotically stable since (3-35) holds. Indeed, for symmetric matrix A and square matrix B, the quadratic form

$$L \;=\; \mathbf{x}A\mathbf{x}^T \tag{3-38}$$

behaves as a strictly decreasing Lyapunov function for any linear dynamical system,

$$\dot{\mathbf{x}} \;=\; \mathbf{x}B \tag{3-39}$$

if and only if the matrix $AB^T + BA$ is negative definite, since

$$
\begin{aligned}
\dot{L} \;&=\; \mathbf{x}A\dot{\mathbf{x}}^T + \dot{\mathbf{x}}A\mathbf{x}^T \\
&=\; \mathbf{x}AB^T\mathbf{x}^T + \mathbf{x}BA\mathbf{x}^T \\
&=\; \mathbf{x}[AB^T + BA]\mathbf{x}^T
\end{aligned}
\tag{3-40}
$$

In (3-29) A represents the identity matrix multiplied by one half. In the decay system (3-30) B represents the negative of the identity matrix. So the right-hand side of (3-40) equals the right-hand side of (3-33).

Asymptotic stability often corresponds to an eigenvalue condition in engineering settings (Parker and Chua [1987]), a practice we shall follow. In particular, a dynamical system is asymptotically stable if and only if the Jacobian matrix of the dynamical system has eigenvalues with negative real parts. The decay system (3-30) satisfies this condition, since its Jacobian equals the negative of the identity matrix. The negative identity matrix also accounts for the exponential convergence of the linear system (3-30).

A general theorem in dynamical systems theory [Hirsch and Smale, 1974] relates convergence rate to eigenvalue sign: A nonlinear dynamical system converges exponentially quickly if its system Jacobian has eigenvalues with negative real parts. Locally such nonlinear systems behave linearly.

Stability ensures only that near an equilibrium no eigenvalue has positive real part. Zero real parts may arise if the system Hessian matrix is degenerate (has zero determinant). Asymptotic stability ensures that all eigenvalues have negative real parts, and hence that the dynamical system converges exponentially fast. We shall use this fact repeatedly in the stability arguments in this chapter and in Chapter 6.

A Lyapunov function summarizes total system behavior. At each moment a single real number represents the entire system. Something like this happens in economics, where a single numerical interest rate summarizes complex microeconomic and macroeconomic phenomena. In statistical mechanics, a single numerical temperature summarizes the interactions of arbitrarily many molecules.

A Lyapunov function often measures the *energy* of a physical system. Hopfield [1982] popularized this neural-network interpretation. A quadratic form in the system state variables (or velocities) approximates the system's energy, whether potential or kinetic.

Consider a physical system of n variables and its potential-energy function E. Suppose the coordinate x_i measures the displacement from equilibrium of the ith unit. So the origin represents a potential-energy equilibrium. Suppose for convenience that the potential energy of the origin equals zero.

The potential energy depends on only coordinates x_i, so $E = E(x_1, \ldots, x_n)$. Since E is a physical quantity, we can assume it is sufficiently smooth to permit a multivariable Taylor-series expansion about the origin:

$$E = E(0, \ldots, 0) + \sum_i \frac{\partial E}{\partial x_i} x_i + \frac{1}{2} \sum_i \sum_j \frac{\partial^2 E}{\partial x_i \, \partial x_j} x_i x_j$$

$$+ \frac{1}{3!} \sum_i \sum_j \sum_k \frac{\partial^3 E}{\partial x_i \, \partial x_j \, \partial x_k} x_i x_j x_k + \cdots \tag{3-41}$$

$$\approx \frac{1}{2} \sum_i \sum_j \frac{\partial^2 E}{\partial x_i \, \partial x_j} x_i x_j \tag{3-42}$$

$$= \frac{1}{2} \mathbf{x} A \mathbf{x}^T \tag{3-43}$$

where matrix A is symmetric, since

$$a_{ij} = \frac{\partial^2 E}{\partial x_i \, \partial x_j} = \frac{\partial^2 E}{\partial x_j \, \partial x_i} = a_{ji} \tag{3-44}$$

The approximation (3-42) follows for three reasons. First, we defined the origin as an equilibrium of zero potential energy; so $E(0, \ldots, 0) = 0$. Second, the origin is an equilibrium only if all first partial derivatives equal zero:

$$\frac{\partial E}{\partial x_i} = 0 \qquad (3\text{-}45)$$

Third, we can neglect higher-order terms for small displacements, since we assume the higher-order products are smaller than the quadratic products.

The quadratic energy (3-43) has the same form as the Lyapunov function (3-38). If we replace the variables x_i with their time derivatives, the same approximation argument yields the kinetic energy of the system as a quadratic form in the velocities [Kosko, 1989], generalizing the familiar $\frac{1}{2}mv^2$ relationship.

The negative of the system energy provides a good guess for a candidate Lyapunov function. We shall follow this approach in the next section.

For neural-network purposes a Lyapunov function need only *decrease* and be *bounded*. The quadratic forms in (3-29), (3-30), and (3-43) can grow without bound as the state variables increase in magnitude. Neural networks avoid this by using the bounded nonlinear signals, instead of the unbounded neuronal activations, as the Lyapunov state variables.

Bounded decreasing Lyapunov functions provide an intuitive way to describe global "computations" in neural networks and other dynamical systems. The dynamical system passes through transient states as the Lyapunov function decreases. Since the Lyapunov function cannot decrease forever—at least not if it attains its lower bound—it must come to a stop. The stopping point corresponds to system equilibration. The stopping point represents the *point of programmability* in neural-network dynamical systems. It may correspond to the learning or recall of pattern information, or to the solution of a combinatorial optimization problem.

Bivalent BAM Theorem

What is a Lyapunov function for a bivalent BAM system? One guess might be the activation energy $-XMY^T$. This function can increase without bound, since activations can range from negative infinity to positive infinity. We require a bounded Lyapunov function.

Another guess might be the average *signal energy* L of the forward pass of the F_X signal state vector $S(X)$ through M, and the backward pass of the F_Y signal state vector $S(Y)$ through M^T:

$$L = -\frac{S(X)MS(Y)^T + S(Y)M^TS(X)^T}{2} \qquad (3\text{-}46)$$

Since each quadratic form is a scalar, and the transpose of a scalar is just the scalar $(3^T = 3)$,

$$S(Y)M^TS(X)^T = [S(Y)M^TS(X)^T]^T$$
$$= S(X)MS(Y)^T \qquad (3\text{-}47)$$

So the two quadratic forms in (3-46) coincide. Then we can rewrite E as the quadratic form

$$L = -S(X)MS(Y)^T \tag{3-48}$$

$$= -\sum_i^n \sum_j^p S_i(x_i)S_j(y_j)m_{ij} \tag{3-49}$$

So the signal-energy Lyapunov function simply sums the products of signals and synaptic connection strengths over all connected pairs.

The signal-energy Lyapunov function is clearly bounded below (and above). For binary or bipolar signals, the matrix coefficients define the attainable lower bound:

$$L \geq -\sum_i \sum_j |m_{ij}| \tag{3-50}$$

The attainable upper bound of L equals the negative of this expression.

The signal-energy Lyapunov function for the general BAM system $(F_X, F_Y, M, I, J, U, V)$ takes the form

$$L = -S(X)MS(Y)^T - S(X)[I - U]^T - S(Y)[J - V]^T \tag{3-51}$$

for constant vectors of inputs $I = [I_1, \ldots, I_n]$ and $J = [J_1, \ldots, J_p]$ and constant vectors of thresholds $U = [U_1, \ldots, U_n]$ and $V = [V_1, \ldots, V_p]$. The attainable lower bound of this signal-energy Lyapunov function equals

$$L \geq -\sum_i \sum_j |m_{ij}| - \sum_i [|I_i - U_i|] - \sum_j [|J_j - V_j|] \tag{3-52}$$

We now show that L behaves as a Lyapunov function. We must show that L decreases along discrete state trajectories for either synchronous or subset asynchronous state changes.

Bivalent BAM theorem.   Every matrix is bidirectionally stable for synchronous or asynchronous state changes.

Proof. Consider the signal state changes that occur from time k to time $k+1$. Define the vectors of signal state change as

$$\triangle S(X) = S(X_{k+1}) - S(X_k)$$

$$= (\triangle S_1(x_1), \ldots, \triangle S_n(x_n)) \tag{3-53}$$

$$\triangle S(Y) = S(Y_{k+1}) - S(Y_k)$$

$$= (\triangle S_1(y_1), \ldots, \triangle S_p(y_p)) \tag{3-54}$$

and define the individual state changes as

$$\triangle S_i(x_i) = S_i(x_i^{k+1}) - S_i(x_i^k) \tag{3-55}$$

$$\triangle S_j(y_j) = S_j(y_j^{k+1}) - S_j(y_j^k) \tag{3-56}$$

We assume at least one neuron changes state from time k to time $k+1$: $\triangle S_i(x_i) \neq 0$. Any subset of neurons in a field can change state, but in only one field at a time. This allows us to model synchronous, simple asynchronous, and the more general subset asynchronous state-change policies. Note that for binary threshold signal functions if a state change is nonzero, then $\triangle S_i(x_i) = 1 - 0 = 1$ or $\triangle S_i(x_i) = 0 - 1 = -1$. For bipolar threshold signal functions, $\triangle S_i(x_i) = 2$ or $\triangle S_i(x_i) = -2$.

The "energy" change $\triangle L$,

$$\triangle L = L_{k+1} - L_k \tag{3-57}$$

differs from zero because of changes in field F_X or in field F_Y. Suppose first that change occurs only in field F_X. Then by (3-51),

$$\triangle L = -\triangle S(X)MS(Y_k)^T - \triangle S(X)[I - U]^T \tag{3-58}$$

$$= -\triangle S(X)[S(Y_k)M^T + I - U]^T \tag{3-59}$$

$$= -\sum_i \sum_j \triangle S_i(x_i)S_j(y_j^k)m_{ij} - \sum_i \triangle S_i(x_i)I_i + \sum_i \triangle S_i(x_i)U_i \tag{3-60}$$

$$= -\sum_i \triangle S_i(x_i) \sum_j S_j(y_j^k)m_{ij} - \sum_i \triangle S_i(x_i)I_i + \sum_i \triangle S_i(x_i)U_i \tag{3-61}$$

$$= -\sum_i \triangle S_i(x_i)[\sum_j S_j(y_j^k)m_{ij} + I_i - U_i] \tag{3-62}$$

$$= -\sum_i \triangle S_i(x_i)[x_i^{k+1} - U_i] \tag{3-63}$$

$$< 0 \tag{3-64}$$

along trajectories. Equation (3-63) follows from (3-62) by the F_X activation dynamical equation (3-24) above. Equation (3-64) follows from (3-63) by the threshold-signal function equation (3-26) and the following argument. $\triangle S_i(x_i) \neq 0$ implies $x_i^{k+1} \neq U_i$ by Equation (3-26). Then we need check only two cases. First, suppose $\triangle S_i(x_i) > 0$. Then

$$\triangle S_i(x_i) = S_i(x_i^{k+1}) - S_i(x_i^k)$$

$$= 1 - 0$$

By the threshold law (3-26), this implies $x_i^{k+1} > U_i$. So the product in (3-63) is positive:

$$\triangle S_i(x_i)[x_i^{k+1} - U_i] > 0$$

Second, suppose $\triangle S_i(x_i) < 0$. Then

$$
\begin{aligned}
\triangle S_i(x_i) &= S_i(x_i^{k+1}) - S_i(x_i^k) \\
&= 0 - 1
\end{aligned}
$$

In this case (3-26) implies $x_i^{k+1} < U_i$. Again the product terms in (3-63) agree in sign, and so the product is positive:

$$
\triangle S_i(x_i)[x_i^{k+1} - U_i] \;>\; 0
$$

This argument holds by (3-26) for any of the $2^n - 1$ possible subsets of F_X neurons with nonzero state changes at time $k + 1$. The same argument produces the same strict energy decrease for any nonzero state changes in the F_Y field (at, say, time k or time $k + 2$) when we use Equations (3-25) and (3-27).

So $L_{k+1} - L_k < 0$ for every state change. Since L is bounded, L behaves as a Lyapunov function for the additive BAM dynamical system defined by (3-24) through (3-27). Since the matrix M was arbitrary in (3-24)–(3-25), every matrix is bidirectionally stable. **Q.E.D.**

The bivalent BAM theorem has an intriguing database interpretation: BAM dynamical networks perform $O(1)$, order-one, search for stored patterns. This interpretation depends on how, indeed on whether, we encode pattern information in the synaptic connection matrix M.

Correlation encoding techniques, discussed below and in Chapter 4, tend to store information at local "energy" or Lyapunov minima. Then the time it takes to descend into an attractor basin does not depend on the number m of attractor basins (stored patterns). This *dimension-independent* property holds for all globally stable neural networks and suggests a distributed mechanism for *real-time* pattern recognition and retrieval in biological nervous systems. Figure 3.1 depicts the geometry of this "$O(1)$ search" property of globally stable dynamical systems.

The $O(1)$-search hypothesis of associative neural networks gives one explanation why the time it takes us to recognize our parent's face in a picture is the same for us at age 5 as it is for us at age 25 or age 50. Familiar patterns produce well-dug attractor basins. Other attractor basins affect the width and contour of an attractor basin. They do not appreciably affect the rapid speed with which a state trajectory descends into the attractor basin.

Classification accuracy may change with time. In particular, the number of learned items stored in memory at any time may change. Overcrowding may degrade classification accuracy. Similar learned patterns may have overlapping basins. Spurious basins may occur. Still, classification speed will remain roughly constant over time.

That also makes evolutionary sense. Natural selection has surely favored real-time pattern-recognition mechanisms over alternative non-real-time recognition mechanisms in the evolution of adaptive organisms and the optimization of their time-energy budgets—in the shaping of their foraging, antipredation, and reproduc-

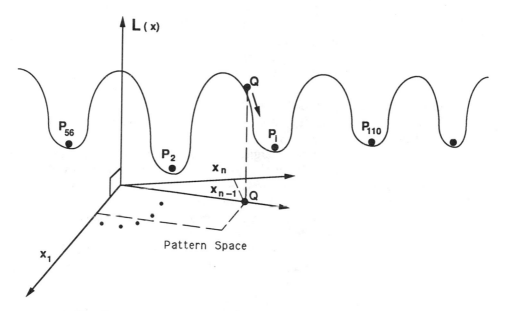

FIGURE 3.1 Convergence in a globally stable neural network, such as a BAM. Encoding sculpts attractor basins in the n-dimensional state space. We can view this as placing rocks on a Lyapunov or "energy" rubber sheet. An input state vector Q quickly rolls into the nearest attractor basin *independently* of the number of other attractor basins ("$O(1)$ search"). If the learning procedure encodes patterns $P_1, P_2, \ldots$ by placing them at local energy minima, the system "classifies" input pattern Q as a pattern of the ith type if Q rolls into P_i's basin. The neural network *misclassifies* input Q if Q is of the ith type but rolls into the basin of another pattern P_j, or if Q rolls into a spurious attractor produced by an imperfect encoding procedure. In the latter case, the network experiences *deja vu*. The network "remembers" a pattern it never learned.

tive behavior [Wilson 1975]. Selection usually favors quick behavior over perfect behavior.

Order-one search provides a central, if only theoretical, relative advantage of computing with dynamical systems instead of with digital computers. (Other advantages are learning from examples and model-free function estimation.) For with computers, serial or parallel, search time increases as we store more items in memory: *memory up, search up*. Often this presents no problem. Computers are fast, and we may need to store only a tractable number of items in memory. Sufficiently many, sufficiently powerful, sufficiently fast digital computers may in principle perform any conceivable computation fast enough to be considered real-time. Whether we judge a neural or nonneural computational approach more efficient depends on several problem-specific factors: training time, computing costs, hardware realizability, etc. At present, dynamical-system computation may prove more efficient for rapidly recognizing large numbers of ill-defined, high-dimensional patterns—images or power spectra, say.

BAM trajectories rapidly converge to fixed-point attractors. In the continuous case we can prove that convergence is exponentially fast under a wide range of conditions. We can study convergence in the discrete case by examining how the Lyapunov function L decreases with neuronal state changes.

The Lyapunov or "energy" decrease in (3-63) equals the sum of the separate energy decreases produced by each F_X neuron that changes state. This yields two insights about the rate of convergence of the BAM dynamical system to the nearest local minimum of L.

First, the individual energies decrease nontrivially. So the overall energy decreases nontrivially. The BAM system does not creep arbitrarily slowly down the Lyapunov or "energy" surface toward the nearest local minimum. The system takes definite hops into the basin of attraction of the fixed point.

The BAM system hops down the attractor basin because the difference $x_i^{k+1} - U_i$ has some small but definite lower bound. This holds because the matrix entries m_{ij}, which appear in the activation difference equation (3-24), are finite constants (often integers). So, for a given synaptic connection matrix M, the inner product

$$\sum_j^p S_j(y_j) m_{ij}$$

and hence the difference $x_i^{k+1} - U_i$, can assume at most 2^p possible values. Moreover, the individual energy change

$$\triangle S_i(x_i)[x_i^{k+1} - U_i]$$

equals an integer multiple of the difference $x_i^{k+1} - U_i$, since, if a state change occurs, the absolute state change $|\triangle S_i(x_i)|$ equals $+1$ in the binary case or $+2$ in the bipolar case.

Second, a synchronous BAM tends to converge faster than an asynchronous BAM. Intuitively, asynchronous updating should take more iterations to converge, especially the fewer the number of neurons updated per iteration. Equation (3-63) grounds this intuition. Since every state change decreases the Lyapunov function, the sum of all state changes tends to produce a comparatively large decrease in the Lyapunov function. The more state changes per iteration, the more the Lyapunov function or "energy" decreases and the quicker the system reaches the nearest local minimum. The network not only hops down the energy surface, it bounds down. So in practice synchronous networks usually converge in two or three iterations.

The index notation and back-and-forth information flow also make clear that state changes occur only in one neuron field at a time. We can drop this restriction in the continuous case. If we update neurons in both fields at the same time, the total "energy" change equals

$$\Delta L = -\Delta S(X) \mathbf{M} S(Y_k)^T - \Delta S(Y) \mathbf{M}^T S(X_k)^T - \Delta S(X) \mathbf{M} \Delta S(Y)^T$$

where for convenience all thresholds and inputs equal zero. The new third term can differ in sign from the other two terms. The product of differences $\Delta S_i(x_i)\,\Delta S_j(y_j)$ need not have the same sign for all neurons at all times. The product of differences defines a two-valued random variable. On average the M-weighted sum of these differences will tend to equal zero as the differences tend to cancel out one another. Then convergence proceeds unabated. Instability may result if the third term differs in sign from the first two terms and exceeds them in magnitude.

BAM CONNECTION MATRICES

Any method of learning, supervised or unsupervised, can produce the matrices M and M^T in the BAM theorem. The matrices can even change gradually—very gradually in general relative to the BAM equilibration rate, or a bit faster if the network uses a Hebbian or competitive learning law, as we establish in the ABAM theorem in Chapter 6.

The most popular method for constructing M, and hence for sculpting the system energy surface, is the **bipolar Hebbian** or **outer-product** learning method. This method sums weighted correlation matrices. We can use binary or bipolar vectors. A Boolean sum (pairwise maximum) sometimes replaces the arithmetic sum when combining binary correlation matrices.

We begin with m many associations of binary vectors (A_i, B_i). A_i denotes a binary vector of length n. B_i denotes a binary vector of length p. Or we begin with m bipolar vector associations (X_i, Y_i). X_i denotes a point in the bipolar n-cube $\{-1, 1\}^n$. Y_i denotes a point in the bipolar p-cube $\{-1, 1\}^p$. We can convert the binary pair (A_i, B_i) to the bipolar pair (X_i, Y_i) if we replace 0s with -1s, and vice versa. Formally this conversion scales and translates the vectors:

$$A_i \;=\; \frac{1}{2}[X_i + \mathbf{1}] \tag{3-65}$$

$$X_i \;=\; 2A_i - \mathbf{1} \tag{3-66}$$

where $\mathbf{1}$ denotes the n-vector of all 1s: $\mathbf{1} = (1, \ldots, 1)$. So the binary vector $A_i = (1\ \ 0\ \ 0\ \ 1)$ corresponds to the bipolar vector $X_i = (1\ \ -1\ \ -1\ \ 1)$.

The **bipolar outer-product law** sums the individual n-by-p bipolar correlation matrices $X_k^T Y_k$:

$$M \;=\; \sum_k^m X_k^T Y_k \tag{3-67}$$

The **binary outer-product law** sums the m binary correlation matrices $A_k^T B_k$:

$$M \;=\; \sum_k^m A_k^T B_k \tag{3-68}$$

The **Boolean outer-product law** combines the m binary correlation matrices $A_k^T B_k$ with pairwise maxima:

$$M \;=\; \bigoplus_{k=1}^{m} A_k^T B_k \tag{3-69}$$

where $\oplus$ denotes the Boolean sum defined pointwise by

$$m_{ij} \;=\; \max(a_1^i b_1^j, \ldots, a_m^i b_m^j) \tag{3-70}$$

So in Boolean arithmetic: $0 \oplus 0 = 0$, $1 \oplus 0 = 0 \oplus 1 = 1$, but $1 \oplus 1 = 1$.

Different associations can have different priority weights. Suppose we assign the real-valued weight w_k to the association (A_k, B_k). If $w_k \geq 0$, we can enter the importance of the kth association multiplicatively in the **weighted outer-product law**, which sums the scaled bipolar correlation matrices $w_k X_k^T Y_k$:

$$M \;=\; \sum_k^m w_k X_k^T Y_k \tag{3-71}$$

A similar multiplicative scaling extends the binary outer product (3-68) and the Boolean outer product (3-69). Though unnecessary, we can normalize the weights w_k to take values in $[0, 1]$ and to sum to unity:

$$\sum_k^m w_k \;=\; 1 \tag{3-72}$$

Normalization converts the weight vector $(w_1, \ldots, w_n)$ to a probability distribution, a set of convex coefficients. The weighted-sum structure (3-71) allows us to recursively update M with weighted future associations, $w_{k+1} X_{k+1}^T Y_{k+1}$, without increasing the dimensions of M.

We can denote the weighted outer product law more compactly in matrix notation as

$$M \;=\; \mathbf{X}^T \mathbf{W} \mathbf{Y} \tag{3-73}$$

The bipolar vector X_k defines the kth row of the m-by-n matrix $\mathbf{X}$. The vector Y_k defines the kth row of the m-by-p matrix $\mathbf{Y}$. The m-by-m diagonal matrix $\mathbf{W}$ displays the weight vector $(w_1, \ldots, w_m)$ along its main diagonal and zeroes elsewhere:

$$\mathbf{X}^T \;=\; [X_1^T | \ldots | X_m^T] \tag{3-74}$$

$$\mathbf{Y}^T \;=\; [Y_1^T | \ldots | Y_m^T] \tag{3-75}$$

$$\mathbf{W} \;=\; \text{Diagonal}[w_1, \ldots, w_m] \tag{3-76}$$

We recover the classical outer-product laws when the weight matrix $\mathbf{W}$ reduces to the m-by-m identity matrix $\mathbf{I}$—when $\mathbf{W} = \mathbf{I}$.

We can arrange coefficients of the weight matrix $\mathbf{W}$ to produce a *recency* or *fading-memory* effect. This gives more weight to more recent associations (those with larger index values) than to less recent associations. This models the inherent exponential fading of first-order unsupervised learning laws, as discussed in Chapter 4. A simple recency effect time-orders the weights: $w_1 < w_2 < \cdots < w_m$. An **exponential fading memory**, constrained by $0 < c < 1$, results if

$$w_k = c^{m-k} \tag{3-77}$$

The neural-network literature has largely overlooked the weighted outer-product laws. Unweighted encoding skews memory-capacity analyses. It suggests that outer-product feedback networks apply to only a small set of association training samples, and overlooks the signal-processing potential of many stable feedback networks. An appropriately weighted fading-memory network yields a "moving window" nonlinear filter for an arbitrary sequence of association training samples.

Optimal Linear Associative Memory Matrices

The **optimal linear associative memory** (OLAM) matrix, studied by Wee [1968, 1971] and Kohonen [1988], provides another data-dependent BAM connection matrix:

$$M = \mathbf{X}^* \mathbf{Y} \tag{3-78}$$

The n-by-m matrix $\mathbf{X}^*$ denotes the **pseudo-inverse** of $\mathbf{X}$. We can define the pseudo-inverse matrix in many ways. We shall say that $\mathbf{X}^*$ is the pseudo-inverse of $\mathbf{X}$ if and only if $\mathbf{X}^*$ behaves as a left-identity and right-identity, and $\mathbf{X}^*\mathbf{X}$ and $\mathbf{XX}^*$ are symmetric (Hermitian in the complex case):

$$\mathbf{XX}^*\mathbf{X} = \mathbf{X} \tag{3-79}$$

$$\mathbf{X}^*\mathbf{XX}^* = \mathbf{X}^* \tag{3-80}$$

$$\mathbf{X}^*\mathbf{X} = (\mathbf{X}^*\mathbf{X})^T, \qquad \mathbf{XX}^* = (\mathbf{XX}^*)^T \tag{3-81}$$

The pseudo-inverse exists, and is unique, for every matrix. For a scalar x, $x^* = 1/x$ if x is nonzero, and $x^* = 0$ if x is zero. For a vector $\mathbf{x}$, $\mathbf{x}^*$ equals the zero vector if $\mathbf{x}$ equals the zero vector. Else $\mathbf{x}^*$ equals the scaled transpose

$$\mathbf{x}^* = \frac{\mathbf{x}^T}{\mathbf{x}\mathbf{x}^T}$$

For rectangular matrix $\mathbf{X}$, if the inverse matrix $(\mathbf{XX}^T)^{-1}$ exists, the pseudo-inverse matrix $\mathbf{X}^*$ equals the product $\mathbf{X}^T(\mathbf{XX}^T)^{-1}$:

$$\mathbf{X}^* = \mathbf{X}^T(\mathbf{XX}^T)^{-1}$$

The recursive Greville algorithm, computationally equivalent to an appropriate time-independent Kalman filter [Kishi, 1964], provides an efficient method for computing pseudo-inverses. The OLAM memory matrix $\widehat{\mathbf{M}} = \mathbf{X}^*\mathbf{Y}$ minimizes the mean-squared error of forward recall in a one-shot, synchronous linear network.

Consider a BAM stripped of feedback (M^T connections) and with linear signal functions: $S_i(x_i) = x_i$ and $S_j(y_j) = y_j$. So forward recall reduces to vector-matrix multiplication. The associations (X_k, Y_k) can consist of arbitrary real n-vectors and p-vectors.

Define the **matrix Euclidean norm** $||M||$ as

$$||M|| = \sqrt{\text{Trace}(MM^T)} \qquad (3\text{-}82)$$

We want to minimize the mean-squared error of forward recall, to find the matrix $\widehat{\mathbf{M}}$ that satisfies the relation

$$||\mathbf{Y} - \mathbf{X}\widehat{\mathbf{M}}|| \leq ||\mathbf{Y} - \mathbf{X}M|| \text{ for all } M \qquad (3\text{-}83)$$

As often happens in mathematics, we can find the general solution if we first find a simple restricted solution, then judiciously remove the restriction. Suppose the matrices $\mathbf{X}$ and $\mathbf{Y}$ are square, and the number of associations m equals the number of neurons in field F_X and field F_Y: $m = n = p$. Suppose further that the inverse matrix $\mathbf{X}^{-1}$ exists. Then

$$
\begin{aligned}
0 &= ||\mathbf{O}|| \\
&= ||\mathbf{Y} - \mathbf{Y}|| \\
&= ||\mathbf{Y} - \mathbf{X}\mathbf{X}^{-1}\mathbf{Y}||
\end{aligned}
$$

So the OLAM matrix $\widehat{\mathbf{M}}$ corresponds to

$$\widehat{\mathbf{M}} = \mathbf{X}^{-1}\mathbf{Y} \qquad (3\text{-}84)$$

The general OLAM solution $\mathbf{X}^*\mathbf{Y}$ follows from (3-84) if we replace the inverse $\mathbf{X}^{-1}$ with the pseudo-inverse $\mathbf{X}^*$ whenever $\mathbf{X}^{-1}$ does not exist. In general the minimizing matrix norm value exceeds zero.

OLAM association resembles least-squares linear regression. They differ because we do not assume the output vector Y_k depends on the input vector X_k. The pair (X_k, Y_k) associates arbitrary vectors.

If the set of vectors $\{X_1, \ldots, X_m\}$ is orthonormal,

$$X_i X_j^T = \begin{cases} 1 & \text{if } i = j \\ 0 & \text{if } i \neq j \end{cases} \qquad (3\text{-}85)$$

then the OLAM matrix reduces to the classical **linear associative memory** (LAM) of Anderson [1972, 1977, 1983]:

$$\widehat{\mathbf{M}} = \mathbf{X}^T\mathbf{Y}. \qquad (3\text{-}86)$$

In (3-86) we recover the outer-product matrix with which we began, though here the Y_k vectors represent arbitrary real p-vectors. The reduction (3-86) follows from the orthonormality condition (3-85), which implies that $(\mathbf{X}\mathbf{X}^T)^{-1}$ exists and equals the identity matrix I:

$$
\begin{aligned}
\widehat{\mathbf{M}} &= \mathbf{X}^*\mathbf{Y} \\
&= \mathbf{X}^T(\mathbf{X}\mathbf{X}^T)^{-1}\mathbf{Y} \\
&= \mathbf{X}^T\mathbf{Y}
\end{aligned}
$$

LAM recall accuracy depends critically on the assumption of orthonormal (linearly independent) input vectors, an assumption seldom satisfied in practice. Even if $X_1, \ldots, X_m$ are orthonormal, LAM recall accuracy degrades rapidly if the input vector X differs from the stored vectors X_k. The "noise" in the input passes through the linear system as faithfully as the "signal" passes through the system. Chapter 4 discusses how signal-function nonlinearity suppresses noise, and restores signal, in BAM systems that use outer-product encoding.

Autoassociative OLAM Filtering

Autoassociative OLAM systems behave as linear filters. We store the m known signal vectors $\mathbf{x}_1, \ldots, \mathbf{x}_m$ in an OLAM matrix M. Each signal vector $\mathbf{x}_i$ represents a point in the real Euclidean signal space R^n. In the autoassociative case the OLAM matrix encodes only the known signal vectors $\mathbf{x}_i$. So the paired heteroassociation $(\mathbf{x}_i, \mathbf{y}_i)$ reduces to the redundant pair $(\mathbf{x}_i, \mathbf{x}_i)$. Then the OLAM matrix equation (3-78) reduces to

$$
M = \mathbf{X}^*\mathbf{X}
$$

M linearly "filters" input measurement $\mathbf{x}$ to the output vector $\mathbf{x}'$ by vector-matrix multiplication: $\mathbf{x}M = \mathbf{x}'$. M behaves as a **linear filter** because the filtered output $\mathbf{x}'$ is a linear function of the input data $\mathbf{x}$. M represents an ordered set of filter coefficients. The OLAM solution gives M as a nonlinear function of the known signals $\mathbf{x}_i$ because the OLAM solution requires a pseudo-inverse computation.

The OLAM matrix $\mathbf{X}^*\mathbf{X}$ behaves as a **projection operator** [Sorenson, 1980]. Algebraically, this means the matrix M is *idempotent*: $M^2 = M$. Since matrix multiplication is associative, pseudo-inverse property (3-80) implies idempotency of the autoassociative OLAM matrix M:

$$
\begin{aligned}
M^2 &= MM \\
&= \mathbf{X}^*\mathbf{X}\mathbf{X}^*\mathbf{X} \\
&= (\mathbf{X}^*\mathbf{X}\mathbf{X}^*)\mathbf{X} \\
&= \mathbf{X}^*\mathbf{X} \\
&= M
\end{aligned}
$$

Let I denote the n-by-n identity matrix. Identity matrices are trivially idempotent. Then (3-80) also implies that the additive dual matrix $I - \mathbf{X}^*\mathbf{X}$ behaves as a projection operator:

$$
\begin{aligned}
(I - \mathbf{X}^*\mathbf{X})^2 &= (I - \mathbf{X}^*\mathbf{X})(I - \mathbf{X}^*\mathbf{X}) \\
&= I^2 - \mathbf{X}^*\mathbf{X} - \mathbf{X}^*\mathbf{X} + \mathbf{X}^*\mathbf{X}\mathbf{X}^*\mathbf{X} \\
&= I - 2\mathbf{X}^*\mathbf{X} + (\mathbf{X}^*\mathbf{X}\mathbf{X}^*)\mathbf{X} \\
&= I - 2\mathbf{X}^*\mathbf{X} + \mathbf{X}^*\mathbf{X} \\
&= I - \mathbf{X}^*\mathbf{X}
\end{aligned}
$$

Geometrically, projection operators project input signals onto some linear subspace $\mathbf{L}$. $\mathbf{L}$ defines a subset of R^n. Projection operator M projects the entire signal space R^n onto $\mathbf{L}$. So we can represent a projection matrix M as the mapping $M: R^n \to \mathbf{L}$.

The Pythagorean theorem underlies projection operators. Suppose we lean a rod against the wall in a room. The rod forms the hypotenuse of a unique right triangle with lines in the wall and the floor as perpendicular legs. A flashlight can project the rod's shadow onto either leg of the triangle. If the flashlight shines squarely from the opposite wall, the rod's shadow coincides with the vertical leg. If the flashlight shines on the rod squarely from the ceiling, the rod's shadow coincides with the horizontal leg on the floor. The horizontal and vertical flashlight positions are perpendicular. They represent dual orthogonal projection operators.

The known signal vectors $\mathbf{x}_1, \ldots, \mathbf{x}_m$ span some unique linear subspace $\mathbf{L}(\mathbf{x}_1, \ldots, \mathbf{x}_m)$ of R^n. We denote $\mathbf{L}(\mathbf{x}_1, \ldots, \mathbf{x}_m)$ as $\mathbf{L}$ for simplicity. So $\mathbf{L}$ equals $\{\sum_i^m c_i\mathbf{x}_i: \text{for all } c_i \in R\}$, the set of all linear combinations of the m known signal vectors. $\mathbf{L}^\perp$ denotes the *orthogonal complement* space $\{\mathbf{x} \in R^n: \mathbf{x}\mathbf{y}^T = 0 \text{ for all } \mathbf{y} \in \mathbf{L}\}$, the set of all real n-vectors $\mathbf{x}$ orthogonal to every n-vector $\mathbf{y}$ in $\mathbf{L}$.

Operator $\mathbf{X}^*\mathbf{X}$ projects R^n onto $\mathbf{L}$. The dual operator $I - \mathbf{X}^*\mathbf{X}$ projects R^n onto $\mathbf{L}^\perp$. Standard Hilbert-space theory [Rudin, 1974] implies that the dual projection operators $\mathbf{X}^*\mathbf{X}$ and $I - \mathbf{X}^*\mathbf{X}$ uniquely decompose every R^n vector $\mathbf{x}$ into a summed *signal* vector $\hat{\mathbf{x}}$ and a noise or *novelty* vector $\tilde{\mathbf{x}}$, the orthogonal "legs" of a Hilbert-space right triangle:

$$
\begin{aligned}
\mathbf{x} &= \mathbf{x}\mathbf{X}^*\mathbf{X} + \mathbf{x}(I - \mathbf{X}^*\mathbf{X}) \\
&= \hat{\mathbf{x}} + \tilde{\mathbf{x}}
\end{aligned}
$$

The same theorem implies that the dual projection operators "annihilate" vectors—project them to the null vector—if the vectors belong to the operator's orthogonal complement. They behave as identity operators on their own subspaces. So $\mathbf{x}\mathbf{X}^*\mathbf{X} = \mathbf{O}$, and thus $\mathbf{x}(I - \mathbf{X}^*\mathbf{X}) = \mathbf{x}$, if $\mathbf{x} \in \mathbf{L}^\perp$. Similarly, $\mathbf{x}\mathbf{X}^*\mathbf{X} = \mathbf{x}$, and thus $\mathbf{x}(I - \mathbf{X}^*\mathbf{X}) = \mathbf{O}$, if $\mathbf{x} \in \mathbf{L}$.

The unique additive decomposition $\hat{\mathbf{x}} + \tilde{\mathbf{x}}$ obeys a generalized Pythagorean theorem:

$$||\mathbf{x}||^2 = ||\hat{\mathbf{x}}||^2 + ||\tilde{\mathbf{x}}||^2$$

where $||\mathbf{x}||^2 = x_1^2 + \cdots + x_n^2$ defines the squared Euclidean or l^2 norm. Chapter 7 presents a new and unique l^p extension of the Pythagorean theorem.

Kohonen [1988] calls $I - \mathbf{X}^*\mathbf{X}$ the **novelty filter** on R^n. Projection $\hat{\mathbf{x}}$ measures what we know about input $\mathbf{x}$ relative to stored signal vectors $\mathbf{x}_1, \ldots, \mathbf{x}_m$ since, by definition of $\mathbf{L}$,

$$\hat{\mathbf{x}} = \sum_i^m c_i \mathbf{x}_i$$

for some constant vector $(c_1, \ldots, c_n)$. The novelty vector $\tilde{\mathbf{x}}$ measures what is maximally unknown or novel in the measured input signal $\mathbf{x}$. The novelty vector provides a slot-by-slot measurement of system uncertainty. The larger the novelty-vector norm $||\tilde{\mathbf{x}}||$, the less certain we are of the "explained" or filtered signal $\hat{\mathbf{x}}$. Kohonen has applied novelty filtering to image-subtraction problems, such as tumor detection in brain scans. In these problems nonnull novelty vectors may indicate disorders or anomalies.

Suppose we model a random measurement vector $\mathbf{x}$ as a random signal vector $\mathbf{x}_s$ corrupted by an additive, independent random-noise vector $\mathbf{x}_N$:

$$\mathbf{x} = \mathbf{x}_s + \mathbf{x}_N$$

We can estimate the unknown signal $\mathbf{x}_s$ as the OLAM-filtered output $\hat{\mathbf{x}} = \mathbf{x}\mathbf{X}^*\mathbf{X}$.

Kohonen [1988] has shown that if the multivariable noise distribution is radially symmetric, such as a multivariable Gaussian distribution, then the OLAM capacity m and pattern dimension n scale the variance of the random-variable estimator-error norm $||\hat{\mathbf{x}} - \mathbf{x}_s||$:

$$V[||\hat{\mathbf{x}} - \mathbf{x}_s||] = \frac{m}{n}||\mathbf{x} - \mathbf{x}_s||^2$$

$$= \frac{m}{n}||\mathbf{x}_N||^2$$

The autoassociative OLAM filter suppresses noise if $m < n$, when memory capacity does not exceed signal dimension. The OLAM filter amplifies noise if $m > n$, when capacity exceeds dimension. More "neurons" allow us to more reliably store, filter, detect, and recall more patterns. This vividly illustrates how dimensionality limits memory capacity in associative networks.

BAM Correlation Encoding Example

The above data-dependent encoding schemes add outer-product correlation matrices. Even the pseudo-inverse $\mathbf{X}^*$ equals an infinite sum of matrices [Kohonen,

1988]. We conclude this section with a simple example of outer-product encoding. The example illustrates a complete nonlinear feedback neural network in action, with data deliberately encoded into the system dynamics.

Suppose the data consists of two unweighted ($w_1 = w_2 = 1$) binary associations (A_1, B_1) and (A_2, B_2) defined by the nonorthogonal binary signal vectors:

$$A_1 = (1\ 0\ 1\ 0\ 1\ 0), \qquad B_1 = (1\ 1\ 0\ 0)$$
$$A_2 = (1\ 1\ 1\ 0\ 0\ 0), \qquad B_2 = (1\ 0\ 1\ 0)$$

These binary associations correspond to the two bipolar associations (X_1, Y_1) and (X_2, Y_2) defined by the bipolar signal vectors:

$$X_1 = (1\ -1\ 1\ -1\ 1\ -1), \qquad Y_1 = (1\ 1\ -1\ -1)$$
$$X_2 = (1\ 1\ 1\ -1\ -1\ -1), \qquad Y_2 = (1\ -1\ 1\ -1)$$

These vectors represent signal, not activation state vectors of binary and bipolar signals. For convenience we use the binary A_i and bipolar X_i notation instead of the more accurate, but more cumbersome, notation of signal functions, activations, and neuron fields. We compute the BAM memory matrix M by adding the bipolar correlation matrices $X_1^T Y_1$ and $X_2^T Y_2$ pointwise. The first correlation matrix $X_1^T Y_1$ equals

$$X_1^T Y_1 = \begin{pmatrix} 1 \\ -1 \\ 1 \\ -1 \\ 1 \\ -1 \end{pmatrix} (1\ 1\ -1\ -1)$$

$$= \begin{pmatrix} 1 & 1 & -1 & -1 \\ -1 & -1 & 1 & 1 \\ 1 & 1 & -1 & -1 \\ -1 & -1 & 1 & 1 \\ 1 & 1 & -1 & -1 \\ -1 & -1 & 1 & 1 \end{pmatrix}$$

Observe that the ith row of the correlation matrix $X_1^T Y_1$ equals the bipolar vector Y_1 multiplied by the ith element of X_1. Similarly the jth column equals the bipolar vector X_1 multiplied by the jth element of Y_1. So $X_2^T Y_2$ equals

$$X_2^T Y_2 = \begin{pmatrix} 1 & -1 & 1 & -1 \\ 1 & -1 & 1 & -1 \\ 1 & -1 & 1 & -1 \\ -1 & 1 & -1 & 1 \\ -1 & 1 & -1 & 1 \\ -1 & 1 & -1 & 1 \end{pmatrix}$$

Adding these matrices pairwise gives M:

$$M = X_1^T Y_1 + X_2^T Y_2$$

$$= \begin{pmatrix} 1 & 1 & -1 & -1 \\ -1 & -1 & 1 & 1 \\ 1 & 1 & -1 & -1 \\ -1 & -1 & 1 & 1 \\ 1 & 1 & -1 & -1 \\ -1 & -1 & 1 & 1 \end{pmatrix} + \begin{pmatrix} 1 & -1 & 1 & -1 \\ 1 & -1 & 1 & -1 \\ 1 & -1 & 1 & -1 \\ -1 & 1 & -1 & 1 \\ -1 & 1 & -1 & 1 \\ -1 & 1 & -1 & 1 \end{pmatrix}$$

$$= \begin{pmatrix} 2 & 0 & 0 & -2 \\ 0 & -2 & 2 & 0 \\ 2 & 0 & 0 & -2 \\ -2 & 0 & 0 & 2 \\ 0 & 2 & -2 & 0 \\ -2 & 0 & 0 & 2 \end{pmatrix}$$

Suppose, first, we use binary state vectors. All thresholds and inputs equal zero. All update policies are synchronous. Suppose we present binary vector A_1 as input to the system—as the current signal state vector at F_X. Then applying the threshold law (3-26) synchronously gives

$$A_1 M = (4 \ 2 \ -2 \ -4) \ \longrightarrow \ (1 \ 1 \ 0 \ 0) = B_1$$

where the arrow $\longrightarrow$ indicates application of the nonlinear threshold signal function. Passing B_1 through the backward filter M^T, and applying the bipolar version of the threshold law (3-27), gives back A_1:

$$B_1 M^T = (2 \ -2 \ 2 \ -2 \ 2 \ -2) \ \longrightarrow \ (1 \ 0 \ 1 \ 0 \ 1 \ 0) = A_1$$

So (A_1, B_1) is a fixed point of the BAM dynamical system. It has Lyapunov "energy" $L(A_1, B_1) = -A_1 M B_1^T = -6$, which equals the backward value $-B_1 M^T A_1^T = -6$. Similarly, passing A_2 through M, and passing B_2 through M^T, give

$$A_2 M = (4 \ -2 \ 2 \ -4) \ \longrightarrow \ (1 \ 0 \ 1 \ 0) = B_2$$
$$B_2 M^T = (2 \ 2 \ 2 \ -2 \ -2 \ -2) \ \longrightarrow \ (1 \ 1 \ 1 \ 0 \ 0 \ 0) = A_2$$

establishing (A_2, B_2) as a fixed point with "energy" $-A_2 M B_2^T = -6$. So the two deliberately encoded fixed points reside in equally "deep" attractors. This does not imply that the attractors have the same width in the binary state space $\{0, 1\}^n \times \{0, 1\}^p$. For pattern recognition, we learn more from the width of an attractor basin than from its Lyapunov "depth." In general we must exhaustively compute the width of the attractor basin.

Recall that **Hamming distance** H equals ℓ^1 distance. $H(A_i, A_j)$ counts the number of slots in which binary vectors A_i and A_j differ:

$$H(A_i, A_j) \;=\; \sum_k^n |a_i^k - a_j^k|$$

Consider for example the input $A = (0\ 1\ 1\ 0\ 0\ 0)$, which differs from A_2 by 1 bit, or, equivalently, has unity Hamming distance from A_2: $H(A, A_2) = 1$. Then

$$AM \;=\; (2\ -2\ 2\ -2) \;\longrightarrow\; (1\ 0\ 1\ 0) \;=\; B_2$$

and so the BAM dynamical system recalls the resonant pair (A_2, B_2). The initial Lyapunov "energy" equals $-AMB_2^T = -4$.

Consider next the input $A = (0\ 0\ 1\ 1\ 0\ 0)$. This input is closer to A_1 than A_2: $H(A, A_1) = 3 < 5 = H(A, A_2)$. A will more likely "roll" into the (A_1, B_1) fixed point instead of the (A_2, B_2) fixed point. But A may roll into another attractor:

$$AM \;=\; (-2\ 2\ -2\ 2) \;\longrightarrow\; (0\ 1\ 0\ 1) \;=\; B_2^c$$

B^c denotes the complement bit vector of B, with energy $-AM(B_2^c)^T = -4$. In turn B_2^c leads to

$$B_2^c M^T \;=\; (-2\ -2\ -2\ 2\ 2\ 2) \;\longrightarrow\; (0\ 0\ 0\ 1\ 1\ 1) \;=\; A_2^c$$

which leads to

$$A_2^c M \;=\; (-4\ 2\ -2\ 4) \longrightarrow (0\ 1\ 0\ 1) \;=\; B_2^c$$

establishing the complement pair (A_2^c, B_2^c) as another fixed point with energy $-A_2^c M (B_2^c)^T = -6$. We refer to these unintended fixed points as **spurious attractors**. For simple correlation (Hebbian) encoding, spurious attractors tend to increase in frequency as the network dimensionality increases.

The above example illustrates an artifact of bipolar correlation encoding: we add $(X_k^c)^T Y_k^c$ to M in (3-67) if we add $X_k^T Y_k$ to M. For the complement X^c of a bipolar vector X equals the vector multiplied by -1:

$$X_k^T Y_k \;=\; (-X_k)^T (-Y_k)$$
$$=\; (X_k^c)^T Y_k^c$$

For the same reason the energies coincide: $-X_k M Y_k^T = -X_k^c M (Y_k^c)^T$. So the correlation sum (3-67) of m associations (X_k, Y_k) equals the weighted sum of $2m$ associations:

$$M = \frac{1}{2} \sum_{k'=1}^{2m} X_{k'}^T Y_{k'}$$

where $X_{k'} = X_k$ and $Y_{k'} = Y_k$ for $k < m + 1$, and $X_{k'} = -X_k = X_k^c$ and $Y_{k'} = Y_k^c$ for $k > m$.

We can also use bipolar signal state vectors in the above example. On average, bipolar signal state vectors produce more accurate recall than binary signal state vectors when we use bipolar outer-product encoding (see Kosko [1988]). Intuitively, binary signals implicitly favor 1s over 0s, whereas bipolar signals are not biased in favor of 1s or -1s: $1 + 0 = 1$, whereas $1 + (-1) = 0$.

For completeness we rework the above example with bipolar signal state vectors:

$$X_1 M \; = \; (8 \; 4 \; -4 \; -8) \; \longrightarrow \; (1 \; 1 \; -1 \; -1) \; = \; Y_1$$

$$X_2 M \; = \; (8 \; -4 \; 4 \; -8) \; \longrightarrow \; (1 \; -1 \; 1 \; -1) \; = \; Y_2$$

$$Y_1 M^T \; = \; (4 \; -4 \; 4 \; -4 \; 4 \; -4) \; \longrightarrow \; (1 \; -1 \; 1 \; -1 \; 1 \; -1) \; = \; X_1$$

$$Y_2 M^T \; = \; (4 \; 4 \; 4 \; -4 \; -4 \; -4) \; \longrightarrow \; (1 \; 1 \; 1 \; -1 \; -1 \; -1) \; = \; X_2$$

So (X_1, Y_1) and (X_2, Y_2) are fixed points, both with Lyapunov energy -24, in the bipolar product space $\{-1, 1\}^n \times \{-1, 1\}^p$. The 1-bit "noisy" input $X = (-1 \; 1 \; 1 \; -1 \; -1 \; -1)$ evokes the fixed point (X_2, Y_2) with initial energy $-XMY_2^T = -16$, since

$$XM \; = \; (4 \; -4 \; 4 \; -4) \; \longrightarrow \; (1 \; -1 \; 1 \; -1) \; = \; Y_2$$

The still "noisier" input $X = (-1 \; -1 \; -1 \; 1 \; 1 \; -1)$, which differs from X_2^c by 1 bit, evokes the complement fixed point (X_2^c, Y_2^c), since

$$XM \; = \; (-4 \; 4 \; -4 \; 4) \; \longrightarrow \; (-1 \; 1 \; -1 \; 1) = -Y_2 \; = \; Y_2^c$$

$$Y_2^c M^T \; = \; (-4 \; -4 \; -4 \; 4 \; 4 \; 4) \; \longrightarrow \; X_2^c$$

with initial energy $-XMY_2^{cT} = -16$ decreasing at the next iteration to $-X_2^c M Y_2^{cT} = -24$.

Figure 3.2 illustrates asynchronous recall. Snapshots of subset asynchronous updates illustrate the BAM recall process. The BAM network stores three spatial alphabetic associations—(S, E), (M, V), and (G, N)—with bipolar outer-product encoding. Field F_X contains $n = 14 \times 10 = 140$ neurons. Field F_Y contains $p = 12 \times 9 = 108$ neurons. We generate a 40% noise-corrupted version of the bipolar association (S, E) by randomly flipping 99 of the bits. We present the noisy (S, E) pattern to the BAM as the input-signal state. The BAM dynamical system quickly recalls the original uncorrupted (S, E) pattern. We randomly chose, on average, subsets of six neurons to make state-change "decisions" at each iteration. Different random subset asynchronous inputs usually produce the same result.

The example shows how a dynamic equilibrium can be rapidly reached and indefinitely maintained by individual "selfish" neurons operating with their own characteristics and on their own time scales, how computational order can emerge from nonlinear chaos. The neurons do not know that a global pattern "error" has occurred. They do not know that they should correct the error, or whether their current behavior helps correct it. The network does not provide the neurons with a global error signal. The network also does not provide neurons with global Lyapunov "energy" information, though state-changing neurons decrease the energy. Instead

FIGURE 3.2 Subset asynchronous recall in a bivalent bidirectional associa-tive memory with bipolar correlation encoding of the spatial association patterns (S, E), (M, V), and (G, N). There are 140 (14×10) neurons in field F_X, 108 (12×9) neurons in field F_Y. Every neuron in F_X synaptically connects to every neuron in F_Y and vice versa. The BAM dynamical system recalls pattern (S, E) when presented with a 40% noise-corrupted version of (S, E). *Source:* Kosko, B., "Adaptive Bidirectional Associative Memories," *Applied Optics*, Volume 26, Number 23, p. 4952, December 1987.

the system dynamics guide the local behavior to a global reconstruction (recollection) of a learned pattern. Similar "invisible hand" equilibria occur in predator-prey ecological, and supply-demand economic, modeling. Such equilibrating systems "self-organize" their internal parameters in response to sampled stimuli.

Memory Capacity: Dimensionality Limits Capacity

Synaptic connection matrices encode limited information. Consider for example the binary Boolean outer-product matrix (3-69). As we sum more binary correlation matrices, $m_{ij} = 1$ holds more frequently. After a point, adding additional associations (A_k, B_k) does not significantly change the connection matrix. The system "forgets" some patterns. This limits the **memory capacity**, the number of distinct patterns the network can accurately learn and recall, encode and decode. The network tends to exceed the memory capacity as the number m of patterns approaches the number n or p, or $\min(n, p)$, of neurons in the network. We already saw this for autoassociative OLAM filtering.

The dimensionality-capacity trade-off reflects a general property of neural networks: *Dimensionality limits capacity*. Dimensionality limitation holds in different ways for different networks, but some form of it always holds.

Supervised networks, especially feedforward multilayer networks trained with the backpropagation algorithm or other estimated-gradient-descent algorithms, can usually accurately store more patterns m than the number n of network neurons. The price paid may be hundreds, perhaps thousands or hundreds of thousands, of training iterations. Still, capacity improves with network size. Networks tend to learn more patterns as the number of neurons, and hence synapses, increases. A two-neuron network has less capacity than a comparably configured hundred-neuron network.

In unsupervised feedforward networks we pay a dimensionality price by using a large number of laterally inhibitive neurons in the competitive field(s). Small-dimensional input patterns project onto a much higher-dimensional competitive layer of neurons. Grossberg's **sparse coding theorem** [1976] then says, for *deterministic* encoding, that pattern dimensionality must exceed pattern number to prevent learning some patterns at the expense of forgetting others.

The number of competing or "coding" neurons similarly constrains the capacity of the *adaptive resonance theory* (ART) models of Carpenter and Grossberg [1987(a), (b)]. The ART models use coding neurons as "grandmother" cells. One neuron, and its synaptic fan-in vector, codes for one pattern or pattern class.

In discussing unsupervised feedback networks, such as BAMs and Hopfield networks, the neural literature sometimes asserts that network dimensionality severely bounds memory capacity. These analyses invariably assume simple Hebbian encoding, as in the McEliece et al. [1987] capacity bound

$$\frac{n}{2 \log_2 n}$$

for bipolar correlation encoding in the Amari-Hopfield network below.

For Boolean encoding (3-69) of binary associations, Haines and Hecht-Nielsen [1988] showed that the memory capacity of bivalent additive BAMs can greatly exceed $\min(n, p)$ if we judiciously choose the thresholds U_i and V_j. The new upper bound becomes $\min(2^n, 2^p)$. Different thresholds produce different hyperplane partitions in the geometry of the product cube $\{0, 1\}^n \times \{0, 1\}^p$. Different sets of thresholds should also improve capacity in the bipolar case, including bipolar Hebbian encoding. Haines and Hecht-Nielsen used heuristics and exhaustive search to determine high-capacity sets of thresholds. They left as an open research problem the development of data-dependent algorithms for manipulating thresholds in the high-capacity direction.

The Hopfield Model

Many neural researchers credit the **Hopfield model** [1982] with reinvigorating, indeed legitimizing, the field of neural networks in the 1980's. Historically the Hopfield model offers a stochastic reinterpretation of the earlier Amari [1977] model that Hopfield cited in his original 1982 paper. For this reason the neural literature sometimes refers to the model as the **Amari-Hopfield model**.

Taxonomically the Hopfield model illustrates an autoassociative additive bivalent BAM operated serially with simple asynchronous state changes. An autoassociative BAM has a symmetric connection matrix, since the BAM forward flow of information through M, and backward flow through M^T, symmetrizes any matrix M.

Autoassociativity means the network topology reduces to only one field, F_X, of neurons: $F_X = F_Y$. The synaptic connection matrix M symmetrically *intra*connects the n neurons in field F_X: $M = M^T$, or $m_{ij} = m_{ji}$. We set the diagonal of M equal to the null vector: $m_{ii} = 0$. This improves recall accuracy (for bipolar autocorrelation encoding) and reflects the observed absence of self-feedback in biological neurons.

The autoassociative version of Equation (3-24) describes the additive neuronal activation dynamics:

$$x_i^{k+1} = \sum_j S_j(x_j^k)m_{ji} + I_i \qquad (3\text{-}87)$$

for constant input I_i, with threshold signal function

$$S_i(x_i^{k+1}) = \begin{cases} 1 & \text{if } x_i^{k+1} > U_i \\ S_i(x_i^k) & \text{if } x_i^{k+1} = U_i \\ 0 & \text{if } x_i^{k+1} < U_i \end{cases} \qquad (3\text{-}88)$$

for constant threshold U_i.

We precompute the Hebbian synaptic connection matrix M by summing bipolar outer-product (autocorrelation) matrices and zeroing the main diagonal:

$$M = \sum_{k=1}^{m} X_k^T X_k - mI \tag{3-89}$$

where I denotes the n-by-n identity matrix. For example, we encode the bipolar pattern $(1 \ -1)$ as the zero-diagonal matrix

$$\begin{pmatrix} 0 & -1 \\ -1 & 0 \end{pmatrix}$$

Zeroing the main diagonal tends to improve recall accuracy by helping the system transfer function $S(XM)$ behave less like the identity operator. For suppose state vectors are random bipolar (or binary) vectors. Then, for symmetric probability distributions, the off-diagonal terms in each column of $X_k^T X_k$ tend to contain as many 1s as -1s. The diagonal term m_{ii} always equals 1. Adding m autocorrelation matrices produces $m_{ii} = m$. The larger m, the more such a sum of autocorrelation matrices M tends to behave as the identity matrix I, since (3-88) thresholds activations (by sign if $U_i = 0$). Every random input pattern X tends to threshold to itself: $S(XM) = X$. So every pattern tends to be a fixed point. Zeroing the main diagonal removes this encoding bias.

So far the Hopfield model restates the Amari [1977] model. The Hopfield model departs from the synchronous Amari model in its assumption of simple asynchrony. The Hopfield model updates only one neuron at a time. Simple asynchrony reduces the network's parallel vector-matrix operation to serial operation.

Simple asynchrony allows us to stochastically *interpret* the Hopfield network. We interpret each neuron as a scalar random process. So the entire network defines a vector random process. Each neuron randomly decides whether to threshold its current activation. Hopfield [1982] suggested that a mean update value and an update variance characterize each "random" neuron. Presumably if we discretize time finely enough, at most one neuron changes state per cycle. In general, arbitrary subsets of neurons change state per cycle.

The stochastic interpretation of simple asynchrony makes a virtue of a necessity. For, in general, if more than one neuron changes state at a time, the Hopfield network is unstable. It is bidirectionally stable, but not unidirectionally stable.

Consider again the 2-by-2 matrix with zero diagonal that encodes the bipolar state vector $(1 - 1)$:

$$M = \begin{pmatrix} 0 & -1 \\ -1 & 0 \end{pmatrix}$$

Now pass the unit vector $X = (1 \ \ 1)$ through M. This gives $XM = (-1 \ \ -1)$. Synchronously applying the bipolar threshold signal function gives back the same vector: $S(XM) = -X = (-1 \ \ -1)$. Passing $-X$ through M, and synchronously thresholding, gives $S(-XM) = X$.

So the Hopfield network is unidirectionally unstable. It oscillates in a two-step limit cycle, $\{X, -X\}$. Note, though, that this defines a bidirectional fixed point. The BAM structure governs since $M^T = M$ in the backward pass of the signal state vector.

Now suppose we pass the unit vector (1 1) through M but we update only the first neuron. Then the new signal state vector is $(-1\ \ 1)$. Passing this vector through M yields the same vector if we update any neuron. Simple asynchrony has produced stability.

ADDITIVE DYNAMICS AND THE NOISE-SATURATION DILEMMA

Additive dynamics resemble linear dynamics. The change in neuronal activation equals a sum of activations and signals, not a product, logarithm, square, or other nonlinear function of these variables. Additivity begets simplicity. We can more easily solve activation differential equations, and Kirchoff's law naturally holds. But activations may saturate in the presence of large inputs.

Usually we, or the environment, restrict the activation x_i to an obtainable finite upper bound x_i^u and finite lower bound x_i^l. The interval $[x_i^l, x_i^u]$ defines the activation's **operating range**. In theory we have allowed activation x_i to arbitrarily range from negative infinity to positive infinity. We have required boundedness only of nonlinear signal functions. In practice the passive decay term $-A_i x_i$ imposes a finite dynamical range on x_i. Infinite activations, like infinite voltages, provide convenient theoretical fictions. Physical systems "burn out" long before they reach infinite values.

Will activations x_i saturate to their upper bounds x_i^u when the system confronts arbitrarily large inputs I_i? Suppose the **dynamical range** equals $[0, \infty]$. How can a neuron, with limited operating range, remain sensitive to an unlimited range of inputs?

We have arrived at the **noise-saturation dilemma** of Grossberg [1982]. If the network is sensitive to large inputs, the network tends to ignore small inputs as noise. If the network is sensitive to small inputs, large inputs tend to saturate the system. A limited operating range seems incompatible with an unlimited dynamical range of inputs.

The noise-saturation dilemma resembles a measurement uncertainty principle. It applies to any signal-processing system, not just to neural networks. A signal-processing system cannot, it seems, perform well for signals of both arbitrarily large and arbitrarily small intensity.

Grossberg's central result shows that, for feedforward networks, additive networks saturate for large inputs, but multiplicative or shunting networks do not saturate. Grossberg claims that this result also resolves the noise part of the noise-saturation dilemma. The result does resolve the noise part of the dilemma, but only if we narrowly interpret noise and noise suppression, and carefully choose network parameters. In Chapter 6 we present the RABAM noise suppression theorem as a more full-blooded resolution of the noise half of the dilemma. Still, Grossberg's result provides the premiere theoretical motivation for using the multiplicative activation models discussed below.

Grossberg's Saturation Theorem

Grossberg's **saturation theorem** states that additive activation models saturate for large inputs, but multiplicative models do not. Moreover, the multiplicative or shunting models remain sensitive to the pattern information in the input—*if* a discrete probability distribution $P = (p_1, \ldots, p_n)$ defines "pattern information."

The saturation theorem applies to the early stages of visual information processing in the brain and eye. These networks are essentially feedforward. A nonnegative background intensity $I(t)$ stimulates the network system. The intensity or energy $I(t)$ represents background light, the illuminant the system must "discount."

Autoassociative field F_X contains n neurons. Each neuron has the limited operating range $[0, B_i]$. In his early works, Grossberg interprets the ith neuron in F_X as a *population* of on-off neurons. Then activation $x_i(t)$ measures the proportion of firing neurons in the population at time t. Grossberg builds his framework out of "grandmother" cells, dedicated feature detectors that reside more frequently in sensory-processing areas of the brain than in cognitive-processing areas. The population interpretation of x_i magnifies this difficulty. We shall continue to view activation x_i as the potential difference across a single neuronal cell membrane.

The stationary "reflectance pattern" $P = (p_1, \ldots, p_n)$ confronts the system amid the background illumination $I(t)$. The pattern P defines a probability distribution. So $p_i \geq 0$ and $p_1 + \cdots + p_n = 1$. Intuitively we see the same reflectance pattern in a room as the lights dim or increase.

On this planet, natural selection has evolved adaptive organisms whose behavior does not vary with wide fluctuations in light intensity. (Cave-dwelling creatures, and bottom-dwelling marine life in the deep sea, experience minimal and effectively constant light intensity. Sight atrophies to blindness, or may never evolve.) The adaptive system must adapt, equilibrate, or "self-organize" its visual parameters in real time to survive in an arbitrary photonic environment. We do not go blind when a cloud blocks the sun.

The ith neuron receives input I_i. Convex coefficient p_i defines the "reflectance" I_i:

$$I_i \;=\; p_i I$$

The neuron must remain sensitive to the pattern information p_i as the background intensity I fluctuates arbitrarily.

The simplest additive activation model combines the feedforward input I_i, the passive decay rate A, and the activation bound $[0, B]$, where we have dropped subscripts for convenience. This results in the *additive Grossberg model*:

$$\dot{x}_i = -Ax_i + (B - x_i)I_i \tag{3-90}$$

$$= -(A + I_i)x_i + BI_i$$

We can solve this linear differential equation to yield

$$x_i(t) = x_i(0)e^{-(A+I_i)t} + \frac{BI_i}{A + I_i}[1 - e^{-(A+I_i)t}] \tag{3-91}$$

For initial condition $x_i(0) = 0$, as time increases the activation converges to its steady-state value:

$$x_i = \frac{BI_i}{A + I_i} \tag{3-92}$$

We can obtain (3-92) directly if we set (3-90) equal to zero and solve for x_i. Using $I_i = p_iI$, we can rewrite (3-92) to give

$$x_i = B\frac{p_iI}{A + p_iI} \longrightarrow B \tag{3-93}$$

as $I \longrightarrow \infty$. So the additive model saturates. The value of p_i does not affect the equilibrium activation value.

Grossberg [1982] suggests that the $n - 1$ inputs I_j should inhibit x_i. This gives an **on-center off-surround** flow of input values, the kind found in biological vision systems. In the simplest case we could add the other inhibitive I_j terms to the activation equation (3-90). But saturation would still result. Instead Grossberg advances the *multiplicative activation model*:

$$x_i = -Ax_i + (B - x_i)I_i - x_i\sum_{j \neq i}I_j \tag{3-94}$$

$$= -(A + I_i + \sum_{j \neq i}I_j)x_i + BI_i$$

$$= -(A + I)x_i + BI_i$$

since $I_i = p_iI$ and $p_1 + \cdots + p_n = 1$. For initial condition $x_i(0) = 0$, the solution to this differential equation becomes

$$x_i = p_iB\frac{I}{A + I}(1 - e^{-(A+I)t}) \tag{3-95}$$

As time increases, the neuron reaches steady state exponentially fast:

$$x_i = p_iB\frac{I}{A + I} \tag{3-96}$$

$$\longrightarrow p_iB$$

as $I \longrightarrow \infty$. So at equilibrium the neuronal activation measures the scaled pattern information p_i. Pattern sensitivity replaces activation saturation. This proves the

Grossberg saturation theorem: *Additive models saturate, multiplicative models do not.*

The equilibrium condition (3-96) and the probability structure of the reflectance pattern $P = (p_1, \ldots, p_n)$ implies a dimension-independent constraint:

$$\sum_{i=1}^{n} x_i = \frac{BI}{A+I}$$

for any number n of neurons in F_X. Grossberg calls this constraint a *normalization rule* or conservation law. He uses it to analyze problems of brightness contrast. Grossberg [1982] also observes that this equilibrium ratio has the form of a **Weber law** of sensation——of "just measurable" sensation to a change in sensory stimulus—as found in psychology.

In general the activation variable x_i can assume negative values. Then the operating range equals $[-C_i, B_i]$ for $C_i > 0$. In the neurobiological literature the lower bound $-C_i$ is usually smaller in magnitude than the upper bound B_i: $C_i << B_i$.

This leads to the slightly more general shunting activation model:

$$\dot{x}_i = -Ax_i + (B - x_i)I_i - (C + x_i)\sum_{j \neq i} I_j \qquad (3\text{-}97)$$

Setting the right-hand side of (3-97) equal to zero and solving gives the equilibrium activation value. An easy exercise rearranges the equilibrium value in a form that directly generalizes (3-96):

$$x_i = [p_i - \frac{C}{B+C}]\frac{(B+C)I}{A+I} \qquad (3\text{-}98)$$

which reduces to (3-96) if $C = 0$. Grossberg refers to the threshold $C/(B+C)$ as an *adaptation level index*. The neuron generates nonnegative activations only if the pattern piece p_i exceeds this threshold. Equation (3-98) still says that x_i remains proportional to p_i as background intensity I grows large. Again the neuron avoids saturation, but maintains sensitivity.

Grossberg argues that (3-98) resolves the noise part of the noise-saturation dilemma. The argument depends on a particular, and limited, interpretation of noise and a particular choice of constants: $B = n - 1$, and $C = 1$.

Grossberg interprets noise as a uniform probability distribution: $p_i = 1/n$ for all i. For instance, in Grossberg's original conception [1982] of adaptive resonance models, he believes an incoming signal pattern and a mismatched stored pattern add together, as two out-of-phase sawtooth patterns, to produce a net constant signal pattern. In such mismatched contexts, which approximate uniform input distributions, Grossberg wants all neurons to shut off, to generate zero activation values.

The parameter choices $B = n - 1$ and $C = 1$ reflect the property $B >> C$. Grossberg gives a further justification based on a laterally inhibitive connection topology for the F_X neurons. The ith neuron excites 1 neuron—namely itself—and inhibits the remaining $n - 1$ neurons.

Now suppose we confront the network with a "noise" pattern, the uniform distribution $P = (1/n, \ldots, 1/n)$. An exercise shows that the parameter choices $B = n - 1$ and $C = 1$ imply the equality

$$\frac{C}{B+C} = \frac{1}{n} \tag{3-99}$$

Substituting these two conditions into (3-98) gives

$$x_i = [\frac{1}{n} - \frac{1}{n}]\frac{(B+C)I}{A+I} \tag{3-100}$$
$$= 0$$

So the neurons shut off. Intuitively, if a perfectly uniform pattern illuminates our retina, we see black not white.

The saturation theorem underlies the more general shunting activation models Grossberg uses, for instance:

$$\dot{x}_i = -A_i x_i + (B_i - x_i)(S_i(x_i) + I_i)$$
$$-(C_i + x_i)(\sum_{j \neq i} S(x_j) + J_i) \tag{3-101}$$

where I_i denotes a general excitatory input, and J_i denotes a general inhibitory input, to the ith neuron in F_X. The right-hand side of a Grossberg activation differential equation always has this structure: DECAY + EXCITATION − INHIBITION.

The saturation theorem also underlies the Grossberg-Mingolla [1985] theory of vision, though here its explanatory power seems to compete with the observed logarithmic transduction of light intensity, at least in the early stages of visual processing. Mead and Mahowald [1988] describe this logarithmic transduction in the context of their silicon retina:

> The photoreceptor transduces an image focused on the retina into an electrical potential proportional to the logarithm of the local light intensity. The logarithmic nature of the response has two important system-level consequences:
>
> 1. An intensity range of many orders of magnitude is compressed into a manageable excursion in signal level.
>
> 2. The voltage difference between two points is proportional to the *contrast ratio* between the two corresponding points in the image, independent of incident light intensity.

Mead's first point offers a simple neuronal mechanism to avoid saturation while the neuron stays sensitive to a wide range of possible inputs. The second point says the same simple mechanism "discounts the illuminant." Grossberg's saturation theorem, though, holds for general neural situations, including those with feedback. Ironically, the saturation theorem seems most informative for nonvisual processing.

Other criticisms concern Grossberg's interpretation of signal and noise. Grossberg [1982] maintains that, first, the unit of learning and perception is the *spatial pattern* (state vector) and, second, this spatial pattern defines a probability distribution or reflectance pattern. This seems reasonable for visual processing and for the signal interpretation of relative muscle contraction rates. But does it make sense for arbitrary regions of the brain? Pattern-vector components might correspond to fuzzy-set elements, which do not obey a sum-to-unity normalization constraint. Or negative pattern components may better fit the facts, especially in the frequency domain.

More controversially, Grossberg interprets noise as a uniform probability distribution. We usually model noise as an additive but independent random variable. The random variable's probability distribution need not be uniform. Often it is Gaussian or unknown.

Why cannot a uniform pattern represent an informative signal? It may indicate a type of consensus within the neural field F_X. And why cannot a nonuniform distribution represent noise? We can interpret as noise many chaotically generated pattern vectors. This interpretation would seem especially natural if the "chaotic" pattern varied with time in the model. The assumption of a constant parameter vector P prevents this stochastic-process interpretation. This reveals another limitation of the model.

Grossberg represents both signal and noise with the same probability-distribution variable p_i. But in general signal and noise occur simultaneously and independently. That helps explain why so many scientists and engineers use the signal-plus-noise model, $S + N$, to model noisy systems. Noise tends to make a signal jiggle or vibrate. Feedback can amplify this vibration. To make the point with an example, Grossberg's framework seems unable to model a noisy uniform signal.

Noise suppression presents another problem. Grossberg simply shuts off neurons to suppress noise. That mechanism fails when signal and noise coexist in the neural network. For then the network would never operate. The network must ceaselessly suppress noise as it processes signals. Chapter 6 develops the additive-noise RABAM model to represent, and suppress, noise in general adaptive feedback networks. Part of the generality of the RABAM model stems from its stochastic use of the deterministic Cohen-Grossberg model of neuronal activation.

GENERAL NEURONAL ACTIVATIONS: COHEN-GROSSBERG AND MULTIPLICATIVE MODELS

Consider the symmetric unidirectional or autoassociative case when $F_X = F_Y$, $M = M^T$, and M is constant. Then a neural network possesses **Cohen-Grossberg** [1983] **activation dynamics** if its activation equations have the form

$$\dot{x}_i = -a_i(x_i)[b_i(x_i) - \sum_{j=1}^{n} S_j(x_j)m_{ij}] \qquad (3\text{-}102)$$

The nonnegative function $a_i(x_i) \geq 0$ represents an abstract **amplification function**. We will assume it has an upper bound. This will allow us to translate a_i when we wish it to assume negative values. The b_i functions are essentially arbitrary. We will require only that the b_i functions behave sufficiently well to keep the integrals bounded in the Lyapunov functions presented below and in Chapter 6. S_j denotes a bounded monotone nondecreasing signal function. In many cases we will assume *strict positivity*: $a_i > 0$ and $S_i' = dS_i/dx_i > 0$.

In Chapter 6 we discuss the global stability of Cohen-Grossberg systems and more general adaptive systems. The Cohen-Grossberg theorem [Grossberg, 1988] ensures the global stability of (3-102). This theorem corresponds—in the sense that $R^n \times R^p = R^{n+p}$—to the continuous BAM theorem discussed below for nonlearning heteroassociative networks, and represents a special case of the ABAM and RABAM theorems in Chapter 6.

Perhaps the most important special cases of (3-102) are additive and shunting networks. The popular versions of these models correspond respectively to the Hopfield circuit and the Hodgkin-Huxley [1952] membrane equation. Grossberg [1988] has also shown that (3-102) reduces to the additive **brain-state-in-a-box** model of Anderson [1977, 1983] and the shunting masking-field model [Cohen, 1987] upon appropriate change of variables.

An autoassociative system has *additive* activation dynamics if the amplification function a_i is constant, and the b_i function is linear. For instance, if $a_i = 1/C_i$, $b_i = (x_i/R_i) - I_i$, $S_i(x_i) = g_i(x_i) = V_i$, and constant $m_{ij} = m_{ji} = T_{ij} = T_{ji}$, where C_i and R_i are positive constants, and input I_i is constant or varies slowly relative to fluctuations in x_i, then (3-102) reduces to the Hopfield circuit [1984]:

$$C_i \dot{x}_i = -\frac{x_i}{R_i} + \sum_j V_j T_{ij} + I_i \qquad (3\text{-}103)$$

An autoassociative network has **shunting** or **multiplicative** activation dynamics when the amplification function a_i is linear, and b_i is nonlinear. For instance, if $a_i = -x_i$, $m_{ii} = 1$ (self-excitation in lateral inhibition), and

$$b_i = \frac{1}{x_i}[-A_i x_i + B_i(S_i + I_i^+) - x_i(S_i + I_i^+) - C_i(\sum_{j \neq i} S_j m_{ij} + I_i^-)]$$

then (3-104) describes the distance-dependent $(m_{ij} = m_{ji})$ unidirectional shunting network:

$$\dot{x}_i = -A_i x_i + (B_i - x_i)[S_i(x_i) + I_i^+]$$
$$-(C_i + x_i)[\sum_{j \neq i} S_j(x_j) m_{ij} + I_i^-] \qquad (3\text{-}104)$$

A_i denotes a positive decay constant. B_i and C_i denote positive saturation constants. The first term on the right-hand side of (3-104) represents a passive decay term. The second and third terms represent respectively positive-feedback and negative-feedback terms. (Strictly speaking, we must keep $a_i(x_i)$ positive. We can always translate a bounded x_i to achieve this.) If we scale to zero the shunting x_i terms in the positive-feedback and negative-feedback terms, (3-104) reduces to an additive model. As we saw above, Grossberg showed that shunting models do not saturate when presented with arbitrarily large positive inputs. They remain sensitive to the relative pattern information in $(I_1, \ldots, I_n)$.

Perhaps more important for neurobiologists, Grossberg [1982, 1988] observed that the shunting model (3-104) represents a special case of the celebrated **Hodgkin-Huxley membrane equation**:

$$c\frac{\partial V_i}{\partial t} = (V^p - V_i)g_i^p + (V^+ - V_i)g_i^+ + (V^- - V_i)g_i^- \qquad (3\text{-}105)$$

V^p, V^+, and V^- denote respectively passive (chloride Cl$^-$), excitatory (sodium Na$^+$), and inhibitory (potassium K$^+$) saturation upper bounds with corresponding shunting conductances g_i^p, g_i^+, and g_i^-, $g^+ >> g^-$. The constant capacitance $c > 0$ scales time. The shunting model (3-104) becomes the Hodgkin-Huxley membrane equation (3-105) if $V_i = x_i$, $V^p = 0$, $V^+ = B_i$, $V^- = -C_i$, $g_i^p = A_i$, $g_i^+ = S_i(x_i) + I_i^+$, and $g_i^- = \sum_{j \neq i} S_j m_{ij} + I_i^-$.

At equilibrium, when the current equals zero, the Hodgkin-Huxley model has the *resting potential* V_{rest}:

$$V_{\text{rest}} = \frac{g_t^p V^p + g^+ V^+ + g^- V^-}{g^p + g^+ + g^-}$$

Hodgkin and Huxley, and others, usually neglect chloride-based passive terms. This gives the resting potential of the shunting model as

$$V_{\text{rest}} = \frac{g^+ V^+ + g^- V^-}{g^+ + g^-}$$

Shunting inhibition refers to a decrease in a neuron's electrical sensitivity caused by increasing its conductances. Thus the multiplicative model gives rise to a quotient or shunting-inhibition equilibrium condition.

Continuous bidirectional associative memories (BAMs) arise when two (or more) neural fields F_X and F_Y are connected in the forward direction, from F_X to F_Y, by an arbitrary n-by-p synaptic matrix M and connected in the backward direction, from F_Y to F_X, by the p-by-n matrix $N = M^T$. BAM activations also possess Cohen-Grossberg dynamics, and their extensions:

$$\dot{x}_i = -a_i(x_i)[b_i(x_i) - \sum_{j}^{p} S_j(y_j)m_{ij}] \qquad (3\text{-}106)$$

$$\dot{y}_j = -a_j(y_j)[b_j(y_j) - \sum_{i}^{n} S_i(x_i)m_{ij}] \qquad (3\text{-}107)$$

with corresponding Lyapunov function L, as we show in Chapter 6:

$$L = -\sum_i \sum_j S_i S_j m_{ij} + \sum_i \int_0^{x_i} S_i'(\theta_i) b_i(\theta_i) \, d\theta_i$$

$$+ \sum_j \int_0^{y_j} S_j'(\varepsilon_j) b_j(\varepsilon_j) \, d\varepsilon_j$$

where we must in principle suitably constrain the functions b_i and b_j to keep L bounded.

The **continuous BAM theorem** states that the BAM system (3-106)–(3-107) is globally stable for *all* matrices M, since $\dot{L} \leq 0$, or $\dot{L} < 0$ if the positivity assumptions hold, along system trajectories. The continuous BAM theorem follows immediately from the ABAM theorem in Chapter 6.

The quadratic form in L is bounded because the signal functions S_i and S_j are bounded. Boundedness of the integral terms requires additional technical hypotheses to avoid pathologies, as Cohen and Grossberg [1983] discuss. For our purpose we simply assume bounded integral terms.

All BAM results extend to any number of BAM-connected fields. Complex topologies are possible and, in theory, will equilibrate as rapidly as the two-layer BAM system. BAM back-and-forth information flow facilitates natural large-scale optical implementations [Guest, 1987; Kinser, 1988].

The BAM model (3-106)–(3-107) clearly reduces to the Cohen-Grossberg model if both neural fields collapse into one, $F_X = F_Y$, and if the constant matrix M is symmetric ($M = M^T$). Conversely, as discussed above, the BAM system symmetrizes an arbitrary matrix M. For if we concatenate the two BAM fields into the new field F_Z, $F_Z = F_X \cup F_Y$, with zero block-diagonal synaptic matrix W that contains M and M^T as respective upper and lower blocks, then the BAM dynamical system (3-106)–(3-107) corresponds to the autoassociative system (3-102).

The BAM system (3-106)–(3-107) includes additive and shunting models. If $a_i = 1 = a_j$, $b_i = x_i - I_i$, and $b_j = y_j - J_j$, for relatively constant inputs I_i and J_j, then an *additive* BAM (3-22)–(3-23) results:

$$\dot{x}_i = -x_i + \sum_j S_j(y_j) m_{ij} + I_i \qquad (3\text{-}108)$$

$$\dot{y}_j = -y_j + \sum_i S_i(x_i) m_{ij} + J_j \qquad (3\text{-}109)$$

Again we can add or multiply constants as desired. More generally, if $a_i = -x_i$, $a_j = -y_j$,

$$b_i = \frac{1}{x_i}[-x_i + (B_i - x_i)[S_i(x_i) + I_i^+] - x_i I_i^-]$$

and

$$b_j = \frac{1}{y_j}[-y_j + (B_j - y_j)[S_j(y_j) + J_j^+] - y_j J_j^-]$$

then a *shunting* BAM [Kosko, 1987] results:

$$\dot{x}_i = -x_i + (B_i - x_i)[S_i + I_i^+] - x_i[\sum_j S_j m_{ij} + I_i^-] \tag{3-110}$$

$$\dot{y}_j = -y_j + (B_j - y_j)[S_j + J_j^+] - y_j[\sum_i S_i m_{ij} + J_j^-] \tag{3-111}$$

The shunting BAM (3-110)–(3-111) reminds us that, in general, distance-dependent competition occurs within fields F_X and F_Y. Suppose the n-by-n matrix R and the p-by-p matrix S describe the distance-dependent ($R = R^T$, $S = S^T$) lateral inhibition within F_X and F_Y respectively. Then we must augment the general BAM model (3-106)–(3-107) to a competitive BAM:

$$\dot{x}_i = -a_i(x_i)[b_i(x_i) - \sum_j^p S_j(y_j)m_{ij} - \sum_k^n S_k(x_k)r_{ki}] \tag{3-112}$$

$$\dot{y}_j = -a_j(y_j)[b_j(y_j) - \sum_i^n S_i(x_i)m_{ij} - \sum_l^p S_l(y_l)s_{lj}] \tag{3-113}$$

So far the synaptic connections have not changed with time, while the neuronal activations have changed with time. Such systems only recall stored patterns. They do not simultaneously learn new ones. In the next two chapters we study the dynamics of synapses that change with time—the dynamics of learning. Chapter 4 studies unsupervised learning. Chapter 5 studies supervised learning. Chapter 6 returns to the activation models we have developed in this chapter, extends them to the random-process domain, allows the synaptic connections to change with time while the neuronal activations change with time, and studies the global behavior of the resulting stochastic adaptive dynamical systems.

REFERENCES

Anderson, B. D. O., and Moore, J. B., *Optimal Filtering*, Prentice Hall, Englewood Cliffs, NJ, 1979.

Anderson, J. A., "A Simple Neural Network Generating an Associative Memory," *Mathematical Biosciences*, vol. 14, 197-220, 1972.

Anderson, J. A., Silverstein, J. W., Ritz, S. R., and Jones, R. S., "Distinctive Features, Categorical Perception, and Probability Learning: Some Applications of a Neural Model," *Psychological Review*, vol. 84, 413-451, 1977.

Anderson, J. A., "Cognitive and Psychological Computations with Neural Models," *IEEE Transactions on Systems, Man, and Cybernetics*, vol. SMC-13, 799-815, September/October 1983.

Amari, S., "Neural Theory of Association and Concept Formation," *Biological Cybernetics*, vol. 26, 175-185, 1977.

Carpenter, G. A., and Grossberg, S., "A Massively Parallel Architecture for a Self-Organizing Neural Pattern Recognition Machine," *Computer Vision, Graphics, and Image Processing*, vol. 37, 54-115, 1987.

Carpenter, G. A., and Grossberg, S., "ART 2: Self-Organization of Stable Category Recognition Codes for Analog Input Patterns," *Applied Optics*, vol. 26, no. 23, 4919-4930, 1 December 1987.

Cohen, M. A., and Grossberg, S., "Absolute Stability of Global Pattern Formation and Parallel Memory Storage by Competitive Neural Networks," *IEEE Transactions on Systems, Man, and Cybernetics*, vol. SMC-13, 815-826, September 1983.

Cohen, M. A., and Grossberg, S., "Masking Fields: A Massively Parallel Neural Architecture for Learning, Recognizing, and Predicting Multiple Groupings of Patterned Data," *Applied Optics*, vol. 26, 1866, 1987.

Elbert, T. F., *Estimation and Control of Systems*, Van Nostrand Reinhold, New York, 1984.

Grossberg, S., "Adaptive Pattern Classification and Universal Recoding: I. Parallel Development and Coding of Neural Feature Detectors," *Biological Cybernetics*, vol. 23, 187-202, 1976.

Grossberg, S.,*Studies of Mind and Brain*, Reidel, Boston, 1982.

Grossberg, S., and Mingolla, E., "Neural Dynamics of Form Perception: Boundary Completion, Illusory Figures, and Neon Color Spreading," *Psychological Review*, vol. 92, 173-211, 1985.

Grossberg, S., and Mingolla, E., "Neural Dynamics of Perceptual Grouping: Textures, Boundaries, and Emergent Segmentation," *Perception and Psychophysics*, vol. 38, 141-171, 1985.

Grossberg, S., "Nonlinear Neural Networks: Principles, Mechanisms, and Architectures," *Neural Networks*, vol. 1, no. 1, 17-61, 1988.

Guest, C. C., and Tekolste, R., "Designs and Devices for Optical Bidirectional Associative Memories," *Applied Optics*, vol. 26, no. 23, 5055-5060, 1 December 1987.

Haines, K., and Hecht-Nielsen, R., "A BAM with Increased Information Storage Capacity," *Proceedings of the IEEE International Conference on Neural Networks (ICNN-88)*, vol. I, 181-190, July 1988.

Hirsch, M. W., and Smale, S., *Differential Equations, Dynamical Systems, and Linear Algebra*, Academic Press, New York, 1974.

Hodgkin, A. L., and Huxley, A. F., "A Quantitative Description of Membrane Current and Its Application to Conduction and Excitation in a Nerve," *Journal of Physiology*, vol. 117, 500-544, 1952.

Hopfield, J. J., "Neural Networks and Physical Systems with Emergent Collective Computational Abilities," *Proceedings of the National Academy of Science*, vol. 79, 2554-2558, 1982.

Hopfield, J. J., "Neural Networks with Graded Response Have Collective Computational

Properties Like Those of Two-State Neurons," *Proceedings of the National Academy of Science*, vol. 81, 3088-3092, May 1984.

Kinser, J. M., Caulfield, H. J., and Shamir, J., "Design for a Massive All-Optical Bidirectional Associative Memory: The Big BAM," *Applied Optics*, vol. 27, no. 16, 3442-3443, 15 August 1988.

Kishi, F. H., "On Line Computer Control Techniques," in *Advances in Control Systems: Theory and Applications*, vol. I, C. T. Leondes (ed.), Academic Press, New York, 1964.

Kohonen, T., *Self-Organization and Associative Memory*, 2nd ed., Springer-Verlag, New York, 1988.

Kosko, B., "Adaptive Bidirectional Associative Memories," *Applied Optics*, vol. 26, no. 23, 4947-4960, December 1987.

Kosko, B., "Bidirectional Associative Memories," *IEEE Transactions on Systems, Man, and Cybernetics*, vol. SMC-18, 49-60, January 1988.

Kosko, B., "Feedback Stability and Unsupervised Learning," *Proceedings of the 1988 IEEE International Conference on Neural Networks (ICNN-88)*, vol. I, 141-152, July 1988.

Kosko, B., "Unsupervised Learning in Noise," *Proceedings of the 1989 International Joint Conference on Neural Networks (IJCNN-89)*, vol. I, 7-17, June 1989; reprinted in *IEEE Transactions on Neural Networks*, vol. 1, no. 1, 44-57, March 1990.

Mathai, G., and Upadhaya, B. R., "Performance Analysis and Application of the Bidirectional Associative Memory to Industrial Spectral Signatures," *Proceedings of the 1989 International Joint Conference on Neural Networks (IJCNN-89)*, vol. I, 33-37, June 1989.

McCulloch, W. S., and Pitts, W., "A Logical Calculus of the Ideas Immanent in Nervous Activity," *Bulletin of Mathematical Biophysics*, vol. 5, 115-133, 1943.

McEliece, R. J., Posner, E. C., Rodemich, E. R., and Venkatesh, S. S., "The Capacity of the Hopfield Associative Memory," *IEEE Transactions on Information Theory*, vol. IT-33, 461-482, July 1987.

Mead, C., and Mahowald, M. A., "A Silicon Model of Early Visual Processing," *Neural Networks*, vol. 1, no. 1, 91-97, 1988.

Parker, T. S., and Chua, L. O., "Chaos: A Tutorial for Engineers," *Proceedings of the IEEE*, vol. 75, no. 8, 982-1008, August 1987.

Perkel, D. H., Mulloney, B., and Budelli, R. W., "Quantitative Methods for Predicting Neuronal Behavior," *Neuroscience*, vol. 6, no. 5, 823-837, 1981.

Rudin, W., *Real and Complex Analysis*, 2nd ed., McGraw-Hill, New York, 1974.

Sorenson, H. W., *Parameter Estimation: Principles and Problems*, Marcel Dekker, New York, 1980.

Wang, Y. F., Cruz, J. B., and Mulligan, J. H., "Two Coding Strategies for Bidirectional Associative Memory," *IEEE Transactions on Neural Networks*, vol. 1, no. 1, 81-92, March 1990.

Wee, W. G., "Generalized Inverse Approach to Adaptive Multiclass Pattern Classification," *IEEE Transactions on Computers*, vol. C-17, no. 12, 1157-1164, December 1968.

Wee, W. G., "Generalized Inverse Approach to Clustering, Feature Detection and Classification," *IEEE Transactions on Information Theory*, vol. IT-17, no. 3, 262-269, May 1971.

Wilson, E. O, *Sociobiology: The New Synthesis*, Harvard University Press, Cambridge, MA, 1975.

PROBLEMS

3.1. Solve the differential equation

$$C\dot{x} = -Ax + P$$

with positive constants A, C, and P. Find the steady-state solution.

3.2. Suppose the potential function $P: R^n \rightarrow R$ defines the n-dimensional *gradient system*

$$\dot{x}_i = -\frac{\partial P}{\partial x_i}$$

or, in vector notation,

$$\dot{\mathbf{x}} = -\nabla P$$

Find a Lyapunov function to prove the system globally stable.

3.3. Suppose the scalar variables x_1 and x_2 are bounded above by the conditions $x_1 < 1$ and $x_2 < 1$ and change in time according to

$$\dot{x}_1 = -x_1 + x_1 x_2$$
$$\dot{x}_2 = -x_2 + x_1 x_2$$

Show that this dynamical system is globally stable.

3.4. Use bipolar outer-product encoding to encode the following binary pairs (A_i, B_i) in a (discrete additive) BAM matrix:

$$
\begin{array}{llll}
A_1 & = & (1\ 0\ 1\ 0\ 1\ 0\ 1\ 0), & B_1 & = & (1\ 1\ 1\ 0\ 0\ 0) \\
A_2 & = & (1\ 1\ 0\ 0\ 1\ 1\ 0\ 0), & B_2 & = & (1\ 0\ 1\ 0\ 1\ 0) \\
A_3 & = & (1\ 1\ 1\ 0\ 0\ 0\ 1\ 1), & B_3 & = & (1\ 1\ 0\ 0\ 1\ 1)
\end{array}
$$

Verify that the pairs (A_i, B_i) are bidirectional fixed points. Use synchronous recall. Verify that the corresponding bipolar associations (X_i, Y_i) are also fixed points. Compute all Lyapunov "energy" values at each iteration.

3.5. Rework Problem 3.4 but use binary outer-product encoding with bipolar state vectors. Are the bipolar pairs (X_i, Y_i) still fixed points?

3.6. Find the pseudo-inverse of the matrix M:

$$
M = \begin{pmatrix}
6 & 0 & 0 & 0 \\
0 & -2 & 0 & 0 \\
0 & 0 & 0 & 0 \\
0 & 0 & 0 & \frac{1}{3}
\end{pmatrix}
$$

3.7. Find the optimal linear associative memory (OLAM) matrix $\widehat{\mathbf{M}}$ given the associations

$$
\begin{array}{llll}
A_1 & = & (1\quad 2\quad 3), & B_1 & = & (4\quad 3\quad 2) \\
A_2 & = & (2\quad 3\quad 4), & B_2 & = & (3\quad 5\quad 2) \\
A_3 & = & (3\quad 4\quad 6), & B_3 & = & (2\quad 2\quad 1)
\end{array}
$$

Determine whether $A_i \widehat{\mathbf{M}} = B_i$.

3.8. Find the linear associative memory (LAM) matrix M given the associations

$$
\begin{array}{lllll}
A_1 & = & (1 \quad 0 \quad 0 \quad 0), & B_1 & = & (\ 1 \quad 2 \quad\quad 3) \\
A_2 & = & (0 \quad 1 \quad 0 \quad 0), & B_2 & = & (-2 \quad 3 \quad -1) \\
A_3 & = & (0 \quad 0 \quad 1 \quad 0), & B_3 & = & (\ 4 \quad 0 \quad\quad 4)
\end{array}
$$

Use outer-product encoding. Determine whether $A_i M = B_i$.

3.9. Suppose $\mathbf{x}_i \in R^n$ defines the ith row of matrix $\mathbf{X}$, and $\mathbf{y}_i \in R^p$ defines the ith row of matrix $\mathbf{Y}$. Suppose there are m associations $(\mathbf{x}_1, \mathbf{y}_1), \dots, (\mathbf{x}_m, \mathbf{y}_m)$ and $m \leq n$. Suppose the input vectors $\mathbf{x}_1, \dots, \mathbf{x}_m$ are linearly independent. Define the projection $\hat{\mathbf{x}}$ as

$$
\begin{aligned}
\hat{\mathbf{x}} &= \mathbf{x}\mathbf{X}^*\mathbf{X} \\
&= \sum_{i=1}^{m} c_i \mathbf{x}_i
\end{aligned}
$$

where $\mathbf{X}^*$ denotes the pseudo-inverse of $\mathbf{X}$. Prove that

$$
\begin{aligned}
\hat{y} &= \mathbf{x}\hat{\mathbf{M}} \\
&= \sum_{i=1}^{m} c_i \mathbf{y}_i
\end{aligned}
$$

where $\hat{\mathbf{M}}$ denotes the system OLAM matrix. So, for each input pattern x, the *same* set of expansion coefficients $\{c_1, \dots, c_m\}$ scales the associated input and output vectors $(\mathbf{x}_i, \mathbf{y}_i)$.

3.10. Store the bipolar pattern vectors

$$
\begin{array}{lllllll}
X_1 & = & (1 & -1 & 1 & -1 & 1) \\
X_2 & = & (1 & 1 & 1 & -1 & -1) \\
X_3 & = & (1 & 1 & -1 & -1 & 1)
\end{array}
$$

in a Hopfield associative memory matrix. Are the pattern vectors X_i unidirectional fixed points? Use synchronous recall. Characterize the (discrete additive) dynamical system's bipolar state space $\{-1, 1\}^5$, which contains 2^5 bipolar pattern vectors. Use each of the 32 possible bipolar vectors as input vectors in synchronous operation. How many fixed points and limit cycles does the dynamical system contain? Compute all Lyapunov "energy" values. Is the unit vector $(1\ 1\ 1\ 1\ 1)$ unidirectionally stable with synchronous recall? With asynchronous recall?

3.11. Solve the differential equation

$$
\dot{x}_i = -Ax_i + (B - x_i)I_i - x_i \sum_{j \neq i} I_j
$$

3.12. Show that the differential equation

$$\dot{x}_i = -Ax_i + (B - x_i)I_i - (C + x_i)\sum_{j \neq i} I_j$$

has the equilibrium solution

$$x_i = [p_i - \frac{C}{B + C}]\frac{(B + C)I}{A + I}$$

where $I_i = p_i I$ and $p_1 + \cdots + p_n = 1$.

3.13. (Continuous BAM theorem) Show that the nonadaptive dynamical system

$$\dot{x}_i = -a_i(x_i)[b_i(x_i) - \sum_{j=1}^{p} S_j(y_j)m_{ij}]$$

$$\dot{y}_j = -a_j(y_j)[b_j(y_j) - \sum_{i=1}^{n} S_i(x_i)m_{ij}]$$

is globally stable. Use the candidate Lyapunov function

$$L = -\sum_{i=1}^{n}\sum_{j=1}^{p} S_i(x_i)S_j(y_j)m_{ij} + \sum_{i=1}^{n}\int_0^{x_i} S_i'(\theta_i)b_i(\theta_i)d\theta_i$$

$$+ \sum_{j=1}^{p}\int_0^{y_j} S_j'(\varepsilon_j)b_j(\varepsilon_j)d\varepsilon_j$$

Show that $\dot{L} < 0$ along trajectories if $a > 0$ and $S' > 0$.

SOFTWARE PROBLEMS

Part I: Discrete Additive Bidirectional Associative Memory (BAM)

The accompanying C-software allows you to construct (discrete additive) BAMs of different dimensions. The number of neurons in field F_X can differ from the number in F_Y. For simplicity the problems that follow use the same number of neurons in both fields.

Construct a 36-by-36 discrete additive BAM as follows:

1. Run BAM.

2. Use Set Load Parameters to set the x-dimension and y-dimension of both layers to 6.

3. Select "Open Network" to create the BAM network.

4. Use the "Weights" menu to clear the synaptic weight array.

5. Use "Set State Mode," under the "Run" menu, to set the network's operating mode to "Binary."

Steps 1 through 5 should have created the desired BAM network with a zero synaptic weight matrix (array).

6. Enter an X pattern. Set to 1 (white) all nodes in the first two rows of the X display.

7. Enter the same pattern for Y in the Y display.

8. Use the "Weights" menu to select "Add Pattern" *exactly once*.

9. Select "Run" from the "Run" menu to verify the pattern's presence. The pattern should remain unchanged.

Problem 1: Does the procedure in step 9 actually verify that the pattern is "in" the BAM weight matrix?

10. Use the "Nodes" menu to clear the X and Y patterns.

11. Enter a new pattern pair (X, Y) as in steps 6 through 8. The newly activated nodes should be in only the third and fourth rows.

12. Use "Run" to verify the pattern in the BAM memory.

Problem 2: Does the procedure in step 12 verify that the pattern is "in" the BAM weight matrix this time?

13. Repeat steps 10 through 12. The last two rows of each side should be the only active rows.

Problem 3: The BAM should now know at least three pattern pairs. What happens if you switch the state mode to "Bipolar"?

Part II

Construct a 4-by-4 BAM as follows:

1. Use "Close Network" to dismantle the existing BAM network.

2. Set the state mode to "Binary."

3. Use Set Load Parameters to set the x-dimension of both layers to 4 and the y-dimension of both layers to 1.

4. Repeat the encoding process of Part I to add patterns to the BAM synaptic weight array. Use the three associations $(U(1, 4), U(1, 4))$, $(U(2, 4), U(2, 4))$, $(U(3, 4), U(3, 4))$. Here $U(i, j)$ denotes a j-dimensional bit vector with a 1 in the ith slot and 0s elsewhere. So $U(3, 5) = (0\ \ 0\ \ 1\ \ 0\ \ 0)$.

Problem 4:   Does the BAM dynamical system accurately recall the three associations? Add the association $(U(4, 4), U(4, 4))$ to the memory. What happens?

Problem 5:   Explain why the 4-by-4 BAM can accurately encode and recall four associations while the 36-by-36 BAM may have difficulty with three associations.

SYNAPTIC DYNAMICS I: UNSUPERVISED LEARNING

The most lively thought still is inferior to the dullest sensation.

David Hume
On Human Nature and the Understanding

When one cell repeatedly assists in firing another, the axon of the first cell develops synaptic knobs (or enlarges them if they already exist) in contact with the soma of the second cell.

D. O. Hebb
The Organization of Behavior

LEARNING AS ENCODING, CHANGE, AND QUANTIZATION

Learning encodes information. A system learns a pattern if the system encodes the pattern in its structure. The system structure changes as the system learns the information.

We may use a behavioristic encoding criterion. Then the system has learned the stimulus-response pair $(\mathbf{x}_i, \mathbf{y}_i)$ if it responds with $\mathbf{y}_i$ when $\mathbf{x}_i$ stimulates the system. $\mathbf{x}_i$ may represent a spectral distribution, and $\mathbf{y}_i$ a pattern-class label. Or $\mathbf{x}_i$ may represent the musical notes in a bar of music text, and $\mathbf{y}_i$ the corresponding control vector of depressed piano keys.

The input-output pair $(\mathbf{x}_i, \mathbf{y}_i)$ represents a sample from the function $f: R^n \longrightarrow R^p$. The function f maps n-vectors $\mathbf{x}$ to p-vectors $\mathbf{y}$. The system has learned the function f if the system responds with $\mathbf{y}$, and $\mathbf{y} = f(\mathbf{x})$, when $\mathbf{x}$ confronts the system for all $\mathbf{x}$. The system has partially learned, or approximated, the function f if the system responds with $\mathbf{y}'$, which is "close" to $\mathbf{y} = f(\mathbf{x})$, when presented with an $\mathbf{x}'$ "close" to $\mathbf{x}$. The system maps similar inputs to similar outputs, and so estimates a continuous function.

Learning involves change. A system learns or adapts or "self-organizes" when sample data changes system parameters. In neural networks, learning means *any change in any synapse*. We do not identify learning with change in a neuron, though in some sense a changed neuron has learned its new state.

Synapses change more slowly than neurons change. We learn more slowly than we "think." This reflects the conventional interpretation of learning as *semi-permanent change*. We have learned calculus if our calculus-exam-taking behavior has changed from failing to passing, and has stayed that way for some time. In the same sense we can say that the artist's canvas learns the pattern of paints the artist smears on its surface, or that the overgrown lawn learns the well-trimmed pattern the lawnmower imparts. In these cases the system learns when pattern stimulation changes a memory medium and leaves it changed for some comparatively long stretch of time.

Learning also involves quantization. Usually a system learns only a small proportion of all patterns in the sampled pattern environment. The number of possible samples may well be infinite. The discrete index notation "$(\mathbf{x}_i, \mathbf{y}_i)$" reflects this sharp disparity between the number of actually learned patterns and the number of possibly learned patterns. In general there are continuum many function samples, or random-vector realizations, $(\mathbf{x}, \mathbf{y})$.

Memory capacity is scarce. An adaptive system must efficiently, and continually, replace the patterns in its limited memory with patterns that accurately represent the sampled pattern environment. Learning replaces old stored patterns with new patterns. Learning forms "internal representations" or prototypes of sampled patterns.

Learned prototypes define quantized patterns. Neural network models represent prototype patterns as vectors of real numbers. So we can alternatively view learning as a form of **adaptive vector quantization (AVQ)** [Pratt, 1978].

The AVQ perspective embeds learning in a dynamic geometry. Learned prototype vectors define synaptic points $\mathbf{m}_i$ in some sufficiently large pattern space R^n. Then the system *learns* if and only if some point $\mathbf{m}_i$ *moves* in the pattern space R^n. The prototypes $\mathbf{m}_i$ gradually wiggle about R^n as learning unfolds. Figure 4.1 illustrates a two-dimensional snapshot of this dynamical n-dimensional process.

Vector quantization may be optimal according to different criteria. The prototypes may spread themselves out so as to minimize the mean-squared error of vector quantization or to minimize some other numerical performance criterion. More generally, the quantization vectors should estimate the underlying unknown probability distribution of patterns. The distribution of prototype vectors should statistically resemble the unknown distribution of patterns.

Uniform sampling probability provides an information-theoretic criterion for an optimal quantization. If we choose a pattern sample at "random"—according to

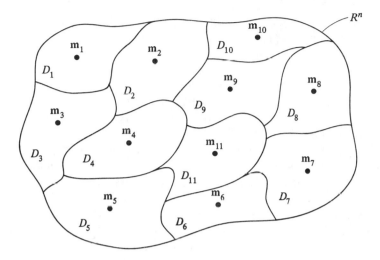

FIGURE 4.1 Quantization of pattern space R^n into k regions of quantization or decision classes. In the simplest case exactly one quantization vector occupies a single region. In the unsupervised case we do not know the class boundaries. Clusters of the k quantization vectors estimate the unknown pattern classes.

thc underlying unknown probability density function $p(\mathbf{x})$—then pattern $\mathbf{x}$ should belong with equal probability to each quantization-estimated pattern class. These equiprobable partitions are not unique.

Supervised and Unsupervised Learning in Neural Networks

The distinction between supervised and unsupervised learning depends on information. The learning adjectives "supervised" and "unsupervised" stem from pattern-recognition theory. The distinction depends on whether the learning algorithm uses pattern-class information. Supervised learning uses pattern-class information; unsupervised learning does not.

An *unknown* probability density function $p(\mathbf{x})$ describes the continuous distribution of patterns $\mathbf{x}$ in the pattern space R^n. Learning seeks only to accurately estimate $p(\mathbf{x})$. The supervision in supervised learning provides information about the pattern density $p(\mathbf{x})$. This information may be inaccurate.

The supervisor assumes a pattern-class structure and perhaps other $p(\mathbf{x})$ properties. The supervisor may assume that $p(\mathbf{x})$ has few or equally likely modes or maximum values. Or the supervisor may assume that $p(\mathbf{x})$ does not change with time, that random quantities are *stationary* in some sense. Or the supervisor may assume that $p(\mathbf{x})$-distributed random vectors have finite covariances or resemble Gaussian-distributed random vectors. Unsupervised learning makes no $p(\mathbf{x})$ assumptions. It uses minimal information.

Supervised learning algorithms depend on the class membership of each training sample $\mathbf{x}$. Suppose the disjoint pattern or decision classes $D_1, \ldots, D_k$ exhaustively partition the pattern space R^n. Then pattern sample $\mathbf{x}$ belongs to some class D_j and not to any other class D_i: $\mathbf{x} \in D_j$ and $\mathbf{x} \notin D_i$ for all $i \neq j$. (With zero probability $\mathbf{x}$ can belong to two or more classes.)

Class-membership information allows supervised learning algorithms to detect pattern misclassifications and perhaps compute an *error* signal or vector. Error information reinforces the learning process. It rewards accurate classifications and punishes misclassifications.

Unsupervised learning algorithms use *unlabelled* pattern samples. They "blindly" process the pattern sample $\mathbf{x}$. Unsupervised learning algorithms often have less computational complexity and less accuracy than supervised learning algorithms. Unsupervised learning algorithms learn rapidly, often on a single pass of noisy data. This makes unsupervised learning practical in many high-speed real-time environments, where we may not have enough time, information, or computational precision to use supervised techniques.

We similarly distinguish supervised and unsupervised learning in neural networks. Supervised learning usually refers to estimated gradient descent in the space of all possible synaptic-value combinations. We estimate the gradient of an unknown mean-squared performance measure that depends on the unknown probability density function $p(\mathbf{x})$. The supervisor uses class-membership information to define a numerical error signal or vector, which guides the estimated gradient descent. Chapter 5 presents supervised and unsupervised (competitive) learning in the statistical framework of *stochastic approximation*.

Unsupervised synaptic learning refers to how biological synapses modify their parameters with physically local information about neuronal signals. The synapses do not use the class membership of training samples. They process raw unlabelled data.

Unsupervised learning systems adaptively cluster patterns into clusters or decision classes D_j. Competitive learning systems evolve "winning" neurons in a neuronal competition for activation induced by randomly sampled input patterns. Then synaptic fan-in vectors tend to estimate pattern-class centroids. The centroids depend explicitly on the unknown underlying pattern density $p(\mathbf{x})$. This illustrates why we do not need to learn if we know $p(\mathbf{x})$. Then only numerical, combinatorial, or optimization tasks remain. Learning is a means to a computational end.

Other unsupervised neural systems evolve attractor basins in the pattern state space R^n. Attractor basins correspond to pattern classes. Feedback dynamics allocate the basins in width, position, and number. Chapter 6 studies unsupervised basin formation in the ABAM and RABAM theorems.

First-order difference or differential equations define unsupervised learning laws. In general, stochastic differential equations define unsupervised learning laws. The differential equations describe how synapses evolve with *locally available information*.

Local information is information physically available to the synapse. The synapse has access to this information only briefly. Local information usually involves synaptic properties or neuronal signal properties. In mammalian brains or opto-electronic integrated circuits, synapses may have local access to other types of information: glial cells, inter-cellular fluids, specific and nonspecific hormones, electromagnetic effects, light pulses, and other "interference" sources. We shall lump these phenomena together and model them as net random unmodelled effects. These noisy unmodelled effects give a Brownian-motion nature to synaptic equilibria but usually do not affect the structure of global network computations.

Biological synapses learn locally and without supervision. Neuroscientist Richard Thompson [1986] summarized biological synaptic learning as follows: "All evidence to date indicates that the mechanisms of memory storage are local and do not involve the formation of new projection pathways. Local changes could include the formation of new synapses, structural and chemical alterations in neurons and synapses, and alterations in membrane properties that influence functional properties of preexisting synapses."

Locality allows asynchronous synapses to learn in real time. The synapses need not wait for a global error message, a message that may never come. Similarly, air molecules vibrate and collide locally without global temperature information, though their behavior produces this global information. Economic agents locally modify their behavior without knowing how they daily, or hourly, affect their national interest rate or their gross national product.

Locality also shrinks the function space of feasible unsupervised learning laws. Synapses have local access to very limited types of information. Learning laws contain only synaptic, neuronal, and noise (unmodelled effects) terms.

Associativity further shrinks the function space. Neural networks associate patterns with patterns. They associate vector pattern **y** with vector pattern **x** as they learn the association $(\mathbf{x}, \mathbf{y})$. Neural networks estimate continuous functions $f: X \longrightarrow Y$. More generally they estimate the unknown joint probability density function $p(\mathbf{x}, \mathbf{y})$.

Locally unsupervised synapses associate signals with signals. This leads to conjunctive, or correlation, learning laws constrained by locality. In the simplest case the signal product $S_i(x_i)S_j(y_j)$ drives the learning equation. Neuroscientist Donald Hebb [1949] first developed the mechanistic theory of conjunctive synaptic learning in his book *The Organization of Behavior.*

FOUR UNSUPERVISED LEARNING LAWS

We shall examine four unsupervised learning laws: signal Hebbian, competitive, differential Hebbian, and differential competitive. We first state their deterministic forms. We then formally review probability theory, random processes, Brownian motion, and white noise. This review allows us to state rigorously the stochastic version of the learning laws, which we actually use in practice and which we study throughout this book. For instance, stochastic synaptic vectors wander in a Brownian motion about their equilibrium values. In simulations the Brownian wandering corresponds to the synaptic vectors jittering slightly after convergence. Finally we study in detail the distinct properties of the four types of unsupervised learning.

Chapter 6 examines the dynamical behavior of stochastic competitive learning in feedforward networks in the AVQ convergence theorem. It examines the dynamical behavior of all four types of learning in feedback networks in the RABAM theorem.

Four Deterministic Unsupervised Learning Laws

The deterministic **signal Hebbian learning law** correlates local neuronal signals:

$$\dot{m}_{ij} \;=\; -m_{ij} + S_i^X(x_i)S_j^Y(y_j) \tag{4-1}$$

or more simply,

$$\dot{m}_{ij} \;=\; -m_{ij} + S_i(x_i)S_j(y_j) \tag{4-2}$$

The overdot denotes time differentiation.

m_{ij} denotes the synaptic efficacy of the synapse along the directed axonal edge from the ith neuron in the input neuron field F_X to the jth neuron in field F_Y. The ijth synaptic junction is *excitatory* if $m_{ij} > 0$, *inhibitory* if $m_{ij} < 0$.

The monotone-nondecreasing signal function S_i transduces the real-valued activation, or membrane potential, x_i to the bounded signal $S_i(x_i)$.

The deterministic **competitive learning law** [Grossberg, 1969] modulates the signal-synaptic difference $S_i(x_i) - m_{ij}$ with the zero-one competitive signal $S_j(y_j)$:

$$\dot{m}_{ij} \;=\; S_j(y_j)[S_i(x_i) - m_{ij}] \tag{4-3}$$

A steep logistic signal function,

$$S_j(y_j) \;=\; \frac{1}{1 + e^{-cy_j}} \tag{4-4}$$

with large $c > 0$, can approximate a binary win-loss indicator. The jth F_Y neuron "wins" the competition at time t if $S_j(y_j(t)) = 1$, "loses" if $S_j(y_j(t)) = 0$. So competitive learning means that synapses learn only if their postsynaptic neurons win. Postsynaptic neurons code for presynaptic signal patterns.

The F_Y neurons compete for the activation induced by signal patterns $\mathbf{S(x)}$ at input field F_X [Von der Malsburg, 1973]. $\mathbf{S(x)}$ denotes the F_X signal vector $(S_1(x_1), \ldots, S_n(x_n))$. The F_Y neurons excite themselves and inhibit one another in a *laterally inhibitive* within-field connection topology.

In practice this F_Y feedback competition reduces to nearest-neighbor pattern matching. The user compares the input pattern to the p synaptic vectors $\mathbf{m}_1, \ldots, \mathbf{m}_p$.

$\mathbf{m}_j$ denotes the synaptic row vector $(m_{1j}, \ldots, m_{nj})$. The jth neuron "wins" the competition if synaptic vector $\mathbf{m}_j$ yields the smallest distance $d(\mathbf{S}(\mathbf{x}), \mathbf{m}_j)$ between F_X signal vector $\mathbf{S}(\mathbf{x})$ and synaptic vector $\mathbf{m}_j$. So the competitive signal $S_j(y_j)$ becomes a metaphor for the metrical indicator function

$$S_j(y_j) = \begin{cases} 1 & \text{if } d(\mathbf{S}(\mathbf{x}), \mathbf{m}_j) = \min_k d(\mathbf{S}(\mathbf{x}), \mathbf{m}_k) \\ 0 & \text{if } d(\mathbf{S}(\mathbf{x}), \mathbf{m}_j) > \min_k d(\mathbf{S}(\mathbf{x}), \mathbf{m}_k) \end{cases} \tag{4-5}$$

In practice F_X neurons behave linearly: $S_i(x_i) = x_i$. The input pattern vector $\mathbf{x}$ represents the output of the F_X neurons. Then the competitive learning law reduces to the **linear competitive learning law**:

$$\dot{\mathbf{m}}_j = S_j(y_j)[\mathbf{x} - \mathbf{m}_j] \tag{4-6}$$

in vector notation.

The deterministic **differential Hebbian learning law** [Kosko, 1988] correlates signal velocities as well as neuronal signals:

$$\dot{m}_{ij} = -m_{ij} + S_i(x_i)S_j(y_j) + \dot{S}_i(x_i)\dot{S}_j(y_j) \tag{4-7}$$

The signal velocity $\dot{S}_i$ decomposes with the chain rule as

$$\frac{dS_i(x_i)}{dt} = \frac{dS_i}{dx_i}\frac{dx_i}{dt} \tag{4-8}$$

$$= S'_i \dot{x}_i \tag{4-9}$$

In principle synapse m_{ij} has access to signal velocities if it has access to signals. Below we show how pulse-coded signal functions, discussed in Chapter 2, greatly simplify computing signal time derivatives and reduce differential Hebbian learning to Hebbian learning. The simplest differential Hebbian learning law omits the signal Hebbian product $S_i S_j$ in (4-7).

The signal velocities may be positive or negative even though the signals may take only nonnegative values. This helps provide synaptic modification with an "arrow of time." It also allows us to use differential Hebbian learning laws for real-time causal inference.

The deterministic **differential competitive learning law** [Kosko, 1990] combines competitive and differential Hebbian learning:

$$\dot{m}_{ij} = \dot{S}_j(y_j)[S_i(x_i) - m_{ij}] \tag{4-10}$$

Differential competition means *learn only if change*. The signal velocity $\dot{S}_j$ provides local reward-punish reinforcement. The $\dot{S}_j$ factor resembles the reward-punish behavior of supervised competitive learning systems, as we discuss below and illustrate in Chapter 1 of the companion volume [Kosko, 1991].

Pulse-coded signal functions reduce differential competitive learning to competitive learning in many cases. Then the presence or absence of a binary pulse

$y_j(t)$ estimates the instantaneous sign of the signal velocity, and hence estimates whether the signal function increases or decreases at that instant.

In practice we use the **linear differential competitive learning law** for pattern recognition and probability-density-function estimation:

$$\dot{m}_j = \dot{S}_j(y_j)[\mathbf{x} - \mathbf{m}_j] \qquad (4\text{-}11)$$

Again metrical comparison picks the "winning" synaptic vector $\mathbf{m}_j$ for modification. In discrete simulations, as in Chapters 8 through 11 and in Chapter 1 of [Kosko, 1991], we approximate the signal velocity with the sign—1, 0, or -1—of the signal difference $S_j(y_j^{k+1}) - S_j(y_j^k)$ or the linear activation difference $y_j^{k+1} - y_j^k$. This resembles the *delta modulation* technique used in communication theory [Hambly, 1990].

The stochastic version of the above unsupervised learning laws relate time changes in random processes. They also perturb learning with Brownian-motion changes or white noise. For instance, the stochastic differential version of the signal Hebbian learning law (4-2) takes the form

$$dm_{ij} = -m_{ij}\, dt + S_i S_j\, dt + dB_{ij} \qquad (4\text{-}12)$$

where $\{B_{ij}(t)\}$ represents a Brownian-motion diffusion process. The less rigorous "noise" form of (4-12) takes the form

$$\dot{m}_{ij} = -m_{ij} + S_i S_j + n_{ij} \qquad (4\text{-}13)$$

where $\{n_{ij}(t)\}$ represents a zero-mean Gaussian white-noise random process.

Intuitively the stochastic differential equation (4-13) simply adds noise to the signal Hebbian learning law (4-2). But mathematically we cannot "add" random functions to deterministic differential equations. The following sections provide motivation and partial justification for this application of the stochastic calculus. The complete justification [Skorokhod, 1982] of our use of the stochastic calculus lies beyond the mathematical scope of this book. Still, we can identify enough of the stochastic-calculus framework to prove theorems about how several neural dynamical systems behave in stochastic equilibrium and on their way to stochastic equilibrium.

We now review the concepts we will need from the theory of probability and random processes. We will apply these concepts extensively in Chapters 5 and 6.

Brownian Motion and White Noise

Brownian motion and white noise are related random processes. They model many physical and information systems. A sample from a Brownian process resembles a continuous jittery curve drawn with kinks everywhere.

White noise represents the idealized time derivative of Brownian motion. Ideal white noise has a flat spectrum across the infinite band of frequencies. A flat spectrum implies that the white-noise process $\{n_t\}$ has infinite *average power* $E[n_t^2]$ at each time instant t:

$$E[n_t^2] \;=\; \infty \qquad\qquad (4\text{-}14)$$

as we show below.

   White noise is a mathematical fiction in a special sense. White noise represents a limiting extreme of observed behavior: thermal-noise jitter in circuits, vibration about stock-price trends, static on a radio, "snow" on a television screen. The ideal gas law, $PV = nrT$, also represents limiting extremes of observed behavior. It does not "really" hold in spacetime reality. Neither does white noise. But unlike the ideal gas law, white noise does not exist in mathematical reality, except as a pseudolimiting case.

   Brownian motion is continuous but *nowhere* differentiable, since it has kinks everywhere. Yet we define white noise as the derivative of Brownian motion. This forces us to approximate white noise as a pseudoderivative in the mean-squared stochastic calculus. As we will see, white noise involves the equally elusive, equally useful, Dirac delta function.

PROBABILITY SPACES AND RANDOM PROCESSES

   Random processes are sequences or families of random variables. More generally they are sequences or families of random vectors, multidimensional random variables.

   Random processes are *indexed* random variables. Different index sets define different species of random processes. A finite index set defines a **random vector**, such as $\mathbf{x} = (x_1, \ldots, x_n)$. A finite or countable index set defines a **random sequence**. A continuous or uncountable index set defines a **random process**. Subintervals of the real line R define continuous or timelike index sets. Connected subsets of R^n define spacelike index sets, which index **random fields**. An infinite sequence of slot-machine plays defines a random sequence. An oscilloscope voltage display defines a random process. The random pixels on a television screen define a discrete random field.

   Random-process properties depend on probability-space properties. Probability spaces are special cases of measure spaces. Intuitively a *measure* generalizes length, area, and volume in abstract mathematical settings. The following section reviews basic measure theory.

Measurability and Sigma-Algebras

   Formally we define the random vector process $\{\mathbf{x}_t\}$ as a function $\mathbf{x}\colon R^n \times T \longrightarrow R^m$ from the *product space* $R^n \times T$ to the space R^m of vector realizations. At each time t, the random process $\mathbf{x}(t)$ can assume different outcome R^m vectors $\mathbf{r}$ with different probabilities. R^n defines the *sample space* of point events. T denotes the index set, perhaps the interval $[0, T]$ or the half-line $[0, \infty)$.

The notation $\mathbf{x}(., .)$ indicates that a random process is a function of two variables. For a fixed realization $\mathbf{r} \in R^n$, $\mathbf{x}(\mathbf{r}, .)$ defines a **sample function** from the random process. For a fixed index value t, $\mathbf{x}(., t)$ defines a random vector, or a random variable in the scalar case.

Technically the mapping $\mathbf{x}$ must be "measurable." Measurability involves which R^n subsets possess probabilities. This depends on R^m subsets. Suppose the R^m subset B consists of interval products. In general this means

$$B = (-\infty, y_1] \times \cdots \times (-\infty, y_m] \tag{4-15}$$

$$= \{(b_1, \ldots, b_p) \in R^p : b_1 \leq y_1, \ldots, b_p \leq y_p\} \tag{4-16}$$

Suppose the R^n subset A consists of n-vectors $\mathbf{r}$ that $\mathbf{x}$ maps into B:

$$A = \mathbf{x}^{-1}(B) = \{\mathbf{r} \in R^n : \mathbf{x}(\mathbf{r}) \in B\} \tag{4-17}$$

The set A is a $\mathbf{B}(R^n)$-*measurable* set, or a **Borel set**, if A belongs to the set collection $\mathbf{B}(R^n)$. If A belongs to $\mathbf{B}(R^n)$, then the probability $P(A)$ is defined. If the inverse-image set $\mathbf{x}^{-1}(B)$ belongs to $\mathbf{B}(R^n)$ for every interval-product subset B, then the mapping $\mathbf{x} \colon R^n \longrightarrow R^m$ is **Borel measurable**, or simply **measurable**. More generally, a function or mapping is measurable if and only if inverse images of measurable sets are measurable sets.

The triple $(R^n, \mathbf{B}(R^n), P)$ defines the underlying **probability space**. The sample space R^n provides the probability space with points or elementary events. The set collection $\mathbf{B}(R^n)$ provides it with sets of points, with *events*. The *probability measure* $P \colon \mathbf{B}(R^n) \longrightarrow [0, 1]$ weights set events with numbers in the unit interval $[0, 1]$. $\mathbf{B}(R^n)$ denotes the **Borel field**, the topological sigma-algebra [Rudin, 1974], on R^n. $\mathbf{B}(R^n)$ contains the Borel-measurable subsets of R^n. $\mathbf{B}(R^m)$ contains the Borel sets of R^m.

A **sigma-algebra**, or **sigma-field**, is a collection of subsets of a sample space. A sigma-algebra contains enough sets to endow it with an algebraic structure. The "sigma" in sigma-algebra means that in general $\mathbf{A}$ contains infinitely many subsets of the sample space. Suppose R^n defines the sample space. Then 2^{R^n} denotes the **power set** of R^n, the collection of all subsets of R^n. So the sigma-algebra $\mathbf{A}$ defines a subcollection of the power set $2^{R^n} \colon \mathbf{A} \subset 2^{R^n}$.

Formally set collection $\mathbf{A}$ is a **sigma-algebra** on the sample space R^n if $\mathbf{A}$ satisfies three properties. First, $\mathbf{A}$ must contain a *unit*:

$$R^n \in A \tag{4-18}$$

The sample space R^n behaves as a unit or identity operator with the operation of set intersection: $S \cap R^n = S$ for all $S \subset R^n$. Second, $\mathbf{A}$ must be *closed* under complementation. If set S belongs to $\mathbf{A}$, then the complement set $S^c = \{\mathbf{r} \in R^n : \mathbf{r} \notin S\}$ must belong to $\mathbf{A}$:

$$S \in \mathbf{A} \longrightarrow S^c \in \mathbf{A} \tag{4-19}$$

Third, **A** must be closed under *countable* unions of its subsets:

$$S_1 \in \mathbf{A}, \; S_2 \in \mathbf{A}, \; \dots \; \longrightarrow \; \bigcup_{i=1}^{\infty} S_i \in \mathbf{A} \qquad (4\text{-}20)$$

The countability in the third property formally accounts for the "sigma" in sigma-algebra. If **A** satisfies the first two properties but contains only finite unions of its subsets, **A** is simply an *algebra*.

The first two properties imply that **A** contains the empty set, because $(R^n)^c = \emptyset$. The empty set $\emptyset$ behaves as a unit or identity operator with set union: $S \cup \emptyset = S$ for all $S \subset R^n$. The second and third properties imply that **A** contains all countable intersections of its subsets. The proof uses DeMorgan's law for countable set collections. The power set 2^{R^n} satisfies trivially all three properties. 2^{R^n} is the "largest" sigma-algebra on R^n, even though uncountably many other sigma-algebras share its 2^c cardinality. c denotes the cardinality of the continuum, or of R^n—the transfinite "number" of elements in R^n.

The set of all sigma-algebras on R^n is closed under intersection. The intersection of any number of sigma-algebras produces another sigma-algebra. This includes uncountable intersections. We can see this with the two extreme sigma-algebras, the power set 2^{R^n} and the smallest sigma-algebra, $\{\emptyset, R^n\}$. Their intersection gives back $\{\emptyset, R^n\}$.

Not all collections **C** of R^n subsets define sigma-algebras. But every collection **C** *generates* a sigma-algebra $g(\mathbf{C})$. The generated sigma-algebra $g(\mathbf{C})$ equals the intersection of all sigma-algebras $\mathbf{A}_c$ that contain the set collection **C**:

$$g(\mathbf{C}) = \bigcap_c \mathbf{A}_c \quad \text{and} \quad \mathbf{C} \subset \mathbf{A}_c \qquad (4\text{-}21)$$

The intersection index c usually ranges over an uncountable set.

The interval-product subsets of R^n generate the Borel sigma-algebra $\mathbf{B}(R^n)$:

$$\mathbf{B}(R^n) \;=\; g(\{O \subset R^n : O = (-\infty, r_1] \times \cdots \times (-\infty, r_n]\}) \qquad (4\text{-}22)$$

Intuitively, Borel sets are made up of products of intervals. If a finite set X defines the sample space, then the Borel sigma-algebra equals the finite power set 2^X: $\mathbf{B}(X) = 2^X$. This does not hold in general if X is infinite.

Other set "bases" generate the Borel field $\mathbf{B}(R^n)$. For example, the set of all open balls with rational radii centered at rational-number coordinates provides a countable base for $\mathbf{B}(R^n)$. This implies [Rudin, 1974] that $\mathbf{B}(R^n)$ has the cardinality c of the continuum. We can place the sets in $\mathbf{B}(R^n)$ in one-to-one correspondence with the real numbers or, equivalently, with the points in R^n.

We cannot place the Borel sets in one-to-one correspondence with 2^{R^n}, all subsets of R^n, which has cardinality 2^c. 2^{R^n} contains vastly more R^n subsets than the Borel field $\mathbf{B}(R^n)$ contains—so many more that in general [Chung, 1974] we cannot define probability measures on 2^{R^n}. In practice all "infinite" sets we model are countable, or are intervals or interval products, and so they are Borel sets. We treat the Borel sigma-algebra $\mathbf{B}(R^n)$ *as if* it equaled the set of all subsets 2^{R^n}.

The range space R^p admits the Borel field $\mathbf{B}(R^p)$. In general a set X and a sigma-algebra $\mathbf{A}$ on X define a **measurable space** $(X, \mathbf{A})$. A measurable function from R^n to R^p actually defines a mapping between the measurable spaces $(R^n, \mathbf{B}(R^n))$ and $(R^p, \mathbf{B}(R^p))$. We formally define a random vector $\mathbf{x}: R^n \longrightarrow R^p$ as a Borel-measurable mapping. So inverse images $\mathbf{x}^{-1}(B)$ of Borel sets B in R^p are Borel sets in R^n:

$$\mathbf{x}^{-1}(B) \in \mathbf{B}(R^n) \quad \text{for all } B \in \mathbf{B}(R^p) \tag{4-23}$$

If we replace the domain space R^n with R and the range space R^p with R, the random vector reduces to the **random variable** $x: R \longrightarrow R$.

In practice we often assume that random variables and vectors define the identity function. In pattern recognition, the random pattern vector $\mathbf{x}$ equals the measurable identity function $\mathbf{x}: R^n \longrightarrow R^n$ such that $\mathbf{x}(\mathbf{v}) = \mathbf{v}$ for all $\mathbf{v}$ in R^n. Then a probability density function $p(\mathbf{x})$ measures the occurrence probability of pattern vectors in different R^n subsets.

The measurable space $(R^n, \mathbf{B}(R^n))$ becomes a **measure space** $(R^n, \mathbf{B}(R^n), M)$ when we define a nonnegative measure M on $\mathbf{B}(R^n)$. If M assigns only finite values to Borel sets, then M equals a probability measure P, perhaps unnormalized, and we arrive at the probability space $(R^n, \mathbf{B}(R^n), P)$.

Probability Measures and Density Functions

Probability measures assign numbers to sets. They assign nonnegative numbers to sets in the Borel sigma-algebra $\mathbf{B}(R^n)$. Probability measures are *finite* measures. They assign a finite positive number to the sample space R^n. We take this number $P(R^n)$ to be unity: $P(R^n) = 1$.

Formally the triple $(R^n, \mathbf{B}(R^n), P)$ defines a probability space if the set function P defines a probability measure. The set function $P : \mathbf{B}(R^n) \longrightarrow [0, 1]$ defines a **probability measure** if P is **countably additive** on disjoint subsets $A_1, A_2, \ldots$ of the Borel sigma-algebra $\mathbf{B}(R^n)$:

$$P(\bigcup_{i=1}^{\infty} A_i) = \sum_{i=1}^{\infty} A_i \tag{4-24}$$

$$A_i \cap A_j = \emptyset \quad \text{if } i \neq j \tag{4-25}$$

The random vector $\mathbf{x}: R^n \longrightarrow R^p$ induces the **cumulative probability function** $P_{\mathbf{x}}$ on R^p vectors $\mathbf{y}$:

$$P_{\mathbf{x}}(\mathbf{y}) = P(\{\mathbf{r} \in R^n: \mathbf{x}(\mathbf{r}) \leq \mathbf{y}\}) \tag{4-26}$$

$$= P(\{\mathbf{r} \in R^n: x_1(r) \leq y_1, \ldots, \leq x_p(r)\}) \tag{4-27}$$

Many texts denote $P_{\mathbf{x}}(y)$ **as** $F_{\mathbf{x}}(y)$, or simply as $F(y)$. Since most random vectors $\mathbf{x}$ we encounter are identity functions, $F_{\mathbf{x}}(\mathbf{x})$ denotes the cumulative distribution function. For simplicity we shall also refer to $F_{\mathbf{x}}(\mathbf{x})$, or simply $F(\mathbf{x})$, instead of $F_{\mathbf{x}}(\mathbf{y})$.

Suppose $F(\mathbf{x})$ has continuous partial derivatives of all orders. Then we define the **probability density function** $p(\mathbf{x})$ as the multiple partial derivative

$$p(\mathbf{x}) \;=\; \frac{\partial^n}{\partial x_1 \dots \partial x_n}\, F(\mathbf{x}) \tag{4-28}$$

where $\mathbf{x}$ refers to the row vector $(x_1, \dots, x_n)$.

The probability density function $p(\mathbf{x})$ need not exist. In practice we assume it does. The cumulative distribution function $F(\mathbf{x})$ always exists, since random vector $\mathbf{x}$ is Borel measurable. We shall focus on $p(\mathbf{x})$ instead of $F(\mathbf{x})$ throughout this book.

The probability density function $p(\mathbf{x})$ cannot assume negative values, and it sums (integrates) to unity:

$$p(\mathbf{x}) \;\geq\; 0 \quad \text{for all } \mathbf{x} \in R^n \tag{4-29}$$

$$\int_{R^n} p(\mathbf{x})\mathbf{dx} \;=\; \int_{-\infty}^{\infty} \dots \int_{-\infty}^{\infty} p(\mathbf{x})\,dx_1 \dots dx_n \;=\; 1 \tag{4-30}$$

Continuous partial derivatives of $F(\mathbf{x})$ allows us to integrate $p(\mathbf{x})$. Strictly speaking, $p(\mathbf{x}) = 0$ for every $\mathbf{x}$ in R^n, since the sample space R^n is continuous. When we write $p(\mathbf{x}) > 0$, we mean that the infinitesimal volume $\mathbf{dx}$ has nonzero probability: $p(\mathbf{x})\,\mathbf{dx} > 0$.

The **Gaussian density** $p_G(\mathbf{x})$ represents one of the most important probability density functions. We define $p_G(\mathbf{x})$ relative to our original probability density function $p(\mathbf{x})$:

$$p_G(\mathbf{x}) \;=\; \frac{1}{(2\pi)^{n/2}|\mathbf{K}|^{1/2}} \exp\{-\frac{1}{2}(\mathbf{x} - \mathbf{m_x})\mathbf{K}^{-1}(\mathbf{x} - \mathbf{m_x})^T\} \tag{4-31}$$

$|\mathbf{K}|$ denotes the (positive) determinant of the **covariance matrix** K:

$$\mathbf{K} \;=\; E[(\mathbf{x} - \mathbf{m_x})^T(\mathbf{x} - \mathbf{m_x})] \tag{4-32}$$

which we define componentwise below. $\mathbf{m_x}$ denotes the **mean vector** of the random vector $\mathbf{x}$:

$$\mathbf{m_x} \;=\; E[\mathbf{x}] \tag{4-33}$$

$$=\; \int_{R^n} \mathbf{x}p(\mathbf{x})\,\mathbf{dx} \tag{4-34}$$

$$=\; \int_{-\infty}^{\infty} \dots \int_{-\infty}^{\infty} \mathbf{x}p(\mathbf{x})\,dx_1 \dots dx_n \tag{4-35}$$

$$=\; (\int_{-\infty}^{\infty} \dots \int_{-\infty}^{\infty} x_1 p(\mathbf{x})\,\mathbf{dx}, \dots, \int_{-\infty}^{\infty} \dots \int_{-\infty}^{\infty} x_n p(\mathbf{x})\,\mathbf{dx}) \tag{4-36}$$

$$=\; (E[x_1], \dots, E[x_n]) \tag{4-37}$$

We have defined the **mathematical expectation** $E[x]$ of scalar random variable x as the Riemann (or Lebesgue) integral, if it exists, of x over R^n with weight function $p(\mathbf{x})$:

$$E[x] \quad = \quad \int_{R^n} x p(\mathbf{x}) \, d\mathbf{x} \tag{4-38}$$

We define the **correlation** $R(\mathbf{x}, \mathbf{z})$ between two R^n vectors as the average inner product

$$R(\mathbf{x}, \mathbf{z}) \quad = \quad E[\mathbf{x} \mathbf{z}^T] \tag{4-39}$$

We define the **variance** $V[x]$ of scalar random variable x as

$$V[x] \quad = \quad E[(x - m_x)^2] \tag{4-40}$$

$$= \quad E[x^2] - m_x^2 \tag{4-41}$$

where $m_x = E[x]$. If $V(x) < \infty$, we often denote $V(x)$ as σ_x^2. We define the **covariance** $\text{Cov}[x, y]$ of two scalar random variables x and z as

$$\text{Cov}[x, z] \quad = \quad E[(x - m_x)(z - m_z)] \tag{4-42}$$

$$= \quad E[xz] - m_x m_z \tag{4-43}$$

which generalizes the variance relationship (4-41).

We define the **covariance matrix K** of random vector $\mathbf{x}$ as

$$\mathbf{K} \quad = \quad E[(\mathbf{x} - \mathbf{m_x})^T (\mathbf{x} - \mathbf{m_x})] \tag{4-44}$$

Each component k_{ij} of $\mathbf{K}$ equals the covariance term $\text{Cov}[x_i, x_j]$. The diagonal terms k_{ii} equal the variances $V[x_i]$.

More generally, suppose $\mathbf{x}$ denotes a random vector on R^n, and $\mathbf{z}$ denotes a random vector on R^p. Then the joint probability density function $p(\mathbf{x}, \mathbf{z})$ describes the continuous distribution of vector pairs $(\mathbf{x}, \mathbf{z})$ in the product sample space $R^n \times R^p$. We define the **cross covariance matrix $\mathbf{K_{xz}}$** as

$$\mathbf{K_{xz}} \quad = \quad E_{\mathbf{xz}}[(\mathbf{x} - \mathbf{m_x})^T (\mathbf{z} - \mathbf{m_z})] \tag{4-45}$$

The subscript on $E_{\mathbf{xz}}$ indicates that we integrate with respect to the joint density $p(\mathbf{x}, \mathbf{z})$. We form the mean vectors $\mathbf{m_x}$ and $\mathbf{m_z}$ with the individual or "marginal" densities $p(\mathbf{x})$ and $p(\mathbf{z})$.

Random vectors $\mathbf{x}$ and $\mathbf{z}$ are **independent** if the joint density $p(\mathbf{x}, \mathbf{z})$ factors:

$$p(\mathbf{x}, \mathbf{z}) \quad = \quad p(\mathbf{x}) p(\mathbf{z}) \tag{4-46}$$

For instance, the random vector $\mathbf{x}$ decomposes into n separate random variables x_i if the joint density $p(\mathbf{x})$ factors into the product $p(x_1)p(x_2)\ldots p(x_n)$.

Scalar random variables x and z are **uncorrelated** if their joint expectation factors:

$$E[xz] \quad = \quad E[x] E[z] \tag{4-47}$$

Independent random variables are uncorrelated, but most uncorrelated random variables are dependent. Uncorrelated Gaussian random variables are independent. Vector quantization reduces to scalar quantization when vector components are uncorrelated or independent [Pratt, 1978].

For scalar random variables x and z we define the **conditional probability density function** $p(x|z)$ as

$$p(x|z) \quad = \quad \frac{p(x,\, z)}{p(z)} \tag{4-48}$$

when $p(z) > 0$ locally. We define a realization of the **conditional expectation** random variable $E[x|z]$ as

$$E[x|z = \xi] \quad = \quad \int_{-\infty}^{\infty} xp(x|z = \xi)\, dx \tag{4-49}$$

when random variable z assumes or "realizes" the value ξ. The set of all realizations $E[x|z = \xi]$ defines the random variable $E[x|z]$. This definition directly extends to the random-vector case. Random variables x and y are **conditionally independent** of random variable z if the conditional expectation $E[xy|z]$ factors for all realizations of z:

$$E[xy|z] \quad = \quad E[x|z]E[y|z] \tag{4-50}$$

Suppose set $D \subset R^n$ belongs to the Borel sigma-algebra $\mathbf{B}(R^n)$. We can equivalently define D as the **indicator function** I_D:

$$I_D(\mathbf{x}) \quad = \quad \begin{cases} 1 & \text{if} \quad \mathbf{x} \in D \\ 0 & \text{if} \quad \mathbf{x} \notin D \end{cases} \tag{4-51}$$

The indicator function I_D indicates whether element $\mathbf{x}$ belongs to set D. I_D is Borel measurable because the inverse-image sets equal D or D^c, and both belong to $\mathbf{B}(R^n)$ and admit the probabilities $P(D)$ and $P(D^c)$. So we can take the mathematical expectation of the random variable I_D. This gives a functional representation of the **class probability** $P(D)$:

$$E[I_D] \quad = \quad \int_{R^n} I_D(\mathbf{x})p(\mathbf{x})\, \mathbf{dx} \tag{4-52}$$

$$= \quad \int_D p(\mathbf{x})\, \mathbf{dx} \tag{4-53}$$

$$= \quad P(D) \tag{4-54}$$

Equality (4-53) follows because D and D^c *partition* the sample space R^n—$R^n = D \cup D^c$, and $D \cap D^c = \emptyset$—and because $I_D(\mathbf{x}) = 1 - I_D^c(\mathbf{x}) = 0$ for all $\mathbf{x}$ in D^c. This property rigorously distinguishes supervised from unsupervised learning in pattern recognition. Supervised learning algorithms depend explicitly on indicator functions I_{D_i} of pattern classes D_i. Unsupervised learning algorithms do not depend on indicator functions.

Random vectors converge componentwise as their constituent random variables converge. So consider a random-variable sequence $x_1, x_2, \ldots$ defined on the scalar probability space $(R, \mathbf{B}(R), P)$. This defines a discrete random process $\{\mathbf{x}_n\}$ in the sense that the index n assumes only integer values. Each random variable $x_i: R \to R$ assigns the real number $x_i(r)$ to the real number r.

The random sequence $\{x_t\}$ can converge to a random variable x in many ways. The sequence converges "everywhere" if it converges in the traditional sense of nonrandom functions. In general we cannot attain convergence everywhere in a probability space. We state the definition for completeness. $\{x_n\}$ **converges everywhere** to x if for all $r \in R$ and for all $\varepsilon > 0$, there exists an integer n_0 such that if $n > n_0$, then

$$|x_n(r) - x(r)| \;<\; \varepsilon \tag{4-55}$$

So for every real r the limit of $x_n(r)$ exists:

$$\lim_{n \to \infty} x_n(r) \;=\; x(r) \tag{4-56}$$

The choice of n_0 may depend on ε and r. Uniform convergence results when it does not depend on r. The random sequence $\{x_n\}$ **converges uniformly** to random variable x if for all $\varepsilon > 0$, there exists an integer n_0 such that if $n > n_0$, then

$$|x_n(r) - x(r)| \;<\; \varepsilon \quad \text{for all } r \in R \tag{4-57}$$

Conditions (4-55) and (4-57) differ by where the universal quantifier "for all r" occurs relative to the existential quantifier "there exists an n_0." SOME-ALL implies ALL-SOME, but not conversely. If someone knows everyone, then everyone is known by someone. But everyone can be known by someone without some one person knowing everyone. So uniform convergence everywhere implies convergence everywhere, but not conversely. If the sequence $\{x_n\}$ converges uniformly to x, then in principle we can always find an n_0 to make terms x_n, where $n > n_0$, at least ε close to x, no matter how small ε and for every r.

A random sequence can *stochastically* converge to a random variable in at least four ways. $\{x_n\}$ **converges with probability one** to random variable x if

$$P(\{r \in R: \lim_{n \to \infty} x_n(r) = x(r)\}) \;=\; 1 \tag{4-58}$$

Equivalently, $P(x_n \longrightarrow x) = 1$ if $\{x_n\}$ converges to x except possibly on a Borel set S of probability measure zero: $P(S) = 0$. For this reason we synonymously say that $\{x_n\}$ **converges almost everywhere** or **converges almost surely** to x with respect to probability measure P.

Convergence in probability provides a weaker type of stochastic convergence. Random sequence $\{x_n\}$ **converges in probability** to random variable x if for all $\varepsilon > 0$:

$$\lim_{n \to \infty} P(\{r \in R: |x_n(r) - x(r)| > \varepsilon\}) \;=\; 0 \tag{4-59}$$

The limit occurs outside the probability measure $P(.)$ in convergence in probability, but inside it in convergence with probability one.

Convergence with probability one implies convergence in probability, but not conversely. In this sense convergence in probability is "weaker" than convergence with probability one. For instance, the *weak law of large numbers* refers to convergence in probability of the sample mean to the population mean of a sequence of independent identically distributed random variables. The *strong law of large numbers* refers to convergence with probability one of the same sample mean to the same population mean.

Random sequence $\{x_n\}$ **converges in the mean-square sense** to random variable x if

$$\lim_{n \to \infty} E[(x_n - x)^2] \; = \; 0 \qquad (4\text{-}60)$$

Mean-squared convergence also implies convergence in probability, but not conversely. In general mean-squared convergence and convergence with probability one do not imply each other. Both are "strong" types of convergence.

Convergence in distribution provides the weakest type of stochastic convergence. It involves convergence everywhere of cumulative distribution functions, not random variables. The random sequence $\{x_n\}$ **converges in distribution** to the random variable x if the corresponding deterministic sequence of evaluated cumulative distribution functions $\{F_{x_n}(r)\}$ converges everywhere to the cumulative-distribution-function value $F_x(r)$ of the random variable x at all points of continuity r of F_x:

$$\lim_{n \to \infty} F_{x_n}(r) \; = \; F_x(r) \qquad (4\text{-}61)$$

Convergence with probability one, convergence in probability, and convergence in the mean-squared sense each imply convergence in distribution. These three stronger types of stochastic convergence directly involve random variables. As the random variables approach convergence, in the "tail" of the sequence, the random variables grow more dependent in the sense that the error between them tends to decrease. This need not happen in convergence in distribution. The random variables in the tail of the sequence can stay independent. This happens in *central limit theorems*, where sums of independent random variables, normalized by $\sqrt{n}$ not n, converge in distribution to a random variable with a Gaussian probability distribution. (Normalizing by n yields either the weak or strong laws of large numbers, and the sequence converges to the constant population mean, not a random variable.)

We now apply these probabilistic concepts to the Brownian-motion and white-noise processes. We shall assume a continuous time index t throughout.

Gaussian White Noise as a Brownian Pseudoderivative Process

$\{B_t\}$ denotes a continuous **Brownian-motion diffusion** or **Wiener process**. With probability one, $\{B_t\}$ is continuous everywhere but differentiable nowhere

[Maybeck, 1982]. Yet $\{B_t\}$ has a pseudoderivative, a time derivative in a limiting mean-squared sense. This pseudoderivative random process $\{dB/dt\}$ equals ideal Gaussian white noise $\{n_t\}$:

$$\frac{dB}{dt} = n \qquad (4\text{-}62)$$

We will heuristically argue this relationship with autocorrelation functions and power spectra. A Gaussian density function describes the joint probability density function of any collection of the random variables in the process $\{n_t\}$. In general a noise process is any process independent or uncorrelated in time, no matter how small the time difference.

The white-noise process $\{n_t\}$ is *zero-mean* and *uncorrelated in time*:

$$E[n(t)] = 0 \quad \text{for all } t \qquad (4\text{-}63)$$

$$E[n(t)n(t+s)] = 0 \quad \text{for all } s \neq 0 \qquad (4\text{-}64)$$

The mathematical expectation in (4-64) uses the joint probability density function of the random variables $n(t)$ and $n(t+s)$. The white-noise process $\{n_t\}$ has finite variance σ^2 for each time t:

$$V[n(t)] = E[n^2] = \sigma^2 < \infty \quad \text{for all } t \qquad (4\text{-}65)$$

Equivalently, the noise process $\{n_t\}$ has a deterministic **autocorrelation function** R_n equal to a variance-scaled delta pulse:

$$R_n(t, t+s) = E[n(t)n(t+s)] \qquad (4\text{-}66)$$

$$= R_n(s) \qquad (4\text{-}67)$$

$$= \sigma^2 \delta(s) \qquad (4\text{-}68)$$

In the discrete case, when a countable set indexes the random sequence $\{n_t\}$, a Kroenecker delta function replaces the Dirac delta function in (4-68) (defined below in (4-78)–(4-79)).

Relations (4-63) and (4-67)–(4-68) imply that white noise is a **wide-sense stationary** (WSS) random process. A random process is WSS if and only if time shifts do not affect its first two moments. This means the process has a constant mean function $E[x(t)]$ for all t and its correlation function $R(t, t+s)$ depends only on the time difference $t + s - t$, or s:

$$R(t, t+s) = R(s) \quad \text{for all } s \qquad (4\text{-}69)$$

This traditional notation emphasizes that R depends only on the time difference s. It wrongly suggests that R is a function of one variable, not two.

We can assume that WSS processes are zero-mean. For the mean function is constant, and we can subtract it out wherever it occurs. Then covariance functions equal correlation functions.

A relationship holds in the **mean-squared sense** if it is the mean-squared limit of a sequence of convergent functions. For instance, the random process $\{x_t\}$ has a **mean-squared derivative at t** if the following mean-squared limit exists:

$$\lim_{s \to 0} E\left[\left(\frac{x(t+s) - x(t)}{s} - \frac{dx(t)}{dt}\right)^2\right] = 0 \qquad (4\text{-}70)$$

The derivative random process $\{dx/dt\}$ exists if the original random process $\{x(t)\}$ has a mean-squared derivative at t for all t.

The mean-squared stochastic calculus resembles superficially the familiar differential calculus. But we must remember that every equality relation, $X = Y$, indicates the mean-squared or probability-one limit of a convergent sequence of random functions. These limits may not exist, or they may behave differently than comparable quantities in the differential calculus behave. In general the chain rule of differential calculus does not apply. We will use linearly scaled additive noise to avoid this problem and retain the chain rule [Maybeck, 1982].

The continuous Brownian-motion process $\{B_t\}$ is Gaussian, has zero-mean and autocorrelation function R_B [Stark, 1986]:

$$R_B(t, s) = \sigma^2 \min(t, s) \qquad (4\text{-}71)$$

Then, in the mean-squared sense, the time-derivative process $\{dB_{ij}/dt\}$ has autocorrelation function $R_{dB/dt}$ that depends on the original autocorrelation function R_B through partial differentiation:

$$R_{dB/dt}(t, s) = \frac{\partial^2}{\partial t\, \partial s} R_B(t, s) \qquad (4\text{-}72)$$

$$= \frac{\partial^2}{\partial t\, \partial s} \sigma^2 \min(t, s) \qquad (4\text{-}73)$$

$$= \frac{\partial}{\partial t} \sigma^2\, \text{STEP}[t - s] \qquad (4\text{-}74)$$

$$= \sigma^2 \delta(t - s) \qquad (4\text{-}75)$$

$$= \sigma^2 \delta(s) \qquad (4\text{-}76)$$

which equals (4-68), if we use the time coordinates $(t+s, t)$ instead of (t, s). STEP denotes the threshold or **unit step function**:

$$\text{STEP}[t - s] = \begin{cases} 1 & \text{if } t > s \\ 0 & \text{if } t < s \end{cases} \qquad (4\text{-}77)$$

with discontinuity when $t = s$. So $R_{dB/dt} = R_n$. This justifies (4-62) in a WSS sense.

The partial derivative of a minimum equals a constant except at the single point of discontinuity. We can approximate the derivative of a step function with the derivative of a threshold linear signal function, as defined in Chapter 2. The steeper the ramp in the signal function, the more its derivative resembles an infinite pulse, the **Dirac delta function** $\delta(x)$:

$$\delta(x) = \begin{cases} 0 & \text{if } x \neq 0 \\ \infty & \text{if } x = 0 \end{cases} \tag{4-78}$$

and

$$\int_{-\infty}^{\infty} \delta(x)\, dx = 1 \tag{4-79}$$

The Dirac delta function does not have a Riemann integral, since it equals zero except at a single point. (It has a Lebesgue integral, which equals zero.) The integral in (4-79) is figurative.

The **power spectral density** $\mathbf{S}_x(\omega)$ of WSS random process $\{x(t)\}$ equals the *Fourier transform* of the autocorrelation function R_X:

$$\mathbf{S}_x(\omega) = \int_{-\infty}^{\infty} R_x(s)e^{-iws}\, ds \tag{4-80}$$

where $i = \sqrt{-1}$ and ω denotes a frequency integration variable. Technically we require that the original WSS process $\{x_t\}$ have finite **average energy** $E[x^2(t)]$ at all times t:

$$E[x^2(t)] < \infty \tag{4-81}$$

But, as we shall see, ideal white noise violates this integrability condition.

Each spectral density value $\mathbf{S}_x(\omega)$ depends on the *entire* shape of the random autocorrelation function R_x. Transforms of R_x depend on its global behavior. Functions of R_x depend on its local behavior. For example, a single value $e^{R_x(t)}$ of the exponentiated autocorrelation e^{R_x} depends only on the particular value $R_x(t)$.

In general [Rudin, 1974] we can recover the time function $R_x(s)$ from its frequency transform $\mathbf{S}_x(\omega)$:

$$R_x(s) = \frac{1}{2\pi} \int_{-\infty}^{\infty} \mathbf{S}_x(\omega)e^{iws}\, d\omega \tag{4-82}$$

at all points of continuity. So if $s = 0$, this integral equals the average energy:

$$\frac{1}{2\pi} \int_{-\infty}^{\infty} \mathbf{S}_x(\omega)\, d\omega = R_x(0) \tag{4-83}$$

$$= E[x^2(t)] \tag{4-84}$$

We can compute the power spectral density $\mathbf{S}_n$ of the Gaussian white-noise autocorrelation function R_n if we use the "sifting" convolution property of the Dirac delta function:

$$\mathbf{S}_n(\omega) \quad = \quad \int_{-\infty}^{\infty} R_n(s) e^{-i\omega s}\, ds \tag{4-85}$$

$$= \quad \sigma^2 \int_{-\infty}^{\infty} \delta(s) e^{-i\omega s}\, ds \tag{4-86}$$

$$= \quad \sigma^2 e^0 \tag{4-87}$$

$$= \quad \sigma^2 \tag{4-88}$$

So the power spectral density of Gaussian white noise equals the constant variance σ^2. The power spectral density's magnitude increases as the process variance σ^2 increases, as the process vibrates in time more "randomly." In the extreme case, when $\sigma^2 = 0$, white noise does not vibrate at all, since the process is deterministic.

A flat line across the entire infinite frequency band defines the power spectrum of ideal Gaussian white noise. A time-domain pulse transforms to a frequency-domain flat line, and vice versa. As in ideal white light, all frequencies are equally present in a white-noise signal.

The constant power-spectral-density relation $\mathbf{S}_n(\omega) = \sigma^2$ and (4-82) imply that the white-noise process $\{n_t\}$ has infinite instantaneous average signal power:

$$E[n_t^2] \quad = \quad R_n(0) \tag{4-89}$$

$$= \quad \frac{\sigma^2}{2\pi} \int_{-\infty}^{\infty} d\omega \tag{4-90}$$

$$= \quad \infty \tag{4-91}$$

We do not escape this conclusion if we assume only a "little" noise. For $E[n_t^2] = \infty$ holds if and only if the process $\{n_t\}$ is random or, equivalently, $\sigma^2 > 0$.

Instead we must assume our noise processes are white over some large but finite frequency bandwidth interval $[-B, B]$. Noise with nonflat spectrum defines **colored noise**. The "color" of band-limited noise increases as the bandwidth interval $[-B, B]$ shrinks. The color whitens as B approaches infinity.

STOCHASTIC UNSUPERVISED LEARNING AND STOCHASTIC EQUILIBRIUM

The **random-signal Hebbian learning law** relates random processes in the stochastic differential equation

$$dm_{ij} \quad = \quad -m_{ij}\, dt + S_i(x_i) S_j(y_j)\, dt + dB_{ij} \tag{4-92}$$

$\{B_{ij}(t)\}$ denotes a Brownian-motion diffusion process with autocorrelation function $R_B(t, s)$ given by (4-71). Each term in (4-92) denotes a separate random process.

For instance, the activation random process $\{x_i(t)\}$ gives rise to the signal random process $\{S_i(x_i(t))\}$ when we apply the monotone-nondecreasing signal function S_i to every random variable $x_i(t)$ in the activation random process $\{x_i(t)\}$.

If we divide the stochastic differentials in (4-92) by the time differential dt and use the noise relationship (4-62), we can rewrite (4-92) in the less rigorous, more familiar "noise notation"

$$\dot{m}_{ij} \quad = \quad -m_{ij} + S_i(x_i)S_j(y_j) + n_{ij} \tag{4-93}$$

We assume the zero-mean, Gaussian white-noise process $\{n_{ij}(t)\}$ is independent of the "signal" process $-m_{ij} + S_i S_j$. We shall work with the noise version of stochastic differential equations throughout this book. In Chapter 6 we allow independent noise processes to corrupt each activation stochastic differential equation.

The stochastic nature of the differential equation (4-93) complicates its equilibrium behavior. This holds for all first-order stochastic differential equations that define unsupervised learning noise. So we consider first the general case. A noisy random unsupervised learning law takes the abstract form

$$\dot{m}_{ij} \quad = \quad f_{ij}(\mathbf{x}, \mathbf{y}, \mathbf{M}) + n_{ij} \tag{4-94}$$

$\{n_{ij}(t)\}$ still denotes a zero-mean, Gaussian white-noise process independent of the synaptic "signal" process $\{f_{ij}(t)\}$.

Since independence implies uncorrelatedness, we have the following general lemma. We shall use a version of this lemma repeatedly in Chapter 6. The lemma places a *lower bound* on how much synaptic processes randomly vibrate in equilibrium and on their way to equilibrium.

Lemma

$$E[\dot{m}_{ij}^2] \quad \geq \quad \sigma_{ij}^2 \tag{4-95}$$

Proof. Square both sides of (4-94):

$$\dot{m}_{ij}^2 \quad = \quad f_{ij}^2 + n_{ij}^2 + 2f_{ij}n_{ij} \tag{4-96}$$

We can define an expectation operator E with the general joint density $p(\mathbf{x}, \mathbf{y}, \mathbf{M}, \mathbf{n})$. Independence factors this density into the product $p(\mathbf{x}, \mathbf{y}, \mathbf{M})p(\mathbf{n})$, and so the expectation and (4-30) "integrates out" the noise density $p(\mathbf{n})$.

We take expectations of both sides of (4-96) and exploit the independence of the noise process $\{n_{ij}(t)\}$, its zero-mean property (4-63), and its finite-variance property (4-65):

$$E[\dot{m}_{ij}^2] \quad = \quad E[f_{ij}^2] + E[n_{ij}^2] + 2E[f_{ij}n_{ij}] \tag{4-97}$$

$$= \quad E[f_{ij}^2] + \sigma_{ij}^2 + 2E[f_{ij}]E[n_{ij}] \tag{4-98}$$

$$= \quad E[f_{ij}^2] + \sigma_{ij}^2 \tag{4-99}$$

$$\geq \quad \sigma_{ij}^2 \tag{4-100}$$

Inequality (4-100) follows because $f_{ij}^2 \geq 0$ holds everywhere. $\{f_{ij}^2(t)\}$ is an *increasing* process. The expectation $E[f_{ij}^2]$ integrates a nonnegative function weighted with a nonnegative probability density function. So $E[f_{ij}^2] \geq 0$. **Q.E.D.**

The lemma says that stochastic synapses vibrate in equilibrium, and they vibrate at least as much as the driving noise process vibrates. $\sigma_{ij}^2 > 0$ implies $E[\dot{m}_{ij}^2] > 0$. The mean-squared velocity is *always* positive. On average the synaptic vector $\mathbf{m}_j$ changes or vibrates at every instant t. The synaptic vector $\mathbf{m}_j$ can equal a constant value, such as a pattern-class centroid, only on average. Then $\mathbf{m}_j$ wanders in a Brownian motion about the constant value $E[\mathbf{m}_j]$.

Stochastic Equilibrium

Synaptic equilibrium in the deterministic signal Hebbian law (4-2) occurs in "steady state" when the synaptic state vector $\mathbf{m}_j$ stops moving:

$$\dot{\mathbf{m}}_j \;=\; \mathbf{O} \tag{4-101}$$

$\mathbf{O}$ denotes the R^n null vector $(0, \ldots, 0)$.

Suppose the steady-state condition (4-101) holds for the random signal Hebbian learning law (4-93) or, more generally, for the stochastic synaptic differential equation (4-94). Then the randomly vibrating noise vector $\mathbf{n}_j$ cancels the random signal vector $\mathbf{f}_j$ at all future times:

$$-\mathbf{f}_j \;=\; \mathbf{n}_j \tag{4-102}$$

This unlikely event has zero probability at every time t—$P(-\mathbf{f}_j = \mathbf{n}_j) = 0$—because $\{\mathbf{f}_j(t)\}$ and $\{\mathbf{n}_j(t)\}$ are independent vector processes. The positive noise variance σ_{ij}^2 alters our deterministic picture of equilibrium.

The synaptic vector $\mathbf{m}_j$ reaches **stochastic equilibrium** when only the random-noise vector $\mathbf{n}_j$ changes $\mathbf{m}_j$:

$$\dot{\mathbf{m}}_j \;=\; \mathbf{n}_j \tag{4-103}$$

Then $\mathbf{f}_j = \mathbf{O}$ holds with probability one. So $f_{ij}^2 = 0$ holds with probability one in the lemma. Then the lemma confirms componentwise what (4-103) implies: $E[\dot{m}_{ij}^2] = \sigma_{ij}^2$. At stochastic equilibrium the mean-squared synaptic velocity obtains its theoretical lower bound.

Stochastic equilibrium corresponds to deterministic equilibrium on average. Here average means ensemble or spatial average, the $p(\mathbf{x})$-weighted average of all realizations of the synaptic random vector $\mathbf{m}_j(t)$ at time t. It does not mean time average. The zero-mean noise property (4-63) and the stochastic-equilibrium condition (4-103) imply the average equilibrium condition

$$E[\dot{\mathbf{m}}_j(t)] \;=\; \mathbf{O} \tag{4-104}$$

for all times t after $\mathbf{m}_j$ reaches equilibrium.

In practice we observe only samples of the synaptic random process $\{\mathbf{m}_j(t)\}$. At each time t we observe only one of infinitely many possible realizations of the random vector $\mathbf{m}_j(t)$. So (4-103) better describes observation than (4-101) does.

In practice the equilibrium process $\mathbf{m}_j$ often behaves as an *ergodic* process. Then synaptic time averages approximate instantaneous ensemble averages. Time-summed noise vectors $\mathbf{n}_j(t)$ tend to add to the null vector $\mathbf{O}$. Then time-summed synaptic vectors $\mathbf{m}_j(t)$ accurately estimate the stationary constant vector $E[\mathbf{m}_j]$.

Stochastic equilibrium (4-103) implies the equivalent differential Brownian relationship

$$\mathbf{dm}_j \;=\; \mathbf{n}_j \, dt \tag{4-105}$$

$$=\; \mathbf{dB}_j \tag{4-106}$$

We can stochastically integrate [Maybeck, 1982] these stochastic differentials to give the random-process relationship

$$\mathbf{m}_j(t) \;=\; \mathbf{B}_j(t) + \mathbf{B}_j(0) \tag{4-107}$$

$\mathbf{B}_j(0)$ defines the first random vector in the vector Brownian motion process $\{\mathbf{B}_j(t)\}$. Mathematicians often stipulate that $\mathbf{B}_j(0)$ equal the null vector $\mathbf{O}$ with probability one. This "centers" the zero-mean Brownian motion about the R^n origin. If we follow that convention, then (4-107) states that the synaptic vector process $\mathbf{m}_j$ equals the Brownian process $\mathbf{B}_j$.

We can center the Brownian motion about any R^n vector. This just translates the origin. Indeed the average stochastic-equilibrium condition (4-104) centers the Brownian motion for us. It centers it about the stationary equilibrium synaptic vector $E[\mathbf{m}_j]$:

$$\mathbf{B}_j(0) \;=\; E[\mathbf{m}_j] \tag{4-108}$$

with probability one.

The stochastic equilibrium condition (4-103) implies that $\mathbf{f} = \mathbf{O}$ with probability one. For the stochastic signal Hebbian learning law (4-93) this implies that synapses equal signal products with probability one:

$$m_{ij} \;=\; S_i S_j \tag{4-109}$$

or in vector notation,

$$\mathbf{m}_j \;=\; \mathbf{S}(\mathbf{x}) S_j(y_j) \tag{4-110}$$

Below we see that in general the Hebbian synapses learn an exponentially weighted average of sampled signal products. Yet the tendency to converge exponentially toward (4-109) remains.

The random learning laws are first-order stochastic differential equations. The first-order structure drives synaptic vectors toward stochastic equilibrium exponentially quickly. This holds equally for competitive, differential Hebbian, and differential competitive learning. We now state general stochastic version of these unsupervised learning laws.

The **random competitive learning law** modulates the random vector difference $\mathbf{S}(\mathbf{x}) - \mathbf{m}_j$ with the random competitive signal $S_j(y_j)$ and appends an independent Gaussian white-noise vector $\mathbf{n}_j$ to model unmodelled effects:

$$\dot{\mathbf{m}}_j = S_j(y_j)[\mathbf{S}(\mathbf{x}) - \mathbf{m}_j] + \mathbf{n}_j \tag{4-111}$$

The **random linear competitive learning law** replaces the nonlinear signal vector $\mathbf{S}(\mathbf{x})$ with the untransformed random vector $\mathbf{x}$:

$$\dot{\mathbf{m}}_j = S_j(y_j)[\mathbf{x} - \mathbf{m}_j] + \mathbf{n}_j \tag{4-112}$$

As we discuss below, competitive dynamics in effect replace the competitive signal $S_j(y_j)$ with the pattern-class indicator function I_{D_j}. The decision classes $D_1, \ldots, D_k$ **partition** the pattern sample space R^n into disjoint pieces:

$$R^n = \bigcup_{i=1}^{k} D_i \tag{4-113}$$

$$D_i \cap D_j = \varnothing \quad \text{if } i \neq j \tag{4-114}$$

Then (4-111) and (4-112) simplify to

$$\dot{\mathbf{m}}_j = I_{D_j}(\mathbf{x})[\mathbf{S}(\mathbf{x}) - \mathbf{m}_j] + \mathbf{n}_j \tag{4-115}$$

$$\dot{\mathbf{m}}_j = I_{D_j}(\mathbf{x})[\mathbf{x} - \mathbf{m}_j] + \mathbf{n}_j \tag{4-116}$$

The synaptic vector process in (4-116) approximates a linear system [Kailath, 1980] in D_j.

In (4-115) and (4-116) the jth synaptic vector $\mathbf{m}_j$ locally samples D_j in the sense of nearest-neighbor classification. Every input sample $\mathbf{x}$ is closest to some quantizing synaptic vector $\mathbf{m}_j$. (We break ties arbitrarily if $\mathbf{x}$ is closest to multiple synaptic vectors. Since the pattern vectors $\mathbf{x}$ are continuously distributed in R^n, borderline cases occur with probability zero.) The synaptic vectors adaptively define decision-class partitions of R^n. In this sense learning remains unsupervised in (4-115) and (4-116) even though the single indicator function I_{D_j} appears explicitly.

In contrast, the **random supervised competitive learning law** depends explicitly on all class indicator functions $I_{D_1}, \ldots, I_{D_k}$ through the reinforcement function r_j:

$$\dot{\mathbf{m}}_j = r_j(\mathbf{x})S_j(y_j)[\mathbf{S}(\mathbf{x}) - \mathbf{m}_j] + \mathbf{n}_j \tag{4-117}$$

or, in the linear case with indicatorlike competitive signal,

$$\dot{\mathbf{m}}_j = r_j(\mathbf{x})I_{D_j}(\mathbf{x})[\mathbf{x} - \mathbf{m}_j] + \mathbf{n}_j \tag{4-118}$$

The supervised **reinforcement function** r_j rewards correct classifications with a $+1$ and punishes misclassifications with a -1 or negative learning:

$$r_j(\mathbf{x}) = I_{D_j}(\mathbf{x}) - \sum_{i \neq j} I_{D_i}(\mathbf{x}) \tag{4-119}$$

The reinforcement function requires that we *label* every sample pattern $\mathbf{x}$ with its class membership.

If we ignore zero-probability boundary cases, r_j equals only 1 or -1. $r_j = 1$ if $\mathbf{x} \in D_j$, and $r_j = -1$ if $\mathbf{x} \notin D_j$. So we can rewrite (4-117) in the more familiar "supervised" form

$$\dot{\mathbf{m}}_j = \begin{cases} S_j(y_j)[\mathbf{S}(\mathbf{x}) - \mathbf{m}_j] + \mathbf{n}_j & \text{if } \mathbf{x} \in D_j \\ -S_j(y_j)[\mathbf{S}(\mathbf{x}) - \mathbf{m}_j] + \mathbf{n}_j & \text{if } \mathbf{x} \notin D_j \end{cases} \qquad (4\text{-}120)$$

In practice neural engineers often use the supervised competitive learning algorithm, first studied by Tspykin [1974] and other early learning theorists, as the system of discrete stochastic differential equations

$$\mathbf{m}_j(k+1) = \mathbf{m}_j(k) + c_k[\mathbf{x}_k - \mathbf{m}_j(k)] \quad \text{if } \mathbf{x}_k \in D_j \qquad (4\text{-}121)$$

$$\mathbf{m}_j(k+1) = \mathbf{m}_j(k) - c_k[\mathbf{x}_k - \mathbf{m}_j(k)] \quad \text{if } \mathbf{x}_k \notin D_j \qquad (4\text{-}122)$$

$$\mathbf{m}_i(k+1) = \mathbf{m}_i(k) \qquad\qquad\qquad \text{if } i \neq j \qquad (4\text{-}123)$$

The sequence of learning coefficients $\{c_k\}$ defines a decreasing sequence of positive numbers in accord with stochastic-approximation restrictions, as discussed in Chapter 5. The index notation means that $\mathbf{m}_j$ has won the "competition" at time k, and all other synaptic vectors $\mathbf{m}_i$ have lost at time k. The competition is the metrical classification summarized in (4-5) above.

Kohonen [1988] calls the supervised stochastic system (4-121)–(4-123) a **learning vector quantization** system. Kohonen has modified this algorithm slightly by adding a "window" restriction on pattern samples, as discussed in Chapter 1 of the companion volume [Kosko, 1991].

Kohonen calls the simpler unsupervised stochastic system (4-124)–(4-125) a **self-organizing map** system:

$$\mathbf{m}_j(k+1) = \mathbf{m}_j(k) + c_k[\mathbf{x}_k - \mathbf{m}_j(k)] \qquad (4\text{-}124)$$

$$\mathbf{m}_j(k+1) = \mathbf{m}_j(k) \quad \text{if } i \neq j \qquad (4\text{-}125)$$

provided the jth F_Y neuron wins the metrical "competition" at time k. Hecht-Nielsen [1987] calls the same system a **counterpropagation** system when applied heteroassociatively to two concatenated neural fields to estimate a measurable function $f \colon R^n \longrightarrow R^p$ or an unknown joint probability density $p(\mathbf{x}, \mathbf{y})$.

Neural theorists have comparatively only recently *interpreted* the above difference laws as competitive learning laws. Grossberg [1969] appears to have first equated the deterministic learning law (4-3) with competitive learning. Von der Malsburg [1973] and Rumelhart [1985] have also studied this deterministic equation and its variants.

Pattern-recognition theorists have studied the stochastic unsupervised learning system (4-124)–(4-125) for decades. MacQueen [1967] popularized the random linear competitive learning algorithm (4-124)–(4-125) as the **adaptive k-means**

clustering algorithm. Sebestyn [1962] and many other pattern-recognition theorists studied the same algorithm but named it differently.

Kohonen and Hecht-Nielsen allow multiple winners per iteration in their neural interpretations of the competitive learning equations. In analogy with feature-detecting neurons in a lateral-inhibition planar topology, if the jth competing neuron wins, the neurons nearest the jth neuron also win. Kohonen calls these winning subsets "bubbles." Hecht-Nielsen calls them "block parties."

The **random differential Hebbian learning law** [Kosko, 1990] takes the form

$$\dot{m}_{ij} = -m_{ij} + S_i(x_i)S_j(y_j) + \dot{S}_i\dot{S}_j + n_{ij} \tag{4-126}$$

or, in vector notation,

$$\dot{\mathbf{m}}_j = -\mathbf{m}_j + \mathbf{S}(\mathbf{x})S_j(y_j) + \dot{S}(\mathbf{x})\dot{S}_j(y_j) + \mathbf{n}_j \tag{4-127}$$

The **random classical differential Hebbian law** omits the Hebbian product and correlates only signal velocities:

$$\dot{m}_{ij} = -m_{ij} + \dot{S}_i\dot{S}_j + n_{ij} \tag{4-128}$$

Since we can multiply or add scaling constants anywhere in our neural models, (4-126) reduces to (4-128) if we scale the Hebbian signal product S_iS_j in (4-126) with a small constant. Without the Hebbian signal product, feedback differential Hebbian systems can oscillate. The RABAM theorems in Chapter 6 guarantee adaptive global stability only for the complete laws (4-126) and (4-127).

The **random differential competitive learning law** takes the form

$$\dot{m}_{ij} = \dot{S}_j(y_j)[S_i(x_i) - m_{ij}] + n_{ij} \tag{4-129}$$

or, in vector notation,

$$\dot{\mathbf{m}}_j = \dot{S}_j(y_j)[\mathbf{S}(\mathbf{x}) - \mathbf{m}_j] + \mathbf{n}_j \tag{4-130}$$

In practice we often use the random linear differential competitive learning law

$$\dot{\mathbf{m}}_j = \dot{S}_j(y_j)[\mathbf{x} - \mathbf{m}_j] + \mathbf{n}_j \tag{4-131}$$

Chapter 1 of the companion volume [Kosko, 1991] discusses and applies a simple discrete version of (4-131) and applies this stochastic algorithm to centroid estimation and phoneme recognition. Differential competitive learning underlies the product-space clustering method used in Chapters 8 through 11 to generate fuzzy-associative-memory rules from sample data.

We now examine these four types of learning in detail. For convenience we shall work with the noiseless deterministic versions of the learning laws. The derived properties carry over to the noisy stochastic case, though in some cases we must replace random variables with their expectations to suppress noise effects.

SIGNAL HEBBIAN LEARNING

We can solve the deterministic first-order signal Hebbian learning law,

$$\dot{m}_{ij} \;=\; -m_{ij}(t) + S_i(x_i(t))S_j(y_j(t)) \tag{4-132}$$

to yield the integral equation

$$m_{ij}(t) \;=\; m_{ij}(0)e^{-t} + \int_0^t S_i(s)S_j(s)e^{s-t}\,ds \tag{4-133}$$

The integral remains in the solution because in general x_i and y_j depend on m_{ij} through the activation differential equations discussed in Chapter 3.

The bounded signal functions S_i and S_j produce a bounded integral in (4-133). The unbounded activation product $x_i y_j$, found in early models of Hebbian learning, can produce an unbounded integral.

Recency Effects and Forgetting

Hebbian synapses learn an exponentially weighted average of sampled patterns. The solution to an inhomogenous first-order ordinary differential equation dictates the average structure and the exponential weight. This holds for all four types of learning we have examined.

The exponential weight in the solution (4-133) induces a **recency effect** on the unsupervised learning. We learn and forget equally fast. Every day, and every night, we experience an exponential decrease in the information we retain from our day's experience. This well-known recency effect underlies the quote by philosopher David Hume at the beginning of this chapter. In this epistemological sense we remember more vividly our sensory experiences in the last twenty minutes than anything we have experienced in our past. Nothing is more vivid than now.

The exponential weight e^{-t} on the prior synaptic "knowledge" $m_{ij}(0)$ in (4-133) arises from the **forgetting term** $-m_{ij}$ in the differential equation (4-132). Indeed the *forgetting law* provides the simplest local unsupervised learning law:

$$\dot{m}_{ij} \;=\; -m_{ij} \tag{4-134}$$

The forgetting law illustrates the two key properties of biologically motivated unsupervised learning laws. It depends on only local information, the current synaptic strength $m_{ij}(t)$. And it equilibrates exponentially quickly, which allows real-time operation.

Asymptotic Correlation Encoding

Equation (4-133) implies that the synaptic matrix $\mathbf{M}$ of long-term memory traces m_{ij} asymptotically approaches the bipolar correlation matrix $X_k^T Y_k$ discussed in Chapter 3:

$$\mathbf{M} \;=\; X_k^T Y_k \tag{4-135}$$

Equation (4-135) uses the abbreviated notation from Chapter 3. Here state vectors X and Y denote the bipolar signal vectors $\mathbf{S}(\mathbf{x})$ and $\mathbf{S}(\mathbf{y})$, and the signal values S_i and S_j equal 1 or -1.

Suppose the signal functions S_i and S_j are bipolar. Then the signal product $S_i S_j$ is also bipolar:

$$-1 \;\leq\; S_i S_j \;\leq\; 1 \tag{4-136}$$

The signal product equals -1 iff one signal equals 1 and the other equals -1. The signal product equals 1 iff both signals equal 1 or -1. So the signal product $S_i S_j$ behaves as an equivalence operator in a multivalued logic, as discussed in Chapter 1.

Consider the extreme case $S_i S_j = 1$. Then (4-132) reduces to the simple first-order linear ordinary differential equation

$$\dot{m}_{ij} + m_{ij} \;=\; 1 \tag{4-137}$$

This differential equation has solution

$$m_{ij}(t) \;=\; m_{ij}(0)e^{-t} + \int_0^t e^{s-t}\, ds \tag{4-138}$$

$$=\; m_{ij}(0)e^{-t} + 1 - e^{-t} \tag{4-139}$$

$$=\; 1 + [m_{ij}(0) - 1]e^{-t} \tag{4-140}$$

So

$$m_{ij}(t) \longrightarrow 1 \quad \text{as} \quad t \longrightarrow \infty \tag{4-141}$$

for *any* finite initial synaptic value $m_{ij}(0)$.

Similarly in the other extreme, when $S_i S_j = -1$, $m_{ij}(t)$ converges to -1 exponentially quickly. So the signal Hebbian learning law approaches asymptotically, and exponentially quickly, the bipolar outer-product matrix (4-135) discussed in Chapter 3.

This asymptotic analysis assumes we present only one pattern to the learning system for an infinite length of time. The pattern washes away all prior learned pattern information $m_{ij}(0)$ exponentially quickly.

In general, unsupervised synapses learn on a single pass of noisy data. They sample a training set representative of some continuous pattern cluster. Each sample tends to resemble the preceding and succeeding samples.

In practice discrete signal Hebbian learning requires that either we cycle through a training set, presenting each associative pattern sample $(\mathbf{S}(\mathbf{x}_k), \mathbf{S}(\mathbf{y}_k))$ for a minimal period of learning, or we present the more recent of the m associative patterns for shorter periods of learning than we presented previous patterns. Mathematically this means we must use a diagonal fading-memory exponential matrix $\mathbf{W}$, as discussed in Chapter 3:

$$\mathbf{X}^T \mathbf{W} \mathbf{Y} \;=\; \sum_{k=1}^{m} w_k X_k^T Y_k \tag{4-142}$$

The weighted outer-product matrix $\mathbf{X}^T\mathbf{W}\mathbf{Y}$ compensates for the inherent exponential decay of learned information. $\mathbf{X}$ and $\mathbf{Y}$ denote matrices of bipolar signal vectors.

Hebbian Correlation Decoding

Signal Hebbian learning represents the simplest form of correlation learning. The correlation structure admits a direct analysis in the discrete case. The analysis holds approximately in the continuous case.

Consider the bipolar correlation encoding of the m bipolar vector associations (X_i, Y_i). As discussed in Chapter 3, X_i denotes a point in the bipolar n-cube $\{-1, 1\}^n$, and Y_i denotes a point in the bipolar p-cube $\{-1, 1\}^p$.

The bipolar associations (X_i, Y_i) correspond to the binary vector associations (A_i, B_i). We simply replace -1s with 0s in (X_i, Y_i) to produce (A_i, B_i). A_i denotes a point in the Boolean n-cube $\{0, 1\}^n$. B_i denotes a point in the p-cube $\{0, 1\}^p$.

Then Hebbian encoding of the bipolar associations (X_i, Y_i) corresponds to the weighted Hebbian encoding scheme (4-142) if the weight matrix $\mathbf{W}$ equals the m-by-m identity matrix I:

$$\mathbf{M} = \sum_i^m X_i^T Y_i \tag{4-143}$$

We can use the Hebbian synaptic matrix $\mathbf{M}$ for bidirectional processing of F_X and F_Y neuronal signals. We pass neuronal signals through $\mathbf{M}$ in the forward direction, through $\mathbf{M}^T$ in the backward direction. For our purposes we need examine only the forward direction with synchronous thresholding of activations to bipolar signals. The same analysis applies to backward passes of signals.

Suppose we present the bipolar vector X_i to the neural system. In general a different bipolar vector X confronts the system. Suppose X is closer to X_i than to all other X_j vectors. The accuracy of the following decoding approximation argument increases as the resemblance of X to X_i increases.

Vector X_i passes through the filter $\mathbf{M}$. Synchronous thresholding yields the output bipolar vector Y. To what extent will Y resemble Y_i? The analysis depends on the **signal-noise decomposition**

$$X_i M = (X_i X_i^T)Y_i + \sum_{j \neq i}^m (X_i X_j^T)Y_j \tag{4-144}$$

$$= nY_i + \sum_{j \neq i}^m (X_i X_j^T)Y_j \tag{4-145}$$

$$= \sum_j c_{ij} Y_j \tag{4-146}$$

where $c_{ij} = X_i X_j^T = X_j X_i^T = c_{ji}$. In this context Y_i behaves as a "signal" vector and the Y_j vectors behave as "noise" vectors.

The c_{ij} terms are **correction coefficients**. They "correct" the noise terms Y_j in sign and magnitude to make each $c_{ij}Y_j$ vector resemble Y_i *in sign* as much as possible. (Note that the maximum positive weight, $c_{ii} = n$, scales the "signal" vector Y_i.) Then the nonlinear threshold operation tends to produce Y_i. The same correction property holds in the backward direction when Y_i, or some close vector Y, passes through M^T.

The correlation argument shows why dimensionality tends to limit Hebbian BAM storage capacity:

$$m \ < \ \min(n, p) \tag{4-147}$$

The corrected noise terms in (4-146) can swamp the amplified signal vector nY_i in (4-145). The heuristic capacity bound (4-147) depends on the "continuity" of the sample associations (X_i, Y_i), as we discuss next. Randomly generated associations erode capacity: $m << \min(n, p)$.

We now examine how the correction coefficients c_{ij} explain rapid BAM stability. Our analysis will not use the global Lyapunov techniques discussed in Chapter 3. The correlation analysis allows us to estimate where a Hebbian BAM system equilibrates relative to learned pattern information. Most Lyapunov arguments show only that these feedback systems equilibrate.

We must assume that the heteroassociative samples (X_i, Y_i) arise from a sampled *continuous* function f. The function f must map small changes in inputs to small changes in outputs.

Accurate decoding—and high memory capacity—implicitly assumes that if stored input patterns are close to one another, then the corresponding stored output patterns are also close to one another. We formally state this **continuity assumption** in terms of Hamming distance or l^1 distance in binary n-cubes and p-cubes:

$$\frac{1}{n}H(A_i, A_j) \ \approx \ \frac{1}{p}H(B_i, B_j) \tag{4-148}$$

We define the *Hamming distance* $H(A_i, A_j)$ between binary vectors A_i and A_j as the l^1 distance

$$H(A_i, A_j) \ = \ \sum_k^n |a_i^k - a_j^k| \tag{4-149}$$

for binary n-vector $A_i = (a_i^1, \ldots, a_i^n)$, and similarly for binary p-vectors B_i. So the Hamming distance $H(A_i, A_j)$ counts the number of bits or slots where the two bit vectors differ. Bit vectors (1 0 1 0) and (1 0 1 1) have Hamming distance 1.

We implicitly assume the continuity assumption when we use neural networks to estimate continuous functions f from "representative" training data. The continuity assumption (4-148) holds with equality for autoassociative networks, since they encode the redundant pairs (A_i, A_i). When a heteroassociative training set substantially violates the continuity assumption, the network poorly estimates the sampled function f. Memory capacity shrinks. We readily observe this capacity/estimation deterioration if we encode randomly generated associations (X_i, Y_i) in (4-143). Fortunately, training sets derived from real-world problems tend to satisfy the continuity assumption, since most sampled processes are continuous.

The correction coefficients c_{ij} behave as desired because, ultimately, they connect the underlying binary and bipolar pattern spaces through the **correction inequality** [Kosko, 1988a]:

$$c_{ij} \underset{<}{\overset{>}{=}} 0 \quad \text{if and only if} \quad H(A_i, A_j) \underset{>}{\overset{<}{=}} \frac{n}{2} \qquad (4\text{-}150)$$

The correction inequality follows from the equalities

$$c_{ij} \;=\; X_i X_j^T \qquad (4\text{-}151)$$

$$=\; [\text{number of common bits}] - [\text{number of different bits}] \qquad (4\text{-}152)$$

$$=\; [n - H(A_i, A_j)] - H(A_i, A_j) \qquad (4\text{-}153)$$

$$=\; n - 2H(A_i, A_j) \qquad (4\text{-}154)$$

The continuity assumption enhances the corrective power of the coefficients c_{ij}. For, suppose binary vector A_i is close to A_j. Then, geometrically, the two patterns are less than half their space away from each other:

$$H(A_i, A_j) \;<\; \frac{n}{2} \qquad (4\text{-}155)$$

So $c_{ij} > 0$ by the correction inequality (4-150). The smaller $H(A_i, A_j)$ is, the larger c_{ij} is, and the more the "noise" vector Y_j resembles the "signal" vector Y_i by the continuity assumption (4-148). In the extreme case $Y_j = Y_i$; so $c_{ij} = n$.

The rare case that $H(A_i, A_j) = n/2$ results in $c_{ij} = 0$. The correction coefficients discard the metrically ambiguous vector A_j from the superimposing sum (4-146).

Finally, suppose A_i is far away from A_j. So

$$H(A_i, A_j) \;>\; \frac{n}{2} \qquad (4\text{-}156)$$

and thus $c_{ij} < 0$ by the correction inequality. The larger $H(A_i, A_j)$ is, the larger is the magnitude $|c_{ij}| > 0$, and the less Y_j resembles the signal vector Y_i by the continuity assumption. But the less Y_j resembles Y_i in this case, the more $-Y_j$ resembles Y_i, since $-Y_j$ equals the complement bipolar vector Y_j^c. The correction inequality (4-150) and $c_{ij} < 0$ produce the desired sign change of Y_j to $-Y_j$ in the Hebbian-encoding sum. In the extreme case $H(A_i, A_j) = n$, and so $c_{ij} = -n$. So $c_{ij}Y_j = -nY_j = nY_i$.

In all three cases the correction coefficients c_{ij} scale the noise vectors Y_j in sign and magnitude to produce a new vector $c_{ij}Y_j$ that tends componentwise to agree in sign with the signal vector Y_i. Threshold signal functions then tend to reproduce Y_i, since they depend on only the sign of F_Y activations.

Hence *threshold nonlinearity suppresses noise*. The nonlinearities that make system analysis difficult help make decoding accurate or "fault tolerant." Linear signal functions do not suppress noise.

Orthogonality improves decoding accuracy and thus improves capacity. Decoding accuracy increases the more the F_X signal vectors X_i, and the more the F_Y signal vectors Y_i, resemble mutually orthogonal vectors. Equation (4-144) reduces to

$$X_i\mathbf{M} \quad = \quad nY_i \tag{4-157}$$

if $X_1, \ldots, X_m$ are mutually orthogonal: $X_i X_j^T = 0$ if $i \neq j$. The amplified activation vector nY_i agrees in sign with Y_i in all p slots. So it thresholds to Y_i. Similarly, if $Y_1, \ldots, Y_m$ are mutually orthogonal, then a backward pass yields

$$Y_i\mathbf{M}^T \quad = \quad pX_i \tag{4-158}$$

which thresholds to the signal vector X_i. So the BAM synchronously converges in one iteration. The nonorthogonal case may require multiple iterations. In both cases a correlation decomposition of $\mathbf{M}$ accounts for system stability, and correlation correction accounts for rapid convergence to stability.

The correlation-correction argument extends to **temporal associative memories** (TAMs) [Kosko, 1988a]. An n-neuron autoassociative feedback network tends to encode an *ordered sequence* of n-dimensional bit vectors $A_1, A_2, \ldots, A_m, A_1$ as a stable limit cycle in the signal state space $\{0, 1\}^n$. The square synaptic matrix $\mathbf{T}$ contains asymmetric coefficients (else a BAM argument ensures two-step **limit cycles**). We omit the local Lyapunov stability argument, which extends the discrete BAM stability argument in Chapter 3.

The Hebbian encoding method first converts each binary vector A_i to a bipolar vector X_i and then sums contiguous correlation-encoded associations:

$$\mathbf{T} \quad = \quad X_m^T X_1 + \sum_{i}^{m-1} X_i^T X_{i+1} \tag{4-159}$$

if the TAM continuity assumption holds:

$$H(A_i, A_j) \quad \approx \quad H(A_{i+1}, A_{j+1}) \tag{4-160}$$

Then presenting X_i to the synchronous TAM, which synchronously thresholds activations to signals, tends to yield X_{i+1}:

$$X_i\mathbf{T} \quad = \quad nX_{i+1} + \sum_{j \neq i} c_{ij} X_{j+1} \tag{4-161}$$

$$\longrightarrow \quad X \approx X_{i+1} \tag{4-162}$$

The arrow $\longrightarrow$ indicates that the TAM synchronously thresholds the activation vector $\mathbf{X}_i\mathbf{T}$ to yield the signal vector X. The feedforward version of TAM encoding and decoding resembles the Grossberg [1982] **avalanche**, which uses hormonal control neurons to modulate the pattern's *rhythm* or speed as well as its order.

The backward TAM matrix $\mathbf{T}^T$ plays the limit-cycle tune in reverse. Passing X_i through the backward TAM matrix $\mathbf{T}^T$ tends to produce X_{i-1}:

$$X_i \mathbf{T}^T \quad = \quad nX_{i-1} + \sum_{j \neq i} c_{ij} X_{j-1} \tag{4-163}$$

$$\longrightarrow \quad X \approx X_{i-1} \tag{4-164}$$

TAM networks illustrate how to encode discretized time-varying patterns in associative networks. But TAMs inefficiently encode time-varying patterns. They require impractically large TAM matrices to encode most temporal patterns of interest.

Most temporal patterns also violate the TAM continuity assumption. Suppose we represent English alphabet letters as long binary vectors. Then we can accurately encode and decode the sequence NETWORK but not the sequence BABY. Each letter in NETWORK abuts a unique letter pattern. But the B in BABY abuts both A and Y. In principle we can always add tag bits to the A_i bit vectors to distinguish contiguous associations. But the price paid may well be computational intractability.

Consider the three-step limit cycle $A_1 \to A_2 \to A_3 \to A_1$ defined by the bit vectors

$$\begin{aligned} A_1 &= (1 \quad 0 \quad 1 \quad 0) \\ A_2 &= (1 \quad 1 \quad 0 \quad 0) \\ A_3 &= (1 \quad 0 \quad 0 \quad 1) \end{aligned}$$

We convert these bit vectors to bipolar vectors:

$$\begin{aligned} X_1 &= (1 \quad -1 \quad 1 \quad -1) \\ X_2 &= (1 \quad 1 \quad -1 \quad -1) \\ X_3 &= (1 \quad -1 \quad -1 \quad 1) \end{aligned}$$

We then add the contiguous correlation matrices $X_i^T X_i$ to produce the asymmetric TAM matrix $\mathbf{T}$:

$$\begin{aligned} \mathbf{T} &= X_1^T X_2 + X_2^T X_3 + X_3^T X_1 \\[6pt] &= \begin{pmatrix} 1 & 1 & -1 & -1 \\ -1 & -1 & 1 & 1 \\ 1 & 1 & -1 & -1 \\ -1 & -1 & 1 & 1 \end{pmatrix} + \begin{pmatrix} 1 & -1 & -1 & 1 \\ 1 & -1 & -1 & 1 \\ -1 & 1 & 1 & -1 \\ -1 & 1 & 1 & -1 \end{pmatrix} \\[6pt] &\quad + \begin{pmatrix} 1 & -1 & 1 & -1 \\ -1 & 1 & -1 & 1 \\ 1 & -1 & 1 & -1 \\ 1 & -1 & 1 & -1 \end{pmatrix} \\[6pt] &= \begin{pmatrix} 3 & -1 & -1 & -1 \\ -1 & -1 & -1 & 3 \\ -1 & 3 & -1 & -1 \\ -1 & -1 & 3 & -1 \end{pmatrix} \end{aligned}$$

Passing the bit vectors A_i through **T** in the forward direction produces

$$
\begin{array}{rclcl}
A_1\mathbf{T} & = & (2 \quad 2 \quad -2 \quad -2) & \longrightarrow & (1 \quad 1 \quad 0 \quad 0) \quad = \quad A_2 \\
A_2\mathbf{T} & = & (2 \quad -2 \quad -2 \quad 2) & \longrightarrow & (1 \quad 0 \quad 0 \quad 1) \quad = \quad A_3 \\
A_3\mathbf{T} & = & (2 \quad -2 \quad 2 \quad -2) & \longrightarrow & (1 \quad 0 \quad 1 \quad 0) \quad = \quad A_1
\end{array}
$$

So the TAM dynamical system recalls the forward limit cycle $A_1 \rightarrow A_2 \rightarrow A_3 \rightarrow A_1 \rightarrow$ if given any bit vector in the limit cycle. The TAM system will tend to reconstruct this limit cycle if given bit vectors A that resemble some A_i.

Passing the bit vectors A_i through $\mathbf{T}^T$ in the backward direction produces

$$
\begin{array}{rclcl}
A_1\mathbf{T} & = & (2 \quad -2 \quad 2 \quad 2) & \longrightarrow & (1 \quad 0 \quad 0 \quad 1) \quad = \quad A_3 \\
A_3\mathbf{T} & = & (2 \quad 2 \quad -2 \quad -2) & \longrightarrow & (1 \quad 1 \quad 0 \quad 0) \quad = \quad A_2 \\
A_2\mathbf{T} & = & (2 \quad -2 \quad 2 \quad -2) & \longrightarrow & (1 \quad 0 \quad 1 \quad 0) \quad = \quad A_1
\end{array}
$$

So the TAM dynamical system recalls the backward limit cycle $A_1 \rightarrow A_3 \rightarrow A_2 \rightarrow A_1 \rightarrow$ given any A_i or any bit vector A that sufficiently resembles A_i.

COMPETITIVE LEARNING

The deterministic competitive learning law

$$\dot{m}_{ij} \;=\; S_j[S_i - m_{ij}] \tag{4-165}$$

resembles the signal Hebbian learning law (4-132). If we distribute the competitive signal S_j in (4-165),

$$\dot{m}_{ij} \;=\; -S_j m_{ij} + S_i S_j \tag{4-166}$$

we see that the competitive learning law uses the nonlinear forgetting term $-S_j m_{ij}$. The signal Hebbian learning law uses the linear forgetting term $-m_{ij}$.

So competitive and signal Hebbian learning differ in how they forget, not in how they learn. In both cases when $S_j = 1$—when the jth competing neuron wins— the synaptic value m_{ij} encodes the forcing signal S_i and encodes it exponentially quickly. But, unlike a Hebbian synapse, a competitive synapse does not forget when its postsynaptic neuron loses, when $S_j = 0$. Then (4-165) reduces to the no-change relationship

$$\dot{m}_{ij} \;=\; 0 \tag{4-167}$$

while the signal Hebbian law (4-132) reduces to the forgetting law (4-134).

Signal Hebbian learning is distributed. Hebbian synapses encode pieces of every sampled pattern. They also forget pieces of every learned pattern as they learn fresh patterns.

In competitive learning, only the winning synaptic vector $(m_{1j}, \ldots, m_{nj})$ encodes the presently sampled pattern $\mathbf{S}(\mathbf{x})$ or $\mathbf{x}$. If the sample patterns $\mathbf{S}(\mathbf{x})$ or $\mathbf{x}$ persist long enough, competitive synapses behave as "grandmother" synapses. The synaptic value m_{ij} soon equals the pattern piece $S_i(x_i)$ or x_i. No other synapse in the network encodes this pattern piece.

So persistent deterministic competitive learning is not distributed. We can omit the learning metaphor and directly copy the sample pattern **x** into the appropriate memory registers, directly into the jth column of the synaptic matrix **M**. Neural engineers often do this in practice. The adaptive-resonance-theory architectures ART-1 and ART-2 [Carpenter, 1987] encode binary and continuous n-vector patterns in just this way.

Random competitive learning laws distribute sample patterns in the weak, but important, sense of a statistical average. Synaptic vectors learn the centroid of the pattern class they locally and randomly sample. More generally, synaptic vectors only equal centroids on average, as we show below in (4-186). They equal Brownian motions centered at the unknown centroids, as in (4-107).

Competition as Indication

Centroid estimation requires that the competitive signal S_j approximate the indicator function I_{D_j} of the locally sampled pattern class D_j:

$$S_j(y_j) \quad \approx \quad I_{D_j}(\mathbf{x}) \tag{4-168}$$

So if sample pattern **x** comes from region D_j, the jth competing neuron in F_Y should win, and all other competing neurons should lose.

In practice we usually use the random linear competitive learning law (4-112) and a simple additive model, as discussed in Chapter 3, of F_Y neuronal activation:

$$S_j(\mathbf{x}\mathbf{m}_j^T + f_j) \quad \approx \quad I_{D_j}(\mathbf{x}) \tag{4-169}$$

for random row vectors **x** and $\mathbf{m}_j$. f_j denotes the inhibitive within-field feedback the jth neuron receives from its F_Y competitors.

Usually the inhibitive-feedback term f_j equals the additive sum of synapse-weighted signals:

$$f_j \quad = \quad \sum_{k=1}^{p} s_{kj} S_k(y_k) \tag{4-170}$$

S denotes the symmetric within-field inhibition matrix: $\mathbf{S} = \mathbf{S}^T$. So $s_{kj} = s_{jk}$. The symmetry assumption arises from the distance dependence of lateral inhibition strength found in fields of competing biological neurons. In the simplest case diagonal entries are positive and equal, $s_{11} = \ldots = s_{pp} > 0$, and off-diagonal entries are negative and equal, $s_{kj} < 0$ for distinct k and j. For signal functions S_j that approximate binary thresholds, (4-170) reduces to $f_j = s_{jj}$ if the jth neuron wins, and to $f_j = -s_{kj}$ if instead the kth neuron wins. Competing neurons excite themselves (or near neighbors) and inhibit others (distant neighbors).

Competition as Correlation Detection

In practice we estimate the competitive signal function S_j as the metrical indicator function defined in (4-5). The fictitious jth competing neuron wins the imaginary competition if the input vector $\mathbf{x}$ is closer to stored synaptic vector $\mathbf{m}_j$ than to all other stored synaptic vectors $\mathbf{m}_k$. We say all other neurons lose, and we break ties arbitrarily. This gives the metrical indicator function

$$S_j(y_j) \;=\; \begin{cases} 1 & \text{if } ||\mathbf{x} - \mathbf{m}_j||^2 = \min_k ||\mathbf{x} - \mathbf{m}_k||^2 \\[2mm] 0 & \text{if } ||\mathbf{x} - \mathbf{m}_j||^2 > \min_k ||\mathbf{x} - \mathbf{m}_k||^2 \end{cases} \qquad (4\text{-}171)$$

where the squared Euclidean or l^2 vector norm $||\mathbf{z}||^2$ equals the sum of squared entries:

$$\begin{aligned} ||\mathbf{z}||^2 \;&=\; z_1^2 + \cdots + z_n^2 \qquad (4\text{-}172) \\ &=\; \mathbf{z}\mathbf{z}^T \end{aligned}$$

Then competitive learning reduces to classical *correlation detection* of signals [Cooper, 1986]—provided all synaptic vectors $\mathbf{m}_j$ have approximately the same norm value $||\mathbf{m}_j||$. Kohonen [1988] has suggested that synaptic vectors are asymptotically equinorm. The centroid argument below shows, though, that the equinorm property implicitly includes an equiprobable synaptic allocation or vector quantization of R^n as well as other assumptions. We need the equinorm property along the way to asymptotic convergence as well as after the synaptic vectors converge.

In practice competitive learning systems tend to exhibit the equiprobable quantization and equinorm properties after training. Hecht-Nielsen [1987] and others simply require that all input vectors $\mathbf{x}$ and synaptic vectors $\mathbf{m}_j$ be normalized to points on the unit R^n sphere: $||\mathbf{x}||^2 = ||\mathbf{m}_j||^2 = 1$. Such normalization requires that we renormalize the winning synaptic vector $\mathbf{m}_j$ at each iteration in the stochastic-difference algorithm (4-124).

To see how metrical competitive learning reduces to correlation detection, suppose at each moment the synaptic vectors $\mathbf{m}_j$ have the same positive and finite norm value:

$$||\mathbf{m}_1||^2 \;=\; \ldots \;=\; ||\mathbf{m}_p||^2 \qquad (4\text{-}173)$$

Our argument will not require that the norms in (4-173) equal the same constant at every instant, though they often do. From (4-171) the jth competing neuron wins the competition if and only if

$$||\mathbf{x} - \mathbf{m}_j||^2 \;=\; \min_k ||\mathbf{x} - \mathbf{m}_k||^2 \qquad (4\text{-}174)$$

$$\text{iff } (\mathbf{x} - \mathbf{m}_j)(\mathbf{x} - \mathbf{m}_j)^T \;=\; \min_k (\mathbf{x} - \mathbf{m}_k)(\mathbf{x} - \mathbf{m}_k)^T \qquad (4\text{-}175)$$

$$\text{iff } \mathbf{x}\mathbf{x}^T + \mathbf{m}_j\mathbf{m}_j^T - 2\mathbf{x}\mathbf{m}_j \;=\; \min_k (\mathbf{x}\mathbf{x}^T + \mathbf{m}_k\mathbf{m}_k^T - 2\mathbf{x}\mathbf{m}_k^T) \qquad (4\text{-}176)$$

$$\text{iff } ||\mathbf{m}_j||^2 - 2\mathbf{x}\mathbf{m}_j^T \;=\; \min_k (||\mathbf{m}_k||^2 - 2\mathbf{x}\mathbf{m}_k^T) \qquad (4\text{-}177)$$

We now subtract $||\mathbf{m}_j||^2$ from both sides of (4-177) and invoke the equinorm property (4-173) to get the equivalent equality

$$-2\mathbf{xm}_j^T \;\; = \;\; 2 \min_k -\mathbf{xm}_k^T \qquad (4\text{-}178)$$

The DeMorgan's law $\max(a, b) = -\min(-a, -b)$ further simplifies (4-178) to produce the correlation-detection equality

$$\mathbf{xm}_j^T \;\; = \;\; \max_k \mathbf{xm}_k^T \qquad (4\text{-}179)$$

The jth competing neuron wins iff the input signal or pattern x correlates maximally with $\mathbf{m}_j$. We minimize Euclidean distance when we maximize correlations or inner products, and conversely, if the equinorm property (4-173) holds. From a neural-network perspective, this argument reduces neuronal competition to a winner-take-all linear additive activation model.

Correlations favor optical implementation. Optical systems cannot easily compute differences. Instead they excel at computing large-scale parallel multiplications and additions. Small-scale software or hardware implementations can more comfortably afford the added accuracy achieved by using the metrical classification rule (4-170).

The **cosine law**

$$\mathbf{xm}_j^T \;\; = \;\; ||\mathbf{x}||\,||\mathbf{m}_j||\cos(\mathbf{x}, \mathbf{m}_j) \qquad (4\text{-}180)$$

provides a geometric interpretation of metrical competitive learning when the equinorm property (4-173) holds. The term $\cos(\mathbf{x}, \mathbf{m}_j)$ denotes the cosine of the angle between the two R^n vectors x and $\mathbf{m}_j$. The cosine law implies that the jth neuron wins iff the input pattern x is more parallel to synaptic vector $\mathbf{m}_j$ than to any other $\mathbf{m}_k$. Parallel vectors maximize the cosine function and yield $\cos(\mathbf{x}, \mathbf{m}_j) = 1$.

Suppose all pattern vectors x and synaptic vectors $\mathbf{m}_j$ lie on the R^n **unit sphere** $\mathbf{S}^n = \{\mathbf{z} \in R^n: ||\mathbf{z}||^2 = 1\}$. So $||\mathbf{x}||^2 = ||\mathbf{m}_j||^2 = 1$. Then the cosine law confirms the intuition that the jth neuron should maximally win the competition if the current sample x equals $\mathbf{m}_j$, and should maximally lose if x equals $-\mathbf{m}_j$ and thus lies on the opposite side of $\mathbf{S}^n$. All other samples x on $\mathbf{S}^n$ produce win intensities between these extremes.

Asymptotic Centroid Estimation

Approximation (4-169) reduces the linear competitive learning law (4-112) to the simpler competitive law (4-116):

$$\dot{m}_j \;\; = \;\; I_{D_j}(\mathbf{x})[\mathbf{x} - m_j] + \mathbf{n}_j \qquad (4\text{-}181)$$

Most stochastic competitive learning laws approximate (4-181). As a result, as discussed in Chapter 6 in considering the AVQ equilibrium theorem, synaptic vectors

m_j tend to equal the unknown centroid of D_j, at least on average. We can see a simple version of this if we use the (unrealistic) equilibrium condition (4-101):

$$O = \dot{m}_j \tag{4-182}$$

$$= I_{D_j}(x)[x - m_j] + n_j$$

We take expectations (probabilistic averages over all possible realizations x) of both sides of these equations, and use the zero-mean property (4-63) of the noise process $\{n_t\}$, to give

$$O = \int_{R^n} I_{D_j}(x)[x - m_j]p(x)\,dx + E[n_j] \tag{4-183}$$

$$= \int_{D_j} [x - m_j]p(x)\,dx \tag{4-184}$$

$$= \int_{D_j} xp(x)\,dx - m_j \int_{D_j} p(x)\,dx \tag{4-185}$$

Solving for the equilibrium synaptic vector m_j shows that it equals the centroid of D_j:

$$m_j = \frac{\displaystyle\int_{D_j} xp(x)\,dx}{\displaystyle\int_{D_j} p(x)\,dx} \tag{4-186}$$

Chapter 6 discusses this result in more detail and, in the AVQ convergence theorem, guarantees that average synaptic vectors converge to centroids exponentially quickly.

We use the centroid-estimation result repeatedly in competitive learning. Unfortunately, it does not hold for deterministic versions of competitive learning, as found throughout the neural-network literature.

The centroid-estimation result has antecedents in pattern-recognition theory and no doubt elsewhere. In the 1960s MacQueen [1967] proved that, with probability one, the adaptive k-means clustering algorithm drives m_j vectors to centroids. Though this result contains the central asymptotic idea, and central advantage, of competitive learning, MacQueen used a discrete real-analytic approach. MacQueen did not consider a continuous stochastic-differential formulation or how noise affects the learning process. The continuous stochastic-differential formulation allows us to prove not only that average synaptic vectors converge to centroids, but that they converge exponentially quickly to centroids. We use this property in applications to justify learning with only one or a few passes of sample data.

Competitive Covariance Estimation

Parameter estimators have variances or covariances. Centroids provide a first-order estimate of how the unknown probability density function $p(x)$ behaves in the regions D_j. Local covariances provide a second-order description.

We can extend the competitive learning laws to asymptotically estimate the local conditional covariance matrices $\mathbf{K}_j$:

$$\mathbf{K}_j \;=\; E[(\mathbf{x} - \bar{\mathbf{x}}_j)^T(\mathbf{x} - \bar{\mathbf{x}}_j)|D_j] \tag{4-187}$$

$\bar{x}_{\mathbf{j}}$ denotes the D_j centroid in (4-186). If we use the denominator constant in (4-186) to scale the joint density $p(\mathbf{x})$ in the numerator, we can rewrite the centroid as a particular realization $E[\mathbf{x}|D_j]$ of the *conditional expectation* random vector $E[\mathbf{x}|.]$:

$$\bar{\mathbf{x}}_j \;=\; \int_{D_j} \mathbf{x} p(\mathbf{x}|D_j)\, d\mathbf{x} \tag{4-188}$$

$$=\; E[\mathbf{x}|\mathbf{x}\varepsilon D_j] \tag{4-189}$$

So every decision class D_j has exactly one centroid.

The conditional expectation $E[\mathbf{y}|\mathbf{x}]$ provides the mean-squared-error optimal estimate of a Borel-measurable random vector $\mathbf{y}$. Estimation theorists sometimes refer to this result as the **fundamental theorem of estimation theory** [Mendel, 1987]:

$$E[(\mathbf{y} - E[\mathbf{y}|\mathbf{x}])(\mathbf{y} - E[\mathbf{y}|\mathbf{x}])^T] \;\leq\; E[(\mathbf{y} - \mathbf{f}(\mathbf{x}))(\mathbf{y} - \mathbf{f}(\mathbf{x}))^T] \tag{4-190}$$

for *all* Borel-measurable random vector functions $\mathbf{f}$ of $\mathbf{x}$.

At each iteration we estimate the unknown centroid $\bar{\mathbf{x}}_j$ as the current synaptic vector $\mathbf{m}_j$. In this sense $\mathbf{K}_j$ becomes an *error* conditional covariance matrix. So we propose the stochastic difference-equation algorithm

$$\mathbf{m}_j(k+1) \;=\; \mathbf{m}_j(k) + c_k[\mathbf{x}_k - \mathbf{m}_j(k)], \tag{4-191}$$

$$\mathbf{K}_j(k+1) \;=\; \mathbf{K}_j(k) + d_k[(\mathbf{x}_k - \mathbf{m}_j(k))^T(\mathbf{x}_k - \mathbf{m}_j(k)) - \mathbf{K}_j(k)] \tag{4-192}$$

for winning synaptic vector $\mathbf{m}_j$.

We can initialize $\mathbf{m}_j(0)$ as $\mathbf{x}(0)$, and $\mathbf{K}_j(0)$ as the null matrix $\mathbf{O}$. Initializing $\mathbf{K}_j(0)$ as the identity matrix implies that the random vector $\mathbf{x}$ has uncorrelated components x_i (and so reduces the adaptive vector quantization process to scalar quantization [Pratt, 1978]) and might skew the learning procedure for small sets of training samples. We can also scale the difference terms in (4-191) and (4-192) to produce supervised-competitive and differential-competitive learning versions.

If the ith neuron loses the F_Y metrical competition, then

$$\mathbf{m}_i(k+1) \;=\; \mathbf{m}_i(k) \tag{4-193}$$

$$\mathbf{K}_i(k+1) \;=\; \mathbf{K}_i(k) \tag{4-194}$$

$\{d_k\}$ denotes an appropriately decreasing sequence of learning coefficients in (4-192).

The computational algorithm (4-192) corresponds to the stochastic differential equation

$$\dot{\mathbf{K}}_j \;=\; I_{D_j}(\mathbf{x})[(\mathbf{x} - \mathbf{m}_j)^T(\mathbf{x} - \mathbf{m}_j) - \mathbf{K}_j] + \mathbf{N}_j \tag{4-195}$$

$\{\mathbf{N}_j\}$ denotes an independent Gaussian white-noise matrix process. Consider first the noiseless equilibrium condition (4-101):

$$\mathbf{O} = \dot{\mathbf{K}}_j \tag{4-196}$$

$$= I_{D_j}(\mathbf{x})[(\mathbf{x} - \bar{\mathbf{x}}_j)^T(\mathbf{x} - \mathbf{x}_j) - \mathbf{K}_j] + \mathbf{N}_j \tag{4-197}$$

where we have used (4-186) to replace $\mathbf{m}_j$ with its equilibrium centroid value. Now take expectations on both sides of (4-196) and (4-197):

$$\mathbf{O} = \int_{D_j} (\mathbf{x} - \bar{\mathbf{x}}_j)^T(\mathbf{x} - \bar{\mathbf{x}}_j)p(\mathbf{x})\, d\mathbf{x} - \mathbf{K}_j \int_{D_j} p(\mathbf{x})\, d\mathbf{x} \tag{4-198}$$

Now solve for the equilibrium covariance value $\mathbf{K}_j(\infty)$:

$$\mathbf{K}_j = \frac{\displaystyle\int_{D_j} (\mathbf{x} - \bar{\mathbf{x}}_j)^T(\mathbf{x} - \bar{\mathbf{x}}_j)p(\mathbf{x})\, d\mathbf{x}}{\displaystyle\int_{D_j} p(\mathbf{x})\, d\mathbf{x}} \tag{4-199}$$

$$= \int_{D_j} (\mathbf{x} - E[\mathbf{x} \,|\, D_j])^T(\mathbf{x} - E[\mathbf{x} \,|\, D_j])p(\mathbf{x} \,|\, D_j)\, d\mathbf{x} \tag{4-200}$$

$$= E[(\mathbf{x} - E[\mathbf{x} \,|\, D_j])^T(\mathbf{x} - E[\mathbf{x} \,|\, D_j]) \,|\, D_j] \tag{4-201}$$

as desired.

The general stochastic-equilibrium condition takes the form

$$\dot{\mathbf{K}}_j = \mathbf{N}_j \tag{4-202}$$

Then, taking expectations of both sides of (4-202) gives (4-196). Since the asymptotic $\mathbf{K}_j$ is no longer constant—$\mathbf{K}_j$ wanders in a Brownian motion about the $\mathbf{K}_j$ in (4-187)—we cannot directly factor $\mathbf{K}_j$ out of the second integral in (4-198). So we must take expectations a second time and rearrange. This gives the more general equilibrium result

$$E[\mathbf{K}_j(\infty)] = E[(\mathbf{x} - \bar{\mathbf{x}}_j)^T(\mathbf{x} - \bar{\mathbf{x}}_j) \,|\, D_j] \tag{4-203}$$

The special case $D_j = R^n$ arises frequently in estimation theory and signal processing. The Kalman filter [Kalman, 1960] estimates the conditional expectation and conditional error covariance matrix of a Gauss-Markov process with a linear state dynamical model [Kailath, 1980]. Then the current random sample $\mathbf{x}_k$ depends at most, and linearly, on the previous state sample $\mathbf{x}_{k-1}$.

More generally, as discussed in Chapter 4 of [Kosko, 1991], we can use higher-order correlation tensors, and their Fourier transforms or *polyspectra* [Nikias, 1987], to estimate nonlinear non-Gaussian system behavior from noisy sample data. In these cases we add the appropriate third-order, fourth-order, and perhaps higher-order difference equations to the stochastic system (4-191) and (4-192).

DIFFERENTIAL HEBBIAN LEARNING

The deterministic differential Hebbian learning law

$$\dot{m}_{ij} = -m_{ij} + S_i S_j + \dot{S}_i \dot{S}_j \qquad (4\text{-}204)$$

and its simpler version

$$\dot{m}_{ij} = -m_{ij} + \dot{S}_i \dot{S}_j \qquad (4\text{-}205)$$

arose from attempts to dynamically estimate the causal structure of fuzzy cognitive maps from sample data [Kosko, 1985, 1986, 1987(c)]. This led to abstract mathematical analyses, which in turn led to neural interpretations of these signal-velocity learning laws, as we shall see. Intuitively, Hebbian correlations promote spurious causal associations among concurrently active units. Differential correlations estimate the concurrent, and presumably causal, variation among active units.

Fuzzy Cognitive Maps

Fuzzy cognitive maps (FCMs) are fuzzy signed directed graphs with feedback [Kosko, 1986, 1988b]. The directed edge e_{ij} from causal concept C_i to concept C_j measures how much C_i causes C_j. The time-varying concept function $C_i(t)$ measures the nonnegative occurrence of some fuzzy event, perhaps the strength of a political sentiment, historical trend, or military objective. FCMs model the world as a collection of classes and causal relations between classes.

Taber [1987, 1991] has used FCMs to model gastric-appetite behavior and popular political developments. Styblinski [1988] has used FCMs to analyze electrical circuits. Zhang [1988] has used FCMs to analyze and extend graph-theoretic behavior. Gotoh and Murakami [1989] have used FCMs to model plant control.

The edges e_{ij} take values in the fuzzy causal interval $[-1, 1]$. $e_{ij} = 0$ indicates no causality. $e_{ij} > 0$ indicates causal increase: C_j increases as C_i increases, and C_j decreases as C_i decreases. $e_{ij} < 0$ indicates causal decrease or negative causality: C_j decreases as C_i increases, and C_j increases as C_i decreases. For instance, freeway congestion and bad weather increase automobile accidents on freeways. Auto accidents decrease the patroling frequency of highway patrol officers. Increased highway-patrol frequency decreases average driving speed.

Simple FCMs have edge values in $\{-1, 0, 1\}$. Then, if causality occurs, it occurs to maximal positive or negative degree. Simple FCMs provide a quick first approximation to an expert's stated or printed causal knowledge. For instance, a syndicated article on South African politics by political economist Walter Williams [1986] led to the simple FCM depicted in Figure 4.2.

Causal feedback loops abound in FCMs in thick tangles. Feedback precludes the graph-search technique used in artificial-intelligence expert systems and causal trees [Pearl, 1988]. These inferencing algorithms tend to get stuck in infinite loops in cyclic knowledge networks.

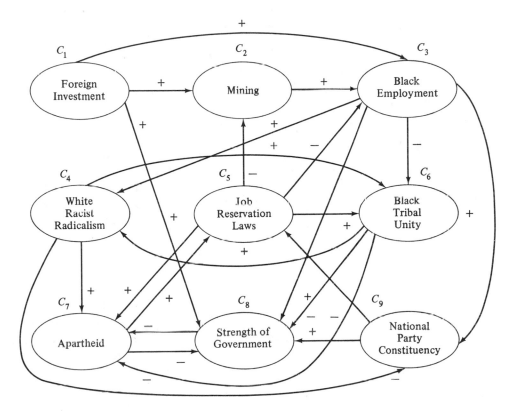

FIGURE 4.2 Simple trivalent fuzzy cognitive map of South African politics. Massive feedback precludes graph search but favors TAM limit-cycle analysis.

Perhaps more important for "knowledge acquisition," for constructing AI search trees, in general we cannot combine two or more trees and still produce a tree. The problem increases with the number of trees combined. This limits the number of knowledge sources or experts who can build the search tree. Larger expert sample sizes should produce more, not less, reliable knowledge structures.

FCM feedback allows experts to freely draw causal pictures of their problem and allows causal adaptation laws, such as (4-204) and (4-205), to infer causal links from sample data. FCM feedback forces us to abandon graph search, forward and especially backward "chaining" [Winston, 1984]. Instead we view the FCM as a dynamical system and take its equilibrium behavior as a forward-evolved inference. Synchronous FCMs behave as temporal associative memories (TAMs), discussed above. We can also always add two or more FCMs to produce a new FCM. The strong law of large numbers ensures in some sense that knowledge reliability increases with expert sample size.

We reason with FCMs as we recall with TAMs. We pass state vectors **C** repeatedly through the FCM connection matrix **E**, thresholding or nonlinearly transforming the result after each pass. Independent of the FCM's size, it quickly settles down to a TAM limit cycle [Kosko, 1988a] or "hidden pattern." The limit-cycle inference summarizes the joint effects of all the interacting fuzzy knowledge. The inference does not merely enumerate a branch in a tree.

Consider the 9-by-9 causal connection matrix $\mathbf{E}$ that represents the political FCM in Figure 4.2:

$$
\mathbf{E} \;=\; \begin{array}{c} \\ C_1 \\ C_2 \\ C_3 \\ C_4 \\ C_5 \\ C_6 \\ C_7 \\ C_8 \\ C_9 \end{array}
\begin{array}{ccccccccc}
C_1 & C_2 & C_3 & C_4 & C_5 & C_6 & C_7 & C_8 & C_9 \\
\left(\begin{array}{ccccccccc}
0 & 1 & 1 & 0 & 0 & 0 & 0 & 1 & 1 \\
0 & 0 & 1 & 0 & 0 & 0 & 0 & 1 & 0 \\
0 & 0 & 0 & 1 & 0 & -1 & 0 & 1 & 1 \\
0 & 0 & 0 & 0 & 0 & 1 & 1 & 0 & -1 \\
0 & -1 & -1 & 0 & 0 & 1 & 1 & 0 & 0 \\
0 & 0 & 0 & 1 & 0 & 0 & -1 & -1 & 0 \\
0 & 0 & 0 & 0 & 1 & 0 & 0 & -1 & 0 \\
0 & 0 & 0 & 0 & 0 & 0 & -1 & 0 & 0 \\
0 & 0 & 0 & 0 & -1 & 0 & 0 & 1 & 0
\end{array}\right)
\end{array}
$$

Concept nodes can represent processes, events, values, or policies. The foreign investment policy corresponds to persistently firing the first node: $C_1 = 1$. We hold or "clamp" C_1 on during the TAM recall process. Threshold signal functions synchronously update each concept after each pass through the connection matrix.

We start with the foreign investment policy

$$\mathbf{C_1} \;=\; (1\ 0\ 0\ 0\ 0\ 0\ 0\ 0\ 0)$$

Then

$$\mathbf{C_1 E} \;=\; (0\ 1\ 1\ 0\ 0\ 0\ 0\ 1\ 1)$$
$$\longrightarrow\; (1\ 1\ 1\ 0\ 0\ 0\ 0\ 1\ 1) \;=\; \mathbf{C_2}$$

The arrow indicates the threshold operation with, say, $1/2$ as the threshold value. So zero causal input produces zero causal output. $\mathbf{C_2}$ contains $C_1 = 1$ because we are testing the foreign-investment policy option. Next

$$\mathbf{C_2 E} \;=\; (0\ 1\ 2\ 1\ -1\ -1\ -1\ 4\ 1)$$
$$\longrightarrow\; (1\ 1\ 1\ 1\ 0\ 0\ 0\ 1\ 1) \;=\; \mathbf{C_3}$$

Next

$$\mathbf{C_3 E} \;=\; (0\ 1\ 2\ 1\ -1\ 0\ 0\ 4\ 1)$$
$$\longrightarrow\; (1\ 1\ 1\ 1\ 0\ 0\ 0\ 1\ 1) \;=\; \mathbf{C_3}$$

So $\mathbf{C_3}$ is a fixed point of the FCM dynamical system. The FCM has associatively inferred the answer $\{C_1, C_2, C_3, C_4, C_8, C_9\}$ given the policy what-if question $\{C_1\}$. This simple FCM predicts that sustained foreign investment maintains a balance of government stability and racism (without apartheid laws).

Now consider the disinvestment policy. We turn off the first concept in C_3 to produce the new state vector

$$\mathbf{D}_1 \;=\; (0\ 1\ 1\ 1\ 0\ 0\ 0\ 1\ 1)$$

Then, passing $\mathbf{D}_1$ through $\mathbf{E}$ and repeating the above recall process leads to the two-step limit cycle $\{C_4, C_5\} \rightarrow \{C_6, C_7\} \rightarrow \{C_4, C_5\} \rightarrow \ldots$. As Williams's analysis predicts, disinvestment has collapsed the government ($C_8 = 0$ in equilibrium) and ended in social chaos.

This example illustrates the strengths and weaknesses of FCM analysis. The FCM allows experts to represent factual and evaluative concepts in an interactive framework. Experts can quickly draw FCM pictures or respond to questionnaires. Knowledge engineers can likewise transcribe interviews or printed documents. Experts can consent or dissent to the local causal structure and perhaps the global equilibrations. The FCM knowledge representation and inferencing structure reduces to simple vector-matrix operations, favors integrated-circuit implementation, and allows extension to neural, statistical, or dynamical systems techniques. Yet an FCM equally encodes the expert's knowledge or ignorance, wisdom or prejudice. Worse, different experts differ in how they assign causal strengths to edges and in which concepts they deem causally relevant. The FCM seems merely to encode its designers' biases, and may not even encode them accurately.

FCM combination provides a partial solution to this problem. We can additively superimpose each expert's FCM in associative-memory fashion, even though the FCM connection matrices $\mathbf{E}_1, \ldots, \mathbf{E}_k$ may not be conformable for addition. Combined conflicting opinions tend to cancel out and, assisted by the strong law of large numbers, a consensus emerges as the sample opinion approximates the underlying population opinion. FCM combination allows knowledge engineers to construct FCMs with iterative interviews or questionnaire mailings.

Laws of large numbers require that the random samples be independent identically distributed random variables with finite variance. Independence models each expert's individuality. Identical distribution models a particular problem-domain focus.

We combine arbitrary FCM connection matrices $\mathbf{E}_1, \ldots, \mathbf{E}_k$ by adding *augmented* FCM matrices $\mathbf{F}_1, \ldots, \mathbf{F}_k$. Each augmented matrix $\mathbf{F}_i$ has n rows and n columns. n equals the total number of distinct concepts used by the experts. We permute the rows and columns of the augmented matrices to bring them into mutual coincidence. Then we add the $\mathbf{F}_i$ pointwise to yield the combined FCM matrix $\mathbf{F}$:

$$\mathbf{F} \;=\; \sum_i \mathbf{F}_i \qquad (4\text{-}206)$$

We can then use $\mathbf{F}$ to construct the combined FCM digraph.

Even if each expert gives trivalent descriptions in $\{-1, 0, 1\}$, the combined (and normalized) FCM entry f_{ij} tends to be in $[-1, 1]$. The strong law of large numbers ensures that f_{ij} provides a rational approximation to the underlying unknown population opinion of how much C_i affects C_j. We can normalize f_{ij} by the number k of experts.

Experts tend to give trivalent evaluations more readily and more accurately than they give weighted evaluations. When transcribing interviews or documents, a knowledge engineer can more reliably determine an edge's sign than its magnitude.

Some experts may be more credible than others. We can weight each expert with a nonnegative credibility weight w_i by multiplicatively weighting the expert's augmented FCM matrix:

$$\mathbf{F} = \sum_i w_i \mathbf{F}_i \qquad (4\text{-}207)$$

The weights need not be in $[0, 1]$. Since we use threshold TAM recall, the weights need only be nonnegative. Different weights may produce different equilibrium limit-cycles or "hidden patterns." We can also weight separately any submatrix of each expert's augmented FCM matrix.

Augmented FCM matrices imply that every expert causally discusses every concept $C_1, \ldots, C_n$. If an expert does not include C_j in his FCM model, the expert implicitly says that C_j is not causally relevant. So the jth row and jth column of his augmented connection matrix contains only zeroes.

Consider the four simple FCMs in Figure 4.3. Four experts discuss a total of six concepts. No expert explicitly discusses more than four concepts.

The four FCMs in Figure 4.3 give rise to the 6-by-6 augmented FCM matrices

$$\mathbf{F}_1 = \begin{pmatrix} 0 & 1 & -1 & 1 & 0 & 0 \\ 0 & 0 & 0 & -1 & 0 & 0 \\ -1 & 1 & 0 & -1 & 0 & 0 \\ 0 & 0 & 1 & 0 & 0 & 0 \\ 0 & 0 & 0 & 0 & 0 & 0 \\ 0 & 0 & 0 & 0 & 0 & 0 \end{pmatrix}$$

$$\mathbf{F}_2 = \begin{pmatrix} 0 & 1 & -1 & 0 & 1 & 0 \\ 1 & 0 & -1 & 0 & 1 & 0 \\ -1 & -1 & 0 & 0 & 1 & 0 \\ 0 & 0 & 0 & 0 & 0 & 0 \\ 0 & 1 & -1 & 0 & 0 & 0 \\ 0 & 0 & 0 & 0 & 0 & 0 \end{pmatrix}$$

$$\mathbf{F}_3 = \begin{pmatrix} 0 & 1 & -1 & 0 & 0 & 0 \\ 1 & 0 & 1 & 0 & 0 & -1 \\ -1 & -1 & 0 & 0 & 0 & 1 \\ 0 & 0 & 0 & 0 & 0 & 0 \\ 0 & 0 & 0 & 0 & 0 & 0 \\ 1 & -1 & -1 & -1 & 0 & 0 \end{pmatrix}$$

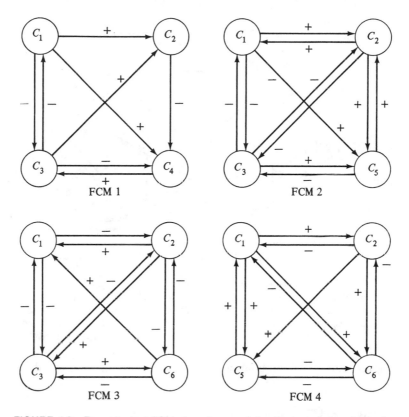

FIGURE 4.3 Four trivalent FCMs from four experts. Each expert explicitly discusses only four of the six concepts. *Source:* Kosko, B., "Hidden Patterns in Combined and Adaptive Knowledge Networks," *International Journal of Approximate Reasoning*, vol. 2, no. 4, October 1988.

$$\mathbf{F}_4 = \begin{pmatrix} 0 & 1 & 0 & 0 & 1 & -1 \\ -1 & 0 & 0 & 0 & 1 & -1 \\ 0 & 0 & 0 & 0 & 0 & 0 \\ 0 & 0 & 0 & 0 & 0 & 0 \\ 1 & 0 & 0 & 0 & 0 & -1 \\ 1 & -1 & 0 & 0 & -1 & 0 \end{pmatrix}$$

Note that each matrix contains two zero rows and two zero columns corresponding to the expert's causally irrelevant concepts. Then the combined FCM matrix $\mathbf{F}$ equals the unweighted sum $\mathbf{F}_1 + \mathbf{F}_2 + \mathbf{F}_3 + \mathbf{F}_4$:

$$\mathbf{F} = \begin{pmatrix} 0 & 4 & -3 & 1 & 2 & -1 \\ 1 & 0 & 0 & -1 & 2 & 0 \\ -3 & -1 & 0 & -1 & 1 & 1 \\ 0 & 0 & 1 & 0 & 0 & 0 \\ 1 & 1 & -1 & 0 & 0 & -1 \\ 2 & -2 & -1 & 0 & -1 & 0 \end{pmatrix}$$

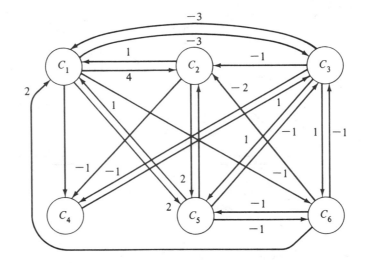

FIGURE 4.4 Additively combined FCM arising from the four trivalent FCMs in Figure 4.3. Though unnormalized, the FCM causal edge strengths are "fuzzier" than the trivalent edge strengths of the original four simple FCMs. *Source:* Kosko, B., "Hidden Patterns in Combined and Adaptive Knowledge Networks," *International Journal of Approximate Reasoning*, vol. 2, no. 4, October 1988.

Figure 4.4 shows the combined FCM digraph that corresponds to **F**.

Adaptive Causal Inference

Adding FCM matrices represents a simple form of causal learning. In general how should we grow causal edge strengths from raw concept data? How do we make adaptive causal inferences?

Empiricist philosopher John Stuart Mill [1843] suggested that we infer causality between variables when we observe *concomitant variation* or lagged variation between them. If B changes when A changes, we suspect a causal relationship. The more correlated the changes, the more we suspect a causal relationship, or, more accurately, we suspect a strong causal relationship.

Time derivatives measure changes. Products of derivatives correlate changes. This leads to the simplest differential Hebbian learning law [Kosko, 1985, 1986]:

$$\dot{e}_{ij} \;=\; -e_{ij} + \dot{C}_i \dot{C}_j \tag{4-208}$$

The passive decay term $-e_{ij}$ forces zero causality between unchanging concepts. (We also require that $e_{ii} = 0$ so that no concept causes itself.)

The concomitant-variation law (4-208) corresponds to the simple differential Hebbian learning law (4-205) if we interpret concept functions as signal functions. Many concept functions may not be sigmoidal or monotone nondecreasing. The more general differential Hebbian learning law (4-204) arises from a fixed-point stability analysis [Kosko, 1988c].

The concomitant-variation term $\dot{C}_i\dot{C}_j$ indicates causal increase or decrease according to joint concept movement. The derivatives take on positive and negative real values. The concepts take on only nonnegative values. If C_i and C_j both increase or both decrease, the product of derivatives is positive. If they move in opposite directions, the product of derivatives is negative. The concomitant-variation term $\dot{C}_i\dot{C}_j$ provides a simple causal "arrow of time."

Klopf's Drive Reinforcement Model

Harry Klopf [1986, 1988] independently proposed the following discrete variant of differential Hebbian learning:

$$\Delta m_{ij}(t) \quad = \quad \Delta S_j(y_j(t)) \sum_{k=1}^{\tau} c_j |m_{ij}(t-k)| \Delta S_i(x_i(t-k)) \qquad (4\text{-}209)$$

where the synaptic difference $\Delta m_{ij}(t)$ updates the current synaptic efficacy $m_{ij}(t)$ in the first-order difference equation

$$m_{ij}(t+1) \quad = \quad m_{ij}(t) + \Delta m_{ij}(t) \qquad (4\text{-}210)$$

Klopf imposes several constraints on the **drive-reinforcement model** (4-209). The neuronal membrane potentials y_j and x_i must obey additive activation laws:

$$y_j(t) \quad = \quad \sum_{i=1}^{n} m_{ij}(t) S_i(x_i(t)) - T_j \qquad (4\text{-}211)$$

for some positive membrane threshold T_j. Neurons transduce activations to signals with threshold-linear signal functions:

$$S_j(y_j) \quad = \quad \min[U_j, \max(0, y_j(t))] \qquad (4\text{-}212)$$

for some positive upper bound U_j. The threshold-linear signal function sets $S_j = y_j$ within the interval $[0, U_j]$, sets $S_j = 0$ if $y_j < 0$, and sets $S_j = U_j$ if $y_j > U_j$. Klopf requires that presynaptic signals S_i cannot decrease in time. Then the presynaptic signal differences $\Delta S_i(t)$ must be nonnegative:

$$\Delta S_i(x_i(t-k)) \quad = \quad S_i(x_i(t-k)) - S_i(x_i(t-k-1)) \quad \geq \quad 0 \qquad (4\text{-}213)$$

So Klopf replaces the presynaptic differences $\Delta S_i(x_i)$ in (4-209) with the threshold terms $\max[0, \Delta S_i(x_i(t-k))]$. Then the synapse learns only if the frequency of presynaptic pulses increases.

Klopf empirically chooses the learning coefficients c_i in (4-211) to model an exponential recency effect: $c_1 = 5$, $c_2 = 3.0$, $c_3 = 1.5$, $c_4 = 0.75$, and $c_5 = 0.25$. He includes the synaptic magnitudes $|m_{ij}(t-k)|$ in (4-209) to "account for the initial positive acceleration in the acquisition curves of classical conditioning." Klopf uses this drive-reinforcement model to successfully retrodict several types of Pavlovian conditioning at the level of the individual neuron. Hebbian synaptic models fail to retrodict the same learning behavior.

The term drive reinforcement arises from variables and their velocities. Klopf defines a *neuronal drive* as the weighted signal $m_{ij}S_i$, and a *neuronal reinforcer* as the weighted difference $m_{ij}\Delta S_i$. Sutton [1988] has developed a related recursive reinforcement model for supervised learning with *temporal differences* instead of with desired-less-actual errors, which extends earlier adaptive-heuristic-critic reinforcement models [Barto, 1983].

A differentiable version of the drive-reinforcement model (4-209) would take the form

$$\dot{m}_{ij} \quad = \quad -m_{ij} + |m_{ij}|\dot{S}_i\dot{S}_j \tag{4-214}$$

The synaptic magnitude $|m_{ij}|$ amplifies the synapse's plasticity. In particular, suppose the ijth synapse is excitatory: $m_{ij} > 0$. Then we can rewrite (4-214) as

$$\dot{m}_{ij} \quad = \quad m_{ij}(\dot{S}_i\dot{S}_j - 1) \tag{4-215}$$

Implicitly the passive decay coefficient A_{ij} scales the $-m_{ij}$ term in (4-214). A_{ij} will usually be much smaller than unity to prevent rapid forgetting. Then (4-215) becomes

$$\dot{m}_{ij} \quad = \quad m_{ij}\dot{S}_i\dot{S}_j \tag{4-216}$$

So drive-reinforcement synapses are exponentially sensitive to concomitant variation in signals. This allows drive-reinforcement synapses to rapidly encode neuronal signal information. Moreover, signal velocities or directions tend to be more robust, more noise tolerant, than the underlying signals, as we show below in terms of delta modulation.

Unfortunately, drive-reinforcement synapses tend to zero as they equilibrate, and they equilibrate exponentially quickly. This holds for both excitatory and inhibitory synapses. For instance, in (4-216) the equilibrium condition $\dot{m}_{ij} = 0$ implies that

$$m_{ij}\dot{S}_i\dot{S}_j \quad = \quad 0 \tag{4-217}$$

or $m_{ij} = 0$ in general. This would hold equally in a signal Hebbian model if we replaced the signal product S_iS_j with the magnitude-weighted product $|m_{ij}|S_iS_j$. Klopf apparently overcomes this tendency in his simulations by forbidding zero synaptic values: $|m_{ij}(t)| \geq 0.1$.

In contrast, the simple differential Hebbian learning law (4-205) or (4-208) equilibrates to

$$m_{ij} \quad = \quad \dot{S}_i\dot{S}_j \tag{4-218}$$

More generally the differential Hebbian law (4-205) learns an exponentially weighted average of sampled concomitant variations, since it has the solution

$$m_{ij}(t) \quad = \quad m_{ij}(0)e^{-t} + \int_0^t \dot{S}_i(s)\dot{S}_j(s)e^{s-t} \, ds \tag{4-219}$$

in direct analogy to the signal-Hebbian integral equation (4-133).

Concomitant Variation as Statistical Covariance

The very term *concomitant variation* resembles the term *covariance*. In differential Hebbian learning we interpreted variation as time change, and concomitance as conjunction or product. Alternatively we can interpret variation spatially as a statistical variance or covariance—as an ensemble's dispersion about its average value. We shall see that these two interpretations relate in an approximate ergodic sense of time averages estimating space averages.

Sejnowski [1977] has cast synaptic modification as a mean-squared optimization problem and derived a covariance-based solution. After some simplifications the optimal solution takes the form of the **covariance learning law**

$$\dot{m}_{ij} = -m_{ij} + \text{Cov}[S_i(x_i), S_j(y_j)] \qquad (4\text{-}220)$$

where (4-42) defines the deterministic covariance $\text{Cov}[x, z]$ of random variables x and z. In (4-220) we have added the passive decay term $-m_{ij}$ to keep the synaptic model bounded and to model forgetting. Sejnowski's model does not include $-m_{ij}$.

The covariance term in (4-220) implies that we have returned to the stochastic-calculus framework, and (4-220) describes the time evolution of the synaptic random process $\{m_{ij}(t)\}$. We can add an independent Gaussian white-noise term n_{ij} to the right-hand side of (4-220). For notational convenience we omit this term in the discussion below.

With (4-43) we can rewrite (4-220) as

$$\dot{m}_{ij} = -m_{ij} + E_{xy}[S_i S_j] - E_x[S_i]E_y[S_j] \qquad (4\text{-}221)$$

The subscripts on E_{xy} indicate that E_{xy} averages random variables with the joint probability density function $p(S_i(x_i), S_j(y_j))$.

Sejnowski invokes the stochastic-approximation principle [Tsypkin, 1974] to estimate expected random variables as the observed realizations of the random variables. Chapter 5 discusses this technique in detail. In its simplest form the technique replaces an unknown average $E[z]$ with the observed realization of random variable z.

At time t we observe the single realization product $S_i(t, \omega)S_j(t, \omega)$, one of infinitely many possible realizations of the random-variable product $S_i(t)S_j(t)$ from the product random process $\{S_i(t)S_j(t)\}$. The expectation $E_{xy}[S_i(t)S_j(t)]$ weights all possible realizations of the random-variable product $S_i(t)S_j(t)$ with the unknown joint probability $p(S_i(t, \cdot), S_j(t, \cdot))$. The stochastic-approximation approach estimates the unknown expectation with the observed realization product:

$$E_{xy}[S_i S_j] \approx S_i S_j \qquad (4\text{-}222)$$

So we estimate a random process with its observed time samples. Sejnowski uses this stochastic-approximation argument to rewrite (4-221) as

$$\dot{m}_{ij} = -m_{ij} + S_i S_j - E_x[S_i]E_y[S_j] \qquad (4\text{-}223)$$

and has since searched for biological support for this covariance learning law.

The covariance law (4-223) involves a conceptual difficulty. It both uses and estimates the unknown joint density $p(S_i, S_j)$. We should apply the stochastic-approximation estimation technique to the *original* covariance term $\text{Cov}[S_i, S_j]$ in (4-220). In equation (4-221) we have already distributed the joint expectation E_{xy} and simplified:

$$\text{Cov}[S_i, S_j] = E_{xy}\{S_iS_j + E_x[S_i]E_y[S_j] - S_iE_y[S_j] - E_x[S_i]S_j\} \quad (4\text{-}224)$$

$$= E_{xy}[S_iS_j] + E_x[S_i]E_y[S_j] - E_{xy}[S_i]E_y[S_j] - E_x[S_i]E_{xy}[S_j] \quad (4\text{-}225)$$

$$= E_{xy}[S_iS_j] - E_x[S_i]E_y[S_j] \quad (4\text{-}226)$$

in accord with (4-43). We have used the unknown joint-density information to replace $E_{xy}[S_i]$ and $E_{xy}[S_j]$ with $E_x[S_i]$ and $E_y[S_j]$.

Suppose instead that we estimate the unknown joint-expectation term

$$E_{xy}[(S_i - E_x[S_i])(S_j - E_y[S_j])]$$

as the observed time samples in the E_{xy} integrand:

$$\text{Cov}[S_i, S_j] \approx (S_i - E_x[S_i])(S_j - E_y[S_j]) \quad (4\text{-}227)$$

This leads to the new covariance learning law

$$\dot{m}_{ij} = -m_{ij} + (S_i - E_x[S_i])(S_j - E_y[S_j]) \quad (4\text{-}228)$$

Covariance learning law (4-228) differs from (4-223), since, in general,

$$2E_x[S_i]E_y[S_j] \neq S_iE_y[S_j] + E_x[S_i]S_j \quad (4\text{-}229)$$

We trivially eliminate the difference between (4-223) and (4-228) if we assume that $E_x[S_i] = S_i$ and that $E_y[S_j] = S_j$, because then both laws collapse to the forgetting law (4-134). So we cannot appeal directly to a stochastic approximation similar to (4-222) or (4-227). Then how should a synapse estimate the unknown averages $E_x[S_i(t)]$ and $E_y[S_j(t)]$ at each time t?

We can lag slightly the stochastic-approximation estimate in time to make a *martingale assumption* [Stark, 1986]. A martingale assumption estimates the immediate future as the present, or the present as the immediate past:

$$E_x[S_i(t)] \approx E_x[S_i(t) \,|\, S_i(s)] \qquad \text{for } 0 \le s < t] \quad (4\text{-}230)$$

$$\approx S_i(s) \quad (4\text{-}231)$$

for some time instant s arbitrarily close to t.

The martingale assumption (4-230) increases in accuracy as s approaches t. Approximation (4-231) represents a "rational expectations" hypothesis [Muth, 1961; Samuelson, 1973], as when we estimate tomorrow's stock price as today's, even though we know all past stock prices. This approximation assumes that the signal processes $\{S_i(t)\}$ are well-behaved: continuous (differentiable), have finite variance, and are at least approximately wide-sense stationary.

We can rewrite the martingale assumption in discrete notation as

$$E_x[S_i(t)] = S_i(t-1) \tag{4-232}$$

Then the covariance term reduces to a timelike concomitant-variation term, and the learning law becomes a classical differential Hebbian learning law:

$$
\begin{aligned}
m_{ij}(t+1) &= m_{ij}(t) + (S_i(t) - E_x[S_i(t)])(S_j(t) - E_y[S_j(t)]) & (4\text{-}233) \\
&= m_{ij}(t) + (S_i(t) - S_i(t-1))(S_j(t) - S_j(t-1)) & (4\text{-}234) \\
&= m_{ij}(t) + \Delta S_i(t)\Delta S_j(t) & (4\text{-}235)
\end{aligned}
$$

which corresponds to a discrete form of (4-205). If the recursion (4-233) begins with $m_{ij}(0) = 0$, then (4-235) has the form of an unnormalized sample conditional covariance [Kosko, 1985], since

$$
\begin{aligned}
m_{ij}(t+1) &= \sum_{k=1}^{t-1} \Delta S_i(k)S_j(k) + \Delta S_i(t)\Delta S_j(t) & (4\text{-}236) \\
&= \sum_{k=1}^{t} \Delta S_i(k)\Delta S_j(k) & (4\text{-}237) \\
&= \sum_{k=1}^{t} (S_i(k) - E_x[S_i(k)\,|\,S_i(k-1)])(S_j(k) \\
&\quad - E_y[S_j(k)\,|\,S_j(k-1)]) & (4\text{-}238)
\end{aligned}
$$

In an approximate sense when time averages resemble ensemble averages, differential Hebbian learning and covariance learning coincide.

Pulse-Coded Differential Hebbian Learning

In Chapter 2 we introduced the *pulse-coded* signal functions [Gluck, 1988, 1989]:

$$S_i(t) = \int_{-\infty}^{t} x_i(s)e^{s-t}\,ds \tag{4-239}$$

$$S_j(t) = \int_{-\infty}^{t} y_j(s)e^{s-t}\,ds \tag{4-240}$$

for pulse functions

$$
x_i(t) = \begin{cases} 1 & \text{if pulse present at } t \\ 0 & \text{if pulse absent at } t \end{cases} \tag{4-241}
$$

and similarly for the F_Y binary pulse functions y_j. We now view the pulse trains $\{x_i(t)\}$ and $\{y_j(t)\}$ as random point processes and view the exponentially windowed integrals (4-239) and (4-240) as stochastic integrals.

In Chapter 2 we proved the *velocity-difference property* for pulse-coded signal functions:

$$\dot{S}_i(t) = x_i(t) - S_i(t) \tag{4-242}$$

$$\dot{S}_j(t) = y_j(t) - S_j(t) \tag{4-243}$$

So synapses can estimate signal velocities in real time by the mere presence or absence of the pulses $x_i(t)$ and $y_j(t)$.

The **pulse-coded differential Hebbian law** replaces the signal velocities in the usual differential Hebbian law with the differences (4-242) and (4-243):

$$\dot{m}_{ij} = -m_{ij} + \dot{S}_i \dot{S}_j + n_{ij} \tag{4-244}$$

$$= -m_{ij} + (x_i - S_i)(y_j - S_j) + n_{ij} \tag{4-245}$$

$$= -m_{ij} + S_i S_j + [x_i y_j - x_i S_j - y_j S_i] + n_{ij} \tag{4-246}$$

The pulse-coded differential Hebbian learning law reduces to the random-signal Hebbian law (4-93) exactly when no pulses are present, when $x_i = y_j = 0$, a frequent event. This suggests that signal-Hebbian behavior represents *average* pulse-coded differential-Hebbian behavior. When no pulses impinge on the synapse at an instant, the synapse still changes but fills in the missing pulses with the expected pulse frequencies S_i and S_j in (4-239) and (4-240).

Suppose pulses are sufficiently infrequent to be zero-mean:

$$E[x_i] = E[y_j] = 0 \tag{4-247}$$

Alternatively we could replace the binary pulse functions in (4-241) with the bipolar pulse functions

$$x_i(t) = \begin{cases} 1 & \text{if pulse present at } t \\ -1 & \text{if pulse absent at } t \end{cases} \tag{4-248}$$

and similarly for y_j. Then zero-mean assumption (4-247) can model even frequent or dense pulse trains.

Next, suppose the $x_i(t)$ and $y_j(t)$ pulses, and the expected pulse frequencies $S_i(t)$ and $S_j(t)$, are pairwise independent. The ijth synapse blindly associates the independent signals or pulses that impinge on it. Then the average behavior of (4-246) reduces to

$$E[\dot{m}_{ij}] = -E[m_{ij}] + E[S_i]E[S_j] + E[x_i]E[y_j] - E[x_i]E[S_j]$$
$$-E[y_j]E[S_i] + E[n_{ij}] \tag{4-249}$$

$$= -E[m_{ij}] + E[S_i]E[S_j] \tag{4-250}$$

the ensemble-averaged random signal Hebbian learning law (4-93) or, equivalently, the classical deterministic-signal Hebbian learning law (4-2).

In the language of estimation theory [Sorenson, 1980], both random-signal Hebbian learning and random pulse-coded differential Hebbian learning provide *unbiased estimators* of signal Hebbian learning (4-2). On average they are all equal at every moment of time. Since pulses are infrequent events, this holds approximately in (4-246) even if we drop the independence assumption of associated pulses and signals.

In practice observed behavior tends to equal (ensemble) average behavior, especially if we time-average observed samples. This suggests that simple neurophysiological measurement of synaptic behavior cannot adjudicate between random-signal Hebbian and pulse-coded differential Hebbian learning models.

The expected pulse frequencies S_i and S_j in (4-239) and (4-240) are exponentially weighted time averages. We can interpret them ergodically (time averages equaling space averages) as ensemble averages:

$$S_i(t) = E[x_i(t)|x_i(s), 0 \le s < t] \qquad (4\text{-}251)$$

$$= E[x_i(t)|x_i(s)] \qquad (4\text{-}252)$$

for some time instant s arbitrarily close to t, and similarly for $S_j(t)$. Substituting these martingale assumptions into (4-245) gives

$$\dot{m}_{ij} = -m_{ij} + (x_i - E[x_i(t)|x_i(s)])(y_j - E[y_j(t)|y_j(s)]) + n_{ij} \quad (4\text{-}253)$$

which has the same conditional covariance structure as (4-238).

Equation (4-253) suggests that random pulse-coded differential Hebbian learning provides a real-time stochastic approximation to covariance learning:

$$\dot{m}_{ij} = -m_{ij} + \text{Cov}[S_i, S_j] + n_{ij} \qquad (4\text{-}254)$$

or, more fundamentally, at the level of pulse trains,

$$\dot{m}_{ij} = -m_{ij} + \text{Cov}[x_i, y_j] + n_{ij} \qquad (4\text{-}255)$$

In general the point processes $\{x_i(t)\}$ and $\{y_j(t)\}$ are not statistically independent in (4-254) and (4-255). At those times t when x_i and y_j represent causally unrelated or "independent" activity, (4-254) and (4-255) reduce to the noisy forgetting law

$$\dot{m}_{ij} = -m_{ij} + n_{ij} \qquad (4\text{-}256)$$

The covariance laws (4-254) and (4-255) show again how differential Hebbian learning and covariance learning coincide when appropriate time averages resemble ensemble averages.

DIFFERENTIAL COMPETITIVE LEARNING

The differential competitive learning law [Kosko, 1990]

$$\dot{m}_{ij} = \dot{S}_j(y_j)[S_i(x_i) - m_{ij}] + n_{ij} \qquad (4\text{-}257)$$

admits a pulse-coded interpretation. If the velocity-difference property (4-243) replaces the competitive signal velocity $\dot{S}_j$ in (4-257), then

$$\dot{m}_{ij} = (y_j - S_j)[S_i - m_{ij}] + n_{ij} \qquad (4\text{-}258)$$

$$= y_j[S_j - m_{ij}] - S_j[S_i - m_{ij}] + n_{ij}. \qquad (4\text{-}259)$$

Equation (4-259) reduces to the random competitive learning law (4-111) when the jth F_Y neuron first wins the F_Y competition. Then $y_j = 1$ and $S_j = 0$. If the jth neuron continues to win, S_j rapidly approaches unity, and learning ceases. The synapse learns the time-averaged pulse-train information S_i exponentially fast. Most learning occurs in the first few instances. This approximation to competitive learning frequently holds because at each instant most competing neurons lose, or have lost, the intrafield competition.

The rapid burst of learning as S_j approaches unity helps prevent the jth neuron from winning too frequently. When the jth neuron wins too frequently, it prematurely encodes a new synaptic pattern in $\mathbf{m}_j$ at the expense of the current $\mathbf{m}_j$ pattern. Carpenter [1987] and Grossberg [1976] call this overwriting problem *code instability*, and propose their adaptive resonance theory (ART) model as a solution to it. In nondifferential competitive learning, winning neurons may tend to keep winning. The constant win signal $S_j = 1$ gives the jth neuron an activation edge over its F_Y competitors when a new input pattern stimulates the system.

In differential competitive learning, the win signal S_j rapidly stops changing once the jth neuron has secured its competitive victory. This does not give the jth neuron a competitive edge when a fresh pattern stimulates the system. Indeed in the pulse-coded framework, the signal velocity equals zero because $y_j = S_j = 1$. So if the signal velocity differs from zero, it must be negative. The winner can only lose.

Differential competitive learning punishes losing with a sign change, that is, $\dot{S}_j(y_j) < 0$. In the pulse-coded formulation, the jth neuron loses when $S_j = 1$ and when, over some nondegenerate time interval $[s, t]$, $y_j(s) = 0$. Then S_j rapidly falls to zero, and learning again ceases. Before S_j reaches zero, (4-259) reduces to the noisy anticompetitive law

$$\dot{m}_{ij} = -S_j[S_i - m_{ij}] + n_{ij} \qquad (4\text{-}260)$$

The nonzero factor S_j rapidly moves the misclassifying synaptic vector $\mathbf{m}_j$ away from the local region of R^n that contains the stimulating signal vector $\mathbf{S}(\mathbf{x})$ or $\mathbf{x}$. This decreases the future probability that the jth neuron will win when input patterns resembling $S(x)$ or $\mathbf{x}$ stimulate the network.

The competitive signal velocity $\dot{S}_j$ approximates the supervised reinforcement function r_j in the supervised competitive learning system (4-117)–(4-120). Both terms reward with a positive signal and punish with a negative signal. Both types of learning tend to rapidly estimate unknown pattern-class centroids. But the reinforcement function r_j requires that in real time synapses "know" and use shifting class memberships of pattern samples.

The unsupervised signal velocity $\dot{S}_j$ does not depend on unknown class memberships. In effect it estimates this information with instantaneous win-rate information. The reinforcement function r_j ignores this win-rate information. This suggests that in many cases differential competitive learning will perform comparably to supervised competitive learning, even though differential competitive learning uses less information. Chapter 1 of [Kosko, 1991], and [Kong, 1991], confirm this conjecture for problems of centroid estimation and phoneme recognition.

Pulse-coded differential competitive learning presents a biologically plausible model of real-time local synaptic modification. Consider the ijth synapse with synaptic efficacy m_{ij}. An axon runs from the ith neuron in the F_X field to the ijth synaptic knob. Pulse trains $\{x_i(t)\}$ propagate down the axon and arrive at the synapse. The synapse abuts the cell membrane of the jth competing neuron in the F_Y field. Intercellular fluid separates the synaptic knob from the jth neuron's cell membrane. This decreases the synapse's ability to estimate the postsynaptic neuron's behavior.

The postsynaptic signal velocity $\dot{S}_j$ presents computational difficulties. How can a noisy synapse compute a nonlinear time derivative in real time? How can the synapse even sample the neuron's *output* signal $S_j(y_j)$?

The velocity-difference property (4-243) answers both questions. The nonlinear derivative $\dot{S}_j$ reduces to the locally available difference $y_j - S_j$. Except when $y_j = S_j = 0$ or $y_j = S_j = 1$ for long stretches of time, S_j lies between its saturation values: $0 < S_j < 1$. Then $\dot{S}_j = 1 - S_j > 0$, and $\dot{S}_j = 0 - S_j < 0$. So the mere presence or absence of the postsynaptic pulse $y_j(t)$ estimates the signal velocity at time t. The bound $0 < S_j < 1$ will tend to hold more in high-speed sensory environments, where stimulus patterns shift constantly, than in slower, stabler pattern environments.

The synapse can physically detect the presence or absence of pulse y_j as a change in the postsynaptic neuron's polarization. The synapse clearly detects the presynaptic pulse train $\{x_i(t)\}$, and thus the pulse-train's pulse count $S_j(t)$ in the most recent 30 milliseconds or so. The differential-competitive synaptic conjecture then states that the synapse electrochemically correlates the incoming pulse train with the detected postsynaptic pulse.

The pulse-coded differential Hebbian learning law (4-245) can also exploit the presence or absence of pulses x_i and y_j to estimate in real time presynaptic and postsynaptic signal velocities. But do synapses always encode pulse-train velocities instead of pulse-train patterns?

Klopf [1988] and Gluck [1988] suggest that input signal velocities provide pattern information for the broader behavior patterns involved in animal learning. In these cases biological synapses may employ pulse-coded differential Hebbian learning, which may appear under the microscope as classical signal Hebbian learning. But in many other cases, perhaps most, synapses process signals and store, recognize, and recall patterns. In these cases pulse-coded differential competitive learning may operate. Then noisy synaptic vectors can locally estimate pattern centroids in real time without supervision.

Differential Competitive Learning as Delta Modulation

The discrete (or pulse-coded) differential competitive learning law

$$\mathbf{m}_j(k+1) \quad = \quad \mathbf{m}_j(k) + \Delta S_j(y_j(k))[\mathbf{x}_k - \mathbf{m}_j(k)] \tag{4-261}$$

represents a neural version of adaptive **delta modulation**.

In communication theory [Cooper, 1986], delta-modulation systems transmit consecutive sampled amplitude *differences* instead of the sampled amplitude values themselves. A delta-modulation system may transmit only $+1$ and -1 signals, indicating local increase or decrease in the underlying sampled waveform. As we shall see, delta modulation works well for signals sampled at high frequency or for signals, such as speech signals, with inherent correlation between samples. The system must sample the waveform frequently enough to result in high positive correlation between consecutive differences.

In Chapter 1 of [Kosko, 1991] we approximate the signal difference ΔS_j as the activation difference Δy_j:

$$\Delta S_j(y_j(t)) \quad \approx \quad \Delta y_j(t) \tag{4-262}$$

$$= \quad \text{sgn}[y_j(t+1) - y_j(t)] \tag{4-263}$$

where the *signum* operator sgn(.) behaves as a modified threshold function:

$$\text{sgn}(x) \quad = \quad \begin{cases} 1 & \text{if} \quad x > 0 \\ 0 & \text{if} \quad x = 0 \\ -1 & \text{if} \quad x < 0 \end{cases} \tag{4-264}$$

The signum operator fixes the step size of the delta modulation. A variable step size, as in (4-261), results in *adaptive* delta modulation. The activation difference (4-263) stays sensitive to changes in activation long after signal (sigmoid) differences saturate.

Consecutive differences are more informative than consecutive samples if the consecutive samples are sufficiently correlated. To show this we define the **statistical correlation** $\rho(x, z)$ between random variables x and z as the deterministic ratio

$$\rho(x, z) \quad = \quad \frac{\text{Cov}[x, z]}{\sqrt{V[x]} \sqrt{V[z]}} \tag{4-265}$$

$$= \quad \frac{\text{Cov}[x, z]}{\sqrt{\sigma_x^2} \sqrt{\sigma_z^2}} \tag{4-266}$$

$$= \frac{\text{Cov}[x, z]}{\sigma_x \sigma_z} \tag{4-267}$$

for finite nonzero variances σ_x^2 and σ_z^2. The correlation function takes values in the bipolar interval $[-1, 1]$:

$$-1 \le \rho(x, z) \le 1 \tag{4-268}$$

Random variables x and z are positively correlated if $\rho > 0$, negatively correlated if $\rho < 0$.

Let d_k denote the F_Y activation or pulse difference Δy_j:

$$d_k = y_j(k+1) - y_j(k) \tag{4-269}$$

Our derivation will not require that the functions $y_j(k)$ be two-valued. What are the first and second moments, the means and variances, of the random-difference sequence $\{d_k\}$?

For simplicity suppose the wide-sense-stationary random sequence $\{y_j(k)\}$ is zero mean, and each random variable $y_j(k)$ has the same finite variance [Cooper, 1986]:

$$E[y_j(k)] = 0 \quad \text{for all } k \tag{4-270}$$

$$V[y_j(k)] = \sigma^2 \quad \text{for all } k \tag{4-271}$$

Properties (4-270) and (4-271) imply that the consecutive samples $y_j(k+1)$ and $y_j(k)$ have correlation value

$$\rho(y_j(k+1), y_j(k)) = \frac{E[y_j(k+1)y_j(k)]}{\sigma^2} \tag{4-272}$$

We can rewrite (4-272) as

$$E[y_j(k+1)y_j(k)] = \sigma^2 \rho \tag{4-273}$$

where ρ denotes $\rho(y_j(k+1), y_j(k))$.

The zero-mean property (4-270) and the linearity of the expectation (integral) operator E imply that the random sequence $\{d_k\}$ also has zero mean:

$$E[d_k] = E[y_j(k+1)] - E[y_j(k)] \tag{4-274}$$

$$= 0 \tag{4-275}$$

The zero-mean property simplifies the variance

$$V[d_k] = E[(d_k - E[d_k])^2] = E[d_k^2] \tag{4-276}$$

$$= E[(y_j(k+1) - y_j(k))^2] \tag{4-277}$$

$$= E[y_j^2(k+1) + y_j^2(k) - 2y_j(k+1)y_j(k)] \tag{4-278}$$

$$= E[y_j^2(k+1)] + E[y_j^2(k)] - 2E[y_j(k+1)y_j(k)] \tag{4-279}$$

We now use (4-270) and (4-271) to eliminate the first two terms in (4-279), and use (4-273) to eliminate the third term. This gives the variance of each random variable in the difference sequence $\{d_k\}$ as

$$V[d_k] \;=\; \sigma^2 + \sigma^2 - 2\rho\sigma^2 \tag{4-280}$$

$$=\; 2\sigma^2(1 - \rho) \tag{4-281}$$

So if consecutive samples are highly positively correlated, if $\rho > 1/2$, then (4-281) implies that the differences d_k have less variance than the samples $y_j(k+1)$ and $y_j(k)$ have:

$$2\sigma^2(1 - \rho) < \sigma^2 \tag{4-282}$$

Differential competitive learning allows synapses to exploit this generic property of sampled signals. In the pulse-coded case, when the jth F_Y neuron wins, it emits a dense pulse train. This winning pulse frequency may be sufficiently high to satisfy (4-282). Pulse-coded differential Hebbian synapses may similarly exploit (4-282) with high-frequency presynaptic and postsynaptic pulse trains.

REFERENCES

Barto, A. G., Sutton, R. S., and Anderson, C. W., "Neuronlike Elements That Can Solve Difficult Control Problems," *IEEE Transactions on Systems, Man, and Cybernetics*, vol. SMC-13, no. 5, 834-846, September 1983.

Carpenter, G. A., and Grossberg, S., "A Massively Parallel Architecture for a Self-Organizing Neural Pattern Recognition Machine," *Computer Vision, Graphics, and Image Processing*, vol. 37, 54-115, 1987.

Carpenter, G. A., and Grossberg, S., "ART 2: Self-Organization of Stable Category Recognition Codes for Analog Input Patterns," *Applied Optics*, 4919-4930, 1 December 1987.

Chung, K. L., *A Course in Probability Theory*, Academic Press, New York, 1974.

Cooper, G. R., and McGillem, C. D., *Modern Communications and Spread Spectrum*, McGraw-Hill, New York, 1986.

Gluck, M. A., Parker, D. B., and Reifsnider, E. S., "Some Biological Implications of a Differential-Hebbian Learning Rule," *Psychobiology*, vol. 16, no. 3, 298-302, 1988.

Gluck, M. A., Parker, D. B., and Reifsnider, E. S., "Learning with Temporal Derivatives in Pulse-Coded Neuronal Systems," *Proceedings 1988 IEEE Neural Information Processing Systems (NIPS) Conference*, Denver, CO, Morgan Kaufman, 1989.

Gotoh, K., Murakami, J., Yamaguchi, T., and Yamanaka, Y., "Application of Fuzzy Cognitive Maps to Supporting for Plant Control," (in Japanese) *SICE Joint Symposium of 15th Syst. Symp. and 10th Knowledge Engineering Symposium*, 99-104, 1989.

Grossberg, S., "On Learning and Energy-Entropy Dependence in Recurrent and Nonrecurrent Signed Networks," *Journal of Statistical Physics*, vol. 1, 319-350, 1969.

Grossberg, S., "Adaptive Pattern Classification and Universal Recoding, I: Parallel Development and Recoding of Neural Feature Detectors," *Biological Cybernetics*, vol. 23, 121-134, 1976.

Hambly, A. R., *Communication Systems*, Computer Science Press, New York, 1990.

Hebb, D. O., *The Organization of Behavior*, Wiley, New York, 1949.

Hecht-Nielsen, R., "Counterpropagation Networks," *Applied Optics*, vol. 26, no. 3, 4979-4984, 1 December 1987.

Kailath, T., *Linear Systems*, Prentice Hall, Englewood Cliffs, NJ, 1980.

Kalman, R. E., "A New Approach to Linear Filtering and Prediction Problems," *Transactions of the ASME, Journal of Basic Engineering*, vol. 82, 35-46, 1960.

Klopf, A. H., "A Drive-Reinforcement Model of Single Neuron Function: An Alternative to the Hebbian Neuronal Model," *Proceedings American Institute of Physics: Neural Networks for Computing*, 265-270, April 1986.

Klopf, A. H., "Drive-Reinforcement Learning: A Real-Time Learning Mechanism for Unsupervised Learning," *Proceedings IEEE International Conference on Neural Networks (ICNN-88)*, vol. II, 441-445, June 1987.

Klopf, A. H., "A Neuronal Model of Classical Conditioning," *Psychobiology*, vol. 16, no. 2, 85-125, 1988.

Kohonen, T., *Self-Organization and Associative Memory*, 2nd ed., Springer-Verlag, New York, 1988.

Kong, S.-G., and Kosko, B., "Differential Competitive Learning for Centroid Estimation and Phoneme Recognition," *IEEE Transactions on Neural Networks*, vol. 2, no. 1, 118-124, January 1991.

Kosko, B., "Adaptive Inference," monograph, Verac Inc. Technical Report, June 1985.

Kosko, B., and Limm, J. S., "Vision as Causal Activation and Association," *Proceedings SPIE: Intelligent Robotics and Computer Vision*, vol. 579, 104-109, September 1985.

Kosko, B., "Fuzzy Cognitive Maps," *International Journal of Man-Machine Studies*, 65-75, January 1986.

Kosko, B., "Differential Hebbian Learning," *Proceedings American Institute of Physics: Neural Networks for Computing*, 277-282, April 1986.

Kosko, B., "Bidirectional Associative Memories," *IEEE Transactions on Systems, Man and Cybernetics*, vol. SMC-18, 49-60, January 1988.

Kosko, B., "Feedback Stability and Unsupervised Learning," *Proceedings 2nd IEEE International Conference on Neural Networks (ICNN-88)*, vol. I, 141-152, July 1988.

Kosko, B., "Hidden Patterns in Combined and Adaptive Knowledge Networks," *International Journal of Approximate Reasoning*, vol. 2, no. 4, 377-393, October 1988.

Kosko, B., "Unsupervised Learning in Noise," *IEEE Transactions on Neural Networks*, vol. 1, no. 1, 44-57, March 1990.

Kosko, B., *Neural Networks for Signal Processing*, Prentice Hall, Englewood Cliffs, NJ, 1991.

MacQueen, J., "Some Methods for Classification and Analysis of Multivariate Observations," *Proceedings of the 5th Berkeley Symposium on Mathematical Statistics and Probability*, 281-297, 1967.

Maybeck, P. S., *Stochastic Models, Estimation, and Control*, vol. 2, Academic Press, Orlando, FL, 1982.

Mendel, J. M., *Lessons in Digital Estimation Theory*, Prentice Hall, Englewood Cliffs, NJ, 1987.

Mill, J. S., *A System of Logic*, 1843.

Muth, F., "Rational Expectations and the Theory of Price Movements," *Econometrica*, vol. 29, 315-335, 1961.

Nikias, C. L., and Raghuveer, M. R., "Bispectrum Estimation: A Digital Signal Processing Framework," *Proceedings of the IEEE*, vol. 75, 869-891, July 1987.

Pearl, J., *Probability Reasoning in Intelligent Systems: Networks of Plausible Inference*, Morgan-Kaufman, 1988.

Pratt, W. K., *Digital Image Processing*, Wiley, New York, 1978.

Rudin, W., *Real and Complex Analysis*, 2nd ed., McGraw-Hill, New York, 1974.

Rumelhart, D. E., and Zipser, D., "Feature Discovery by Competitive Learning," *Cognitive Science*, vol. 9, 75-112, 1985.

Samuelson, P. A., "Proof that Properly Discounted Present Values of Assets Vibrate Randomly," *Bell Journal of Economics and Management Science*, vol. 4, 41-49, 1973.

Sebestyen, G. S., *Decision Making Processes in Pattern Recognition*, Macmillan: New York, 1962.

Sejnowski, T. J., "Storing Covariance with Nonlinearly Interacting Neurons," *Journal of Mathematical Biology*, vol. 4, 303-321, 1977.

Skorokhod, A. V., *Studies in the Theory of Random Processes*, Addison-Wesley, Reading, MA, 1982.

Sorenson, H. W., *Parameter Estimation*, Marcel Dekker, New York, 1980.

Styblinski, M. A., and Meyer, B. D., "Fuzzy Cognitive Maps, Signal Flow Graphs, and Qualitative Circuit Analysis," *Proceedings of the 2nd IEEE International Conference on Neural Networks (ICNN-87)*, vol. II, 549-556, July 1988.

Stark, H., and Woods, J. W., *Probability, Random Processes, and Estimation Theory for Engineers*, Prentice Hall, Englewood Cliffs, NJ, 1986.

Sutton, R. S., "Learning to Predict by the Methods of Temporal Differences," *Machine Learning*, vol. 3, 9-44, 1988.

Taber, W. R., and Siegel, M., "Estimation of Expert Weights with Fuzzy Cognitive Maps," *Proceedings of the 1st IEEE International Conference on Neural Networks (ICNN-87)*, vol. II, 319-325, June 1987.

Taber, W. R., "Knowledge Processing with Fuzzy Cognitive Maps," *Expert Systems with Applications*, vol. 2, no. 1, 83-87, 1991.

Thompson, R. F., "The Neurobiology of Learning and Memory," *Science*, 941-947, 29 August 1986.

Tspykin, Ya. Z., *Foundations of the Theory of Learning Systems*, Academic Press, Orlando, FL, 1973.

Von der Malsburg, C., "Self-Organization of Orientation Sensitive Cells in Striata Cortex," *Kybernetik*, vol. 14, 85-100, 1973.

Williams, W. E., "South Africa is Changing," *San Diego Union*, Heritage Foundation Syndicate, August 1986.

Winston, P. H., *Artificial Intelligence*, 2nd ed., Addison-Wesley, Reading, MA, 1984.

Zhang, W., and Chen, S., "A Logical Architecture for Cognitive Maps," *Proceedings of the 2nd IEEE International Conference on Neural Networks (ICNN-88)*, vol. I, 231-238, July 1988.

PROBLEMS

4.1. Prove that sigma-algebras are closed under countable intersections:

$$\text{If } A_1 \in \mathbf{A}, \; A_2 \in \mathbf{A}, \; \ldots, \quad \text{then } \bigcap_{i-1}^{\infty} A_i \in \mathbf{A}$$

4.2. Find the power spectral density $\mathbf{S}$ of a WSS random process with correlation function R given by

$$R(s) \;=\; e^{-c|s|}$$

for some constant $c > 0$.

4.3. A *band-limited* function f has a Fourier transform $\mathbf{F}$ that equals zero everywhere outside some finite band or interval of frequencies. We define the Fourier transform of f as

$$\mathbf{F}(\omega) \;=\; \int_{-\infty}^{\infty} f(s)e^{-i\omega s} \, ds$$

and the inverse transform as

$$f(s) \;=\; \frac{1}{2\pi} \int_{-\infty}^{\infty} \mathbf{F}(\omega)e^{i\omega s} \, d\omega$$

where $i = \sqrt{-1}$. Suppose the Fourier transform $\mathbf{F}$ looks like a rectangle centered about the origin:

$$F(\omega) \;=\; \begin{cases} \frac{1}{2B} & \text{if} \quad -B \leq \omega \leq B \\ 0 & \text{if} \quad |\omega| > B \end{cases}$$

for constant $B > 0$. Show that the original function f is proportional to $(\sin ct)/ct$ for some $c > 0$. You may need the trigonometric identities

$$\sin x \;=\; \frac{e^{ix} - e^{-ix}}{2i}$$

$$\cos x \;=\; \frac{e^{ix} + e^{-ix}}{2}$$

4.4. Prove: $V[x + y] = V[x] + V[y] + 2\text{Cov}[x, y]$ for random variables x and y.

4.5. Prove: $\text{Cov}[x, y] = 0$ if x and y are independent.

4.6. Prove: $-1 \leq \rho(x, y) \leq 1$, where

$$\rho(x, y) = \frac{\text{Cov}[x, y]}{\sqrt{V[x]} \sqrt{V[y]}}$$

4.7. Define the conditional-expectation random variable $E[x|y]$ for each realization r of random variable y as

$$E[x|y = r] = \int_{-\infty}^{\infty} xp(x|r)\,dx$$

Prove the following equalities:

(a) $E[x|y] = E[x]$ if x and y are independent;

(b) $E_y[E_x[z|y]] = E_x[z]$ if $z = f(x)$;

(c) $E_{xy}[ax + by|z] = aE_x[x|z] + bE_y[y|z]$ for constants a and b.

4.8. Suppose $\mathbf{x}$ and $\mathbf{y}$ are random row vectors. Suppose $\mathbf{f}$ is any Borel-measurable vector function of $\mathbf{x}$, $\mathbf{f}(\mathbf{x})$. Prove the fundamental theorem of estimation theory:

$$E[(\mathbf{y} - E[\mathbf{y}|\mathbf{x}])(\mathbf{y} - E[\mathbf{y}|\mathbf{x}])^T] \leq E[(\mathbf{y} - \mathbf{f}(\mathbf{x}))(\mathbf{y} - \mathbf{f}(\mathbf{x}))^T]$$

4.9. Define the conditional variance $V[x|y]$ and conditional covariance $\text{Cov}[x, y|z]$ as

$$V[x|y] = E[(x - E[x|y])^2|y] \qquad (4\text{-}283)$$
$$\text{Cov}[x, y|z] = E[(x - E[x|z])(y - E[y|z])|z] \qquad (4\text{-}284)$$

Prove the following equalities:

$$V[x|y] = E[x^2|y] - E^2[x|y]$$
$$V_x[x] = E_y[V[x|y]] + V_y[E[x|y]]$$
$$\text{Cov}[x, y|z] = E[xy|z] - E[x|z]E[y|z]$$

4.10. Construct a temporal associative memory (TAM) that successively encodes and decodes the limit cycle $A_1 \longrightarrow A_2 \longrightarrow A_3 \longrightarrow A_1 \longrightarrow \ldots$ and the limit cycle $A_1 \longrightarrow A_3 \longrightarrow A_2 \longrightarrow A_1 \longrightarrow \ldots$ with the following binary vectors:

$$A_1 = (1 \ \ 0 \ \ 0 \ \ 1 \ \ 0 \ \ 0 \ \ 1 \ \ 0 \ \ 0 \ \ 1)$$
$$A_2 = (1 \ \ 1 \ \ 0 \ \ 0 \ \ 1 \ \ 1 \ \ 0 \ \ 0 \ \ 1 \ \ 1)$$
$$A_3 = (1 \ \ 0 \ \ 1 \ \ 0 \ \ 1 \ \ 1 \ \ 1 \ \ 0 \ \ 1 \ \ 0)$$

Compute the forward "energies" $-A_i \mathbf{T} A_{i+1}^T$ and backward energies $-A_i \mathbf{T}^T A_{i-1}^T$. Test the recall accuracy with binary vectors A that have nonzero Hamming distance from stored binary vectors A_i: $H(A, A_i) = 1$, $H(A, A_i) = 2$, $H(A, A_i) = 3$, and so on.

4.11. Construct trivalent fuzzy cognitive maps (FCMs) from opposing political or economic articles in the editorial section of a Sunday newspaper. Identify the fixed points and limit cycles of each FCM. Combine the two FCMs and identify the new equilibrium states.

4.12. Combine the following three fuzzy cognitive maps and identify the equilibrium states of the combined FCM:

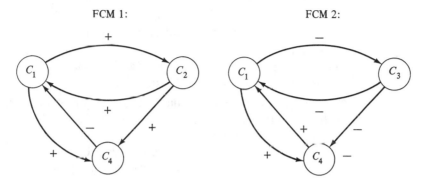

FCM 1:   FCM 2:

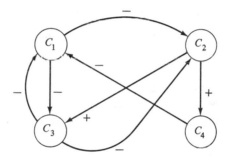

FCM 3:

SOFTWARE PROBLEMS

Part I: Competitive Learning

Construct a feedforward two-layer competitive network. The input field F_X should contain two neurons or nodes. The competitive field F_Y should contain 25 nodes. The following command sequence constructs such a competitive neural network. Dots indicate the 25 quantizing synaptic vectors.

1. Run CL.

2. Select "Activate KOH" from the "Network" menu.

3. Set the Load Parameters as follows:

Nodes in Network	=	25
Inputs to Network	=	2
Weight Scatter Radius	=	1.0
Weight Scatter Bias	=	0.0

4. Set the Distribution to "Uniform" in the "Run" menu. Set the X and Y extents to 1.0. This creates a uniform probability density of pattern vectors over a 2-by-2 square centered at the origin. Setting the "Weight Scatter Radius" to unity in step 3 ensures that the 25 synaptic vectors will be initialized with the first 25 samples drawn from the uniform distribution.

5. Use "Set Run Parameters" in the "Run" menu to set the "Learning Rate" to 0.06 and "Examples per Run" to 10. (This is a slow learning rate.) Do not change the Scatter Radius or the Scatter Bias.

6. Select "Train" from the "Run" menu. Let the network iterate for 500 iterations. While the network trains, occasionally press the SPACE bar to see the network's progress. After very few iterations, the 25 2-dimensional synaptic vectors should be approximately uniformly distributed over the square.

7. After the 500 iterations, press the SPACE bar and let the network iterate for another 200 iterations. Then press the "Q" key to pause the network.

Question 1:   Have the 25 synaptic vectors converged (stopped moving)? Do they approximate the (unknown) uniform pattern distribution?

8. Select the "Restart Network" option in the "Run" menu.

9. Use "Set Run Parameters" and set the "Learning Rate" to 0.3, a much faster learning rate.

10. Let the network iterate for 500 iterations.

11. After the 500 iterations, press the SPACE bar and let the network iterate for another 200 iterations. Then press the "Q" key to pause the network.

Question 2:   Have the 25 synaptic vectors converged this time? How does this competitive network compare with the first competitive network? Which network estimates more accurately the unknown, but sampled, probability density function? Do the two networks exhibit a "stability-plasticity" tradeoff?

Question 3:   Repeat steps 8 through 11 but now set the "Weight Scatter Radius" to 0.5. This initializes the 25 synaptic vectors with points from an interior subsquare centered at the origin. Are 700 iterations sufficient for approximating the uniform density?

Question 4:   Repeat steps 8 through 11 but now set the "Weight Scatter Radius" to 0.0. This degenerate initialization scheme does not depend on the sample distribution. Do the synaptic vectors converge after 500 iterations? After 50,000 iterations?

Part II: Differential Competitive Learning

Construct a feedforward two-layer competitive neural network but train it with differential competitive learning (DCL). The DCL software allows a small number of multiple winners.

1. Run CL.

2. Select "Activate DCL" from the "Network" menu.

3. Set the Load Parameters as follows:

$$
\begin{aligned}
\text{Nodes in Network} \quad &= \quad 25 \\
\text{Inputs to Network} \quad &= \quad 2 \\
\text{Weight Scatter Radius} \quad &= \quad 1.0 \\
\text{Weight Scatter Bias} \quad &= \quad 0.0
\end{aligned}
$$

4. Set the Distribution to "Uniform" in the "Run" menu. Set the X and Y extents to 1.0. This creates a uniform probability density of pattern vectors over the 2-by-2 square, as in Part I above.

5. Use "Set Run Parameters" in the "Run" menu to set the following specifications:

$$
\begin{aligned}
\text{Learning Rate} \quad &= \quad 0.06 \\
\text{Examples Per Run} \quad &= \quad 10 \\
\text{Alpha Value} \quad &= \quad 0.0 \\
\text{Beta Value} \quad &= \quad 1.0 \\
\text{Prorate Exponent} \quad &= \quad 1.0 \\
\text{Winners} \quad &= \quad 1
\end{aligned}
$$

6. Select "Train" from the "Run" menu. Let the network iterate for 500 iterations. Press the SPACE bar and let the network iterate for another 200 iterations. Then press the "Q" key to pause the network.

Question 5:   Have the 25 synaptic vectors converged?

7. Now repeat steps 1 through 6 but set "Learning Rate" to 0.3, "Winners" to 5, and "Prorate Exponent" to 0.01. For multiple winners, you must multiply the original learning rate by the number of winners and you must keep "Prorate Exponent" close to zero.

Question 6:   Have the 25 synaptic vectors converged? Has the rate of convergence changed? Has the equilibrium distribution of synaptic vectors changed?

You can also experiment with the nonuniform distribution "Triangle 2" in the "Run" menu and the "KCON" algorithm in the "Network" menu. KCON supervises competitive learning with a "conscience mechanism" to force the n competing neurons to win approximately $1/n$ percent of the time. (For KCON, set "B" to 0.0001 and "C" to 10.0 when using "Set Run Parameters" in the "Run" menu.)

5

SYNAPTIC DYNAMICS II: SUPERVISED LEARNING

The need for learning arises whenever available a priori information is incomplete. The type of learning depends on the degree of completeness of this a priori information. In learning with supervision, it is assumed that at each instant of time we know in advance the desired response of the learning system, and we use the difference between the desired and actual response, that is, the error of the learning system, to correct its behavior. In learning without supervision, we do not know the desired response of the learning system.

Ya. Z. Tsypkin
Foundations of the Theory of Learning Systems

Operant conditioning shapes behavior as a sculptor shapes a lump of clay. Although at some point the sculptor seems to have produced an entirely novel object, we can always follow the process back to the original undifferentiated lump, and we can make the successive stages by which we return to this condition as small as we wish. At no point does anything emerge which is very different from what preceded it.

B. F. Skinner
Science and Human Behavior

SUPERVISED FUNCTION ESTIMATION

This chapter presents supervised learning as stochastic approximation of an unknown average error surface. We attempt to estimate an unknown function $f: X \rightarrow Y$ from observed random row-vector samples $(\mathbf{x}_1, \mathbf{y}_1), \ldots, (\mathbf{x}_m, \mathbf{y}_m)$ by minimizing an unknown expected error functional $E[J]$.

We define error as *desired* performance minus *actual* performance. "Supervision" determines desired behavior. Supervision provides an ever-present error or "teaching" signal. In pattern recognition, supervision means complete knowledge of pattern-class boundaries.

We search for a locally optimal synaptic vector $\mathbf{m}^*$ that depends on all possible random vector realizations $(\mathbf{x}, \mathbf{y})$ through the unknown joint probability density function $p(\mathbf{x}, \mathbf{y})$. $E[J]$ defines an average error surface "over" the synaptic weight space. At each iteration, the current sample $(\mathbf{x}_i, \mathbf{y}_i)$ and previous initial conditions define an instantaneous error surface in $R^{n \times p} \times R$ if the neural network contains np synapses and if random-vector realizations $\mathbf{x}_i$ and $\mathbf{y}_i$ belong respectively to R^n and R^p. We indirectly search $E[J]$ for a local minimum by using stochastic gradient descent to search the instantaneous error surfaces.

The local minimum $\mathbf{m}^*$ may differ significantly from the unknown global minimum of $E[J]$. Some shallow local minima may be no better than expected error values determined by randomly picking network parameters. Since we do not know $E[J]$, we do not know the depth of its local minima. In general, nonlinear multivariable systems define complicated, and bumpy, average error surfaces. We may overlook this geometric fact when experimenting with small-dimensional problems.

Local minima are the best we can achieve, and then only in certain cases. In the set of all possible supervised learning algorithms, most algorithms define dynamical systems that do not converge at all. They oscillate or chaotically wander through the network parameter space. Even at a local minimum $\mathbf{m}^*$, we do not know what the neural network has learned unless we check all input-output pairs $(\mathbf{x}, \mathbf{y})$, a computationally prohibitive task. Even if we could check all $(\mathbf{x}, \mathbf{y})$ pairs, we still would not know what the neural network will learn or forget in future training. Model-free estimation of $f: X \rightarrow Y$ gives only a computational black box.

The neural network's linear or nonlinear dynamics depend on a synaptic *parameter* vector $\mathbf{m}$. The neural network $\mathbf{N}_m$ behaves as a transformation $\mathbf{N}_m: X \rightarrow Y$ between random sample spaces, usually R^n and R^p. Supervised learning defines a family of neural networks $\{\mathbf{N}_m\}$ indexed by synaptic parameter vectors $\mathbf{m}$. For suitable learning-rate coefficients, stochastic approximation ensures that the network $\mathbf{N}_m$ converges eventually to $\mathbf{N}_{m^*}$. Parameter vector $\mathbf{m}^*$ locally minimizes the unknown mean-squared error functional $E[J]$. At each iteration, the current synaptic vector $\mathbf{m}(k)$ estimates the unknown optimal synaptic vector $\mathbf{m}^*$.

The desired output $\mathbf{y}_i$ minus the actual neural-network output $\mathbf{N}(\mathbf{x}_i)$ defines the *instantaneous* error $\mathbf{y}_i - \mathbf{N}(\mathbf{x}_i)$, a random vector. $E[\mathbf{y}_i - \mathbf{N}(\mathbf{x}_i)]$ defines the instantaneous *average* or expected error, a deterministic vector. We do not know $E[\mathbf{y}_i - N(\mathbf{x}_i)]$ because we do not know the joint probability density function $p(\mathbf{x}, \mathbf{y})$ that defines the linear mathematical expectation operator E. $E[(\mathbf{y}_i - \mathbf{N}(\mathbf{x}_i))(\mathbf{y}_i - \mathbf{N}(\mathbf{x}_i))^T]$ defines the *mean-squared* error, a scalar, of the random sample $(\mathbf{x}_i, \mathbf{y}_i)$. The total sampled mean-squared error sums over all random realizations or "samples" $(\mathbf{x}_i, \mathbf{y}_i)$.

Stochastic approximation estimates expected quantities with observed random quantities, then uses these estimates in a discrete approximation algorithm, usually a stochastic gradient-descent algorithm.

Most supervised neural-network algorithms reduce to stochastic approximation. Rosenblatt's perceptron estimates a linear expected cost or misclassification error with an instantaneous observed misclassification error. Widrow's LMS algorithm estimates the mean-squared error of a linear network with an instantaneous squared error. Recently White [1989] formally reduced the backpropagation algorithm to stochastic approximation. Related reinforcement learning [Barto, 1983, 1985] and learning-automata algorithms explicitly define stochastic-approximation algorithms. Reinforcement learning performs estimated gradient descent with an error scalar instead of an error vector.

Below we review these neural-network algorithms and reductions. First we examine supervised learning in behaviorist psychology and in pattern recognition.

SUPERVISED LEARNING AS OPERANT CONDITIONING

Operant conditioning reinforces *responses*. Environmental stimuli excite or stimulate an organism. The organism "freely" emits behavioral responses. A supervisor ignores or reinforces these responses to increase the intensity or probability of desired responses. The reinforcement can be a reward or a punishment, a bonus or a fine. The supervisor can represent the naturally selecting environment or a loving parent.

Skinner [1953] originally defined *positive* reinforcement as any applied reinforcement. The reinforcement could be either reward or punishment. He defined *negative* reinforcement as absence of reinforcement ("No work, no pay"). These definitions have largely given way to the more intuitive interpretation of positive reinforcement as reward and negative reinforcement as punishment. We continue this interpretation.

Classical conditioning reinforces *stimuli*. The organism has already learned a stimulus-response sequence $S \rightarrow R$. Then reinforcer or "conditioned stimulus" B combines with the unconditioned stimulus S in repeated learning trials. The organism still emits response R but in the new behavioral sequence $(B\&S) \rightarrow R$, or its logical equivalent, $B \rightarrow (S \rightarrow R)$. With continued practice the organism emits R when stimulated with the reinforcer B alone. The organism learns the "conditioned reflex" $B \rightarrow R$. In the spirit of Pavlov's [1927] experiments, we can think of the stimulus S as food, the organism as a dog, the response as saliva, and the reinforcer B as a ringing bell. Initially the dog salivates in the presence of food. After training the dog salivates when only the bell rings.

Operant and classical conditioning correspond to supervised and unsupervised learning. Supervised learning rewards or punishes a neural network by incrementing or decrementing the components of the synaptic vector **m**. The algorithmic supervisor reinforces the neural network based on only the network output $\mathbf{N}(\mathbf{x})$, not directly on the network input **x**. The algorithm reinforces the neural network if, and only if, the desired network output **y** differs from the actual output $\mathbf{N}(\mathbf{x})$. Convergence may require thousands of learning iterations, if it occurs at all.

Unsupervised learning directly couples the input **x** with the output **y**. This usually requires very few learning iterations and can occur asynchronously in real time. Hebb's [1949] synaptic learning displays rapid coupling by encoding the correlation products $x_i y_j$ in the outer-product matrix $\mathbf{x}^T \mathbf{y}$.

Grossberg [1982] observed that the ith input or F_X neuron—the fan-out or **outstar** neuron—in a Hebbian network acts as a conditioned stimulus to the F_Y neurons. The ith neuron generates a real-valued membrane potential or activation x_i. The neuron transduces x_i to the bounded signal $S_i(x_i)$ and sends $S_i(x_i)$ out along its fan-out of axons and synapses. The signal $S_i(x_i)$ synaptically couples with the F_Y inputs or unconditioned stimuli J_j to produce the output neuronal activations y_j, and thus signals $S_j(y_j)$. The synaptic efficacy m_{ij} exponentially equilibrates to the correlation value $S_i y_j$, or $S_i S_j$. Then, for nonnegative input signals S_i, large activations x_i tend to reproduce the F_Y activation pattern $\mathbf{y} = (y_1, \ldots, y_p)$, or at least the relative values of the y_j components, and thus the signal pattern $\mathbf{S}(\mathbf{y}) = (S_1(y_1), \ldots, S_p(y_p))$. Grossberg's [1982] **outstar learning theorems** ensure such classical-conditioning behavior in very general mathematical settings.

Competitive and signal-velocity learning algorithms also behave as classical conditioners. These learning laws, like the Hebb law, learn with only locally available information. Recently Klopf [1988] showed that differential Hebbian learning successfully reproduces a wide range of Pavlov's experimental findings on animal learning.

Most physiologists believe that biological synapses simulate some form of Hebbian learning. They do not know which mathematical model of Hebbian learning best approximates biological synaptic learning. They cannot yet accurately monitor and measure individual synapses in action.

Supervised learning seems implausible at the level of individual synapses. Operant conditioning governs much of the learning of organisms, perhaps even groups of organisms. But that does not imply it governs an organism's neural microstructure. Supervised synaptic learning requires error or pattern-class information that is not locally available. Thousands of synchronized learning iterations preclude real-time performance. Supervised synapses must compute amid potential dynamical instabilities, compound round-off errors, and high-dimensional local error minima.

SUPERVISED LEARNING AS STOCHASTIC PATTERN LEARNING WITH KNOWN CLASS MEMBERSHIPS

Patterns define points in R^n. *Time-varying* patterns x define trajectories in R^n. They define mappings x: $[0, T] \rightarrow R^n$ from the time interval $[0, T]$ to the pattern space R^n. So patterns define time-varying patterns that map $[0, T]$ to a single point in R^n.

The probability density function $p(\mathbf{x})$ describes the distribution of patterns x in R^n. $p(\mathbf{x})$ is nonnegative and integrates to unity over R^n. The patterns x are continuously distributed in R^n.

x denotes a random pattern vector. We observe pattern samples or realizations $\mathbf{x}(1)$, $\mathbf{x}(2)$, ... of the random vector x. We do not know $p(\mathbf{x})$. We estimate $p(\mathbf{x})$ with the observed samples. In stochastic pattern recognition, we estimate the probability that an observed sample comes from a pattern class.

Random vector x is a function [Chung, 1974]. x is a *measurable* function from a sample space to a vector space. (Measurability means inverse images $\mathbf{x}^{-1}(O)$ of measurable or open sets O of the vector space are measurable subsets of the sample space, as discussed in Chapter 4.) In the autoassociative case, both spaces are R^n, so x: $R^n \rightarrow R^n$. The *sigma algebra* of measurable sets refers to the Borel sigma algebra $B(R^n)$, the topological sigma algebra. As discussed in Chapter 4, the open subsets of R^n generate $B(R^n)$. In practice the random vector x: $R^n \rightarrow R^n$ defines the identity function: $\mathbf{x}(\mathbf{v}) = \mathbf{v}$ for all v in R^n.

The cumulative distribution function P: $B(R^n) \rightarrow [0, 1]$ maps open subsets of R^n to numbers in $[0, 1]$ and is countably additive on countably infinite disjoint unions of R^n subsets. $P(O)$ equals the integral of $p(\mathbf{x})$ on the open subset O of R^n. $p(\mathbf{x})$ characterizes the "randomness" of the random vector x. Since x is the identity function on R^n, $p(\mathbf{x})$ characterizes the *occurrence probability* of the observed pattern samples or realizations.

We rigorously define the stochastic pattern-recognition framework when we specify the probability space $(R^n, B(R^n), P)$ and the random pattern vector x. In theory x can assume at least as many forms as there are real numbers. So can $p(\mathbf{x})$, and thus so can P. In the default case x is the identity function.

Pattern clusters or classes are subsets of R^n. Suppose we *partition* the pattern sample space R^n into k disjoint exhaustive subsets or **decision classes** $D_1, \ldots, D_k$:

$$R^n = D_1 \cup \ldots \cup D_k \quad \text{and} \quad D_i \cap D_j = \emptyset \quad \text{if } i \neq j \tag{5-1}$$

Some classes are more probable than others. A multiple integral gives the cumulative or **class probability** $P(D_i)$:

$$P(D_i) \quad = \quad \int_{D_i} p(\mathbf{x}) \, d\mathbf{x} \tag{5-2}$$

$$= \quad \int_{R^n} I_{D_i}(\mathbf{x}) p(\mathbf{x}) \, d\mathbf{x} \tag{5-3}$$

$$= \quad E[I_{D_i}] \tag{5-4}$$

The integral in (5-2) denotes an n-dimensional multiple integral. Equation (5-3) defines $E[x]$, the mathematical expectation of random variable x. The function $I_S: R^n \to \{0, 1\}$ denotes the indicator function of set S:

$$I_S(\mathbf{x}) = \begin{cases} 1 & \text{if } \mathbf{x} \in S \\ 0 & \text{if } \mathbf{x} \notin S \end{cases} \qquad (5\text{-}5)$$

The indicator function indicates class membership. In the probabilistic setting, the indicator function is random or Borel measurable [Chung, 1974], and thus a random variable. S is a *continuous* or multivalued or "fuzzy" set if I_S maps R^n into the unit interval $[0, 1]$ instead of $\{0, 1\}$. Then patterns $\mathbf{x}$ belong to different classes to different degrees.

The partition property (5-1) and $P(R^n) = 1$ imply that

$$P(D_1) + \cdots + P(D_k) = 1 \qquad (5\text{-}6)$$

With probability one, a pattern $\mathbf{x}$ resides in exactly one decision class. With probability zero, pattern $\mathbf{x}$ can lie on the border of two or more decision classes. Technically $p(\mathbf{x}) = 0$ for every $\mathbf{x}$ in R^n.

The distinction between supervised and unsupervised pattern learning depends on the available information and how the learning system uses it. In both cases the system does not know the probability density $p(\mathbf{x})$. If it knew $p(\mathbf{x})$, it would not need to learn. We could numerically determine probability clusters, maxima, and centroids.

Supervised pattern learning uses more information than unsupervised learning uses. Unsupervised learning uses minimal information.

Pattern learning is **supervised** if we know the decision classes $D_1, \ldots, D_k$ and if the learning system uses this information. The user knows, and the algorithm uses, the class membership of every sample pattern $\mathbf{x}$. The user knows that $\mathbf{x} \in D_i$ and that $\mathbf{x} \notin D_j$ for all $j \neq i$. Equivalently, the user knows every indicator function I_{D_i}. Pattern learning is **unsupervised** if we do not know or use class memberships (indicator functions).

Supervision allows us to compute an error measure. The simplest error measure equals the desired outcome minus the actual outcome. The error measure guides the learning process with feedback error correction.

Consider the noisy stochastic competitive learning law:

$$\dot{\mathbf{m}}_j = S_j(y_j)[\mathbf{S}(\mathbf{x}) - \mathbf{m}_j] + \mathbf{n}_j \qquad (5\text{-}7)$$

where $\mathbf{S}(\mathbf{x}) = (S_1(x_1), \ldots, S_n(x_n))$. S_i and S_j denote bounded monotone-nondecreasing signal functions. y_j denotes the real-valued activation of the jth neuron in neuronal field F_Y. Independent Gaussian white-noise vector $\mathbf{n}_j$ is zero-mean and equals vector Brownian motion's pseudotime derivative.

The stochastic differential equation (5-7) does not use knowledge of class memberships to modify the synaptic vector $\mathbf{m}_j$. So learning is unsupervised.

In practice the input neuronal field F_X feeds neuronal information forward through the synaptic connection matrix M to the competitive field F_Y, and F_Y does not feed neuronal information back to F_X. Then the F_X neurons use linear signal functions: $S_i(x_i) = x_i$. The competitive signal function S_j is steep, usually a binary threshold function. S_j measures win status. In biological neurons S_j measures recent pulse frequency.

In practice "competition" means metrical or nearest-neighbor classification. The competition metaphor summarizes a variety of neuronal activation models, most laterally inhibitive. The jth neuron in F_Y wins iff pattern $\mathbf{x}$ falls in D_j, and $\mathbf{m}_j$ codes for D_j. So we can replace S_j in (5-7) with the indicator function I_{D_j}:

$$\dot{\mathbf{m}}_j = I_{D_j}(\mathbf{x})[\mathbf{x} - \mathbf{m}_j] + \mathbf{n}_j \qquad (5\text{-}8)$$

Learning remains unsupervised in (5-8) because "competition" approximates the individual indicator function I_{D_j} and because we do not know in general the class membership of random sample $\mathbf{x}$. In Kosko [1990] and in Chapter 6, we show that in stochastic steady state, when

$$\dot{\mathbf{m}}_j = \mathbf{n}_j \qquad (5\text{-}9)$$

the mean synaptic vector $E[\mathbf{m}_j]$ equals the unknown D_j-class centroid. If we eliminate the additive noise term in (5-8), the synaptic vector $\mathbf{m}_j$ itself equals the class centroid with probability one. This follows by taking expectations of both sides of (5-8) at equilibrium and rearranging.

The **supervised stochastic competitive learning law**,

$$\dot{\mathbf{m}}_j = r_j(\mathbf{x})S_j(y_j)[\mathbf{x} - \mathbf{m}_j] + \mathbf{n}_j \qquad (5\text{-}10)$$

uses the **reinforcement function** r_j, defined by known decision-class indicator functions:

$$r_j = I_{D_j} - \sum_{i \neq j} I_{D_i} \qquad (5\text{-}11)$$

r_j rewards correct pattern classifications with $+1$ and punishes misclassifications with -1. Kohonen's [1988] supervised adaptive-vector-quantization algorithm, discussed in Chapter 6, uses implicitly the reinforcement function (5-11) but in a noiseless stochastic difference equation. Tsypkin [1974] has derived a similar and more general "adaptive Bayes" estimator.

SUPERVISED LEARNING AS STOCHASTIC APPROXIMATION

Supervised learning is goal-directed learning. Supervised learning attempts to minimize an unknown expected-cost or error functional $E[J]$. An unknown probability density function $p(\mathbf{x})$ defines the expected-cost functional $E[J]$.

$p(\mathbf{x})$ summarizes the unknown a priori information Tsypkin refers to in the quote at the beginning of this chapter. If we knew $p(\mathbf{x})$, we would not need learning to estimate it. In principle we could minimize $E[J]$ with numerical or calculus-of-variations techniques [Fleming, 1975].

In practice we do not know $p(\mathbf{x})$. More generally, we do not know the joint probability density function $p(\mathbf{x}, \mathbf{y})$ that characterizes the sampled function $f\colon X \to Y$. $\mathbf{x}$ denotes a random vector in R^n. $\mathbf{y}$ denotes a random vector in R^p. The expected error $E[J]$ weights all possible input-output samples $(\mathbf{x}, \mathbf{y})$ in the sample space $R^n \times R^p$ with $p(\mathbf{x}, \mathbf{y})$.

We know only the m random samples $(\mathbf{x}_1, \mathbf{y}_1), (\mathbf{x}_2, \mathbf{y}_2), \ldots, (\mathbf{x}_m, \mathbf{y}_m)$. So we *estimate* $E[J]$ as simply J. The error measure J depends on the observed random samples $(\mathbf{x}_i, \mathbf{y}_i)$, the observed realizations of augmented random vector $[\mathbf{x} \,|\, \mathbf{y}]$. So J is random, though $E[J]$ is deterministic.

Monro and Robbins [1951] introduced stochastic approximation in 1951. Blum [1954] and Albert [1967] present more general vector formulations. Below we use only the simplest version of this mathematically rich and complicated subject, and we use it for estimated gradient descent.

Stochastic approximation estimates the unknown gradient $\nabla E[J]$ with the known random gradient ∇J at each iteration in a discrete stochastic gradient-descent algorithm:

$$\mathbf{m}_{k+1} \;=\; \mathbf{m}_k - c_k \nabla J_k \qquad (5\text{-}12)$$

The unknown gradient $-\nabla E[J]$ points in the direction of steepest descent on the unknown expected-error surface defined by $E[J]$ in the product sample space $R^{np} \times R$. Here the total synaptic vector $\mathbf{m}$ contains np components.

$\mathbf{m}_k$ denotes a synaptic vector at iteration k. $\mathbf{m}_k$ may list all synaptic connection weights in a neural network. ∇J_k denotes the gradient of J_k with respect to $\mathbf{m}_k$. J_k is random, since the samples $(\mathbf{x}_k, \mathbf{y}_k)$ are random. So ∇J_k is random. So $\mathbf{m}_k$ is random.

The **learning-rate coefficients** $\{c_k\}$ decrease with time k to suppress random disturbances. (In the deterministic case the learning coefficients are constant, or converge to a constant.) Then, with probability one, the random synaptic vector $\mathbf{m}_k$ in (5-12) converges to a local minimum of the unknown expected-error function $E[J]$. Stochastic approximation guarantees this result if we appropriately choose the learning coefficients c_k [Albert, 1967].

The learning coefficients decrease slowly but not too slowly, quickly but not too quickly. We formally constrain the sequence $\{c_k\}$ to decrease slowly in the sense of a divergent sum:

$$\sum_{k=1}^{\infty} c_k \;=\; \infty \qquad (5\text{-}13)$$

The slower c_k decreases, the faster $\mathbf{m}_k$ learns in the stochastic difference equation (5-12). We constrain c_k to decrease quickly in the sense of a convergent sum of squares:

$$\sum_{k=1}^{\infty} c_k^2 \;<\; \infty \qquad (5\text{-}14)$$

The faster c_k decreases, the less $\mathbf{m}_k$ forgets learned pattern information. The harmonic series, $c_k = 1/k$, satisfies (5-13) and (5-14).

The slow-fast constraints (5-13) and (5-14) rigorously summarize, and resolve, one version of Grossberg's **stability-plasticity dilemma**. How, Grossberg [1982] asks, can a learning system be stable enough to remember old learned patterns, and yet plastic enough to learn new patterns? Grossberg pursues an architectural answer to this oak-versus-willow question with his adaptive-resonance theory, discussed in Chapter 6 and illustrated in the software homework problems.

Stochastic approximation interprets and answers the stability-plasticity question in terms of learning *rates*, not system architectures. The learning-rate condition (5-13) permits plasticity while (5-14) constrains plasticity. (5-14) permits "stability" or memory rigidity while (5-13) constrains it. Both neural and non-neural systems can execute the discrete stochastic gradient-descent algorithm (5-12) subject to rate conditions (5-13) and (5-14).

Stochastic approximation shows that suitably supervised learning is asymptotically optimal learning. When learning ceases, we know the previously unknown probability density information $p(\mathbf{x}, \mathbf{y})$. More realistically, we know only a local minimum of some function of $p(\mathbf{x}, \mathbf{y})$.

We now examine the perceptron, *LMS*, and backpropagation algorithms in the context of stochastic approximation.

The Perceptron: Learn Only If Misclassify

The **perceptron** [Anderson, 1988; Rosenblatt, 1962] is a feedforward network with one output neuron that learns a separating hyperplane in a pattern space. n linear F_X neurons feed forward to one threshold output F_Y neuron. The perceptron separates linearly separable sets of patterns. The perceptron need not separate linearly inseparable pattern sets. Most collections of pattern sets are linearly inseparable. Here we consider only the two-class perceptron. Chapter 1 of the companion volume [Kosko, 1991] discusses the multiclass perceptron.

The **perceptron learning theorem** states that a perceptron separates linearly separable pattern sets in *finite* iterations. Stochastic approximation ensures only asymptotic convergence in countably infinite iterations.

Rosenblatt first proved the perceptron learning theorem. In doing so he launched the *first wave* of modern neural-network theory. Nilsson's 1965 book, *Learning Machines*, represents the crest of the perceptron wave. We refer the reader to Nilsson's book for a proof of the perceptron learning theorem.

Minsky and Papert's 1969 book, *Perceptrons*, signaled the end of the perceptron wave. In the *second wave* of neural networks—characterized by backpropagation and Hopfield models, the so-called *BackHop* wave—many neural network enthusiasts blamed Minsky and Papert for crushing the perceptron wave. This was inaccurate.

Minsky and Papert actually extended the perceptron framework with their "computational geometry" topological formulation of supervised learning. In both the 1969 and 1988 editions of *Perceptrons*, Minsky and Papert accurately criticized estimated gradient descent performed on nonlinear systems of high dimension. They illustrated these criticisms with proofs that simple perceptrons could not compute apparently "simple" yet global properties like connectedness or parity (the number of 1s in a bit vector).

Their criticisms remain valid for the backpropagation algorithm discussed below. As they state in the 1988 edition of *Perceptrons*, "We have the impression that many people in the connectionist community do not understand that this [backpropagation] is merely a particular way to compute a gradient and have assumed instead that backpropagation is a new learning scheme that somehow gets around the basic limitations of hill-climbing."

We now develop the preliminaries for the two-class perceptron. Consider two decision classes D_i and D_j, subsets of R^n. Suppose D_i contains the sample patterns $\mathbf{x}_i(1), \ldots, \mathbf{x}_i(k)$, and D_j contains $\mathbf{x}_j(1), \ldots, \mathbf{x}(l)$. Many R^n subsets contain the pattern sets $\{\mathbf{x}_i\}$ and $\{\mathbf{x}_j\}$. In particular, D_i contains $\{\mathbf{x}_i\}$.

Many *convex* R^n subsets contain $\{\mathbf{x}_i\}$ and $\{\mathbf{x}_j\}$. Set S is **convex** if S contains all points on all line segments with endpoints in S. R^n spheres, ellipsoids, and cubes are convex. Tori and disconnected sets are not convex.

$C(\{\mathbf{x}_i\})$ denotes the **convex hull** of pattern set $\{\mathbf{x}_i\}$. $C(\{\mathbf{x}_i\})$ is the smallest convex set in R^n that contains $\{\mathbf{x}_i\}$. Equivalently, $C(\{\mathbf{x}_i\})$ equals the intersection of all convex sets in R^n that contain $\{\mathbf{x}_i\}$. $C(S) = S$ if set S is convex. The convex set R^n always contains $\{\mathbf{x}_i\}$.

Pattern sets $\{\mathbf{x}_i\}$ and $\{\mathbf{x}_j\}$ are **linearly separable** if their convex hulls are disjoint:

$$C(\{\mathbf{x}_i\}) \cap C(\{\mathbf{x}_j\}) \;=\; \emptyset \qquad (5\text{-}15)$$

We ignore the more complicated case where the two convex hulls just touch. The convex hulls are usually closed and bounded (exceptions possible). So some minimum-distance line segment $d_{\min}$ connects them. So some perpendicular hyperplane H bisects $d_{\min}$. (We here invoke the convex **separation theorem** found in optimization theory and functional analysis [Rudin, 1973].) The pattern sets are linearly separable iff H exists. For instance, in R^3 we can pass a plane between any two disjoint balls or cubes. In R^2 we can pass a line between any two disjoint discs or squares.

We define the **linear discriminant function** $g_i: R^n \to R$ as the inner product

$$g_i(\mathbf{x}) \;=\; \mathbf{x}\mathbf{m}_i^T + m_{i0} \qquad (5\text{-}16)$$

where m_{i0} denotes an optional threshold value. $\{\mathbf{x}: g_i(\mathbf{x}) = 0\}$ defines a **hyperplane** H in R^n. H is manifold of dimension $n - 1$. In the real line R, H is a point; in R^2, a line; in R^3, a plane.

Suppose g^i and g^j represent D_i and D_j by sign. So $g_i(\mathbf{x}) > 0$ if $\mathbf{x}$ resides in D_i, and $g_i(\mathbf{x}) < 0$ if $\mathbf{x}$ resides in the complement set D_i^c. $\{\mathbf{x}: g_i(\mathbf{x}) = g_j(\mathbf{x})\}$ defines the **decision surface** between D_i and D_j. Chapter 1 of the companion volume [Kosko, 1991] discusses discriminant functions in more detail.

The perceptron convergence theorem says that if the pattern sets $\{x_i\}$ and $\{x_j\}$ are linearly separable, then the two-class perceptron algorithm finds a separating linear discriminant function g_i—finds a separating synaptic vector m_i^{sep}—in finite iterations for *any* initial synaptic vector $m_i(0)$. In the 1960s, researchers often illustrated this theorem with randomly chosen initial synaptic weights. This mistakenly suggested that the theorem required random or unstructured initial conditions. In fact it illustrates the theorem's robustness.

Consider a two-layer network with one output neuron. The neuron behaves as a bipolar threshold function according to the sign of the discriminant function g_k at iteration k. m_k denotes the synaptic fan in vector to the neuron at time k.

The perceptron learns only if it misclassifies random sample pattern x. The perceptron misclassifies x_k at iteration k if D_i contains x_k but $g_k(x_k) < 0$, or if D_j contains x_k but $g_k(x_k) > 0$.

$E[J_k]$ denotes the unknown mean misclassification error at iteration k. So we can write the ideal **gradient-descent algorithm** as

$$m_{k+1} = m_k - c_k \nabla_m E[J_k] \tag{5-17}$$

for some sequence of learning-rate coefficients $\{c_k\}$. Stochastic approximation [Albert, 1967] requires that the coefficients $\{c_k\}$ decrease in time and obey (5-13) and (5-14). The perceptron convergence theorem does not require that the coefficients decrease. As in the LMS algorithm discussed below, the system can converge even if the learning-rate coefficients are constant. This reflects the independent research paths that led to these algorithms.

We estimate $E[J]$ as J. We identify J with the scalar random linear misclassification function

$$J = -\sum_{x \in \chi} x m^T \tag{5-18}$$

where χ denotes the set of observed misclassified vector realizations or "training samples." We take the gradient of J with respect to m to give the random vector

$$\nabla_m J = -\sum_{x \in \chi} x \tag{5-19}$$

We now substitute the estimate (5-19) into (5-17) to derive the **perceptron learning algorithm**:

$$m_{k+1} = m_k + c_k \sum_{x \in \chi} x \tag{5-20}$$

Equation (5-20) defines an estimated gradient algorithm that need not converge in finite iterations. In general we do not know whether the pattern sets $\{x_i\}$ and $\{x_j\}$ are linearly separable.

We can state the **two-class perceptron algorithm** as iterative discriminant-function test conditions:

$$
\mathbf{m}_{k+1} \;=\; \begin{cases} \mathbf{m}_k + c\mathbf{x}_k & \text{if } \mathbf{x} \in D_i \text{ and } \mathbf{x}_k \mathbf{m}_k^T < 0 \\ \mathbf{m}_k - c\mathbf{x}_k & \text{if } \mathbf{x} \notin D_i \text{ and } \mathbf{x}_k \mathbf{m}_k^T > 0 \end{cases} \tag{5-21}
$$

The supervised reinforcement in (5-21) resembles the competitive supervised reinforcement algorithm (5-10) with reinforcement function r_j in (5-11). But the reinforcement signal in (5-21) depends on more than the indicator-function difference $I_{D_i} - I_{D_j}$. It also depends on the sign behavior of the discriminant function g_k.

Equation (5-21) assumes that we arbitrarily choose the initial synaptic vector $\mathbf{m}_0$. Different choices can produce different convergence rates and perhaps different separating hyperplanes. The learning coefficients c in (5-21) can vary with time k. In practice they are constant, positive, and small.

The LMS Algorithm: Linear Stochastic Approximation

Stochastic approximation describes the *mean-squared* convergence of estimators to minima of $E[J]$. Linear estimation provides perhaps the simplest example. In the form of Widrow's [1960] *least-mean-square* (LMS) algorithm, adaptive linear mean-squared estimation has proved the most popular, and currently most important, practical application of the theory. Widrow [1975, 1985, 1989] has successfully applied the LMS algorithm to adaptive beam forming in antennas, adaptive noise cancellation in long-distance telephone calls and high-speed modems, and several other adaptive filtering tasks.

The LMS algorithm is a close cousin of the adaptive linear mean-squared Kalman filter [Kalman, 1960, 1961]. Both systems recursively estimate the "normal equations" of linear least-squares regression [Mendel, 1987]. The Kalman filter adds a Gauss-Markov state equation to describe system dynamics. Chapter 11 compares a simple Kalman filter to a fuzzy control system for target tracking. The LMS and Kalman-filter algorithms account for a surprisingly large proportion of today's successful prediction and control of high-speed information processes, from deep-space navigation to real-time bloodstream decomposition.

Widrow and Hoff [1960] derived the LMS algorithm in the late 1950s. They were apparently unaware of the linear stochastic-approximation framework. They instead directly approached the adaptive estimation problem as an engineering problem. They wanted to adaptively estimate the Wiener-Hopf or "normal equations" of linear regression theory without complete knowledge of process statistics.

Widrow eventually determined tight bounds on *constant* learning coefficients that ensured convergence, and he tied these bounds to process quantities that engineers could measure. This makes the LMS algorithm easier to apply than the pure stochastic-approximation solution, which arrives only asymptotically at the same place the LMS algorithm arrives—Gauss's normal equations. The deterministic optimal linear associative memory (OLAM) in Chapter 3 also restates the normal equations.

The **linear model** is a feedforward two-layer network with one output "neuron" that has a linear signal function, $S(y) = y$. So an *additive* activation model describes the linear neuron:

$$y_k = \sum_{i=1}^{n} x_k^i m_k^i \tag{5-22}$$

$$= \mathbf{x}_k \mathbf{m}_k^T \tag{5-23}$$

where $\mathbf{m}_k$ denotes the synaptic fan-in vector $(m_k^1, \ldots, m_k^n)$ to the linear neuron at iteration k. We sample random vector realization $\mathbf{x_k}$ in R^n at time k. Input neuron field F_X contains n linear neurons. Output neuron field F_Y contains just one neuron.

d_k denotes the *desired* system output at time k. The linear system's *actual* scalar output y_k usually differs from d_k. We model $\{d_k\}$ and $\{y_k\}$ as random processes indexed by k. So d_k and y_k define random variables at time k. They can assume any one of infinitely many realizations at k. Different realizations have different occurrence probabilities. We observe only the single realizations d_k and y_k from this infinitude. We use supervised learning to estimate the unknown underlying probability structure. We attempt to learn the random function $f: R^n \to R$ from the random samples $(\mathbf{x}_k, d_k)$.

Define the random *instantaneous error* e_k as the desired-minus-actual system response:

$$e_k = d_k - y_k \tag{5-24}$$

The LMS model assumes that the random processes $\{d_k\}$, $\{x_k\}$, and $\{e_k\}$ are **wide-sense stationary**: They have constant mean functions for all k, and their second-order statistics depend only on time differences. This gives the **instantaneous mean-squared error** $E[e_k^2]$ as the average quadratic

$$E[e_k^2] = E[(d_k - y_k)^2] \tag{5-25}$$

$$= E[(d_k - \mathbf{x}_k \mathbf{m}_k^T)^2] \tag{5-26}$$

$$= E[d_k^2] - 2E[d_k \mathbf{x}_k]\mathbf{m}_k^T + \mathbf{m}_k E[\mathbf{x_k}^T \mathbf{x_k}]\mathbf{m}_k^T \tag{5-27}$$

if we assume $\mathbf{m}_k$ deterministic (in general it is random).

Widrow and Stearns [1985] denote the **input autocorrelation matrix** $E[\mathbf{x}_k^T \mathbf{x}_k]$ as $\mathbf{R}$ and the **cross-correlation vector** $E[d_k \mathbf{x}_k]$ as $\mathbf{P}$. So we can rewrite (5-27) as

$$E[e^2] = E[d^2] - 2\mathbf{Pm}^T + \mathbf{mRm}^T \tag{5-28}$$

where we have dropped the subscript k for convenience.

We can derive the functional form of the unknown optimal mean-squared-error synaptic vector $\mathbf{m}^*$ by partially differentiating both sides of (5-28) with respect to the components of $\mathbf{m}$ and setting this gradient vector of partial derivatives equal to the null vector in R^n. This gives the "normal equations":

$$O = 2m^*R - 2P \qquad (5-29)$$

or, solving for m^*,

$$m^* = PR^{-1} \qquad (5-30)$$

if the matrix inverse R^{-1} exists.

All autocorrelation matrices are positive-*semidefinite* symmetric matrices. So R may possess zero eigenvalues. All other eigenvalues must be positive. We assume R is positive *definite*: $R > O$. Then all eigenvalues are positive. Then R^{-1} exists. Positive definiteness and (5-29) imply that the second-order Hessian matrix of $E[e^2]$ is positive definite. So (5-30) gives a local minimum not maximum. Since the system is linear, $E[e^2]$ defines a convex mean-squared-error surface in R^{n+1}, a "bowl." So m^* in (5-30) is also a global-minimum solution.

We replace R^{-1} with the pseudoinverse R^* discussed in Chapter 3 if R does not have an inverse. Then (5-30) takes the form

$$m^* = PR^* \qquad (5-31)$$

Equation (5-31) gives the general form for the least-squares optimal synaptic weight vector. By (5-23), we pass input x through the linear filter m^T. So (5-31) takes the transposed form

$$m^{*T} = R^*P^T \qquad (5-32)$$

since R is symmetric.

We see that (5-32) has the same form as the *optimal linear associative memory* (OLAM) in Chapter 3. Supervised learning arrives asymptotically at the OLAM synaptic matrix X^*Y. The difficulty in computing the pseudoinverse matrix X^* reflects the computational complexity of even linear optimal supervised learning.

The optimal solutions (5-30) and (5-31) require more information than we usually have available. We do not know the expected cross-correlation vector $E[d_k x_k]$ or the expected autocorrelation matrix $E[x_k^T x_k^k]$ at each k because we do not know the underlying joint probability density functions at each k. We know only the observed realizations d_k and x_k. Before we examine this case, we examine iterative behavior with total information.

Suppose we know P and R, and R^{-1} exists. Then (5-29) gives the gradient vector of the scalar mean-squared error as

$$\nabla E[e^2] = 2mR - 2P \qquad (5-33)$$

Post-multiply both sides of (5-33) by $\frac{1}{2}R^{-1}$:

$$\frac{1}{2}\nabla E[e^2]R^{-1} = m - PR^{-1} \qquad (5-34)$$

$$= m - m^* \qquad (5-35)$$

by (5-30). So we can rewrite the optimal synaptic vector $\mathbf{m}^*$ as

$$\mathbf{m}^* = \mathbf{m} - \frac{1}{2}\nabla_m E[e^2]\mathbf{R}^{-1} \tag{5-36}$$

Equation (5-36) suggests a one-iteration, $\mathbf{R}^{-1}$-scaled gradient descent to the optimal solution $\mathbf{m}^*$. Rewriting (5-36) in iterative or adaptive notation, and scaling the gradient with the learning coefficient c_k, gives

$$\mathbf{m}_{k+1} = \mathbf{m}_k - c_k\nabla_m E[e_k^2]\mathbf{R}^{-1} \tag{5-37}$$

We assume the autocorrelation matrix $\mathbf{R}$, and hence $\mathbf{R}^{-1}$, changes little with time. If $c_k = 1/2$ as in (5-36), then (5-37) defines **Newton's approximation method** for finding the solution (5-30), and the algorithm converges to $\mathbf{m}^*$ in one step.

If $\mathbf{R}$ equals the identity matrix, (5-37) reduces to an **ideal gradient-descent algorithm**:

$$\mathbf{m}_{k+1} = \mathbf{m}_k - c_k\nabla_m E[e_k^2] \tag{5-38}$$

The algorithm is "ideal" because it assumes total knowledge of the gradient vector $\nabla_m E[e_k^2]$—because it assumes total knowledge of the underlying probability density functions that define $E[e_k^2]$, $\mathbf{R}_k$, and $\mathbf{P}_k$.

The LMS algorithm estimates the instantaneous mean-squared error $E[e_k^2]$ with the available measurement e_k^2:

$$E[e_k^2] \approx e_k^2 \tag{5-39}$$

We call (5-39) the **LMS assumption**. We might equally call (5-39) the *stochastic-approximation assumption*, since (5-39) reduces (5-38) to linear stochastic approximation if the learning coefficient sequence $\{c_k\}$ obeys (5-13) and (5-14).

More broadly, (5-39) is a *martingale assumption* if we interpret the deterministic scalar $E[e_k^2]$ as the conditional expectation $E[e_k^2 \mid e_0^2, e_1^2, \ldots, e_{k-1}^2]$, a random variable. The scalar random process $\{e_k^2\}$ behaves as a **martingale** [Chung, 1974] if

$$E[e_{k+1}^2 \mid e_0^2, e_1^2, \ldots, e_k^2] = e_k^2 \tag{5-40}$$

Technically in (5-40) we condition e_k^2 on the sigma-algebras or Borel fields that the random variables e_k^2 generate. The sigma-algebras increase with k and represent *knowledge sets* at k. For instance, many economists [Samuelson, 1973] believe that today's stock prices provide our best estimates of tomorrow's stock prices—even if we know all preceding stock prices. Heavily traded stocks vibrate randomly as a vector martingale process. The kth sigma-algebra can correspond to a compact-disc ROM that contains all stock quotes up to the kth day.

The LMS assumption (5-39) implies an **estimated gradient-descent algorithm**:

$$\mathbf{m}_{k+1} = \mathbf{m}_k - c_k\nabla_m e_k^2 \tag{5-41}$$

Equation (5-41) defines a *stochastic* gradient descent, since the random vector $\nabla_m e_k^2$ estimates the deterministic vector $\nabla_m E[e_k^2]$.

We use the chain rule and (5-22) to compute the stochastic gradient of e_k^2 with respect to m:

$$\nabla_m e_k^2 \;=\; [\frac{\partial e_k^2}{\partial m_1}, \ldots, \frac{\partial e_k^2}{\partial m_n}] \tag{5-42}$$

$$=\; [\frac{\partial(d_k - y_k)^2}{\partial m_1}, \ldots, \frac{\partial(d_k - y_k)^2}{\partial m_n}] \tag{5-43}$$

$$=\; [-2(d_k - y_k)\frac{\partial y_k}{\partial m_1}, \ldots, -2(d_k - y_k)\frac{\partial y_k}{\partial m_n}] \tag{5-44}$$

$$=\; -2e_k[\frac{\partial y_k}{\partial m_1}, \ldots, \frac{\partial y_k}{\partial m_n}] \tag{5-45}$$

$$=\; -2e_k[x_k^1, \ldots, x_k^n] \tag{5-46}$$

$$=\; -2e_k \mathbf{x}_k \tag{5-47}$$

Equation (5-47) and some manipulation show that the LMS gradient estimate is *unbiased* if the synaptic vector $\mathbf{m}$ is constant:

$$E[\nabla_m e_k^2] \;=\; \nabla_m E[e_k^2] \tag{5-48}$$

The estimator $\nabla_m e_k^2$ equals the estimated $\nabla_m E[e_k^2]$ on average for all k. The estimator may fluctuate mildly or wildly depending on the second-order statistics of $\{d_k\}$ and $\{\mathbf{x}_k\}$.

We substitute (5-47) in (5-41) to derive the simple, but powerful, **LMS algorithm:**

$$\mathbf{m}_{k+1} \;=\; \mathbf{m}_k + 2ce_k\mathbf{x}_k \tag{5-49}$$

Like Widrow [1985], we assume a constant learning rate $c > 0$. The LMS algorithm (5-49) uses only physically available quantities at each iteration k, and uses them in a computationally trivial formula.

The learning constant c governs convergence to the least-squares optimal synaptic vector $\mathbf{m}^*$. If c is too small, the LMS algorithm crawls needlessly down each estimated squared-error surface. Learning may be prohibitively slow. If c is too large, the algorithm leaps recklessly down the estimated squared-error surfaces. Then learning may never converge. The system state vector $\mathbf{m}_k$ may hop chaotically in R^n, landing indiscriminately at points that correspond to larger and smaller values of the total $E[e_k^2]$ surface.

A decreasing stochastic-approximation gain sequence $\{c_k\}$ ensures convergence to $\mathbf{m}^*$ only if the linear system samples a countably infinite set of samples $\{\mathbf{x}_k\}$. Widrow has restricted his analyses and applications to fixed learning rates. Fixed rates simplify computation and often allow convergence in finite iterations, as we saw above in Newton's method (5-37). Linearity permits this restriction.

The learning rate c should vary inversely with system uncertainty. The more uncertain the sampling environment, the smaller we must choose c to avoid divergence. The less uncertain the sampling environment, the larger we can choose c to speed convergence.

The autocorrelation matrix $\mathbf{R}$ represents much of the random-sampling environment's uncertainty, its second-order fluctuations. The deterministic matrix $\mathbf{R}$ also defines a linear operator $\mathbf{R}: R^n \rightarrow R^n$ with generalized nonnull fixed points or *eigenvectors* $\mathbf{e}_1, \ldots, \mathbf{e}_n$ and associated scale factors or *eigenvalues* $\lambda_1, \ldots, \lambda_n$:

$$\mathbf{e}_i\mathbf{R} = \lambda_i\mathbf{e}_i \tag{5-50}$$

Since $\mathbf{R}$ is symmetric and positive definite, all eigenvalues are real and positive. Suppose we order the eigenvalues as

$$0 < \lambda_1 \leq \lambda_2 \leq \ldots \leq \lambda_n \tag{5-51}$$

Symmetry also implies [Franklin, 1968] that the eigenvectors $\{\mathbf{e}_1, \ldots, \mathbf{e}_n\}$ are mutually orthogonal, even though some of the eigenvalues may be repeated. Normalizing by the Euclidean norm $\|\mathbf{e}_i\|$, we can assume that the eigenvectors are *orthonormal*:

$$\mathbf{o}_i\mathbf{o}_j^T = \begin{cases} 1 & \text{if } i = j \\ 0 & \text{if } i \neq j \end{cases} \tag{5-52}$$

where $\mathbf{o}_i = \mathbf{e}_i/\|\mathbf{e}_i\|$.

Equations (5-50) and (5-52) imply that we can eliminate matrix $\mathbf{R}$ in favor of its eigenvectors and eigenvalues:

$$\mathbf{R} = \mathbf{O}^T\Lambda\mathbf{O} \tag{5-53}$$

$$= \sum_{i=1}^{n} \lambda_i\mathbf{o}_i^T\mathbf{o}_i \tag{5-54}$$

Eigenvector $\mathbf{o}_i$ defines the ith row of the orthogonal matrix $\mathbf{O}$. Eigenvalue λ_i defines the ith element of the diagonal matrix Λ. In (5-53) we use the fact that $\mathbf{O}^T = \mathbf{O}^{-1}$ since the eigenvectors are orthonormal. Observe that the *spectral decomposition* (5-54) is a form of weighted outer-product or Hebbian learning, as discussed in Chapter 3.

Widrow shows that the autocorrelation eigenvalues constrain the learning rate. The argument uses the geometry of n-dimensional ellipsoids. For a symmetric positive-definite matrix $\mathbf{M}$, the set $\{\mathbf{x}: \mathbf{x}\mathbf{M}\mathbf{x}^T = 1\}$ defines an n-dimensional *ellipsoid* in R^n. $\mathbf{M}$'s n orthogonal eigenvectors $\{\mathbf{e}_1, \ldots, \mathbf{e}_n\}$ define the n **principal component axes** of the ellipsoid [Franklin, 1968]. The eigenvalue quantity $1/\sqrt{\lambda_i}$ measures the length or Euclidean norm $\|\mathbf{e}_i\|$ of the ith eigenvector axis. The smaller the eigenvalue, the longer the eigenvector axis. The ellipsoid's minor axis corresponds to $\mathbf{e}_n$, the eigenvector with the largest eigenvalue.

The eigenvector e_n "explains" the most system variance. e_n is the shortest axis of the n-dimensional *uncertainty* ellipsoid. e_n points in the direction of least uncertainty. Image-compression encoding schemes often exploit this fact by ignoring covariance eigenvectors with small eigenvalues.

Widrow [1985] proves that, for wide-sense stationary signals with a convex mean-squared error surface, the LMS algorithm converges to the global solution m^* if and only if we restrict the learning-rate constant c to the **LMS convergence interval**:

$$0 < c < \frac{1}{\lambda_n} \qquad (5\text{-}55)$$

Widrow established this bound with a principal-components analysis. Nonstationary signals need not obey (5-55). In practice we assume stationarity, not prove it. We still lack convergence bounds for unconditional LMS learning.

The LMS convergence interval (5-55) grounds the intuition that the learning rate c should vary inversely with the system uncertainty. As the uncertainty or variance in the sampled signal environment increases, so the largest eigenvalue λ_n increases in magnitude. Then (5-55) implies that the learning constant decreases in size, and the gradient-descent algorithm (5-49) takes smaller steps down each estimated error surface. The reverse holds as the system uncertainty decreases.

Inequality (5-55) implies the more practical bound

$$0 < c < \frac{1}{\text{Trace}(\mathbf{R})} \qquad (5\text{-}56)$$

since the eigenvalues are positive, and the trace, which sums the diagonal entries r_{ii}, equals the sum of eigenvalues λ_i. In many cases of temporal filtering, $\text{Trace}(\mathbf{R}) = nE[x_k^2]$. The scalar $E[x_k^2]$ defines the instantaneous average *signal strength* or power. Often we can more accurately estimate the signal strength than the underlying autocorrelation eigenvalues.

THE BACKPROPAGATION ALGORITHM

The backpropagation algorithm, a nonlinear extension of the LMS algorithm, reawakened the scientific and engineering world to model-free function estimation with neural networks. We can derive the backpropagation algorithm with a few iterated applications of the chain rule of differential calculus and embed it in the stochastic-approximation framework. First we examine the algorithm's history and relationship to neural-network representations of functions.

History of the Backpropagation Algorithm

Rumelhart et al. [1986] popularized the backpropagation algorithm in the *Parallel Distributed Processing* (PDP) edited volumes in the late 1980s. The PDP volumes summarized the neural research of several psychologists and computer scientists at the University of California at San Diego. The PDP volumes helped secure for neural networks a broad interdisciplinary audience.

The PDP researchers suggested that the backpropagation algorithm—or, as they called it, the *generalized delta rule*—overcame the limitations of the perceptron algorithm, limitations that Minsky and Papert [1969] had carefully enumerated. The algorithm seemed to reopen the research door that Minsky and Papert had earlier shut. The algorithm was a triumph of "connectionism," a term many computer and cognitive scientists use to refer to neural-network theory and applications.

About the same time the PDP volumes appeared, Sejnowski used the backpropagation algorithm to successfully synthesize speech from text without expert-system rules in his NETalk simulation [Anderson, 1988]. A tape recorder replayed NETalk's training experience, from babble to baby talk to articulate speech, and galvanized audiences and the press.

The backpropagation algorithm quickly dominated the neural-network literature. Graduate students turned entrepreneurs and sold inexpensive backpropagation software. Others quietly tried to forecast stock prices. Hundreds of academic, industrial, and government researchers reported the results of backpropagation simulations at technical conferences. Venture capitalists and large companies funded the development of backpropagation accelerator boards for personal computers and computer workstations. The technical and popular press heralded backpropagation as the long-awaited breakthrough in machine intelligence, a new learning method that "learned from experience" and promised manufacturing automation, real-time speech recognition and between-language translation, self-repairing robots, and the like. Physiologists looked for the backpropagation algorithm in the human brain and central nervous system. Theorists conjectured that backpropagation could learn any sampled function $f: X \to Y$.

Backpropagation's popularity begot waves of criticism. The first criticisms were algorithmic. The backpropagation algorithm learned with nonlocal information, often failed to converge, and at best converged to local error minima.

Backpropagation modified synaptic connection strengths with nonlocal error information. Nonlocality, synchrony, supervision, and lengthy training cycles precluded biological plausibility.

The backpropagation algorithm did not always converge in discrete simulations. Some choices of initial conditions led to oscillations, even chaotic wandering. Some backpropagation defenders argued that the algorithm had to converge, since it implemented a gradient descent on the error surface in the synaptic "weight space." The state-space ball simply had to roll down the error surface to the nearest error minimum and stop, as a Hopfield or BAM trajectory ended in a Lyapunov local minimum. The backpropagation algorithm would converge, they argued, if we made the simulation discretization small enough.

These arguments overlooked the algorithm's stochastic nature. They confused realizations of random squared-error surfaces with the unknown deterministic mean-squared-error surface. Researchers interpreted stochastic differential and difference equations as ordinary differential and difference equations, even though the PDP presentation arrived at the backpropagation algorithm by starting with Widrow's stochastic LMS algorithm and cited the LMS algorithm as a linear special case of backpropagation. Learning coefficients were assumed constant. At root many researchers, apparently unfamiliar with estimation and approximation theory, believed that "backpropagation is not statistics," a belief that University of California San Diego economist Halbert White [1989] later proved false.

The backpropagation algorithm converged to a local error minimum, if it converged at all. Backpropagation's defenders admitted this, but some argued that, given the large dimension of the synaptic weight space, local minima were "rare." Several variables had to "conspire" to produce a local minimum. Others argued, incorrectly, that all error minima had the same depth. So local minima were global minima. This empirical speculation pointed to successful (small-dimensional) backpropagation simulations. These defenses ignored the vast mathematical and applications literature on gradient-descent methods. The arguments proved too much. If generic local-minima arguments held for backpropagation, they held in general for the many nonlinear problems of large-scale multivariable optimization, problems that ranged from factory and transportation scheduling to economic and weapon allocation. Indeed the higher the dimension, the "rarer" should be local minima. Gradient descent should work best where in fact it works worst.

The next wave of criticism challenged backpropagation's historical priority. Many researchers claimed that they had first derived the backpropagation algorithm. The publication record quickly showed that Parker [1982] had derived the algorithm as "learning logic" in 1981 and that Werbos [1974] had derived the algorithm as "dynamic feedback" in his 1974 Harvard University dissertation in applied mathematics. Werbos statistically formulated backpropagation and went on to apply the algorithm in the 1970s and 1980s to economic forecasting and other problems. Earlier derivations may exist.

The next wave of criticism challenged whether backpropagation learning was new. The question was not who first derived the algorithm, but whether the algorithm differed from existing techniques. White [1989] reduced backpropagation to the stochastic approximation of the 1950s and so gave a startling answer to the what-is-new question: nothing. The algorithm did not offer a new *kind* of learning. Instead it offered only a new and computationally efficient way to implement an estimation method that statisticians had long explored.

Indeed the 1960s learning-theory literature contains many careful analyses of supervised learning with stochastic approximation. Mendel and Fu [1970] and Tsypkin [1974] present excellent summaries. These early learning theorists focused on mathematics, not on computation. Their approach was more analytical than empirical. Their mathematics tended to be rigorous and real-analytical, in the spirit of adaptive and optimal control theory. Perhaps if these researchers possessed the computing power of a later generation, they would have been less concerned about convergence rates, asymptotic optimality, and local minima and more excited about the observed behavior of simpler learning systems. Perhaps they would have discovered and popularized the backpropagation algorithm or other, computationally richer, algorithms. As the study of chaotic dynamical systems shows, even small-dimensional nonlinear systems can behave in surprisingly complex ways. Often computer exploration first reveals this behavior. In some cases system nonlinearities prevent an analytical characterization, and we can seek answers only with computer simulations.

The backpropagation algorithm, as White observed, also focused attention on certain feedforward nonlinear representations of functions. Sigmoidal "squashings" of activations—bounded nonlinear signal functions—define the nonlinearities. Compared to the theory of bounded linear functionals [Rudin, 1974], we know little about bounded nonlinear functionals.

Feedforward Sigmoidal Representation Theorems

Hornik and White [1989] showed that feedforward sigmoidal architectures can in principle represent any Borel-measurable function to any desired accuracy—*if* the network contains enough "hidden" neurons between the input and output neuronal fields. We do not know how accurate an approximation we get with a fixed number of neurons. Uniform convergence ensures only that we can represent a function with finite neurons. (For all $\varepsilon > 0$, there exists *some* integer m_0 such that for all integers $m > m_0$ and all $\mathbf{x}$: $||N_m(\mathbf{x}) - f(\mathbf{x})|| < \varepsilon$ for feedforward sigmoidal neural network $N_m: R^n \to R^p$.) For many functions the number of neurons we need may be physically unrealizable.

The neural representation theorems resemble the Weierstrass approximation theorem. Every continuous function on a finite (compact) interval is the limit of a uniformly convergent sequence of polynomials. Finite polynomials span the space of continuous functions on a closed and bounded set. After all, in principle we can estimate any function with high-order polynomial regression. Polynomial estimation is model-free if feedforward-sigmoidal estimation is model-free. In regression, sample data modifies polynomial coefficients. In networks, sample data modifies inner-product or synaptic coefficients.

Most neural-network practitioners already believed that neural networks could represent any sampled function. This faith underlay their search for more powerful learning algorithms.

In this sense the representation theorems resemble the minimax theorem of game theory, which ensures that an optimal strategy exists in two-person noncooperative games with total information. We have found such optimal strategies in tic-tac-toe and checkers. In chess we know that either white or black can always win or draw. But we still search the chess game tree, and we will continue to search it for a very long time.

Some neural enthusiasts misinterpreted the representation theorems. They claimed that the theorems proved that backpropagation could learn any (Borel-measurable) function. They confused algorithm with architecture. Indeed the neural literature still refers to "backpropagation networks," which confuses a specific nonlinear architecture with one way to train its synapses. The backpropagation algorithm remains a computationally practical algorithm for performing estimated gradient descent, but only one of possibly infinitely many gradient-descent algorithms.

We must be careful how we interpret the feedforward-sigmoidal representation theorems. We might be tempted to say that, among model-free estimators, neural networks, and *only* neural networks, can represent any function. In fact we can also say that AI expert systems and fuzzy systems can represent any continuous function to any desired level of accuracy. The representation theorems benefit AI expert systems and fuzzy systems as much as they benefit neural networks.

Expert systems appear purely symbolic, incapable of function estimation. Indeed AI researchers seldom speak in terms of function estimation when they discuss expert systems. Nevertheless, we can interpret expert systems as feedforward networks. Expert systems sum weighted step functions. They estimate functions that map an input "problem-domain" space to an output action space, questions to answers.

Binary propositional rules define step functions. The rule antecedents or consequents fire, or are true, over part of the input domain. They do not fire, or are false, over the rest of the input domain. Numerical uncertainty factors may weight the rules. The propositional rules may even occur in concatenated "chains": If A, then B; if B, then C and D; if C, then .... The embedded propositional rules correspond to "hidden" threshold neurons in a feedforward neural network. The summed-weighted-step-function architecture, combined with standard results [Rudin, 1974] on estimating Borel-measurable functions on compact sets with "simple" functions, ensures that some finite AI expert system approximates the sampled continuous function to any specified level of accuracy. Of course this may require thousands, or billions, of rules.

The overall expert system behaves as a model-free estimator of a sampled but unknown continuous function $f: X \rightarrow Y$. Knowledge engineers do not assume the functional form of f. They do not, for instance, assume f is linear or a kth-order polynomial. They work with a problem-domain expert to abstract propositional rules from observed condition-action samples $(\mathbf{x}, \mathbf{y})$, where presumably $\mathbf{y} = f(\mathbf{x})$.

In practice AI expert systems estimate unknown, highly nonlinear functions with a small number of propositional rules. Often knowledge engineers do not weight the rules. A few summed step functions poorly estimate a wavy function. But in principle enough summed step functions can accurately approximate any continuous or measurable function.

AI researchers know this, at least implicitly. They refer to such large rule sets as a "combinatorial explosion," since branching rules grow exponentially. When feedforward-sigmoidal neural networks finely approximate complicated functions, the networks may suffer a like "explosion" of hidden neurons and interconnecting synapses.

Fuzzy systems, discussed in Chapters 8 through 11, sum the weighted consequents of set-level "rules" or associations. Output "fuzzy" (multivalued) set vectors

B_i' define consequents as the result of fuzzy vector-matrix composition: $A \circ M_i = B_i'$. The fuzzy matrix, or relation, M_i houses the fuzzy-set association or "rule" (A_i, B_i).

Fuzzy systems "reason" or inference with multivalued sets instead of bivalent propositions, vectors instead of scalars. Thus they generalize AI expert systems. If a fuzzy system sums enough function samples $(\mathbf{x}_i, \mathbf{y}_i)$ as associative rules (A_i, B_i), and if the system uses impulse functions instead of fuzzy sets as inputs (BIO-FAM case, see Chapter 8), then the same sums-of-weighted-simple-functions results [Rudin, 1974] ensure that the fuzzy system can represent any sampled continuous function to any level of accuracy.

Function representability may be necessary for interesting model-free estimation. It is not sufficient. We want to know *how* to efficiently represent a partially sampled function, not just *whether* we can in principle represent it. Function representability becomes sufficient when the representation scheme depends on sample data but the representation architecture does not—when a given network can learn any sampled function.

Multilayer Feedforward Network Architectures

Neural networks with multiple neuronal fields can often outestimate neural networks with only two neuronal fields. Minsky and Papert [1969] proved this for two-layer perceptron threshold networks for some problems. Hornik and White confirmed this with their recent representation theorems.

Consider the *parity function* p: $\{0, 1\}^n \to \{0, 1\}$. The parity function indicates whether a binary n-vector B in $\{0, 1\}^n$ contains an even $(p(B) = 0)$ or odd $(p(B) = 1)$ number of 1s. If we define the parity function on the Boolean 3-cube $\{0, 1\}^3$, we can write the parity function as 2^3 associations:

	B		$p(B)$
0	0	0	0
0	0	1	1
0	1	0	1
0	1	1	0
1	0	0	1
1	0	1	0
1	1	0	0
1	1	1	1

The *exclusive-or* logic function equals the 2-bit parity function on $\{0, 1\}^2$.

Parity changes if any one bit changes. For high dimensions n, the parity function behaves increasingly discontinuously. Small 1-bit changes in output produce total changes in output. Such discontinuity confuses naively encoded associative memories. But as Rumelhart [1986] observes, in principle we can overcome this problem with associative memories if we add extra binary dimensions to the input, and perhaps the output.

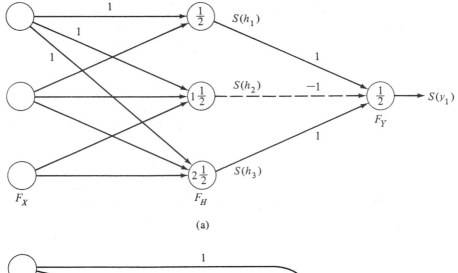

(a)

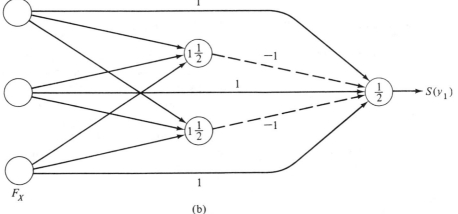

(b)

FIGURE 5.1 (a) Three-bit parity function represented with three "hidden" neurons. (b) Three-bit parity function represented with two hidden neurons. Input F_X neurons are linear. F_H and F_Y neurons have threshold signal functions. Node numbers indicate threshold values. All synaptic paths from F_X have synaptic value $+1$.

Consider the three-layer networks in Figure 5.1. Each network represents the 3-bit parity problem. As in most feedforward networks, the input F_X neurons are linear: $S_i(x_i) = x_i$. The "hidden" neurons in F_H and the output F_Y neurons are threshold neurons. Node numbers indicate threshold values. For instance, $S(y_1) = 1$ if additive activation $y_1 > \frac{1}{2}$, and $S(y_1) = 0$ if $y_1 < \frac{1}{2}$.

How can a multilayer neural network learn? Error correction seems difficult because we do not know how the signal outputs of hidden neurons should behave. We cannot directly compute a hidden-neuron error. We know only the random sample $(\mathbf{x}_i, \mathbf{y}_i)$ and the vector error $\mathbf{y}_i - \mathbf{N}(\mathbf{x}_i)$.

Gradient descent offers a solution. We wish to minimize the total mean-squared error MSE. In practice we do not know the joint density $p(\mathbf{x}, \mathbf{y})$. At best we can minimize the instantaneous squared error $SE(k)$ of sampled associations $(\mathbf{x}_k, \mathbf{y}_k)$. Ideally we should increment or decrement the current synaptic value $m_{ij}(k)$ by the negative of its MSE partial derivative:

$$\Delta m_{ij}(k) \;=\; -c_k \frac{\partial \text{MSE}}{\partial m_{ij}} \tag{5-57}$$

for some learning coefficient $c_k > 0$. Here synaptic value m_{ij} may "connect" an input neuron and a hidden neuron, two hidden neurons, or a hidden and output neuron. In practice we must settle for the *stochastic* gradient estimate

$$\Delta m_{ij}(k) \;=\; -c_k \frac{\partial \text{SE}(k)}{\partial m_{ij}} \tag{5-58}$$

Minsky and Papert [1969] referred to this computational problem as the *credit-assignment problem*.

We can approximate the marginal changes in (5-57) and (5-58) with numerical-analytical techniques. Or we can simply jiggle each synapse about its current value and keep the value that most decreases the squared error. Even if this works, the best we can hope for is a synaptic combination that locally minimizes the unknown MSE function. In many cases this local minimum may be unacceptable. We do not know in advance whether the local minimum will behave well or misbehave. We do not even know whether the apparent local minimum is a local minimum of the MSE function or of just the current SE function. The local minimum may change when we use additional training samples drawn from the sampled process.

The backpropagation algorithm provides a comparatively efficient way to compute the instantaneous partial derivatives in (5-58). The algorithm recursively modifies the synapses between neuronal fields. The algorithm first modifies the synapses between the output field F_Y and the penultimate field F_{H_p} of "hidden" or interior neurons. The algorithm then uses this information to modify the synapses between the hidden fields F_{H_p} and $F_{H_{p-1}}$, then between $F_{H_{p-1}}$ and $F_{H_{p-2}}$, and so on all the way back to the synapses between the first hidden field F_{H_1} and the input field F_X. In practice feedforward multilayer networks usually contain no more than one hidden field, occasionally two. We know no quantitative criteria, or even rules of thumb, to guide our choice of the number of hidden fields and the number of neurons within a hidden field.

Backpropagation Algorithm and Derivation

We present the backpropagation algorithm for the default cases of one and three hidden fields. We can immediately apply the recursive procedure to any number of hidden fields. The presentation follows the PDP presentation [Rumelhart, 1986] but emphasizes the stochastic nature of the learning process.

We update the synaptic connection strength $m_{ij}(k)$ at iteration time k with the first-order stochastic difference equation

$$m_{ij}(k+1) \quad = \quad m_{ij}(k) + c_k \Delta m_{ij}(k) \tag{5-59}$$

The ith and jth neurons reside in contiguous neuronal fields. The learning-coefficient sequence $\{c_k\}$ should decrease in accord with (5-13) and (5-14) to ensure an accurate stochastic approximation of an unknown local MSE minimum.

More generally we can update $m_{ij}(k)$ with the second-order linear stochastic difference equation

$$m_{ij}(k+1) \quad = \quad m_{ij}(k) + c_k \Delta m_{ij}(k) + b_k \Delta m_{ij}(k-1) \tag{5-60}$$

where $\{b_k\}$ should also decrease according to (5-13) and (5-14) to help suppress random disturbances. This second-order difference equation defines a simple type of *conjugate* gradient descent. Rumelhart calls the acceleration term $b_k \Delta m_{ij}(k-1)$ a "momentum" term, presumably because it resembles the mass-velocity product of physical momentum. The acceleration term helps smooth the local curvature between successive squared-error surfaces. Perhaps for this reason, the acceleration term tends to improve simulation results. It adds little computational burden to the first-order update equation (5-59).

We randomly sample the unknown Borel-measurable function $f\colon R^n \to R^p$. This generates a countably infinite sequence of sample associations $(\mathbf{x}_1, \mathbf{d}_1)$, $(\mathbf{x}_2, \mathbf{d}_2)$, ... We desire vector output $\mathbf{d}$ given input vector $\mathbf{x}$: $\mathbf{d} = f(\mathbf{x})$. Realization $\mathbf{x}_i$ is a point $(x_1^i, \ldots, x_n^i)$ in R^n. Desired realization $\mathbf{d}_j$ is a point in R^p. We sample vector association $(\mathbf{x}_k, \mathbf{d}_k)$ at time k. In practice we may have only m training samples or associations. Then we can sample with replacement from this set or simply repeatedly cycle through it.

The current neural network $\mathbf{N}_k$ defines the vector function $\mathbf{N}_k\colon R^n \to R^p$. For simplicity we shall drop the index on $\mathbf{N}_k$ and denote the network as $\mathbf{N}$, even though its synaptic structure changes from iteration to iteration. The actual network output $\mathbf{N}(\mathbf{x})$ corresponds to the output signal vector $\mathbf{S}(\mathbf{y})$:

$$\mathbf{N}(\mathbf{x}) \quad = \quad \mathbf{S}(\mathbf{y}) \tag{5-61}$$

$$= \quad (S_1(y_1), \ldots, S_p(y_p)) \tag{5-62}$$

where we abbreviate the bounded monotone-nondecreasing signal function S_j^Y as S_j. In fact the backpropagation algorithm requires only that signal functions be differentiable with respect to activations. So, for instance, we can use Gaussian or Cauchy signal functions.

The current random sample $(\mathbf{x}_k, \mathbf{d}_k)$ defines the *instantaneous* squared-error vector $\mathbf{e}_k$:

$$\mathbf{e}_k \quad = \quad \mathbf{d}_k - \mathbf{N}(\mathbf{x}_k) \tag{5-63}$$

$$= \quad \mathbf{d}_k - \mathbf{S}(\mathbf{y}_k) \tag{5-64}$$

$$= \quad (d_1^k - S_1(y_1^k), \ldots, d_p^k - S_k(y_p^k)) \tag{5-65}$$

which directly generalizes the LMS linear error scalar (5-24). The backpropagation algorithm uses the instantaneous summed squared error E_k:

$$E_k = \frac{1}{2} \sum_j^p [d_j^k - S_j(y_j^k)]^2 = \frac{1}{2} \mathbf{e}_k \mathbf{e}_k^T \tag{5-66}$$

The total or **cumulative error** E,

$$E = \sum_k E_k \tag{5-67}$$

sums the first k instantaneous errors at iteration k. In general,

$$\frac{\partial E}{\partial m_{ij}} = \sum_k \frac{\partial E_k}{\partial m_{ij}} \tag{5-68}$$

This linear relationship, combined with the linear relationship (5-59), helps justify estimating the gradient of E_k at iteration k instead of estimating the gradient of the cumulative error up to k.

The backpropagation algorithm updates neuronal activations with the simple additive model discussed in Chapter 3:

$$y_j^k = \sum_q S_q(h_q^k) m_{qj} \tag{5-69}$$

if the jth neuron belongs to the output neuronal field F_Y, and

$$h_q^k = \sum_{i=1} x_i^k m_{iq} \tag{5-70}$$

if the qth neuron belongs to the single hidden neuronal field F_H in the feedforward topology $F_X \rightarrow F_H \rightarrow F_Y$. Recall that $S_i(x_i) = x_i$ for input neurons in feedforward networks. If F_G denotes a second, and prior, hidden neuronal field in the feedforward topology $F_X \rightarrow F_G \rightarrow F_H \rightarrow F_Y$, then we replace (5-70) with

$$h_q^k = \sum_r S_r(g_r^k) m_{rq} \tag{5-71}$$

and g_r^k replaces h_q^k in (5-70).

We derive the backpropagation algorithm with repeated application of the chain rule of multivariable differential calculus. First suppose f is a sufficiently smooth function of variables $x_1, \ldots, x_n$ and $y_1, \ldots, y_p$. We denote this as $f = f(x_1, \ldots, x_n; y_1, \ldots, y_p)$. Suppose x_i and y_j are sufficiently smooth functions of t with appropriate domains and ranges. Then the *total derivative* df/dt takes the form (5-71)

$$\frac{df}{dt} = \sum_i^n \frac{\partial f}{\partial x_i} \frac{dx_i}{dt} + \sum_j^p \frac{\partial f}{\partial y_j} \frac{dy_j}{dt} \tag{5-72}$$

Now suppose $f = f(x_1, \ldots, x_n)$ and $x_i = x_i(y_1, \ldots, y_p)$. Then the *partial derivative* $\partial f / \partial y_j$ takes the form

$$\frac{\partial f}{\partial y_j} \; = \; \sum_i^n \frac{\partial f}{\partial x_i} \frac{\partial x_i}{\partial y_j} \tag{5-73}$$

Chain rules (5-72) and (5-73) do not hold in general in the stochastic calculus. For then equalities denote stochastic limits of convergent sequences of random variables.

The backpropagation algorithm uses the stochastic calculus because its variables are random variables or random vectors. But the algorithm uses only simple linear stochastic difference equations without nonlinearly scaled noise (Brownian-motion) terms. So [Maybeck, 1982] we can still use the above chain rules of differential calculus, at least as a very close approximation to how the random variables change.

Errors and activations depend on signal functions. Applying the chain rule to (5-66) gives

$$\frac{\partial E_k}{\partial S_j} \; = \; -[d_j^k - S_j(y_j^k)] \frac{\partial S_j}{\partial S_j} \tag{5-74}$$

$$= \; -[d_j^k - S_j(y_j^k)] \tag{5-75}$$

the negative of the instantaneous error term e_j^k in (5-65). Applying the chain rule to (5-69) and (5-70) gives

$$\frac{\partial y_j^k}{\partial S_q} \; = \; m_{qj} \tag{5-76}$$

$$\frac{\partial h_q^k}{\partial S_i} \; = \; m_{iq} \tag{5-77}$$

where again we use the linearity condition, $S_i(x_i) = x_i$, for input neurons in feed-forward networks. Similarly, applying the chain rule (5-73) to (5-71) gives

$$\frac{\partial h_q^k}{\partial S_r} \; = \; m_{rq} \tag{5-78}$$

Equations (5-69) through (5-71) also imply

$$\frac{\partial y_j^k}{\partial m_{qj}} \; = \; S_q(h_q^k) \tag{5-79}$$

$$\frac{\partial h_q^k}{\partial m_{iq}} \; = \; x_i^k \tag{5-80}$$

$$\frac{\partial h_q^k}{\partial m_{rq}} \; = \; S_r(g_r^k) \tag{5-81}$$

Now we can formally state and derive the backpropagation algorithm. We present two versions. In the first version, the network contains the single hidden neuronal field F_H in the feedforward topology $F_X \rightarrow F_H \rightarrow F_Y$. Again, the neurons in the input neuronal field F_X are linear: $S_i(x_i) = x_i$. This slightly simplifies the algorithm. In the second version, the network contains three hidden neuronal fields in the feedforward topology $F_X \rightarrow F_F \rightarrow F_G \rightarrow F_H \rightarrow F_Y$. This allows direct extension to any number of hidden fields.

The backpropagation algorithm propagates the instantaneous squared error, $E_k = \frac{1}{2}e_k^T e_k$, backward from F_Y through the hidden fields to F_X at each iteration. Random input data $\mathbf{x}_k$ generates the instantaneous squared error. $\mathbf{x}_k$ passes forward from F_X to F_H. Then F_H signals pass to F_Y. Then we subtract the *actual* F_Y output-signal vector $\mathbf{S}(\mathbf{y}_k)$ from the *desired* F_Y vector $\mathbf{d}_k$ to compute the error vector $\mathbf{e}_k$. We repeat the process until we have used all (countably infinite) training associations $(\mathbf{x}_k, \mathbf{d}_k)$ and, ideally, we have reached a local minimum of the unknown mean-squared error surface.

Backpropagation Algorithm

A. If the jth neuron belongs to output field F_Y, then

$$\Delta m_{qj}(k) = [d_j^k - S_j(y_j^k)]S_j'(y_j^k)S_q(h_q^k) \tag{5-82}$$

where $S_j' = dS_j/dy_j$.

B. If the qth neuron belongs to the single hidden neuronal field F_H in the feedforward topology $F_X \rightarrow F_H \rightarrow F_Y$, then

$$\Delta m_{iq}(k) = -[\sum_j^p \frac{\partial E_k}{\partial y_j^k} m_{qj}]S_q'(h_q^k)x_i^k \tag{5-83}$$

$$= [\sum_j^p [d_j^k - S_j(y_j^k)]S_j'(y_j^k)m_{qj}]S_q'(h_q^k)x_i \tag{5-84}$$

C. If the rth neuron belongs to the middle hidden field F_G in the feedforward topology $F_X \rightarrow F_F \rightarrow F_G \rightarrow F_H \rightarrow F_Y$, then

$$\Delta m_{sr}(k) = -[\sum_q \frac{\partial E_k}{\partial h_q^k} m_{rq}]S_r'(g_r^k)S_s(f_s^k) \tag{5-85}$$

In this case we recursively compute the "generalized delta" terms $-\partial E_k/\partial h_q^k$ from (5-82) and (5-83).

Derivation. We derive the backpropagation algorithm with repeated application of the chain rule (5-73).

A. First suppose the jth neuron belongs to the output field F_Y. Then

$$\Delta m_{qj}(k) \;=\; -\frac{\partial E_k}{\partial m_{qj}} \tag{5-86}$$

$$=\; -\frac{\partial E_k}{\partial y_j^k}\frac{\partial y_j^k}{\partial m_{pj}} \tag{5-87}$$

$$=\; -\frac{\partial E_k}{\partial y_j^k} S_q(h_q^k) \tag{5-88}$$

by (5-79),

$$=\; -\frac{\partial E_k}{\partial S_j(y_j^k)}\frac{\partial S_j(y_j^k)}{\partial y_j^k} S_q(h_q^k) \tag{5-89}$$

$$=\; -\frac{\partial E_k}{\partial S_j(y_j^k)} S_j'(y_j^k) S_q(h_q^k) \tag{5-90}$$

since S_j is a signal function and $S_j' = dS_j/dy_j$,

$$=\; [d_j^k - S_j(y_j^k)] S_j'(y_j^k) S_q(h_q^k) \tag{5-91}$$

by (5-75), which establishes (5-82).

B. Next suppose the qth neuron belongs to the single hidden field F_H in the feedforward topology $F_X \to F_H \to F_Y$. So all F_X neurons have linear signal functions: $S_i(x_i) = x_i$. Then

$$\Delta m_{iq}(k) \;=\; -\frac{\partial E_k}{\partial m_{iq}} \tag{5-92}$$

$$=\; -\frac{\partial E_k}{\partial h_q^k}\frac{\partial h_q^k}{\partial m_{iq}} \tag{5-93}$$

$$=\; -\frac{\partial E_k}{\partial h_q^k} x_i^k \tag{5-94}$$

by (5-80),

$$=\; -\frac{\partial E_k}{\partial S_q}\frac{\partial S_q(h_q^k)}{\partial h_q^k} x_i^k \tag{5-95}$$

$$=\; -\frac{\partial E_k}{\partial S_q} S_q'(h_q^k) x_i^k \tag{5-96}$$

$$=\; -[\sum_j^p \frac{\partial E_k}{\partial y_j^k}\frac{\partial y_j^k}{\partial S_q}] S_q'(h_q^k) x_i^k \tag{5-97}$$

$$=\; -[\sum_j^p \frac{\partial E_k}{\partial y_j^k} m_{qj}] S_q'(h_q^k) x_i^k \tag{5-98}$$

by (5-76),

$$= \quad [\sum_{j}^{p}[d_j^k - S_j(y_j^k)]S_j'(y_j^k)m_{qj}]S_q(h_q^k)x_i^k \qquad (5\text{-}99)$$

by (5-88) through (5-91), which establishes (5-84).

C. Finally, suppose the rth neuron belongs to the middle hidden field F_G in the feedforward topology $F_X \to F_F \to F_G \to F_H \to F_Y$. Suppose we have already used the backpropagation algorithm to modify the synapses between F_Y and F_H and between F_H and F_G. So, from (5-94) through (5-99), we know the values of the "generalized delta" terms $\partial E_k/\partial h_q^k$. Then we modify the synapses from F_F to F_G as follows:

$$\Delta m_{sr}(k) \quad = \quad -\frac{\partial E_k}{\partial m_{sr}} \qquad (5\text{-}100)$$

$$= \quad -\frac{\partial E_k}{\partial g_r^k}\frac{\partial g_r^k}{\partial m_{sr}} \qquad (5\text{-}101)$$

$$= \quad -\frac{\partial E_k}{\partial g_r^k}S_s(f_s^k) \qquad (5\text{-}102)$$

since $g_r^k = \sum_{s} S_s(f_s^k)m_{sr}$,

$$= \quad -\frac{\partial E_k}{\partial S_r}\frac{\partial S_r(g_r^k)}{\partial g_r^k}S_s(f_s^k) \qquad (5\text{-}103)$$

$$= \quad -\frac{\partial E_k}{\partial S_r}S_r'(g_r^k)S_s(f_s^k) \qquad (5\text{-}104)$$

$$= \quad -[\sum_{q}\frac{\partial E_k}{\partial h_q^k}\frac{\partial h_q^k}{\partial S_r}]S_r'(g_r^k)S_s(f_s^k) \qquad (5\text{-}105)$$

$$= \quad -[\sum_{q}\frac{\partial E_k}{\partial h_q^k}m_{rq}]S_r'(g_r^k)S_s(f_s^k) \qquad (5\text{-}106)$$

by (5-78), which establishes (5-85) and completes the derivation of the backpropagation algorithm. **Q.E.D.**

The derivation may have seemed to use more indices than necessary. For instance, (5-106) contains the partial derivative $\partial E_k/\partial h_q^k$ instead of the simpler $\partial E_k/\partial h_q$. In the differential calculus h_q^k denotes a scalar at time k, not a "variable." But in the stochastic calculus h_q^k, like E_k, is a random variable at time k. $\{h_q^k\}$ defines a discrete stochastic process, a sequence of random variables (Borel-measurable functions) indexed by k. In this setting the partial derivative $\partial E_k/\partial h_q$ is ambiguous without the index k. It refers to the process but not the particular random variable.

The derivation did not require that neurons process activations with monotone-nondecreasing signal functions. We assumed only that signal functions were differentiable. In practice researchers often use logistic signal functions; then $S' = S(1 - S)$. We can also use Gaussian or other signal functions.

The backpropagation algorithm reduces to the LMS algorithm if all neurons are linear and if the feedforward topology $F_X \to F_Y$ contains no hidden neurons. For then (5-82) reduces to

$$\Delta m_{ij}(k) = [d_j^k - y_j^k] \frac{dy_j^k}{dy_j^k} x_i^k \qquad (5\text{-}107)$$

$$= e_j^k x_i^k \qquad (5\text{-}108)$$

Substituting (5-108) into the stochastic difference equation (5-59) gives

$$m_{ij}(k+1) = m_{ij}(k) + c_k e_j^k x_i^k \qquad (5\text{-}109)$$

which defines a multineuron version of the LMS algorithm (5-49) for appropriate constant learning coefficients c_k.

Many backpropagation variations exist. Williams and Zipser [1989] have extended the algorithm to allow synchronous feedback in discrete time. They use the algorithm to learn some types of sequenced data but observe that "for learning, each weight must have access to both the complete recurrent weight matrix $\mathbf{W}$ and the whole error vector $\mathbf{e}$." Almeida [1987] and Pineda [1989] have found, by linearizing around a fixed point, sufficient conditions for stability for feedback networks with additive activation dynamics and backpropagation-like synaptic dynamics.

Backpropagation as Stochastic Approximation

Recently White [1989] showed that the backpropagation algorithm reduces to a special case of stochastic approximation. We reproduce White's argument for a neural network with a single output F_Y neuron. The network can have a feedforward or feedback topology.

In general stochastic approximation [Robbins, 1951] seeks a solution M^* to the equation

$$E[\phi(M, \mathbf{x}, d)] = 0 \qquad (5\text{-}110)$$

for some arbitrary function ϕ of the random vector $\mathbf{z} = [\mathbf{x} \,|\, d]$.

We do not know the characterizing probability density function $p(\mathbf{z})$. Else we could numerically solve (5-110). We know only the observed random realizations $\mathbf{z}_1, \mathbf{z}_2, \ldots$. Equivalently we know only the observed realizations $(\mathbf{x}_1, d_1), (\mathbf{x}_2, d_2), \ldots$ of the unknown Borel-measurable function $f: R^n \to R$. Stochastic approximation seeks M^* iteratively with the iterative equation

$$M_{k+1} = M_k + c_k \phi(M, \mathbf{z}_k) \qquad (5\text{-}111)$$

If the learning-coefficient sequence $\{c_k\}$ obeys the constraints (5-13) and (5-14), then the parameter sequence $\{M_k\}$ converges with probability one (or in the mean-squared sense) to a locally optimal solution M^* of (5-110).

We can interpret M_k as a high-dimensional vector of all network synaptic values at time k. Then $S(y_k) = f(\mathbf{x}_k, M)$. y_k describes the neuronal activation at time k of the sole neuron in F_Y.

White observed that the identifications

$$\phi(M, \mathbf{z}_k) = [d_k - f(\mathbf{x}_k, M)]\nabla_{\mathbf{M}}f(\mathbf{x}_k, M) \tag{5-112}$$

$$= [d_k - S(y_k)]\nabla_{\mathbf{M}}S(y_k) \tag{5-113}$$

correspond to Equation (5-89) because the term

$$\frac{\partial S(y_k)}{\partial y_k}\frac{\partial y_k}{\partial m_p}$$

defines a term in the gradient $\nabla_{\mathbf{M}}S(y_k)$ and because (5-75) holds. Equations (5-112) and (5-113) are global relationships. They include all synapses. So they include the synaptic connections that arise from hidden neurons in feedforward and feedback topologies.

Robust Backpropagation

White also observed that, since backpropagation embeds in statistics, we can apply the techniques of **robust statistics** [Huber, 1981] to backpropagation function estimation. Robustness refers to insensitivity to small perturbations in the underlying probability distribution. In the statistics literature robustness often means allowable deviation from an assumed Gaussian, Cauchy, or other probability density function $p(\mathbf{x})$.

In practice robust statistical estimators ignore sample "outliers," infrequent observations that differ significantly from the "pattern" of the remaining observations. Robustness resembles the dynamical *structural stability* discussed in Chapter 6.

Mean-squared measures tend to overweight statistical outliers. In contrast, summed absolute error,

$$\sum_j |e_j^k| = \sum_j |d_j^k - S_j(y_j^k)| \tag{5-114}$$

tends to better suppress extreme observations or random input-output samples $(\mathbf{x}_k, \mathbf{d}_k)$. We do not know in advance whether, or to what degree, training samples will resemble statistical outliers. Occasional outliers can skew the estimated gradient descent of nonrobust backpropagation, slowing convergence or leading the system state toward less representative local minima.

Robust backpropagation replaces the scalar error e_k in (5-113) with an **error suppressor** function $s(e_k)$. The linear error suppressor, $s(e_k) = e_k$, is not robust. If the underlying probability density function resembles a logistic density, then

$$s(e_k) = \tanh(\frac{e_k}{2}) \tag{5-115}$$

provides a robust error suppressor [Huber, 1981]. If the underlying probability density function is symmetric but has thick tails, for instance if it is a Cauchy or other infinite-variance density, then

$$s(e_k) = \frac{2e_k}{1 + e_k^2} \tag{5-116}$$

provides a robust error suppressor. Then $s(e_k)$ approaches zero for large-magnitude errors and approximates e_k for small errors. In some Gaussian-like or "least informative" cases the threshold-linear function,

$$s(e_k) = \max[-c, \min(c, x)] \tag{5-117}$$

provides a robust error suppressor. In practice we can choose many neuronal signal functions to approximate a robust error suppressor.

Robust backpropagation takes the general form

$$\phi(M, \mathbf{z}_k) = s(e_k)\nabla_M S(y_k) \tag{5-118}$$

for an appropriate robust error-suppressor function s. In particular, robust Cauchy backpropagation replaces (5-82) with

$$\Delta m_{qj}(k) = \frac{2e_j^k}{1 + (e_j^k)^2} S_j'(y_j^k)S_q(h_q^k) \tag{5-119}$$

and replaces (5-84) with

$$\Delta m_{iq}(k) = [\sum_j^p \frac{2e_j^k}{1 + (e_j^k)^2} S_j'(y_j^k)m_{qj}]S_q'(h_q^k)x_i \tag{5-120}$$

By assumption we do not know the form of the underlying probability density function $p(\mathbf{x})$. So ultimately we must appeal to trial and error, or to an informed guess, to choose an appropriate robust error-suppressor function.

Other Supervised Learning Algorithms

Most known learning algorithms are supervised. The perceptron, LMS, and backpropagation algorithms represent prototypical supervised gradient-descent learning algorithms. There are many more that we have not discussed.

Barto [1985] has used stochastic approximation to combine supervised probabilistic learning with stochastic learning automata in his associative-reward-penalty algorithm. Fukushima [1983] has developed the neocognitron network, a complex multifield system modeled after the visual system, for recognizing distorted input (retinal) patterns with supervised competitive-type learning. Fukushima [1987] has extended the neocognitron to allow synchronous feedback. Inner products of signals propagate forward to recognition or "grandmother" cells, and convolutions propagate backward to provide a dynamical selective-attention mechanism. Nobel-laureate physicist Leon Cooper has built the publicly traded company Nestor around an intuitive, and patented (U.S. patent number 4,326,259), supervised learning algorithm. This geometric algorithm centers pattern spheres around training samples and shrinks sphere radii when mismatching spheres overlap. Reilly [1987] provides a high-level description of the Nestor learning system.

Ideal supervised learning algorithms should quickly learn any sampled function that we can represent with a feedforward-sigmoidal architecture. Multiplicative or other activation dynamics might replace additive neuronal dynamics. Gaussian or other signal functions might replace sigmoidal signal functions. Indeed Gaussian-sum density estimators [Sorenson, 1971] can approximate, in an L^1 sense, any probability density function to any desired level of accuracy. Unfortunately, traditional Gaussian-sum approximators require a bank of "extended" Kalman filters. Each extended Kalman filter estimates a single nonlinear conditional density in the approximating weighted sum. Separate "innovations processes" [Mendel, 1987] determine the weights. The entire process requires extensive computation and round-off accuracy. New supervised learning algorithms may eliminate this computational burden yet retain the approximation accuracy of the Gaussian-sum approach.

Future supervised learning algorithms may not estimate error gradients. Instead they may burrow through, hop over, contract, vibrate, or flatten the unknown mean-squared error surface. Or the algorithms may search estimated error surfaces with irregular combinatorial, topological, or differential-geometric search strategies. Or they may approximate the unknown error surface with shrinking manifolds, stitched-together pattern patches, or locally cooperating search agents. Supervised dynamical systems can marshal tools from any branch of mathematics.

REFERENCES

Albert, A. E., and Gardner, L.A., *Stochastic Approximation and Nonlinear Regression*, M.I.T. Press, Cambridge, MA, 1967.

Almeida, L. B., "A Learning Rule for Asynchronous Perceptrons with Feedback in a Combinatorial Environment," *Proceedings of the First IEEE International Conference on Neural Networks (ICNN-87)*, vol. II, 609-618, June 1987.

Anderson, J. A., and Rosenfeld, E. (eds.), *Neurocomputing: Foundations of Research*, M.I.T. Press, Cambridge, MA, 1988.

Barto, A. C., and Sutton, R. S., "Landmark Learning: An Illustration of Associative Search," *Biological Cybernetics*, vol. 42, 1-8, 1982.

Barto, A. C., and Anandan, P., "Pattern-Recognizing Stochastic Learning Automata," *IEEE Transactions on Systems, Man, and Cybernetics*, vol. SMC-15, no. 3, May-June 1985.

Blum, J. R., "Multidimensional Stochastic Approximation Procedure," *Annals of Mathematical Statistics*, vol. 25, 737-744, 1954.

Chung, K. L., *A Course in Probability Theory*, Academic Press, Orlando, FL, 1974.

Fleming, W. H., and Rishel, R. W., *Deterministic and Stochastic Optimal Control*, Springer-Verlag, New York, 1975.

Franklin, J. N., *Matrix Theory*, Prentice Hall, Englewood Cliffs, NJ, 1968.

Fukushima, K., Miyake, S., and Ito, T., "Neocognition: A Neural Network Model for a Mechanism of Visual Pattern Recognition," *IEEE Transactions on Systems, Man, and Cybernetics*, vol. SMC-13, 826-834, 1983.

Fukushima, K., "Neural-Network Model for Selective Attention in Visual Pattern Recognition and Associative Recall," *Applied Optics*, vol. 26, no. 23, 4985-4992, December 1, 1987.

Grossberg, S., *Studies of Mind and Brain*, Reidel, Boston, 1982.

Hebb, D. O., *The Organization of Behavior*, Wiley, New York, 1949.

Hornik, K., Stinchcombe, M., and White, H., "Multilayer Feedforward Networks Are Universal Approximators," *Neural Networks*, vol. 2, no. 5, 359-366, 1989.

Huber, P. J., *Robust Statistics*, Wiley, New York, 1981.

Kalman, R. E., "A New Approach to Linear Filtering and Prediction Problems," *Transactions ASME J. Basic Eng. Series D*, vol. 82, 35-46, 1960.

Kalman, R. E., and Bucy, R., "New Results in Linear Filtering and Prediction Problems," *Transactions ASME J. Basic Eng. Series D*, vol. 83, 95-108, 1961.

Klopf, H. A., "A Neuronal Model of Classical Conditioning," *Psychobiology*, vol. 16, 85-125, 1988.

Kohonen, T., *Self-Organization and Associative Memory*, 2nd ed., Springer-Verlag, New York, 1988.

Kosko, B., "Stochastic Competitive Learning," *Proceedings of the International Joint Conference On Neural Networks (IJCNN-90)*, vol. II, 215-226, San Diego, June 1990.

Kosko, B., *Neural Networks for Signal Processing*, Prentice Hall, Englewood Cliffs, NJ, 1991.

Maybeck, P. S., *Stochastic Models, Estimation, and Control*, vol. 2, Academic Press, Orlando, FL, 1982.

Mendel, J. M., *Lessons in Digital Estimation Theory*, Prentice Hall, Englewood Cliffs, NJ, 1987.

Mendel, J. M., and Fu, K. S. (eds.), *Adaptive, Learning, and Pattern Recognition Systems: Theory and Applications*, Academic Press, Orlando, FL, 1970.

Minsky, M. L., and Papert, S., *Perceptrons: An Introduction to Computational Geometry*, M.I.T. Press, Cambridge, MA, 1969; 2nd ed., 1988.

Nilsson, N., *Learning Machines*, McGraw-Hill, New York, 1965.

Parker, D. B., "Learning Logic," Invention Report S81-64, File 1, Office of Technology Licensing, Stanford University, Stanford, CA, 1982.

Pavlov, I. P., *Conditioned Reflexes*, Oxford University Press, New York, 1927 (Dover edition, 1960).

Pineda, F. J., "Recurrent Backpropagation and the Dynamical Approach to Adaptive Neural Computation," *Neural Computation*, vol. 1, no. 2, 161-172, Summer 1989.

Reilly, D. L., Scofield, C., Elbaum, C., and Cooper, L. N., "Learning System Architectures Composed of Multiple Learning Modules," *Proceedings of the First IEEE International Conference on Neural Networks (ICNN-87)*, vol. II, 495-503, June 1987.

Robbins, H., and Monro, S., "A Stochastic Approximation Method," *Annals of Mathematical Statistics*, vol. 22, 400-407, 1951.

Rosenblatt, F., *Principles of Neurodynamics*, Spartan Books, New York, 1962.

Rudin, W., *Functional Analysis*, McGraw-Hill, New York, 1973.

Rudin, W., *Real and Complex Analysis*, 2nd ed., McGraw-Hill, New York, 1974.

Rumelhart, D. E., and McClelland, J. L. (eds.), *Parallel Distributed Processing*, vol. I, M.I.T. Press, Cambridge, MA, 1986.

Samuelson, P. A., "Proof That Properly Discounted Present Values of Assets Vibrate Randomly," *Bell Journal of Economics and Management Science*, vol. 4, 41-49, 1973.

Skinner, B. F., *Science and Human Behavior*, The Free Press, New York, 1953.

Sorenson, H. W., and Alspach, D. L., "Recursive Bayesian Estimation Using Gaussian Sums," *Automatica*, vol. 7, no. 4, 465-479, July 1971.

Tsypkin, Y. Z., *Foundations of the Theory of Learning Systems*, trans. Z. J. Nikolic, Academic Press, Orlando, FL, 1973.

Werbos, P. J., *Beyond Regression: New Tools for Prediction and Analysis in the Behavioral Sciences*, Ph.D. Dissertation in Statistics, Harvard University, August 1974.

White, H., "Some Asymptotic Results for Learning in Single Hidden Layer Feedforward Network Models," *Journal of the American Statistical Association*, in press, 1989.

White, H., "Learning in Artificial Neural Networks: A Statistical Perspective," *Neural Computation*, vol. 1, no. 4, 425-469, Winter 1989.

White, H., "Neural Network Learning and Statistics," *AI Expert*, 48-52, December 1989.

Widrow, B., and Hoff, M.E., Jr., "Adaptive Switching Circuits," *IRE WESCON Convention Record*, pt. 4, 96-104, September 1960.

Widrow, B., et al., "Adaptive Noise Canceling: Principles and Applications," *Proceedings of the IEEE*, vol. 63, no. 12, December 1975.

Widrow, B., and Stearns, S. D., *Adaptive Signal Processing*, Prentice Hall, Englewood Cliffs, NJ, 1985.

Widrow, B., and Winter, R., "Neural Nets for Adaptive Filtering and Adaptive Pattern Recognition," *IEEE Computer Magazine*, 25-39, March 1988.

Williams, R. J., and Zipser, D., "A Learning Algorithm for Continually Running Fully Recurrent Neural Networks," *Neural Computation*, vol. 1, no. 2, 270-280, Summer 1989.

PROBLEMS

5.1. Prove that the LMS gradient estimate $\nabla_m e_k^2$ is unbiased if the synaptic vector **m** is constant:

$$E[\nabla_m e_k^2] \quad = \quad \nabla_m E[e_k^2]$$

5.2. (Brain-state-in-a-box learning) Suppose $\mathbf{x}_i \in R^n$ and $\mathbf{y}_i \in R^p$, and we encode the row-vector associations $(\mathbf{x}_1, \mathbf{y}_2), \ldots, (\mathbf{x}_m, \mathbf{y}_m)$ in a Hebbian outer-product matrix **M**:

$$\mathbf{M} \quad = \quad \sum_{i=1}^{m} \mathbf{x}_i^T \mathbf{y}_i$$

Then in general $\mathbf{x}_i\mathbf{M} \neq \mathbf{y}_i$ since the input vectors $\{\mathbf{x}_1, \ldots, \mathbf{x}_m\}$ are not in general orthonormal. This desired-minus-actual recall discrepancy defines an error vector $\mathbf{e}_i$:

$$\mathbf{e}_i = \mathbf{y}_i - \mathbf{x}_i\mathbf{M}$$

Derive a discrete matrix LMS algorithm for updating $\mathbf{M}$.

5.3. (Least squares) Suppose we observe the real scalar pairs $(X_1, Y_1), \ldots, (X_m, Y_m)$. The Y_i are realizations of n independent random variables. The X_i are deterministic variables. Define the translated input variables x_i as

$$x_i = X_i - \frac{1}{m}\sum_{j=1}^{m} X_j$$

So the x_i sum to zero. We model the input-output relationship with a random linear model:

$$Y_i = a + bx_i + e_i$$

The error terms e_i are independent and zero-mean, $E[e_i] = 0$, with finite variance, $V[e_i] = \sigma^2$. So $E[Y_i] = a + bx_i$, and $V[Y_i] = \sigma^2$ (why?). We assume that the observed samples (x_i, y_i) come from the random linear model. We assume that the errors e_i account for all discrepancies between observed and model behavior. So what are our "best" estimates, $\hat{a}$ and $\hat{b}$, of the unknown linear deterministic parameters a and b? Gauss's *least-squares* approach defines "best" as that which minimizes the sum of squared errors J:

$$J = \sum_{i=1}^{m} e_i^2$$

Prove that the least-squares estimates correspond to

$$\hat{a} = \frac{1}{m}\sum_{i=1}^{m} Y_i$$

$$\hat{b} = \frac{\displaystyle\sum_{i=1}^{m} x_i Y_i}{\displaystyle\sum_{i=1}^{m} x_i^2}$$

Be sure to show that $\hat{a}$ and $\hat{b}$ minimize, not maximize, J. Note that $\hat{a}$ equals the random sample mean. $\hat{b}$ is also random. Show that

$$E[\hat{a}] = a, \qquad E[\hat{b}] = b$$
$$V[\hat{a}] = \frac{\sigma^2}{m}, \qquad V[\hat{b}] = \frac{\sigma^2}{\displaystyle\sum_i x_i^2}$$

give the first-order and second-order statistics of these random estimators. Now state the *vector* version of the least-squares linear model. State the mean-squared error. Use the least-squares technique to estimate the mean-squared optimal parameter vector.

5.4. Prove:

$$\text{Trace}(\mathbf{M}) = \sum_i^n \lambda_i$$

$$\text{Determinant}(\mathbf{M}) = \prod_i^n \lambda_i$$

for any n-by-n matrix $\mathbf{M}$, where λ denotes an eigenvalue of $\mathbf{M}$. The trace operator sums the n diagonal entries m_{ii}. The determinant operator sums $n!$ signed and permuted n-products of matrix entries m_{ij}.

5.5. Find the eigenvalues, eigenvectors, inverse, trace, and determinant of matrix

$$M = \begin{pmatrix} 1 & -1 \\ 2 & 4 \end{pmatrix}$$

5.6. Prove: Symmetric matrices have real eigenvalues and orthogonal eigenvectors. Given an example of a matrix $\mathbf{M}$, with real entries m_{ij}, that has complex eigenvalues.

5.7. $\{\mathbf{o}_1, \ldots, \mathbf{o}_n\}$ is a set of orthonormal n-dimensional vectors. Vector $\mathbf{o}_i$ defines the ith row of matrix $\mathbf{O}$. Prove:

$$\mathbf{O}^T = \mathbf{O}^{-1}$$

5.8. A symmetric matrix $\mathbf{M}$ is *positive definite* iff $\mathbf{x}\mathbf{M}\mathbf{x}^T > O$ for all nonnull row vectors $\mathbf{x}$. Prove: A symmetric matrix is positive definite (semidefinite) iff all eigenvalues are positive (nonnegative).

5.9. Suppose $\mathbf{M}$ is positive definite. $\{\mathbf{x}: \mathbf{x}\mathbf{M}\mathbf{x}^T = 1\}$ defines an ellipsoid in R^n. $\mathbf{M}$'s orthogonal eigenvectors $\mathbf{e}_1, \ldots, \mathbf{e}_n$ possess corresponding eigenvalues $\lambda_1 \leq \lambda_2 \leq \ldots \leq \lambda_n$. The eigenvectors represent the principal axes of the ellipsoid. Prove that the Euclidean length of $\mathbf{e}_i$ equals the inverse square-root of its associated eigenvalue:

$$\|\mathbf{e}_i\| = \frac{1}{\sqrt{\lambda_i}}$$

Prove:

$$\lambda_1 = \min \frac{\mathbf{x}\mathbf{M}\mathbf{x}^T}{\mathbf{x}\mathbf{x}^T}, \qquad \lambda_n = \max \frac{\mathbf{x}\mathbf{M}\mathbf{x}^T}{\mathbf{x}\mathbf{x}^T}$$

where the minimum and maximum are taken over all nonnull vectors $\mathbf{x}$.

5.10. Prove: Autocorrelation matrices $\mathbf{M}$ are positive semidefinite, where

$$\mathbf{M} = \sum_{i=1}^m \mathbf{x}_i^T \mathbf{x}_i$$

5.11. The *parity function* $p: \{0, 1\}^n \to \{0, 1\}$ computes whether an n-dimensional bit vector B contains an even or odd number of 1s:

$$p(B) = \begin{cases} 1 & \text{if } B \text{ contains an odd number of 1s} \\ 0 & \text{if } B \text{ contains an even number of 1s} \end{cases}$$

Construct at least two different feedforward multilayer neural networks that implement the parity function for 5-dimensional bit vectors such as (1 0 1 0 1), (0 0 1 1 0), (1 0 0 0 1), etc.

5.12. Carefully state a discrete version of the backpropagation algorithm. Assume a feedforward neural network topology with a single hidden layer. Assume hidden and output neurons transduce activations to signals with a logistic signal function.

SOFTWARE PROBLEMS

The following problems use the accompanying OWL network software.

Part I: Exclusive-OR (XOR)

Construct a feedforward 2-layer (3-layer counting input field) neural network. Use the backpropagation (BKP) algorithm to train the network. Use the following specifications:

1. Run BKP.

2. Select "Activate 2-layer Network" from the "Network" menu.

3. Set the Load Parameters as follows:

$$
\begin{array}{rcl}
\text{Number of Inputs} & = & 2 \\
\text{Number of Hidden Units} & = & 2 \\
\text{Number of Outputs} & = & 1 \\
\text{Connect Inputs to Outputs} & = & \text{"No"} \\
\text{Randomize Half Range} & = & 2.0
\end{array}
$$

When you have implemented steps 1 through 3, the BKP program should have created a feedforward 2-layer (3-layer counting input field) network.

4. Set the Run Parameters as follows:

$$
\begin{array}{rcl}
\text{Layer 1 Learning Rate} & = & 0.70 \\
\text{Layer 2 Learning Rate} & = & 0.70 \\
\text{Layer 1 Momentum} & = & 0.90 \\
\text{Layer 2 Momentum} & = & 0.70 \\
\text{Use Momentum} & = & \text{"Yes"} \\
\text{Batch Size} & = & 0 \\
\text{Randomize Half Range} & = & 2.0 \\
\text{Slow Speed} & = & \text{"No"}
\end{array}
$$

Now you must set up the input-data source.

5. Use the "Data" menu to declare the input file as follows:
   Select "Select Data File" and enter "XORDATA1".
   Select "File Handling" options as follows:

   > Set "Reset After Run" to "No"
   > Set "Reset on EOF" to "Yes"
   > Set "Lesson Size" to 1

6. Set the "Display Mode" to "States".

7. Select the "Train" option of the "Run" menu to train the backpropagation network.

Problem 1:   In the "weights" display, we indicate the desired output in red, the actual output in yellow, and the squared error in white. Does the network learn the XOR (2-bit parity) function?

8. Press the "Q" key to stop the network.

9. Using "Restart Network," randomize the network's weight matrix.

10. Set the input data file to "XORDATA4".

11. Under the File Handling options, set the Lesson Size to 4.

12. Train the network as before.

Problem 2:   Explain why the display differs from that in Software Problem 1.

Part II: Sine Function

Construct a feedforward 2-layer (3-layer counting input field) neural network. Train the network with the backpropagation algorithm. Use the following specifications:

1. Run BKP.

2. Select "Activate 2-layer Network" from the "Network" menu.

3. Set the Load Parameters as follows:

   > Number of Inputs   =   1
   > Number of Hidden Units   =   6
   > Number of Outputs   =   1
   > Connect Inputs to Outputs =   "No"
   > Randomize Half Range   =   2.0

4. Set the Run Parameters the same as in Part I above.

5. Use the "Data" menu to declare the input file as follows:
   Select "Select Data File" and enter "TRUE.SIN".
   Select "File Handling" options as follows:

   > Set "Reset After Run" to "No"
   > Set "Reset on EOF" to "Yes"
   > Set "Lesson Size" to 1

6. Set the "Display Mode" to "States".

7. Select the "Train" option of the "Run" menu to train the network.

Notice the behavior of the output state. Can you explain this? [*Hint:* Consider the range of the sine and logistic-sigmoid functions. Each line of the file "TRUE.SIN" contains a data pair of the form $(x, \sin(x))$.]

Problem 3:   Explain what happens as BKP tries to learn the data in TRUE.SIN.

8. Stop the network.

9. Set the input file to "TRUEPREP.SIN".

10. Reinitialize the network using the "Restart" menu selection.

11. Train the network.

Each line in "TRUEPREP.SIN" contains a data pair of the form $(x, \frac{1}{2}(\sin(x) + 1))$. BKP should more accurately learn the data in this file.

Problem 4:   Would BKP perform better if you used more or less hidden units? More layers? Experiment with different architectures. Allow the network to train adequately each time.

Part III: Training Set versus Test Set

After completing Parts I and II above, use the BKP algorithm to train a feedforward network with the data in data file "HALFPREP.SIN", which is the first half of the file TRUEPREP.SIN. Then do the following:

1. Stop the network. Do not close or restart it.

2. Select "HALFTEST.SIN" as the input file.

3. Run the network, but do not train the network. Does the network produce correct values for HALFTEST.SIN?

Problem 5:   Both files present data pairs from the same function. Does the network perform equally on both files? If not, why would the network perform differently?

ARCHITECTURES
AND EQUILIBRIA

NEURAL NETWORKS AS STOCHASTIC GRADIENT SYSTEMS

We can classify neural-network models by their synaptic connection topologies and by how learning modifies their connection topologies.

A synaptic connection topology is **feedforward** if it contains no closed synaptic loops. The neural network is a **feedback** network if its topology contains closed synaptic loops or feedback pathways. A neural network defines a directed graph with axonal-synaptic pathways as edges and neurons as nodes. The neural network is a feedforward or feedback neural network according as the directed graph is acyclic or cyclic.

Neural networks learn when their synaptic topologies change. *Supervised* learning algorithms use class-membership information of training samples. *Unsupervised* learning algorithms use unlabelled training samples. In practice supervised learning corresponds to stochastic gradient-descent optimization. Supervision allows the gradient-descent algorithm to compute an explicit error term. In this chapter we shall see that unsupervised learning, in both feedforward and feedback networks, also corresponds to stochastic gradient descent.

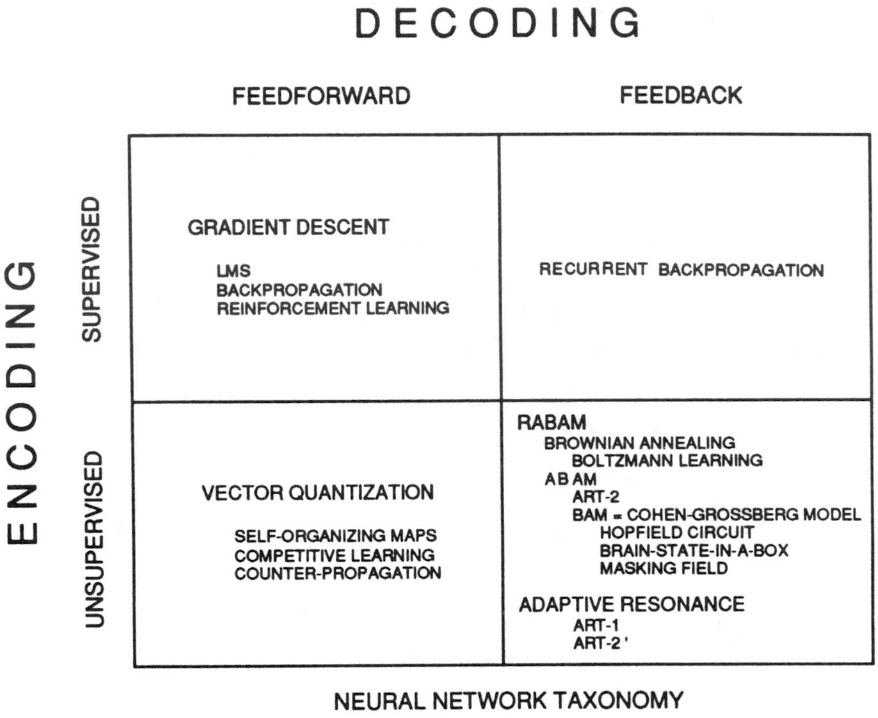

DECODING

FEEDFORWARD   FEEDBACK

ENCODING

SUPERVISED

GRADIENT DESCENT

LMS
BACKPROPAGATION
REINFORCEMENT LEARNING

RECURRENT BACKPROPAGATION

UNSUPERVISED

VECTOR QUANTIZATION

SELF-ORGANIZING MAPS
COMPETITIVE LEARNING
COUNTER-PROPAGATION

RABAM
    BROWNIAN ANNEALING
        BOLTZMANN LEARNING
    ABAM
        ART-2
    BAM = COHEN-GROSSBERG MODEL
        HOPFIELD CIRCUIT
        BRAIN-STATE-IN-A-BOX
        MASKING FIELD

ADAPTIVE RESONANCE
    ART-1
    ART-2'

NEURAL NETWORK TAXONOMY

FIGURE 6.1 Neural-network models classified by how they encode or learn pattern information in their synaptic topologies, and by the cyclic or acyclic structure of the synaptic topology they use to decode or recall information. *Source:* Kosko, B., "Unsupervised Learning in Noise," *IEEE Transactions on Neural Networks*, vol. 1, no. 1, March 1990.

These distinctions partition the set of neural network models into four categories. Figure 6.1 illustrates this taxonomy for several popular neural-network models. Learning encodes information, which passes through the synaptic topology or channel, and the system's nonlinear dynamics recalls or decodes information.

The taxonomy boundaries are fuzzy because the defining terms are fuzzy, and because different versions of some models may or may not contain cycles or learn with supervision. As discussed in Chapter 5, some researchers have recently extended the backpropagation algorithm to allow feedback. And important versions of competitive learning use supervised as well as unsupervised learning. Indeed the "competitive" in competitive learning refers to within-field feedback among competing neurons. In practice inner-product assumptions replace these nonlinear dynamics and preserve the global feedforward structure of competitive learning systems. Neural engineers often train the brain-state-in-a-box (BSB) model [Anderson, 1977, 1983] off-line with the supervised LMS algorithm or with the simpler, and unsupervised, Hebbian learning algorithm. But when the BSB model operates as a dynamical system, when it recalls patterns in its state-space "box," it behaves as a nonadaptive feedback neural network with additive activation dynamics.

Three stochastic gradient systems represent the three main categories: (1) feedforward supervised neural networks trained with the backpropagation (BP) algorithm, (2) feedforward unsupervised competitive learning or adaptive vector quantization (AVQ) networks, and (3) feedback unsupervised random adaptive bidirectional associative memory (RABAM) networks. Each network may contain multiple neuronal fields or "layers." Chapter 5 presented the backpropagation algorithm in the framework of stochastic approximation. Chapters 4 and 5 discussed competitive learning algorithms.

In this chapter we examine the convergence and equilibrium properties of AVQ systems. We also combine the neuronal activation models in Chapter 3 with the unsupervised synaptic learning laws in the RABAM models below and summarize their stochastic equilibrium behavior in the RABAM theorem. If we eliminate the feedback synaptic connections in RABAM models and use a competitive learning law to modify the synaptic values, then the RABAM model reduces to the AVQ model. In this sense the supervised BP (stochastic approximation) algorithm and the unsupervised RABAM models provide two unified frameworks for representing neural networks as stochastic gradient dynamical systems.

GLOBAL EQUILIBRIA: CONVERGENCE AND STABILITY

A neural network's synapses can change, its neurons can change, or both its synapses and neurons can change. These divisions correspond to three dynamical systems: the synaptic dynamical system M, the neuronal dynamical system $\dot{x}$, and the joint neuronal-synaptic dynamical system $(\dot{x}, \dot{M})$. Chapter 3 presented deterministic versions of the major neuronal dynamical systems. Chapters 4 and 5 presented the major unsupervised and supervised synaptic dynamical systems. In this chapter the RABAM model provides a general model of the joint neuronal-synaptic dynamical system $(\dot{x}, \dot{M})$ in the unsupervised stochastic case.

Historically, neural engineers have usually studied only the first or second dynamical systems and usually studied them independently. Either the synapses change or the neurons change, but not both simultaneously. Either the network learns patterns or recalls patterns. Neural engineers usually study learning in feedforward neural networks and neuronal stability in nonadaptive feedback neural networks, such as Hopfield-Amari networks. RABAM networks, and the related adaptive resonance theory (ART) networks, provide two of the few families of feedback neural networks that directly depend on joint equilibration of the synaptic and neuronal dynamical systems.

Equilibrium is steady state (for fixed-point attractors). **Convergence** is synaptic equilibrium:

$$\dot{M} = O \qquad (6\text{-}1)$$

where **O** denotes the null synaptic matrix. **Stability** is neuronal equilibrium:

$$\dot{x} = O \qquad (6\text{-}2)$$

where **O** now denotes the null vector of neuronal activations. More generally neuronal signals reach steady state even though the activations still change. We denote this symbolically as steady state in the neuronal field $\mathbf{F_X}$:

$$\dot{\mathbf{F}}_\mathbf{X} = O \qquad (6\text{-}3)$$

Then **global stability** is joint neuronal-synaptic steady state: both (6-1) and (6-3) hold. When both neurons and synapses change, (6-1) and (6-3) may be incompatible.

Stochastic equilibria vibrate. For stochastic dynamical systems with additive Gaussian white noise, **stochastic global stability** corresponds to the two conditions

$$\dot{x} = n \qquad (6\text{-}4)$$

$$\dot{M} = N \qquad (6\text{-}5)$$

As discussed in Chapter 4, n denotes a random vector from the (independent) Gaussian white random process $\{n_t\}$, and N denotes a random matrix from the Gaussian white random field $\{N_t\}$. In stochastic equilibrium the neuronal state vector and synaptic state matrix hover in a Brownian motion about fixed (deterministic) equilibrium values, which they equal on average.

The convergence-stability terminology reflects usage. Neural engineers sometimes use "stability" and "convergence" interchangeably when they describe neural-network equilibria. Usually, in terms of journal articles and conference proceedings publications, convergence refers to synaptic steady state, especially if the backpropagation or LMS algorithms modify the synapses. The neurons in these feedforward networks stabilize trivially. Usually stability refers to neuronal steady state in non-learning feedback systems, such as Hopfield circuits (or ART-1 or ART-2 systems during usage). In these networks the constant synapses have converged trivially.

In the general, and the biological, case both neurons and synapses change as the feedback system samples fresh environmental stimuli. The neuronal and synaptic dynamical systems ceaselessly approach equilibrium and may never achieve it.

Neurons fluctuate faster than synapses fluctuate. In feedback systems this dynamical asymmetry creates the *stability-convergence dilemma*: learning tends to destroy the neuronal patterns being learned. Stable neuronal outputs represent filtered patterns. The synapses slowly encode these hovering patterns. But when the synapses change in a feedback system, this tends to undo the stable neuronal patterns, and thus the dilemma. Convergence undermines stability.

Below we formally state the stability-convergence dilemma and develop the ABAM theorem as a candidate resolution to it. The RABAM theorem extends this resolution into the stochastic-process and simulated-annealing cases. First we examine separately synaptic convergence in feedforward competitive learning or AVQ systems and neuronal convergence in the feedback activation models developed in Chapter 3. The next section assumes knowledge of competitive learning as presented in Chapter 4.

SYNAPTIC CONVERGENCE TO CENTROIDS: AVQ ALGORITHMS

Competitive learning adaptively quantizes the input pattern space R^n. Probability density function $p(\mathbf{x})$ characterizes the continuous distribution of patterns in R^n, the realizations of random vector $\mathbf{x} : R^n \longrightarrow R^n$. We shall prove that competitive AVQ synaptic vectors $\mathbf{m}_j$ converge exponentially quickly to pattern-class centroids and, more generally, at equilibrium they vibrate about the centroids in a Brownian motion.

Competitive AVQ Stochastic Differential Equations

The decision classes $D_1, \ldots, D_k$ partition R^n into k classes:

$$R^n \quad = \quad D_1 \cup D_2 \cup \ldots \cup D_k \qquad (6\text{-}6)$$

$$D_i \cap D_j = \varnothing \qquad \text{if } i \neq j \qquad (6\text{-}7)$$

The random indicator functions $I_{D_1}, \ldots, I_{D_k}$ equivalently define the decision classes by extension:

$$I_{D_j}(x) \quad = \quad \begin{cases} 1 & \text{if } \mathbf{x} \in D_j \\ 0 & \text{if } \mathbf{x} \notin D_j \end{cases} \qquad (6\text{-}8)$$

Supervised learning algorithms depend explicitly on the indicator functions. Unsupervised learning algorithms do not require this pattern-class information. Centroid $\bar{\mathbf{x}}_j$ defines the deterministic "center of mass" of pattern class D_j:

$$\bar{\mathbf{x}}_j \quad = \quad \frac{\displaystyle\int_{D_j} \mathbf{x}\, p(\mathbf{x})\, d\mathbf{x}}{\displaystyle\int_{D_j} p(\mathbf{x})\, d\mathbf{x}} \qquad (6\text{-}9)$$

The stochastic unsupervised competitive learning law takes the form

$$\dot{m}_j \quad = \quad S_j(y_j)[\mathbf{x} - \mathbf{m}_j] + \mathbf{n}_j \qquad (6\text{-}10)$$

where S_j denotes the steep signal function of the jth competing neuron in field F_Y with activation y_j. $\mathbf{n}_j$ denotes an independent zero-mean Gaussian white-noise vector. More generally we replace the pattern $\mathbf{x}$ in (6-10) with the signal vector $\mathbf{S}(\mathbf{x}) = (S_1(x_1), \ldots, S_n(x_n))$, where S_i denotes the signal function of the ith input neuron in F_X. In feedforward networks input neurons often behave linearly: $S_i(x_i) = x_i$.

We want to show that at equilibrium $\mathbf{m}_j = \bar{\mathbf{x}}_j$, and more generally that $E[\mathbf{m}_j] = \bar{\mathbf{x}}_j$. $E[.]$ denotes the mathematical expectation or average of an n-dimensional random vector with respect to the probability density function $p(\mathbf{x})$.

For instance, $E[\mathbf{m}_j]$ stands for a vector integral:

$$E[\mathbf{m}_j] = \int_{R^n} \mathbf{m}_j p(\mathbf{x})\, d\mathbf{x} \tag{6-11}$$

$$= \int_{-\infty}^{\infty} \cdots \int_{-\infty}^{\infty} \mathbf{m}_j p(x_1, \ldots, x_n)\, dx_1\, dx_2 \ldots dx_n \tag{6-12}$$

$$= (E[m_j^1], \ldots, E[m_j^n]) \tag{6-13}$$

A separate probability density function $p(\mathbf{n})$ describes the independent noise vector $\mathbf{n}$. Technically the expectation in (6-11) uses the joint density $p(\mathbf{x}, \mathbf{n})$, since independence factors $p(\mathbf{x}, \mathbf{n})$ into $p(\mathbf{x})p(\mathbf{n})$ and the vector integration in (6-11) leaves only the marginal density $p(\mathbf{x})$. In the RABAM model below we shall use a joint density $p(\mathbf{x}, \mathbf{y}, M)$ that involves every neuron and synapse, each perturbed by separate independent noise processes. As discussed in Chapter 4, competitive signal functions S_j estimate the indicator function of the local jth pattern class D_j:

$$S_j(y_j) \approx I_{D_j}(\mathbf{x}) \tag{6-14}$$

This property allows us to ignore the complicated nonlinear dynamics of the within-field F_Y competition. The equilibrium and convergence results below depend on approximation (6-14). The property is approximate because the sampled decision classes may vary slightly—or wildly in rare cases—as the synapses converge to their equilibrium values.

Approximation (6-14) reduces the linear stochastic competitive learning law (6-10) to

$$\dot{\mathbf{m}}_j = I_{D_j}(\mathbf{x})[\mathbf{x} - \mathbf{m}_j] + \mathbf{n}_j \tag{6-15}$$

We shall prove the AVQ theorems for this version of unsupervised competitive learning. The results hold approximately for the related supervised competitive learning law and unsupervised differential competitive learning law.

The linear supervised competitive learning law arises from (6-15) when the reinforcement function r_j,

$$r_j(\mathbf{x}) = I_{D_j}(\mathbf{x}) - \sum_{i \neq j} I_{D_i}(\mathbf{x}) \tag{6-16}$$

which depends explicitly on the usually unknown indicator functions, scales the "signal" term in (6-15):

$$\dot{\mathbf{m}}_j = r_j(\mathbf{x})I_{D_j}(\mathbf{x})[\mathbf{x} - \mathbf{m}_j] + \mathbf{n}_j \tag{6-17}$$

The linear differential competitive learning law scales the learning difference $\mathbf{x} - \mathbf{m}_j$ with the competitive signal velocity $\dot{S}_j$:

$$\dot{\mathbf{m}}_j = \dot{S}_j[\mathbf{x} - \mathbf{m}_j] + \mathbf{n}_j \tag{6-18}$$

As discussed in Chapter 1 of the companion volume [Kosko, 1991] and applied in Chapters 8 through 11 of this book, in practice we often use only the sign of the competitive activation velocity:

$$\dot{\mathbf{m}}_j \quad = \quad \text{sgn}[\dot{y}_j][\mathbf{x} - \mathbf{m}_j] + \mathbf{n}_j \qquad (6\text{-}19)$$

since steep signal functions quickly saturate. The signum operator sgn[.] behaves as a three-valued threshold function with zero threshold:

$$\text{sgn}[z] \quad = \quad \begin{cases} 1 & \text{if } z > 0 \\ 0 & \text{if } z = 0 \\ -1 & \text{if } z < 0 \end{cases} \qquad (6\text{-}20)$$

All the above competitive learning laws admit pulse-coded formulations.

We implement these stochastic differential equations as the following stochastic difference equations. We do not add an independent noise term to the difference equations. The noise processes in the above stochastic differential equations model unmodeled effects, round-off errors, or sample-size defects.

Competitive AVQ Algorithms

1. Initialize synaptic vectors: $\mathbf{m}_i(0) = \mathbf{x}(i)$, $i = 1, \ldots, m$. This distribution-dependent initialization scheme avoids the problems that occur when all synaptic vectors initially equal, say, the null vector and that require quasi-supervised support algorithms [De Sieno, 1988].

2. For random sample $\mathbf{x}(t)$, find the closest ("winning") synaptic vector $\mathbf{m}_j(t)$:

$$||\mathbf{m}_j(t) - \mathbf{x}(t)|| \quad = \quad \min_i ||\mathbf{m}_i(t) - \mathbf{x}(t)|| \qquad (6\text{-}21)$$

where $||\mathbf{x}||^2 = x_1^2 + \cdots + x_n^2$ gives the squared Euclidean norm of $\mathbf{x}$.

3. Update the winning synaptic vector(s) $\mathbf{m}_j(t)$ by the UCL, SCL, or DCL learning algorithm.

Unsupervised Competitive Learning (UCL)

$$\mathbf{m}_j(t + 1) \quad = \quad \mathbf{m}_j(t) + c_t[\mathbf{x}(t) - \mathbf{m}_j(t)] \qquad (6\text{-}22)$$

$$\mathbf{m}_i(t + 1) \quad = \quad \mathbf{m}_i(t) \qquad \text{if } i \neq j \qquad (6\text{-}23)$$

where, as in the stochastic approximation of Chapter 5, $\{c_t\}$ defines a slowly decreasing sequence of learning coefficients. For instance, $c_t = 0.1\,[1 - (t/10{,}000)]$ for 10,000 samples $\mathbf{x}(t)$.

Supervised Competitive Learning (SCL)

$$\mathbf{m}_j(t+1) \;=\; \mathbf{m}_j(t) + c_t r_j(\mathbf{x}(t))[\mathbf{x}(t) - \mathbf{m}_j(t)] \tag{6-24}$$

$$= \begin{cases} \mathbf{m}_j(t) + c_t[\mathbf{x}(t) - \mathbf{m}_j(t)] & \text{if } \mathbf{x} \in D_j \\ \mathbf{m}_j(t) - c_t[\mathbf{x}(t) - \mathbf{m}_j(t)] & \text{if } \mathbf{x} \notin D_j \end{cases} \tag{6-25}$$

Equation (6-25) rewrites the reinforcement function r_j into the update algorithm.

Differential Competitive Learning (DCL)

$$\mathbf{m}_j(t+1) \;=\; \mathbf{m}_j(t) + c_t\, \Delta S_j(y_j(t))[\mathbf{x}(t) - \mathbf{m}_j(t)] \tag{6-26}$$

$$\mathbf{m}_i(t+1) \;=\; \mathbf{m}_i(t) \quad \text{if } i \neq j \tag{6-27}$$

where $\Delta S_j(y_j(t))$ denotes the time change of the jth neuron's competitive signal $S_j(y_j)$ in the competition field F_Y:

$$\Delta S_j(y_j(t)) \;=\; S_j(y_j(t+1)) - S_j(y_j(t)) \tag{6-28}$$

In practice we often use only the sign of the signal difference (6-28) or $\text{sgn}[\Delta y_j]$, the sign of the activation difference. In Chapters 8 through 11, we update the F_Y neuronal activations y_j with an *additive* model:

$$y_j(t+1) \;=\; y_j(t) + \sum_{i=1}^{n} S_i(x_i)m_{ij}(t) + \sum_{k=1}^{m} S_k(y_k)w_{kj} \tag{6-29}$$

The fixed competition matrix W defines a symmetric *lateral inhibition* topology within F_Y. In the simplest case, $w_{jj} = 1$ and $w_{ij} = -1$ for distinct i and j.

Stochastic Equilibrium and Convergence

Competitive synaptic vectors $\mathbf{m}_j$ converge to decision-class centroids. The centroids may correspond to local maxima of the sampled but unknown probability density function $p(\mathbf{x})$.

In general, when there are more synaptic vectors than probability maxima, the synaptic vectors cluster about local probability maxima. Comparatively few synaptic vectors may actually arrive at pattern-class centroids. We consider only convergence to centroids. We can view any local connected patch of the sample space R^n as a candidate decision class. Each synaptic vector samples such a local patch and converges to its centroid. Approximation (6-14) reflects this interpretation.

We first prove the AVQ centroid theorem: If a competitive AVQ system converges, it converges to the centroid of the sampled decision class. We prove this equilibrium theorem, anticipated in Chapter 4, only for unsupervised competitive learning, but argue that it holds for supervised and differential competitive learning in most cases of practical interest.

Next we use a Lyapunov argument to reprove and extend the AVQ centroid theorem to the AVQ convergence theorem: Stochastic competitive learning systems are asymptotically stable, and synaptic vectors converge to centroids. So competitive AVQ systems always converge, and converge exponentially fast. Both results hold with probability one.

AVQ Centroid Theorem

$$\text{Prob}(\mathbf{m}_j = \bar{\mathbf{x}}_j) \quad = \quad 1 \qquad \text{at equilibrium} \tag{6-30}$$

Proof. Suppose the jth neuron in F_Y wins the activation competition during the training interval. Suppose the jth synaptic vector $\mathbf{m}_j$ codes for decision class D_j. So $I_{D_i}(\mathbf{x}) = 1$ iff $S_j = 1$ by (6-14). Suppose the synaptic vector has reached equilibrium:

$$\dot{\mathbf{m}}_j \quad = \quad \mathbf{O} \tag{6-31}$$

which holds with probability one (or in the mean-square sense, depending on how we define the stochastic differentials). Take expectations of both sides of (6-31), use the zero-mean property of the noise process, eliminate the synaptic velocity vector $\dot{\mathbf{m}}_j$ with the competitive law (6-15), and expand to give

$$\mathbf{O} \quad = \quad E[\dot{\mathbf{m}}_j] \tag{6-32}$$

$$= \quad \int_{R^n} I_{D_j}(\mathbf{x})(\mathbf{x} - \mathbf{m}_j)p(\mathbf{x})\,d\mathbf{x} + E[\mathbf{n}_j] \tag{6-33}$$

$$= \quad \int_{D_j} (\mathbf{x} - \mathbf{m}_j)p(\mathbf{x})\,d\mathbf{x} \tag{6-34}$$

$$= \quad \int_{D_j} \mathbf{x}p(\mathbf{x})\,d\mathbf{x} - \mathbf{m}_j \int_{D_j} p(\mathbf{x})\,d\mathbf{x} \tag{6-35}$$

since (6-31) implies that $\mathbf{m}_j$ is constant with probability one. Solving for the equilibrium synaptic vector $\mathbf{m}_j$ gives the centroid $\bar{\mathbf{x}}_j$ in (6-9). **Q.E.D.**

In general the AVQ centroid theorem concludes that at equilibrium the *average* synaptic vector $E[\mathbf{m}_j]$ equals the jth centroid $\bar{\mathbf{x}}_j$:

$$E[\mathbf{m}_j] \quad = \quad \bar{\mathbf{x}}_j \tag{6-36}$$

The equilibrium synaptic vector $\mathbf{m}_j$ vibrates in a Brownian motion about the constant centroid $\bar{\mathbf{x}}_j$. $\mathbf{m}_j$ equals $\bar{\mathbf{x}}_j$ on average at each postequilibration instant. The simulated competitive synaptic vectors in [Kong, 1991] and in Chapter 1 of the companion volume [Kosko, 1991] exhibited such random wandering about centroids.

Synaptic vectors learn noise as well as signal. So they vibrate at equilibrium. The independent additive noise process n_j in (6-15) drives the random vibration. The steady-state condition (6-31) models the rare event that noise cancels signal. In general it models stochastic equilibrium in the absence of additive noise.

The equality

$$\dot{\mathbf{m}}_j = \mathbf{n}_j \tag{6-37}$$

models stochastic equilibrium in the presence of additive independent noise.

Taking expectations of both sides of (6-37) still gives (6-32), since the noise process $\mathbf{n}_j$ is zero-mean, and the argument proceeds as before. Taking a second expectation in (6-33) and using (6-32) gives (6-36).

The AVQ centroid theorem applies to the stochastic SCL law (6-17) because the algorithm picks winners metrically with the nearest-neighbor criterion (6-21). The reinforcement function r_j in (6-16) reduces to $r_j(\mathbf{x}) = -I_{D_i}(\mathbf{x}) = -1$ when the jth neuron continually wins for random samples $\mathbf{x}$ from class D_i. This tends to occur once the synaptic vectors have spread out in R^n, and D_i is close, usually contiguous, to D_j. Then $\mathbf{m}_j$ converges to $\bar{\mathbf{x}}_j$, the centroid of D_j, since the steady state condition (6-31) or (6-37) removes the scaling constant -1 that then appears in (6-34).

This argument holds only approximately when, in the exceptional case, $\mathbf{m}_j$ repeatedly misclassifies patterns $\mathbf{x}$ from several classes D_k. Then the difference of indicator functions in (6-16) replaces the single indicator function I_{D_j} in (6-33). The resultant equilibrium $\mathbf{m}_j$ equals a more general ratio than the centroid. For then we must integrate the density $p(\mathbf{x})$ over R^n not just over D_j.

The AVQ centroid theorem applies similarly to the stochastic DCL law (6-18). A positive or negative factor scales the difference $\mathbf{x} - \mathbf{m}_j$. If, as in practice and in (6-26), a constant approximates the scaling factor, the steady state condition (6-31) or (6-37) removes the scaling constant from (6-34) and $\mathbf{m}_j$ or $E[\mathbf{m}_j]$, estimates the centroid $\bar{\mathbf{x}}_j$.

The integrals in (6-32) through (6-35) are *spatial* integrals over R^n or subsets of R^n. Yet in the discrete UCL, SCL, and DCL algorithms, the recursive equations for $\mathbf{m}_j(t + 1)$ define *temporal* integrals over the training interval.

The spatial and temporal integrals are approximately equal. The discrete random samples $\mathbf{x}(0)$, $\mathbf{x}(1)$, $\mathbf{x}(2)$, ... partially enumerate the continuous distribution of equilibrium realizations of the random vector $\mathbf{x}$. The time index in the discrete algorithms approximates the "spatial index" underlying $p(\mathbf{x})$. So the recursion $\mathbf{m}_j(t + 1) = \mathbf{m}_j(t) + \cdots$ approximates the averaging integral. We sample patterns one at a time. We integrate them all at a time.

The AVQ centroid theorem assumes that stochastic convergence occurs. Synapses converge trivially for continuous deterministic competitive learning, at least in feedforward networks. Convergence is not trivial for stochastic competitive learning in noise.

The AVQ convergence theorem below ensures exponential convergence. The theorem does not depend on how the F_Y neurons change in time, provided (6-14) holds. The proof uses a stochastic Lyapunov function. The strictly decreasing deterministic Lyapunov function $E[L]$ replaces the random Lyapunov function L, as in the proof of the RABAM theorem below.

A strictly decreasing Lyapunov function yields asymptotic stability [Parker,

1987] as discussed in Chapter 3. Then the real parts of the eigenvalues of the system Jacobian matrix are strictly negative, and locally the nonlinear system behaves linearly. Synaptic vectors converge [Hirsch, 1974] exponentially quickly to equilibrium points—to pattern-class centroids—in R^n. Technically, nondegenerate Hessian matrix conditions must also hold. Otherwise some eigenvalues may have zero real parts.

AVQ Convergence Theorem. Competitive synaptic vectors converge exponentially quickly to pattern-class centroids.

Proof. Consider the random quadratic form L:

$$L = \frac{1}{2} \sum_i^n \sum_j^m (x_i - m_{ij})^2 \qquad (6\text{-}38)$$

Note that if $\mathbf{x} = \mathbf{x}_j$ in (6-38), then with probability one $L > 0$ if any $\mathbf{m}_j \neq \bar{\mathbf{x}}_j$ and $L = 0$ iff $\mathbf{m}_j = \bar{\mathbf{x}}_j$ for every $\mathbf{m}_j$.

The pattern vectors $\mathbf{x}$ do not change in time. (The following argument is still valid if the pattern vectors $\mathbf{x}$ change slowly relative to synaptic changes.) This simplifies the stochastic derivative of L:

$$\dot{L} = \sum_i \frac{\partial L}{\partial x_i} \dot{x}_i + \sum_i \sum_j \frac{\partial L}{\partial m_{ij}} \dot{m}_{ij} \qquad (6\text{-}39)$$

$$= \sum_i \sum_j \frac{\partial L}{\partial m_{ij}} \dot{m}_{ij} \qquad (6\text{-}40)$$

$$= -\sum_i \sum_j (x_i - m_{ij}) \dot{m}_{ij} \qquad (6\text{-}41)$$

$$= -\sum_i \sum_j I_{D_j}(\mathbf{x})(x_i - m_{ij})^2 - \sum_i \sum_j (x_i - m_{ij}) n_{ij} \qquad (6\text{-}42)$$

L equals a random variable at every time t. $E[L]$ equals a deterministic number at every t. So we use the average $E[L]$ as a Lyapunov function for the stochastic competitive dynamical system. For this we must assume sufficient smoothness to interchange the time derivative and the probabilistic integral—to bring the time derivative "inside" the integral. Then the zero-mean noise assumption, and the independence of the noise process $\mathbf{n}_j$ with the "signal" process $\mathbf{x} - \mathbf{m}_j$, gives

$$\dot{E}[L] = E[\dot{L}] \qquad (6\text{-}43)$$

$$= -\sum_j \int_{D_j} \sum_i (x_i - m_{ij})^2 p(\mathbf{x}) \, d\mathbf{x} \qquad (6\text{-}44)$$

So, on average by the learning law (6-15), $\dot{E}[L] < 0$ iff any synaptic vector $\mathbf{m}_j$ moves along its trajectory. So the competitive AVQ system is *asymptotically* stable, as discussed in Chapter 3, and in general converges exponentially quickly to a local equilibrium. Or note that (6-38), (6-43), and (6-44) imply $\dot{E}[L] = -\frac{1}{2}E[L]$, and so $E[L(t)] = ce^{-t/2}$.

Suppose $\dot{E}[L] = 0$. Then every synaptic vector has reached equilibrium and is constant (with probability one) if (6-31) holds. Then, since $p(\mathbf{x})$ is a nonnegative weight function [Rudin, 1974], the weighted integral of the learning differences $x_i - m_{ij}$ must also equal zero:

$$\int_{D_j} (\mathbf{x} - \mathbf{m}_j)p(\mathbf{x})\, d\mathbf{x} \;=\; \mathbf{O} \tag{6-45}$$

in vector notation. Equation (6-45) is identical to (6-34). So, with probability one, equilibrium synaptic vectors equal centroids. More generally, as discussed above, (6-36) holds. Average equilibrium synaptic vectors are centroids: $E[\mathbf{m}_j] = \bar{\mathbf{x}}_j$. **Q.E.D.**

The sum of integrals (6-44) defines the total mean-squared error of vector quantization for the partition $D_1, \ldots, D_k$, as discussed in [Kong, 1991] and in Chapter 1 of [Kosko, 1991]. The vector integral in (6-45) equals the gradient of $\dot{E}[L]$ with respect to $\mathbf{m}_j$. So the AVQ convergence theorem implies that class centroids—and, asymptotically, competitive synaptic vectors—minimize the mean-squared error of vector quantization.

Then by (6-15), the synaptic vectors perform stochastic gradient descent on the mean-squared-error surface in the pattern-plus-error space R^{n+1}. The difference $\mathbf{x}(t) - \mathbf{m}_j(t)$ behaves as an error vector. The competitive system estimates the unknown centroid $\bar{\mathbf{x}}_j$ as $\mathbf{x}(t)$ at each time t. Learning is unsupervised but proceeds *as if* it were supervised. Competitive learning reduces to stochastic gradient descent.

We now return to the general feedback neural networks discussed in Chapter 3, extend them to allow unsupervised learning, and use the Lyapunov technique to establish their global stability. When the synapses change according to a competitive AVQ model, and when we remove the feedback connections from the network connection topology, we will recover the above AVQ models.

GLOBAL STABILITY OF FEEDBACK NEURAL NETWORKS

Global stability, (6-1) through (6-3), is convergence to fixed points for all inputs and all choices of system parameters. Global stability provides the content-addressable-memory (CAM) property of neural networks, forcing the state ball bearing to roll down the energy surface illustrated in Chapter 3.

Global stability theorems are powerful but limited. Their power comes from their dimension independence, their nonlinear generality, and their exponentially fast convergence to fixed points. They are limited because in general they do not tell us where the equilibria occur in the state space.

Equilibration to fixed points provides pattern recall or decoding. But in general we do not know whether decoded patterns correspond to earlier encoded patterns. Indeed we usually do not even know what patterns learning encoded. In combinatorial optimization networks, as in Chapter 2 of [Kosko, 1991], where we identify the decreasing Lyapunov function with the cost function we seek to minimize, the Lyapunov function will not in general converge to global minima. A pattern-recognition network may fall into spurious attractor basins and suffer a type of *déjà vu*, recognizing patterns it did not "experience" during learning.

Perfect decoding accuracy requires exhaustive search of the state space. In general this is computationally prohibitive. Worse, in an adaptive dynamical system, at each instant the distribution of fixed-point equilibria changes, as does the shape and position of the surrounding attractor basins.

Global stability in AVQ models provides an exception. The AVQ centroid theorem above ensures that the synaptic dynamical system converges and that the synaptic equilibria correspond on average to pattern-class centroids. But even here we do not know in advance the probability density function $p(\mathbf{x})$ of the patterns and hence the distribution of centroids. We know only that the steady-state distribution of synaptic vectors should approximate the unknown underlying pattern distribution.

Still, global stability of neural networks provides a scientific, if not engineering, advance. Global stability offers a mechanism for real-time biological organization. Natural selection may well have exploited architectures that rapidly organize five thousand, or five million, asynchronous neurons into a trainable computational unit. Animal evolution has had hundreds of millions of years to fine-tune the balance between computational speed and approximation accuracy. Achieving practical speed/accuracy balances in distributed silicon, polymer, or other computational media remains an open engineering challenge.

ABAMs and the Stability-Convergence Dilemma

Adaptive bidirectional associative memory (ABAM) dynamical systems [Kosko, 1987, 1989] include several popular feedback neural networks as special cases. This holds in large part because ABAMs generalize the Cohen-Grossberg [1983] activation model:

$$\dot{x}_i \;=\; -a_i(x_i)[b_i(x_i) - \sum_j S_j(x_j)m_{ji}] \qquad (6\text{-}46)$$

described in Chapter 3. Function a_i is nonnegative. b_i denotes an almost arbitrary nonlinear function that satisfies technical boundedness conditions, which we shall assume. Throughout this chapter we will make the **positivity assumptions**:

$$a_i(x_i) \;>\; 0 \qquad (6\text{-}47)$$

$$S_i'(x_i) \;>\; 0 \qquad (6\text{-}48)$$

In the Cohen-Grossberg case the memory matrix $\mathbf{M}$ is constant, and so no learning occurs. It is also symmetric—$\mathbf{M} = \mathbf{M}^T$—so $m_{ij} = m_{ji}$. The Cohen-Grossberg model is autoassociative. All neurons reside in field F_X.

In Chapter 3 we saw how several additive and multiplicative models define special cases of (6-21). For instance, the Hopfield circuit

$$C_i \dot{x}_i = -\frac{x_i}{R_i} + \sum_j T_{ij} V_j + I_i \qquad (6\text{-}49)$$

specifies an additive autoassociative model. The Hopfield circuit follows from the Cohen-Grossberg model if we make the substitutions $a_i = 1/C_i$, $b_i = x_i/R_i - I_i$, $m_{ij} = T_{ij}$, and $S_i(x_i) = V_j$—a controversial subsumption first published as a critical comment in the journal *Science* [Carpenter, 1987b].

Hopfield's [1985] pathbreaking contribution showed how to use (6-49) to solve certain combinatorial optimization problems, as illustrated in Chapters 2, 3, and 8 of the companion volume [Kosko, 1991]. Hopfield equated the governing Lyapunov function L to an unconstrained cost function C, then backsolved for the synaptic connection strengths m_{ij} and input values I_i. When $L = C$, global stability drives the network to a local, and sometimes global, minimum of the cost function C. The Tank-Hopfield [1986] clockless analog-to-binary converter illustrates this technique. Other examples include linear programming, signal decomposition, and speech recognition [Tank, 1987].

ABAM models are adaptive and "multilayered." In the simplest case, ABAM models are heteroassociative or two-layered. Neurons in field F_X pass signals through the adaptive matrix $\mathbf{M}$ to field F_Y. The neurons in F_Y pass signals back through $\mathbf{M}^T$ to F_X. This symmetrizes the nonlinear matrix of functions $\mathbf{M}$, a key property for establishing global stability.

Hebbian ABAMs adapt according to the signal Hebbian learning law discussed in Chapter 4. We can specify a two-layer **Hebbian ABAM** as

$$\dot{x}_i = -a_i(x_i)[b_i(x_i) - \sum_{j=1}^{p} S_j(y_j) m_{ij}] \qquad (6\text{-}50)$$

$$\dot{y}_j = -a_j(y_j)[b_j(y_j) - \sum_{i=1}^{n} S_i(x_i) m_{ij}] \qquad (6\text{-}51)$$

$$\dot{m}_{ij} = -m_{ij} + S_i(x_i) S_j(y_j) \qquad (6\text{-}52)$$

This ABAM model reduces to the Cohen-Grossberg model if no learning occurs, if the two fields F_X and F_Y collapse into the autoassociative field $F_X = F_Y$, and if the constant matrix M is symmetric.

Other unsupervised learning laws can replace the signal Hebbian learning law in an ABAM system. A **competitive ABAM (CABAM)** uses the competitive learning law (6-53) in place of (6-52):

$$\dot{m}_{ij} = S_j(y_j)[S_i(x_i) - m_{ij}] \qquad (6\text{-}53)$$

Similarly, a **differential Hebbian ABAM** uses the differential Hebbian learning law (6-54) in place of (6-52):

$$\dot{m}_{ij} = -m_{ij} + S_i S_j + \dot{S}_i \dot{S}_j \tag{6-54}$$

As discussed in Chapter 4, a **differential competitive ABAM** uses the differential competitive learning law (6-55) in place of (6-52):

$$\dot{m}_{ij} = \dot{S}_j [S_i - m_{ij}] \tag{6-55}$$

These three learning laws each require additional hypotheses to establish global stability. The competitive ABAM, though, requires only that the signal functions S_j be sufficiently steep to approximate a binary threshold function.

Global stability requires joint equilibration of the activation and synaptic dynamical systems. The fast-changing neurons must balance the slow-changing synapses. In principle they must ceaselessly balance in a smooth sequence (manifold) of system equilibria as changing external stimuli continuously perturb the dynamical system. This real-time tug-of-war between short-term and long-term memory leads to the stability-convergence dilemma, and to the ABAM theorem.

Stability-Convergence Dilemma

Neuronal stability must balance synaptic convergence. The neurons must stably represent patterns while the synapses gradually learn them. Learning, though, tends to destabilize the neurons. This describes the **stability-convergence dilemma**.

Global stability requires a delicate dynamical balance between stability and convergence. Achieving such a balance presents arguably the central problem in analyzing, and building, unsupervised feedback dynamical systems. The chief difficulty stems from the dynamical asymmetry between neural and synaptic fluctuations. Neurons fluctuate orders of magnitude faster than synapses fluctuate: Learning is slow. In real neural systems, neurons may fluctuate at the millisecond level, while synapses may fluctuate at the second or even minute level.

The *stability-convergence dilemma* arises from the asymmetry in neuronal and synaptic fluctuation rates. The dilemma unfolds as follows. Neurons change faster than synapses change. Patterns form when neurons stabilize, when $\dot{F}_X = \mathbf{0}$ and $\dot{F}_Y = \mathbf{0}$. The slowly varying synapses in $\mathbf{M}$ try to learn these patterns. Since the neurons are stable for more than a synaptic moment, the synapses begin to adapt to the neuronal patterns, and learning begins. So $\dot{F}_X = \mathbf{0}$ and $\dot{F}_Y = \mathbf{0}$ imply $\dot{\mathbf{M}} \neq \mathbf{0}$. Since there are numerous feedback paths from the synapses to the neurons, the neurons tend to change state. So $\dot{\mathbf{M}} \neq \mathbf{0}$ implies $\dot{F}_X \neq \mathbf{0}$ and $\dot{F}_Y \neq \mathbf{0}$. *Learning tends to undo the very stability patterns to be learned*, and hence the dilemma. In summary, for two fields of neurons, F_X and F_Y, connected in the forward direction by $\mathbf{M}$ and in the backward direction by $\mathbf{M}^T$, the stability-convergence dilemma has four parts:

Stability-Convergence Dilemma

1. **Asymmetry**: Neurons in F_X and F_Y fluctuate faster than the synapses in $\mathbf{M}$.

2. **Stability**: $\dot{F}_X = \mathbf{0}$ and $\dot{F}_Y = \mathbf{0}$ (pattern formation).

3. **Learning**: $\dot{F}_X = \mathbf{0}$ and $\dot{F}_Y = \mathbf{0} \longrightarrow \dot{\mathbf{M}} \neq \mathbf{0}$.

4. **Undoing**: $\dot{\mathbf{M}} \neq \mathbf{0} \longrightarrow \dot{F}_X \neq \mathbf{0}$ and $\dot{F}_Y \neq \mathbf{0}$.

The ABAM theorem offers a general solution to the stability-convergence dilemma. This family of theorems ensures that the neuronal and synaptic dynamical systems rapidly reach equilibrium for different unsupervised learning laws. The RABAM theorem extends this result to stochastic neural processing in the presence of noise.

The ABAM Theorem

The ABAM theorem forms a bridge from the Cohen-Grossberg [1983] theorem to the various RABAM theorems below. We prove the ABAM theorem by finding the appropriate Lyapunov function L for the above ABAM dynamical systems. Here we prove only the ABAM theorem for the signal Hebbian and competitive learning laws. We leave the proof for differential Hebbian learning as an exercise.

ABAM Theorem.   The Hebbian ABAM and competitive ABAM models are globally stable. We define the dynamical systems as

$$\dot{x}_i \quad = \quad -a_i(x_i)[b_i(x_i) - \sum_{j=1}^{p} S_j(y_j)m_{ij}] \qquad (6\text{-}56)$$

$$\dot{y}_j \quad = \quad -a_j(y_j)[b_j(y_j) - \sum_{i=1}^{n} S_i(x_i)m_{ij}] \qquad (6\text{-}57)$$

$$\dot{m}_{ij} \quad = \quad -m_{ij} + S_i S_j \qquad (6\text{-}58)$$

for signal Hebbian ABAMs, or with (6-58) replaced with (6-59) for competitive ABAMs,

$$\dot{m}_{ij} \quad = \quad S_j[S_i - m_{ij}] \qquad (6\text{-}59)$$

If the positivity assumptions $a_i > 0$, $a_j > 0$, $S_i' > 0$, and $S_j' > 0$ hold, then the models are *asymptotically* stable, and the squared activation and synaptic velocities decrease exponentially quickly to their equilibrium values:

$$\dot{x}_i^2 \downarrow 0, \quad \dot{y}_j^2 \downarrow 0, \quad \dot{m}_{ij}^2 \downarrow 0 \qquad (6\text{-}60)$$

Proof. The proof uses the bounded Lyapunov function L:

$$L = -\sum_i \sum_j S_i S_j m_{ij} + \sum_i \int_0^{x_i} S_i'(\theta_i) b_i(\theta_i) \, d\theta_i$$

$$+ \sum_j \int_0^{y_j} S_j'(\varepsilon_j) b_j(\varepsilon_j) \, d\varepsilon_j + \frac{1}{2} \sum_i \sum_j m_{ij}^2 \qquad (6\text{-}61)$$

It takes some real-analytical care [Cohen-Grossberg, 1983] to ensure the boundedness of the integral terms in (6-61). Pathologies can occur, though we shall ignore them. The boundedness of the quadratic form follows from the boundedness of the signal functions S_i and S_j. The sum of squared synaptic value is bounded because the signal functions are bounded.

We time-differentiate L. We can distribute the time-derivative operator across the right-hand side of (6-61). Then we can time-differentiate each term separately. Differentiating the integral terms may appear daunting. But each integral equals a composite function $F(x_i(t))$. So the chain rule of differentiation gives

$$\frac{d}{dt} F(x_i(t)) = \frac{dF}{dx_i} \frac{dx_i}{dt} \qquad (6\text{-}62)$$

$$= S_i' b_i \dot{x}_i \qquad (6\text{-}63)$$

and similarly for the F_Y integrals involving y_j. The term $S_i' \dot{x}_i$ multiplies each additive F_X term. Similarly, $S_j' \dot{y}_j$ multiplies each additive F_Y term. The chain rule produces the velocity factor $\dot{m}_{ij}$ for the **M** terms. This allows us to regroup the F_X, F_Y, and **M** terms as follows:

$$\dot{L} = -\sum_i S_i' \dot{x}_i \sum_j S_j m_{ij} - \sum_j S_j' \dot{y}_j \sum_i S_i m_{ij} - \sum_i \sum_j S_i S_j \dot{m}_{ij}$$

$$+ \sum_i S_i' b_i \dot{x}_i + \sum_j S_j' b_j \dot{y}_j + \sum_i \sum_j m_{ij} \dot{m}_{ij} \qquad (6\text{-}64)$$

$$= \sum_i S_i' \dot{x}_i [b_i - \sum_j S_j m_{ij}] + \sum_j S_j' \dot{y}_j [b_j - \sum_i S_i m_{ij}]$$

$$- \sum_i \sum_j \dot{m}_{ij} [S_i S_j - m_{ij}] \qquad (6\text{-}65)$$

$$= -\sum_i S_i' a_i [b_i - \sum_j S_j m_{ij}]^2 - \sum_j S_j' a_j [b_j - \sum_i S_i m_{ij}]^2$$

$$- \sum_i \sum_j [-m_{ij} + S_i S_j]^2 \qquad (6\text{-}66)$$

upon eliminating the terms in braces with the signal Hebbian ABAM equations (6-56) through (6-58). Since the amplification functions are nonnegative,

$a_i \geq 0$ and $a_j \geq 0$, and the signal functions are monotone nondecreasing, $S_i' \geq 0$ and $S_j' \geq 0$, the right-hand side of (6-66) decreases along system trajectories: $\dot{L} \leq 0$.

This proves global stability for signal Hebbian ABAMs. It also proves global stability for the nonadaptive BAM and Cohen-Grossberg models, since then the squared synaptic terms m_{ij}^2 in (6-61) are constant, and time differentiation zeroes them out. The autoassociative version of the ABAM theorem then reduces to the **Cohen-Grossberg theorem** [Grossberg, 1988].

To prove global stability for the competitive learning law (6-59)—and to outline a general proof strategy for arbitrary candidate first-order learning laws—we eliminate the third term in brackets in (6-66) with (6-59) to give

$$\dot{L} = -\sum_i S_i' a_i [b_i - \sum_j S_j m_{ij}]^2 - \sum_j S_j' a_j [b_j - \sum_i S_i m_{ij}]$$
$$- \sum_i \sum_j S_j [S_i - m_{ij}][S_i S_j - m_{ij}] \tag{6-67}$$

(Equation (6-66) allows for the nonpositive case when $a_i = 0$ or $S_i' = 0$.) We now invoke the **competitive assumption** that S_j is steep and indicates the win-loss status of the jth neuron in the competitive, or laterally inhibitive, neural field F_Y. So we assume that S_j behaves approximately as a zero-one threshold. This holds when S_j equals an appropriately scaled logistic sigmoid signal function. Then the third sum in (6-67) is nonpositive, since its summand is nonnegative:

$$\dot{m}_{ij}[S_i S_j - m_{ij}] = S_j [S_i - m_{ij}][S_i S_j - m_{ij}] \tag{6-68}$$

$$= \begin{cases} 0 & \text{if } S_j = 0 \\ (S_i - m_{ij})^2 & \text{if } S_j = 1 \end{cases} \tag{6-69}$$

So $\dot{L} \leq 0$ along trajectories.

This proves the global stability of the competitive ABAM system (6-56), (6-57), and (6-59). The nonnegativity in (6-67) holds approximately as S_j approximates a zero-one threshold function. We can also argue statistically to further weaken the competitive zero-one assumption since the third term in (6-67) is a large sum. Different summands can be positive or negative so long as the overall sum is nonpositive *on average*.

We prove the stronger asymptotic stability of the ABAM models (Hebbian, competitive, and otherwise) with the positivity assumptions $a_i > 0$, $a_j > 0$, $S_i' > 0$, and $S_j' > 0$, which permit division by a_i and a_j in (6-65). For convenience we shall detail the proof for signal Hebbian learning only. The proof for competitive learning uses (6-67) as above. We discuss below the proof for signal-velocity learning laws, such as the differential Hebb law.

The time derivative of L again equals (6-65), but now we can use the positivity assumptions to eliminate the terms in brackets in a way that differs from (6-66). In particular,

$$\dot{L} = -\sum_i \frac{S_i'}{a_i} \dot{x}_i^2 - \sum_j \frac{S_j'}{a_j} \dot{y}_j^2 - \sum_i \sum_j \dot{m}_{ij}^2 < 0 \qquad (6\text{-}70)$$

along trajectories for any nonzero change in any neuronal activation or any synapse. This proves asymptotic global stability. Trajectories end in equilibrium points, not merely near them [Hirsch, 1974]. Indeed (6-70) implies that

$$\dot{L} = 0 \quad \text{iff} \quad \dot{x}_i^2 = \dot{y}_j^2 = \dot{m}_{ij}^2 = 0 \qquad (6\text{-}71)$$

$$\text{iff} \quad \dot{x}_i = \dot{y}_j = \dot{m}_{ij} = 0 \qquad (6\text{-}72)$$

for all i and j, giving the desired equilibrium condition on squared velocities. The squared velocities decrease *exponentially quickly* [Hirsch, 1974] because of the strict negativity of (6-70) and, to rule out pathologies (system Jacobian eigenvalues with zero real parts), because of the second-order assumption of a nondegenerate Hessian matrix. Asymptotic stability ensures that the real parts of eigenvalues of the ABAM system Jacobian matrix, computed about an equilibrium, are negative (possibly nonpositive). Standard results from dynamical systems theory then ensure that locally the nonlinear system behaves linearly and decreases to its equilibrium exponentially fast. **Q.E.D.**

The ABAM theorem covers a broad class of adaptive feedback neural networks. Three important cases are higher-order signal Hebbian ABAMs, adaptive resonance theory (ART) competitive ABAMs, and differential Hebbian ABAMs. We examine each in turn.

Higher-Order ABAMs

Higher-order ABAMs [Kosko, 1988] learn with higher-order signal-Hebbian correlations. This improves learning accuracy but at the cost of greatly increased storage requirements [Giles, 1987]. In the heteroassociative case higher-order ABAMs tend to give a polynomial approximation of sampled functions.

The second-order *autoassociative* signal Hebbian ABAM corresponds to the dynamical system

$$\dot{x}_i = -a_i(x_i)[b_i(x_i) - \sum_j S_j(x_j)m_{ij} - \sum_j \sum_k S_j(x_j)S_k(x_k)n_{ijk}] \quad (6\text{-}73)$$

$$\dot{m}_{ij} = -m_{ij} + S_i(x_i)S_j(x_j) \qquad (6\text{-}74)$$

$$\dot{n}_{ijk} = -n_{ijk} + S_i(x_i)S_j(x_j)S_k(x_k) \qquad (6\text{-}75)$$

with corresponding Lyapunov function L:

$$L = -\frac{1}{2}\sum_i \sum_j S_i S_j m_{ij} - \frac{1}{3}\sum_i \sum_j \sum_k S_i S_j S_k n_{ijk}$$

$$+ \sum_i \int_0^{x_i} S_i'(\theta_i) b_i(\theta_i)\, d\theta_i + \frac{1}{4}\sum_i \sum_j m_{ij}^2 + \frac{1}{6}\sum_i \sum_j \sum_k n_{ijk}^2 \quad (6\text{-}76)$$

The Lyapunov function remains bounded in the adaptive case. The new terms,

$$\frac{1}{4}\sum_i \sum_j m_{ij}^2 \quad \text{and} \quad \frac{1}{6}\sum_i \sum_j \sum_k n_{ijk}^2 \qquad (6\text{-}77)$$

in (6-76) are bounded because the solutions to (6-74) and (6-75) are bounded since, ultimately, the signal functions S_i are bounded.

If $a_i(x_i) > 0$ and $S_i' > 0$, and if we differentiate (6-76) with respect to time, rearrange terms, and use (6-73)–(6-75) to eliminate terms, then L strictly decreases along trajectories, yielding asymptotic stability (and in general exponential convergence). For

$$\dot{L} = -\sum_i \frac{S_i'}{a_i} \dot{x}_i^2 - \frac{1}{2}\sum_i \sum_j \dot{m}_{ij}^2 - \frac{1}{3}\sum_i \sum_j \sum_k \dot{n}_{ijk}^2 \quad < \quad 0 \qquad (6\text{-}78)$$

if any activation or synaptic velocity is nonzero. The strict monotonicity assumption $S_i' > 0$ and (6-78) further imply that $\dot{L} = 0$ if and only if all state variables stop changing: $\dot{x}_i = \dot{m}_{ij} = \dot{n}_{ijk} = 0$ for all i, j, k. All like higher-order ABAMs are globally stable.

Adaptive Resonance ABAMs

Competitive ABAMs (CABAMs) are topologically equivalent to **adaptive resonance theory** (ART) systems [Grossberg, 1982]. The idea behind ART systems is *learn only if resonate*. "Resonance" means joint stability at F_X and F_Y mediated by the forward connections **M** and the backward connections **N**. When $\mathbf{N} = \mathbf{M}^T$, we describe ART and activation dynamics with some instance of the heteroassociative equations

$$\dot{x}_i = -a_i(x_i)\left[b_i(x_i) - \sum_j^p S_j(y_j)m_{ij} - \sum_k^n S_k(x_k)r_{ki}\right] \qquad (6\text{-}79)$$

$$\dot{y}_j = -a_j(y_j)\left[b_j(y_j) - \sum_i^n S_i(x_i)m_{ij} - \sum_l^p S_l(y_l)s_{lj}\right] \qquad (6\text{-}80)$$

ART models are CABAM models if synapses learn according to the competitive learning law with steep signal functions S_j. This holds in the ART-2 model [Carpenter, 1987] since shunting dynamics describe the activation dynamics of F_X

and F_Y and, in the notation of Carpenter and Grossberg, competitive learning laws describe the synaptic dynamics with threshold F_Y signal functions:

$$\text{top-down } (F_Y \longrightarrow F_X): \qquad \dot{z}_{ji} = g(y_j)[p_i - z_{ji}] \qquad (6\text{-}81)$$

$$\text{bottom-up } (F_X \longrightarrow F_Y): \qquad \dot{z}_{ij} = g(y_j)[p_i - z_{ij}] \qquad (6\text{-}82)$$

where g denotes a threshold signal function, and p_i denotes the signal pattern (itself involving complicated L^2-norm computations) transmitted from F_X. Equations (6-81) and (6-82) imply that matrix $\mathbf{Z}$ contains forward projections, and its transpose $\mathbf{Z}^T$ contains backward connections.

The CABAM model may not extend the earlier binary ART-1 model if the Weber-law structure occurs in the forward "bottom-up" synaptic projections. Then transposition need not identify the forward and backward connection matrices. This explains in part why binary inputs in ART-2 need not produce ART-1 behavior. In general, for large "L" values [Carpenter, 1987a] ART-1 models define CABAMs.

Differential Hebbian ABAMS

Next we state the form of ABAM systems that adapt (and activate) with signal-velocity information as related by the differential Hebb learning law:

$$\dot{x}_i = -a_i(x_i)[b_i(x_i) - \sum_j S_j m_{ij} - \sum_j \dot{S}_j m_{ij}] \qquad (6\text{-}83)$$

$$\dot{y}_j = -a_j(y_j)[b_j(y_j) - \sum_i S_i m_{ij} - \sum_i \dot{S}_i m_{ij}] \qquad (6\text{-}84)$$

$$\dot{m}_{ij} = -m_{ij} + S_i S_j + \dot{S}_i \dot{S}_j \qquad (6\text{-}85)$$

and the further assumptions $\dot{S}_i \approx \ddot{S}_i$, $\dot{S}_j \approx \ddot{S}_j$. In general [Kosko, 1988] we require only that signal velocities and accelerations tend to have the same sign (as in clipped exponentials). The corresponding Lyapunov function now includes a "kinetic energy" term to account for signal velocities:

$$L = -\sum_i \sum_j S_i S_j m_{ij} - \sum_i \sum_j \dot{S}_i \dot{S}_j m_{ij} + \sum_i \int_0^{x_i} S_i'(\theta_i) b_i(\theta_i) \, d\theta_i$$

$$+ \sum_j \int_0^{y_j} S_j'(\varepsilon_j) b_j(\varepsilon_j) \, d\varepsilon_j + \frac{1}{2} \sum_i \sum_j m_{ij}^2 \qquad (6\text{-}86)$$

We leave the proof that $\dot{L} \leq 0$ as an exercise. At equilibrium the signal-velocity terms disappear from (6-83)–(6-85).

Next we study the *qualitative* behavior of global ABAM equilibria. We shall show that ABAM models are *structurally* stable, robust in the face of random perturbations. This requires recasting the ABAM differential equations as stochastic differential equations and perturbing them with Brownian motions, as discusssed in Chapter 4. First we review the nature of structural stability and the stochastic calculus.

STRUCTURAL STABILITY OF UNSUPERVISED LEARNING

How robust are unsupervised learning systems? What happens if local or global effects perturb synaptic mechanisms in real time? Will shaking disturb or prevent equilibria? What effect will thermal-noise processes, electromagnetic interactions, and component malfunctions have on large-scale implementations of unsupervised neural networks? How biologically accurate are unsupervised neural models that do not model the myriad electro-chemical, molecular, and other processes found at synaptic junctions and membrane potential sites?

These questions underly a more general question: Is unsupervised learning *structurally stable*? **Structural stability** [Gilmore, 1981; Thom, 1975] is insensitivity to small perturbations. Structural stability allows us to perturb globally stable feedback systems without changing their qualitative equilibrium behavior. This increases the reliability of large-scale hardware implementations of such networks. It also increases their biological plausibility, since we in effect model the myriad synaptic and neuronal processes missing from neural-network models. But we model them as net random unmodeled effects that do not affect the structure of the global network computations.

Structural stability differs from the global stability, or convergence to fixed points, that endows some feedback networks with content-addressable memory and other computational properties. Initial conditions can affect globally stable systems. Different input states can converge to different limit states; else memory capacity is trivial. Structural stability ignores many small perturbations. Such perturbations preserve qualitative properties. In particular, basins of attractions maintain their basic shape. Structurally stable attractors can include periodic and chaotic equilibria as well as fixed-point equilibria.

The formal approach to structural stability uses the transversality techniques of differential topology [Hirsch, 1976], the study of global properties of differentiable manifolds. Manifolds A and B have nonempty *transversal* intersection in R^n if the tangent spaces of A and B span R^n at every point of intersection, if locally the intersection looks like R^n. Two lines intersect transversely in the plane but not in 3-space, 4-space, or higher n-space. If we shake the lines in 2-space, they still intersect. If we shake them in 3-space, the lines may no longer intersect. In Figure 6.2, manifolds A and B intersect transversely in the plane at points a and b. Manifolds B and C do not intersect transversely at c.

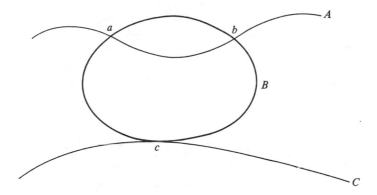

FIGURE 6.2 Manifold intersection in the plane (manifold R^2). Intersection points a and b are transversal. Point c is not: Manifolds B and C need not intersect if even slightly perturbed. No points are transversal in 3-space unless B is a sphere (or other solid). *Source:* Kosko, B., "Unsupervised Learning in Noise," *IEEE Transactions on Neural Networks*, vol. 1, no. 1, March 1990.

We shall approach structural stability indirectly with the calculus of stochastic differential and integral equations. The stochastic-calculus approach abstracts statistically relevant behavior from large sets of functions. The differential-topological approach, in contrast, deals with all possible behavior of all functions (open dense sets of functions). This makes the analysis extremely abstract and makes calculations cumbersome and often impractical.

Below we demonstrate the structural stability of many types of unsupervised learning in the stochastic sense. As in the proof of the AVQ convergence theorem, we use the scalar-valued Lyapunov function of globally stable feedback networks but in a random framework. Then the Lyapunov function defines a random variable at each moment of time t. So we cannot minimize the Lyapunov function as when it was a scalar at each t. Instead we minimize its expectation, its average value, which is a scalar at t.

RANDOM ADAPTIVE BIDIRECTIONAL ASSOCIATIVE MEMORIES

Brownian diffusions perturb **random adaptive bidirectional associative memory** (RABAM) models [Kosko, 1990]. The differential equations in (6-50) through (6-53) now become stochastic differential equations, with random processes as solutions. Every activation and synaptic variable represents a separate stochastic process. The stochastic differential equations relate the time evolution of these stochastic processes. We then add Brownian diffusions, or "noise" processes, to the stochastic differential equations, as discussed in Chapter 4.

Suppose B_i, B_j, and B_{ij} denote Brownian-motion (independent Gaussian increment) processes [Maybeck, 1982; Skorokhod, 1965] that perturb state changes in the ith neuron in F_X, the jth neuron in F_Y, and the synapse m_{ij}, respectively. The Brownian motions may have time-varying diffusion parameters. Then the **diffusion RABAM** corresponds to the adaptive stochastic dynamical system

$$dx_i = -a_i(x_i)[b_i(x_i) - \sum_j S_j(y_j)m_{ij}] \, dt + dB_i \qquad (6\text{-}87)$$

$$dy_j = -a_j(y_j)[b_j(y_j) - \sum_i S_i(x_i)m_{ij}] \, dt + dB_j \qquad (6\text{-}88)$$

$$dm_{ij} = -m_{ij}dt + S_i(x_i)S_j(y_j) \, dt + dB_{ij} \qquad (6\text{-}89)$$

We can replace the signal Hebb diffusion law (6-89) with the stochastic competitive law,

$$dm_{ij} = S_j(y_j)[S_i(x_i) - m_{ij}] \, dt + dB_{ij} \qquad (6\text{-}90)$$

if S_j is sufficiently steep. Alternatively we can replace (6-89) with differential Hebbian or differential competitive diffusion laws if we impose tighter constraints to ensure global stability. For simplicity, we shall formulate the RABAM model in the signal Hebbian case only. The extensions to competitive and differential learning proceed exactly as the above extensions of the ABAM theorem. All RABAM results, like all ABAM results, also extend to high-order systems of arbitrarily high order.

We will develop the RABAM model in the "noise notation" discussed in Chapter 4. Intuitively we add independent zero-mean Gaussian white-noise processes to the ABAM model. The stochastic differential equations then describe the time evolution of network "signals plus noise." This implicitly means that the noise processes are independent of the nonlinear "signal" processes, which we assume. We further assume that the noise processes have finite variances, though these variances may vary with time. Then the signal-Hebbian **noise RABAM model** takes the form

$$\dot{x}_i = -a_i(x_i)[b_i(x_i) - \sum_j S_j(y_j)m_{ij}] + n_i \qquad (6\text{-}91)$$

$$\dot{y}_j = -a_j(y_j)[b_j(y_j) - \sum_i S_i(x_i)m_{ij}] + n_j \qquad (6\text{-}92)$$

$$\dot{m}_{ij} = -m_{ij} + S_i(x_i)S_j(y_j) + n_{ij} \qquad (6\text{-}93)$$

$$E(n_i) = E(n_j) = E(n_{ij}) = 0 \qquad (6\text{-}94)$$

$$V(n_i) = \sigma_i^2 < \infty, \quad \sigma_j^2 < \infty, \quad \sigma_{ij}^2 < \infty \qquad (6\text{-}95)$$

In principle we can add noise within the general b_i and b_j terms, perhaps reflecting random input signals. We discuss this below for additive and shunting activation models. For additive activation models, we can absorb the new noise terms in the noise terms n_i and n_j.

Will so much noise destabilize the system? So much noise with so much feedback might seem to promote chaos, especially since the network dimensions n and p can be arbitrarily large. How can patterns stably form and learning occur?

The RABAM theorem ensures stochastic stability. Nonlinear interactions suppress noise and suppress it exponentially quickly. In effect, RABAM equilibria are ABAM equilibria that randomly vibrate. The noise variances control the range of vibration. Average RABAM behavior equals ABAM behavior. Since noise perturbations do not destroy equilibria, the RABAM theorem states that unsupervised learning is structurally stable in the stochastic sense. The result applies with equal force, though with less theoretical interest, for unsupervised learning in feedforward networks.

We can motivate the RABAM theorem with a simple thought experiment or, better, a few hand calculations. Consider a discrete additive BAM with fixed matrix M, as described in Chapter 3. Find its bipolar fixed points in the product space $\{-1, 1\}^n \times \{-1, 1\}^p$. Now add a small amount of zero-mean noise to each memory element m_{ij}. Since a discrete BAM signal function is a threshold function, it is unlikely that more than very few neurons, if any, change state differently during iterations than they did before. It is even less likely that they will do so as n and p increase. The system tends to reach the same fixed points, and persist in them once reached. This corresponds to adding noise at the synaptic level. Now repeat the computation, but also add zero-mean noise to each neuron's activation at each iteration. Then repeat this computation, adding new noise to the matrix M each time. This allows the synaptic noise processes to be "slower" than the neuronal noise processes. Again the threshold signal functions make it unlikely that the signal patterns will change significantly, if at all, during convergence iterations or in equilibrium. Neuronal nonlinearities suppress noise.

RABAM Theorem. The RABAM model (6-87)–(6-90), or (6-91)–(6-95), is globally stable. If signal functions are strictly increasing and amplification functions a_i and a_j are strictly positive, the RABAM model is asymptotically stable.

Proof. The ABAM Lyapunov function L in (6-61) now defines a random process. At each time t, $L(t)$ equals a random variable. We conjecture that *the expected ABAM Lyapunov function E(L) is a Lyapunov function for the RABAM system*. We take the expectation with respect to all random parameters:

$$E(L) = \int \dots \int L\, p(\mathbf{x}, \mathbf{y}, \mathbf{M})\, d\mathbf{x}\, d\mathbf{y}\, d\mathbf{M} \qquad (6\text{-}96)$$

Recall that each activation and synaptic parameter represents a random process separate from the random process gotten simply by adding noise to a deterministic variable. Since all noise processes are independent of RABAM variables, we have factored the marginal noise densities out of $p(\mathbf{x}, \mathbf{y}, \mathbf{M})$ and integrated them out of the expectation.

The proof replaces the time derivative of the expectation with the expectation of the time derivative of the ABAM Lyapunov function, which we calculated above. Technically we need to assume sufficient smoothness conditions on the RABAM model to bring the time derivative inside the multiple integrals in (6-96). Then

$$\dot{E}(L) \;=\; E(\dot{L}) \tag{6-97}$$

$$= \; E\Big\{\sum_i S_i' \dot{x}_i [b_i - \sum_j S_j m_{ij}] + \sum_j S_j' \dot{y}_j [b_j - \sum_i S_i m_{ij}]$$

$$- \sum_i \sum_j \dot{m}_{ij} [-m_{ij} + S_i S_j]\Big\} \tag{6-98}$$

$$= \; E\Big\{ -\sum_i S_i' a_i [b_i - \sum_j S_j m_{ij}]^2 - \sum_j S_j' a_j [b_j - \sum_i S_i m_{ij}]^2$$

$$- \sum_i \sum_j [-m_{ij} + S_i S_j]^2 \Big\} + \sum_i E\{n_i S_i' [b_i - \sum_j S_j m_{ij}]\}$$

$$+ \sum_j E\{n_j S_j' [b_j - \sum_i S_i m_{ij}]\} - \sum_i \sum_j E\{n_{ij} [-m_{ij} + S_i S_j]\} \tag{6-99}$$

upon eliminating the activation and synaptic velocities in (6-98) with the RABAM dynamical equations (6-91) through (6-93),

$$= E[\dot{L}_{\text{ABAM}}] \; + \; \sum_i E(n_i) E\{S_i' [b_i - \sum_j S_j m_{ij}]\}$$

$$+ \; \sum_j E(n_j) E\{S_j' [b_j - \sum_i S_i m_{ij}]\}$$

$$- \; \sum_i \sum_j E(n_{ij}) E[-m_{ij} + S_i S_j] \tag{6-100}$$

by independence of the "signal" and additive noise terms in the RABAM model, and by the facts that S_i' and S_j' are nonnegative functions of x_i and y_j, respectively, and a_i and a_j are nonnegative essentially arbitrary functions (so $S_i' = a_i$ and $S_j' = a_j$ possible),

$$= E[\dot{L}_{\text{ABAM}}] \tag{6-101}$$

by (6-94). So $\dot{E}(L) \leq 0$ or $\dot{E}(L) < 0$ along trajectories according as $\dot{L}_{\text{ABAM}} \leq 0$ or $\dot{L}_{\text{ABAM}} < 0$. **Q.E.D.**

Noise-Saturation Dilemma and the RABAM Noise-Suppression Theorem

How much do RABAM trajectories and equilibria vibrate? To answer this question we need to examine the *second-order* behavior of the RABAM model. This behavior depends fundamentally on the variances of the additive noise processes. The zero-mean assumption (6-94) implies that the time-varying "variances" σ_i^2, σ_j^2, and σ_{ij}^2 equal the respective instantaneous mean-squared "noises" $E(n_i^2)$, $E(n_j^2)$, and $E(n_{ij}^2)$, since in general $V(x) = E(x^2) - E^2(x)$.

Observed RABAM second-order behavior consists of the observed instantaneous *mean-squared velocities* $E(\dot{x}_i^2)$, $E(\dot{y}_j^2)$, and $E(\dot{m}_{ij}^2)$. The mean-squared velocities measure the magnitude of instantaneous RABAM change. They are at least as large as the underlying instantaneous "variances" of the activation velocity and synaptic velocity processes, since, for example,

$$E(\dot{x}_i^2) \geq E(\dot{x}_i^2) - E^2(\dot{x}_i) = V(\dot{x}_i) \tag{6-102}$$

Intuitively the mean-squared velocities should depend on the instantaneous "variances" of the noise processes in (6-91) through (6-95). The more the noise processes hop about their means, the greater the potential for the activations and synapses to change state. But this intuition seems to run counter to the structural stability we just established with the RABAM theorem. If the magnitudes of the noise fluctuations grow arbitrarily large, the system may reach a point—and perhaps reach it quickly in the midst of massive noisy feedback—where the RABAM system transitions from stability to instability.

The RABAM noise-suppression theorem guarantees that no noise processes can destabilize a RABAM *if* the noise processes have *finite* instantaneous variances (and zero mean). Cauchy noise, for example, could destabilize a RABAM, since it has infinite variance. In practice, even Cauchy variance is finite, and so it will never destabilize a RABAM. Theoretically, for infinite variance, Cauchy RABAMs are only "piecewise convergent."

Grossberg's [1982] **noise-saturation dilemma** motivates the use of the term "noise suppression" in the RABAM corollary below. The noise-saturation dilemma, discussed in Chapter 3, asks how neurons can have an effective infinite dynamical range when they operate between upper and lower bounds and yet not treat small input signals as noise: "If the x_i are sensitive to large inputs, then why do not small inputs get lost in internal system noise? If the x_i are sensitive to small inputs, then why do they not all saturate at their maximum values in response to large inputs?" [Grossberg, 1988].

Grossberg resolves the saturation half of the dilemma by showing, as we saw in Chapter 3, that shunting models remain sensitive to relative pattern information over a wide range of inputs. He also shows that additive models quickly saturate to upper bounds for large inputs.

Grossberg resolves the noise half of the noise-saturation dilemma only partially. As we saw in Chapter 3, Grossberg argues that noisy patterns are uniform input patterns and that, for a particular small threshold value, neurons suppress uniform noise by shutting off. Besides the dependence on a specific noise threshold, this involves two difficulties. First, noise permeates all parameters and all signals and need not be uniform. Grossberg refers to this in his above description of the noise-saturation dilemma when he asks why small inputs do not "get lost in internal system noise." System noise makes everything "jiggle," including relative input pattern values. RABAMs model this noise with additive noise processes. Second, shutting off neurons to suppress noise precludes neural computation in noise. Neurons and synapses must continually process signals and noise.

RABAM Noise-Suppression Theorem

The RABAM noise-suppression theorem directly resolves the noise half of the noise-saturation dilemma. It proves that the instantaneous mean-squared velocities $E[\dot{x}_i^2]$, $E[\dot{y}_j^2]$, and $E[\dot{m}_{ij}^2]$ decrease exponentially quickly to their lower bounds. The lower bounds equal the underlying (unknown) instantaneous noise "variances." The following lemma summarizes these theoretical lower bounds on RABAM second-order behavior:

Lemma

$$E[\dot{x}_i^2] \geq \sigma_i^2, \qquad E[\dot{y}_j^2] \geq \sigma_j^2, \qquad E[\dot{m}_{ij}^2] \geq \sigma_{ij}^2 \qquad (6\text{-}103)$$

This lemma extends the lemma in Chapter 4.

The lemma makes explicit the need for the finite-variance assumption (6-95). The lemma follows if we square both sides of (6-91), (6-92), and (6-93), take expectations on both sides, factor the cross-product expectation in each expression with the independence of "signal" and noise terms, then use the zero-mean assumption (6-63) to eliminate the cross-product terms and to obtain, for example, $E[n_i^2] = V[n_i]$. This proof technique does not depend on a specific learning law if the noise is additive.

The lemma states that noise fluctuations drive system fluctuations. It also eliminates the possibility of ideal noise suppression. At *stochastic* equilibrium the dynamical system still vibrates, still "jiggles" about the deterministic equilibrium. The RABAM noise-suppression theorem guarantees that equality holds in (6-103), and equality holds exponentially quickly. In this sense RABAM models optimally suppress noise in real time. The RABAM theorem and the RABAM noise-suppression theorem imply that we cannot finitely shake an ABAM equilibrium out of equilibrium, or, more accurately, out of its equilibrium shape.

The mean-squared velocities in (6-103) exceed the variances of the activation and synaptic variances, since, for example,

$$V[\dot{x}_i] \;=\; E[\dot{x}_i^2] - E^2[\dot{x}_i] \;\leq\; E[\dot{x}_i^2] \qquad (6\text{-}104)$$

Inequality (6-104) implies that observed system fluctuations, mean-squared velocities, never underestimate the variance of state changes. It also implies that at equilibrium, when equality holds in (6-103), the variances of the activation and synaptic velocities are bounded *above* by the noise variances.

The lemma does not imply that the squared velocity processes must exceed or equal the squared noise processes at every instant. It only implies that the inequality holds *on average* at every instant. The expectation computes the average as a "space" or ensemble average, not a time average. The space corresponds to the sample space Ω consisting of all possible ω in each random realization $x_i(\omega, t)$.

Pathologies can occur if the instantaneous squared noise exceeds the instantaneous squared activation velocity. No problems can arise for synaptic velocities, as we shall see in the proof of the RABAM noise-suppression theorem. The proof also shows that we avoid such pathologies if we require that the random weight functions, S_i'/a_i and S_j'/a_j , are "well-behaved." Well-behavedness means that two integrals stay nonnegative:

$$E\,[\frac{S_i'}{a_i}\,(\dot{x}_i^2 - n_i^2)] \;\geq\; 0 \qquad (6\text{-}105)$$

$$E\,[\frac{S_j'}{a_j}\,(\dot{y}_j^2 - n_j^2)] \;\geq\; 0 \qquad (6\text{-}106)$$

Suppose the random difference function $\dot{x}_i^2 - n_i^2$ is negative for a particular realization. If the positive random weight function S_i'/a_i is sufficiently large, the overall integral, the expectation, could be negative. This would be true, to take an extreme example, if the random weight function behaved as a Dirac delta function. We assume in (6-105) and (6-106) that the weight functions do not place too much weight on the infrequent negative realizations of the difference functions.

RABAM Noise Suppression Theorem. As the above RABAM dynamical systems converge exponentially quickly, the mean-squared velocities of neuronal activations and synapses decrease to their lower bounds exponentially quickly:

$$E[\dot{x}_i^2] \downarrow \sigma_i^2, \quad E[\dot{y}_j^2] \downarrow \sigma_j^2, \quad E[\dot{m}_{ij}^2] \downarrow \sigma_{ij}^2 \qquad (6\text{-}107)$$

for strictly positive amplification functions a_i and a_j, monotone-increasing signal functions, well-behaved weight ratios S_i'/a_i and S_j'/a_j [so (6-105) and (6-106) hold], sufficient smoothness to permit time differentiation within the expectation integral, and nondegenerate Hessian conditions.

Proof. As in the proof of the general RABAM theorem, we use the smoothness condition to expand the time derivative of the bounded Lyapunov function $E[L]$:

$$\dot{E}[L] \;=\; E[\dot{L}] \qquad (6\text{-}108)$$

$$=\; E\{\sum_i S_i'\dot{x}_i[b_i - \sum_j S_j m_{ij}] + \sum_j S_j'\dot{y}_j[b_j - \sum_i S_i m_{ij}]$$

$$-\sum_i \sum_j \dot{m}_{ij}[-m_{ij} + S_i S_j]\} \qquad (6\text{-}109)$$

$$= \quad E\{-\sum_i \frac{S_i'}{a_i}\dot{x}_i^2 - \sum_j \frac{S_j'}{a_j}\dot{y}_j^2 - \sum_i\sum_j \dot{m}_{ij}^2 + \sum_i \frac{S_i'}{a_i}\dot{x}_i n_i$$

$$+ \sum_j \frac{S_j'}{a_j}\dot{y}_j n_j + \sum_i\sum_j n_{ij}\dot{m}_{ij}\} \qquad (6\text{-}110)$$

by using the positivity of the random amplification functions to eliminate the three terms in braces in (6-109) with (6-91), (6-92), and (6-93), respectively,

$$= \quad E[\dot{L}_{\text{ABAM}}] + E[\sum_i \frac{S_i'}{a_i}n_i^2 + \sum_j \frac{S_j'}{a_j}n_j^2 + \sum_i\sum_j n_{ij}^2] \qquad (6\text{-}111)$$

$$= \quad -\sum_i E[\frac{S_i'}{a_i}(\dot{x}_i^2 - n_i^2)] - \sum_j E[\frac{S_j'}{a_j}(\dot{y}_j^2 - n_j^2)]$$

$$- \sum_i\sum_j [E[\dot{m}_{ij}^2] - \sigma_{ij}^2] \qquad (6\text{-}112)$$

We already know from the proof of the general RABAM theorem above that, in the positivity case assumed here, this Lyapunov function strictly decreases along system trajectories. So the system is asymptotically stable and equilibrates exponentially quickly. (Stability holds for different learning laws under appropriate conditions.) This itself, though, does not guarantee that each summand in the first two sums in (6-112) is nonnegative. The well-behavedness of the weight functions S_i'/a_i and S_j'/a_j guarantees this, for then (6-105) and (6-106) hold. The lemma guarantees that the summands in the third sum, the synaptic sum, are nonnegative. Then asymptotic stability ensures that *each* summand in (6-112) decreases to zero exponentially quickly. This gives the desired exponential decrease of the synaptic mean-squared velocities to their lower bounds. For the activation mean-squared velocities, we find at equilibrium:

$$E[\frac{S_i'}{a_i}\dot{x}_i^2] \quad = \quad E[\frac{S_i'}{a_i}n_i^2] \qquad (6\text{-}113)$$

with a like equilibrium condition for each neuron in the F_Y field. Then, because the integrands in (6-113) are nonnegative, we can peel off the expectations almost everywhere [Rudin, 1974] (except possibly on sets of zero probability):

$$\frac{S_i'}{a_i}\dot{x}_i^2 \quad = \quad \frac{S_i'}{a_i}n_i^2 \qquad \text{almost everywhere} \qquad (6\text{-}114)$$

or

$$\dot{x}_i^2 \quad = \quad n_i^2 \qquad \text{almost everywhere} \qquad (6\text{-}115)$$

Taking expectations on both sides now gives the desired equilibrium conditions:

$$E[\dot{x}_i^2] = \sigma_i^2 \quad \text{and} \quad E[\dot{y}_j^2] = \sigma_j^2 \qquad (6\text{-}116)$$

This proves the theorem since the system reaches these equilibrium values exponentially quickly, and since the lemma guarantees they are lower bounds on the mean-squared velocities. **Q.E.D.**

The RABAM noise-suppression theorem generalizes the squared-velocity condition of the ABAM theorem. In the ABAM case, the instantaneous "variances" equal zero. This corresponds to the deterministic case. The probability space has a degenerate sigma-algebra, which contains only two "events," the whole space and the empty set. So expectations disappear and the RABAM squared velocities equal zero everywhere at equilibrium, as in the ABAM case.

The RABAM noise-suppression theorem implies that average RABAM equilibria approach ABAM equilibria exponentially quickly. In estimation-theory terms, RABAM systems define *unbiased* estimators of ABAM systems awash in noise. We can establish unbiasedness directly, and stochastically, if we integrate the RABAM equations, take expectations, and exploit the zero-mean nature of Brownian motion [Maybeck, 1982]. This approach ignores the convergence-rate information that the RABAM noise-suppression theorem provides. We now use the RABAM noise-suppression theorem to characterize the *asymptotic* unbiasedness of RABAM equilibria.

Consider again the ABAM model (6-50)–(6-52). For maximum generality, we can assume the ABAM model defines a set of stochastic differential equations. At equilibrium the activations of the neurons in the field F_X, for example, obey the condition

$$b_i(x_i) \quad = \quad \sum_j S_j(y_j) m_{ij} \qquad (6\text{-}117)$$

since the amplification functions a_i are strictly positive. In the additive case the function $b_i(x_i)$ equals a linear function of x_i. In particular, $b_i(x_i) = x_i - I_i$. Then at equilibrium the neuronal activation in (6-117) equals the sum of the ith neuron's synapse-weighted feedback input signals and external input,

$$x_i \quad = \quad \sum_j S_j m_{ij} + I_i \qquad (6\text{-}118)$$

In the *shunting* case, we restrict the neuronal activation value to a bounded interval. The equilibrium activation value then has the ratio form of a Weber law [Grossberg, 1988]. At equilibrium the synaptic values equal Hebbian products,

$$m_{ij} \quad = \quad S_i S_j \qquad (6\text{-}119)$$

in the signal Hebbian case, or equal presynaptic signals

$$m_{ij} \quad = \quad S_j \qquad (6\text{-}120)$$

in the competitive learning case for the ith "winning" neuron, if S_j behaves approximately as a zero-one threshold signal function. We now prove that RABAM equilibria obey equilibrium conditions with the same average shape as (6-117) through (6-120).

Unbiasedness Corollary. Average RABAM equilibria are ABAM equilibria. Under the assumptions of the RABAM noise-suppression theorem, $b_i(x_i)$ converges to $\sum_j S_j m_{ij}$ in the mean-square sense exponentially quickly. At equilibrium,

$$b_i(x_i) \;=\; \sum_j S_j m_{ij} \quad \text{with probability one} \tag{6-121}$$

and similarly for the b_j functions. The synaptic values similarly converge to the relationships in (6-119) and (6-120) exponentially quickly.

Proof. We reexamine the lemma (6-103). Squaring the RABAM stochastic differential equations (6-91)–(6-93), taking expectations, and invoking the independence of the scaled noise terms and (6-94), gives

$$E[\dot{x}_i^2] \;=\; E[a_i^2[b_i - \sum_j S_j m_{ij}]^2] + \sigma_i^2 \tag{6-122}$$

$$E[\dot{y}_j^2] \;=\; E[a_j^2[b_j - \sum_i S_i m_{ij}]^2] + \sigma_j^2 \tag{6-123}$$

$$E[\dot{m}_{ij}^2] \;=\; E[(-m_{ij} + S_i S_j)^2] + \sigma_{ij}^2 \tag{6-124}$$

or, if the noisy competitive learning law replaces the noisy signal Hebb law (6-93), we replace (6-124) with

$$E[\dot{m}_{ij}^2] \;=\; E[S_j^2(S_i - m_{ij})^2] + \sigma_{ij}^2 \tag{6-125}$$

The right-hand side of each equation is nonnegative, and the RABAM noise-suppression theorem implies that exponentially quickly the mean-squared "signal" terms decrease to zero. Since the weighting functions a_i^2 and a_j^2 are strictly positive, this proves the asserted mean-square convergence. The strict positivity of the weighting functions further implies [Rudin, 1974] the equilibrium conditions

$$E[(b_i - \sum_j S_j m_{ij})^2] \;=\; 0 \tag{6-126}$$

$$E[(b_j - \sum_i S_i m_{ij})^2] \;=\; 0 \tag{6-127}$$

$$E[(-m_{ij} + S_i S_j)^2] \;=\; 0 \tag{6-128}$$

Then the nonnegativity of the integrands allows us to remove the expectations almost everywhere (with probability one). This yields the equilibrium condition. **Q.E.D.**

An important extension of the corollary for stochastic estimation holds in the additive case with random noisy input waveform $[I_1', \ldots, I_n' \mid J_1', \ldots, J_p']$. We can view this random waveform, which may vary slowly with time, as a set of random inputs in the additive random ABAM case perturbed with additive zero-mean finite-variance noise processes:

$$I_i' \;=\; I_i + v_i \tag{6-129}$$

$$J'_j = J_j + w_j \qquad (6\text{-}130)$$

where $E[v_i] = E[w_j] = 0$.

In the additive case, the new noise processes in (6-129) and (6-130) combine additively with the other zero-mean noise processes in the RABAM model. We can similarly scale these new noise terms with different nonnegative deterministic annealing schedules, as discussed below. The new scaled noise terms require that we rescale the variances in the lemma and the RABAM noise-suppression theorem, in addition to appropriately adjusting the variance terms themselves.

Then the additive RABAM model uses constant amplification functions and $b_i^R = x_i - I'_i$ and $b_j^R = y_j - J'_j$, while the additive ABAM uses $b_i^A = x_i - I_i$ and $b_j^A = y_j - J_j$, where again I_i and J_j may be random. Then, at equilibrium, (6-117), (6-118), (6-121), and (6-129) give

$$E[x_i^R - x_i^A] = E[b_i^R + I'_i - b_i^A - I_i] \qquad (6\text{-}131)$$

$$= E[b_i^R - b_i^A] + E[v_i] \qquad (6\text{-}132)$$

$$= 0 \qquad (6\text{-}133)$$

and similarly for the y_j activations. So at equilibrium the RABAM system behaves as an unbiased estimator of the random ABAM system: $E[x_i^R] = E[x_i^A]$.

RABAM Annealing

Gradient systems are globally stable. The above theorems apply this general Lyapunov fact. For example, Cohen and Grossberg [1983] showed that their symmetric nonlearning autoassociative system takes a pseudogradient form for monotone-increasing signal functions and positive amplification functions.

Geman and Hwang [1986] showed that stochastic gradient systems with scaled additive Gaussian white noise perform **simulated annealing** in a weak sense. We form the gradient from a cost function, which we "search" with scaled random hill-climbing. If we scale the initial noise high enough, perhaps to a physically unrealizable size, then gradually decreasing the nonnegative "temperature" scaling factor $T(t)$ can bounce the system state out of local minima and trap it in global minima. Simulated annealing varies from pure random search to "greedy" gradient descent as the temperature $T(t)$ ranges from infinity to zero.

The system must converge exponentially *slowly*. We can achieve convergence only in the weak sense [Rudin, 1972] for measures (analogous to the convergence in distribution found in central limit theorems). Annealing fails for convergence with probability one and convergence in probability. With some probability the system state will bounce out of global or near-global minima as the system "cools."

Szu [1987] showed that using a Cauchy distribution—or symmetric probability density with infinite variance—to govern the random search process leads to a *linear* convergence rate. The occasional infrequent distant samples provide a stochastic fault-tolerance mechanism for bouncing the system out of local cost minima.

We now extend the RABAM theorem and RABAM noise-suppression theorem to include simulated annealing. We introduce the activation "**temperatures**" or **annealing schedules** $T_i(t)$ and $T_j(t)$ and the synaptic schedules $T_{ij}(t)$. The temperatures are nonnegative *deterministic* functions. So we can factor them outside all expectations in proofs. We allow the system to anneal both neurons and synapses.

The **RABAM annealing model** scales the independent Gaussian white-noise processes with the square-root of the annealing schedules:

$$\dot{x}_i = -a_i[b_i - \sum_j S_j m_{ij}] + \sqrt{T_i} n_i \qquad (6\text{-}134)$$

$$\dot{y}_j = -a_j[b_j - \sum_i S_i m_{ij}] + \sqrt{T_j} n_j \qquad (6\text{-}135)$$

$$\dot{m}_{ij} = -m_{ij} + S_i S_j + \sqrt{T_{ij}} n_{ij} \qquad (6\text{-}136)$$

We can replace (6-136) with other unsupervised learning if we impose appropriate additional constraints.

RABAM Annealing Theorem.   The RABAM annealing model is globally stable, and asymptotically stable for monotone-increasing signal functions and positive amplification functions, in which case the mean-squared activation and synaptic velocities decrease to the temperature-scaled instantaneous "variances" exponentially fast:

$$E(\dot{x}_i^2) \downarrow T_i \sigma_i^2, \quad E(\dot{y}_j^2) \downarrow T_j \sigma_j^2, \quad E(\dot{m}_{ij}^2) \downarrow T_{ij} \sigma_{ij}^2 \qquad (6\text{-}137)$$

Proof. The proof largely duplicates the proofs of the RABAM theorem and RABAM noise-suppression theorem. Again $E(L)$ denotes a sufficiently smooth Lyapunov function that allows time differentiation of the integrand. When we use the diffusion or noise RABAM annealing equations to eliminate activation and synaptic velocities in the time-differentiated Lyapunov function, we can factor the temperature functions outside all expectations. The nonnegativity of the temperature functions keeps them from affecting the structure of the expanded time derivative of $E(L)$. We assume the random weight functions S'/a are well-behaved in the sense of (6-105) and (6-106). They keep the expectations in which they occur nonnegative. We can extend the above lemma to show, for instance, that the mean-squared velocity $E(\dot{x}_i^2)$ is bounded below by $T_i \sigma_i^2$. This generalizes (6-112) to

$$\dot{E}(L) = -\sum_i E\left[\frac{S_i'}{a_i}(\dot{x}_i^2 - T_i n_i^2)\right] - \sum_j E\left[\frac{S_j'}{a_j}(\dot{y}_j^2 - T_j n_j^2)\right]$$

$$-\sum_i \sum_j (E(\dot{m}_{ij}^2) - T_{ij}\sigma_{ij}^2) \qquad (6\text{-}138)$$

Q.E.D.

The RABAM annealing theorem provides a nonlinear and continuous generalization of Boltzmann machine learning [Rumelhart, Hinton, 1986] if learning is Hebbian and slow. The **Boltzmann machine** [Fahlman, 1983] uses discrete symmetric additive autoassociative dynamics [Grossberg, 1988]. The system anneals with binary neurons during periods of Hebbian and anti-Hebbian learning. Here Hebbian learning corresponds to (6-136) with $T_{ij}(t) = 0$ for all t. Anti-Hebbian learning further replaces the Hebb product $S_i S_j$ in (6-136) with the negative product $-S_i S_j$.

In principle anti-Hebbian learning (during "free-running" training) can destabilize a RABAM system. This tends not to occur for slow anti-Hebbian learning. The activation terms in the time derivative of $E(L)$ stay negative and can outweigh the possibly positive anti-Hebbian terms, even if learning is fast.

Unsupervised RABAM learning and temperature-supervised annealing learning differ in how they treat noise. Simulated annealing systems search or learn with noise. Unsupervised RABAM systems learn *despite* noise. During "cooling," the continuous annealing schedules define the flow of RABAM equilibria in the product state space of continuous nonlinear random processes. Relation (6-137) implies that no finite temperature value, however large, can destabilize a RABAM.

REFERENCES

Anderson, J. A., Silverstein, J. W., Ritz, S. R., and Jones, R. S., "Distinctive Features, Categorical Perception, and Probability Learning: Some Applications of a Neural Model," *Psychological Review*, vol. 84, 413-451, 1977.

Anderson, J. A., "Cognitive and Psychological Computations with Neural Models," *IEEE Transactions on Systems, Man, and Cybernetics*, vol. SMC-13, 799-815, September/October 1983.

Amari, S., "Neural Theory of Association and Concept Formation," *Biological Cybernetics*, vol. 26, 175-185, 1977.

Carpenter, G. A., and Grossberg, S., "Massively Parallel Architecture for a Self-Organizing Neural Pattern Recognition Machine," *Computer Vision, Graphics, and Image Processing*, vol. 37, 54-115, 1987.

Carpenter, G. A., Cohen, M. A., and Grossberg, S., "Computing with Neural Networks," *Science*, vol. 235, 1226-1227, 6 March 1987.

Carpenter, G. A., and Grossberg, S., "ART 2: Self-Organization of Stable Category Recognition Codes for Analog Input Patterns," *Applied Optics*, vol. 26, no. 23, 4919-4930, 1 December 1987.

Cohen, M. A., and Grossberg, S., "Absolute Stability of Global Pattern Formation and Parallel Memory Storage by Competitive Neural Networks," *IEEE Transactions on Systems, Man, and Cybernetics*, vol. SMC-13, 815-826, September 1983.

Cohen, M. A., and Grossberg, S., "Masking Fields: A Massively Parallel Neural Architecture for Learning, Recognizing, and Predicting Multiple Groupings of Patterned Data," *Applied Optics*, vol. 26, 1866, 1987.

DeSieno, D., "Adding a Conscience to Competitive Learning," *Proceedings of the 2nd IEEE International Conference on Neural Networks (ICNN-88)*, vol. I, 117-124, July 1988.

Fahlman, S. E., Hinton, G. E., and Sejnowski, T. J., "Massively Parallel Architectures for AI: NETL, Thistle and Boltzmann Machines," *Proceedings of the AAAI-83 Conference*, 1983.

Geman, S., and Hwang, C., "Diffusions for Global Optimization," *SIAM Journal of Control and Optimization*, vol. 24, no. 5, 1031-1043, September 1986.

Giles, C. L., and Maxwell, T., "Learning, Invariance, and Generalization in High-Order Neural Networks," *Applied Optics*, vol. 26, no. 23, 4973-4978, December 1, 1987.

Gilmore, R., *Catastrophe Theory for Scientists and Engineers*, Wiley, New York, 1981.

Grossberg, S., "Adaptive Pattern Classification and Universal Recoding: I. Parallel Development and Coding of Neural Feature Detectors," *Biological Cybernetics*, vol. 23, 187-202, 1976.

Grossberg, S., *Studies of Mind and Brain*, Reidel, Boston, 1982.

Grossberg, S., "Nonlinear Neural Networks: Principles, Mechanisms, and Architectures," *Neural Networks*, vol. 1, no. 1, 17-61, 1988.

Hirsch, M. W., and Smale, S., *Differential Equations, Dynamical Systems, and Linear Algebra*, Academic Press, Orlando, FL, 1974.

Hodgkin, A. L., and Huxley, A. F., "A Quantitative Description of Membrane Current and Its Application to Conduction and Excitation in a Nerve," *Journal of Physiology*, vol. 117, 500-544, 1952.

Hopfield, J. J., "Neural Networks and Physical Systems with Emergent Collective Computational Abilities," *Proceedings of the National Academy of Science*, vol. 79, 2554-2558, 1982.

Hopfield, J. J., "Neural Networks with Graded Response Have Collective Computational Properties Like Those of Two-State Neurons," *Proceedings of the National Academy of Science*, vol. 81, 3088-3092, May 1984.

Hopfield, J. J., and Tank, D. W., "Neural Computation of Decisions in Optimization Problems," *Biological Cybernetics*, vol. 52, 141-152, 1985.

Kohonen, T., *Self-Organization and Associative Memory*, 2nd ed., Springer-Verlag, New York, 1988.

Kong, S.-G., and Kosko, B., "Differential Competitive Learning for Centroid Estimation and Phoneme Recognition," *IEEE Transactions on Neural Networks*, vol. 2, no. 1, 118-124, January 1991.

Kosko, B., "Adaptive Bidirectional Associative Memories," *Applied Optics*, vol. 26, no. 23, 4947-4960, December 1987.

Kosko, B., "Bidirectional Associative Memories," *IEEE Transactions on Systems, Man, and Cybernetics*, vol. SMC-18, 49-60, January 1988.

Kosko, B., "Feedback Stability and Unsupervised Learning," *Proceedings of the 1988 IEEE International Conference on Neural Networks (ICNN-88)*, vol. I, 141-152, July 1988.

Kosko, B., "Unsupervised Learning in Noise," *IEEE Transactions on Neural Networks*, vol. 1, no. 1, 44-57, March 1990.

Kosko, B., *Neural Networks for Signal Processing*, Prentice Hall, Englewood Cliffs, NJ, 1991.

Parker, T. S., and Chua, L. O., "Chaos: A Tutorial for Engineers," *Proceedings of the IEEE*, vol. 75, no. 8, 982-1008, August 1987.

Rudin, W., *Real and Complex Analysis*, 2nd ed., McGraw-Hill, New York, 1974.

Szu, H., and Hartley, R., "Fast Simulated Annealing," *Physics Letters*, vol. 1222 (3, 4), 157-162, 1987.

Tank, D. W., and Hopfield, J. J., "Simple 'Neural' Optimization Networks: An A/D Converter, Signal Decision Circuit, and a Linear Programming Circuit," *IEEE Transactions on Circuits and Systems*, vol. CAS-33, no. 5, 533-541, May 1986.

Tank, D. W., and Hopfield, J. J., "Neural Computation by Concentrating Information in Time," *Proceedings of the National Academy of Science*, vol. 84, 1896-1900, April 1987.

Thom, R., *Structural Stability and Morphogenesis*, Addison-Wesley, Reading, MA, 1975.

PROBLEMS

6.1. Show that

$$L = -\frac{1}{2} \sum_i \sum_j T_{ij} V_i V_j + \sum_i \frac{1}{R_i} \int_0^{V_i} S_i^{-1}(W)\, dW - \sum_i I_i V_i$$

is a Lyapunov function for the Hopfield circuit

$$C_i \dot{x}_i = -\frac{x_i}{R_i} + \sum_{j=1}^n T_{ij} V_j + I_i$$

where $T_{ij} = T_{ji}$, $S_i(x_i) = V_i$, $S_i' > 0$, and C_i and R_i are positive constants. Use $S_i^{-1'} > 0$. Suppose $V_i = S_i(cx_i)$, for *gain* constant $c > 0$. Argue why the continuous Hopfield circuit tends to behave as a discrete associate memory for very large gain values.

6.2. Verify that $S_i^{-1'} > 0$ if $S_i(x_i)$ is the logistic signal function

$$S_i(x_i) = \frac{1}{1 + e^{-cx_i}}$$

6.3. Prove that the autoassociative second-order signal Hebbian ABAM

$$\dot{x}_i = -a_i(x_i)[b_i(x_i) - \sum_j S_j(x_j) m_{ij} - \sum_j \sum_k S_j(x_j) S_k(x_k) n_{ijk}]$$

$$\dot{m}_{ij} = -m_{ij} + S_i(x_i) S_j(x_j)$$

$$\dot{n}_{ijk} = -n_{ijk} + S_i(x_i) S_j(x_j) S_k(x_k)$$

is globally stable. Find a global Lyapunov function for an autoassociative kth-order signal Hebbian ABAM.

6.4. Prove that the competitive ABAM

$$\dot{x}_i = -a_i(x_i)[b_i(x_i) - \sum_{j=1}^{p} S_j(y_j)m_{ij} - \sum_{k=1}^{n} S_k(x_k)r_{ki}]$$

$$\dot{y}_j = -a_j(y_j)[b_j(y_j) - \sum_{i=1}^{n} S_i(x_i)m_{ij} - \sum_{l=1}^{p} S_l(y_l)s_{lj}]$$

$$\dot{m}_{ij} = S_j(y_j)[S_i(x_i) - m_{ij}]$$

is globally stable for steep signal functions S_j, $R = R^T$, and $S = S^T$. Extend this theorem by allowing r_{ki} and s_{lj} to adapt according to a stable unsupervised learning law.

6.5. Prove that the differential Hebbian ABAM

$$\dot{x}_i = -a_i(x_i)[b_i(x_i) - \sum_j S_j m_{ij} - \sum_j \dot{S}_j m_{ij}]$$

$$\dot{y}_j = -a_j(y_j)[b_j(y_j) - \sum_i S_i m_{ij} - \sum_i \dot{S}_i m_{ij}]$$

$$\dot{m}_{ij} = -m_{ij} + S_i S_j + \dot{S}_i \dot{S}_j$$

is globally stable if

$$\dot{S}_i \approx \ddot{S}_i, \quad \dot{S}_j \approx \ddot{S}_j$$

6.6. Prove the RABAM annealing lemma:

$$E[\dot{x}_i^2] \geq T_i \sigma_i^2, \quad E[\dot{y}_j^2] \geq T_j \sigma_j^2, \quad E[\dot{m}_{ij}^2] \geq T_{ij} \sigma_{ij}^2$$

6.7. Prove the RABAM annealing theorem:

$$E[\dot{x}_i^2] \downarrow T_i \sigma_i^2, \quad E[\dot{y}_j^2] \downarrow T_j \sigma_j^2, \quad E[\dot{m}_{ij}^2] \downarrow T_{ij} \sigma_{ij}^2$$

6.8. Carefully evaluate the integral

$$\int_{-\infty}^{\infty} \frac{dx}{1 + x^2}$$

Prove that Cauchy-distributed random variables have infinite variance. How does this affect the RABAM annealing theorem?

SOFTWARE PROBLEMS

Part I: Random Adaptive Bidirectional Associative Memory (RABAM)

The accompanying software allows you to construct RABAMs from a wide mix of network parameters. The RABAMs can learn with signal Hebbian, competitive, or differential Hebbian learning laws. You can select several signal functions and noise types. First construct a 4-by-3 signal-Hebb RABAM.

1. Run RABAM.

2. Use Set Load Parameters to set the X-dimension to 4 and the Y-dimension to 3. These values are the program's default dimension values.

3. When the "Run Parameters" dialogue box appears, press ESCAPE to use the Run Parameter default values. Check that the system uses the default Learning Rate of 0.01.

4. Pause briefly. This gives the program time to initialize—to select a random input pattern and compute equilibrium values.

5. Use "Train" to begin learning. You should see the network rapidly suppress noise.

Question 1:   Does the lower-right graph exhibit exponential decrease of the squared error? (The system does not display or compute the *mean*-squared error.)

6. Press "Q" to stop the network.

7. Use "Set Run Parameters" to set the Learning Rate to 0.1.

8. Press ESCAPE to exit "Set Run Parameters." Restart the network. Wait for the program to initialize as before.

9. Use "Train" to begin learning the new pattern.

Question 2:   How does the learning rate affect the squared-error curve?

10. Repeat the above steps with different numbers of F_X and F_Y neurons. Increase the magnitude of the noise on F_X and F_Y neurons and on the input waveform. You may need to decrease the "Step Rate" value for large noise factors.

11. Repeat the above steps with different signal functions.

Question 3:   How does the choice of signal function affect the squared-error curve? The learning rate?

12. Repeat the above steps but with competitive learning and differential Hebbian learning laws. The differential Hebbian law requires the truncated-exponential signal function to ensure stability. The truncated-exponential signal function permits very large noise magnitudes.

13. Repeat the above steps but with uniform and Gaussian noise.

14. Repeat the above steps but with Cauchy noise. Recall that Cauchy-distributed random variables have infinite variance. You might try to prove this.

Question 4:   What convergence behavior does Cauchy noise produce?

When you use the RABAM program, from time to time you should press the SPACE bar to clear scatter diagram in the lower-right corner. This diagram plots two mean-adjusted activation values against each other. As time passes, points should distribute symmetrically about the origin. This reflects the $E[RABAM] = ABAM$ asymptotic unbiasedness property.

Part II: Binary Adaptive Resonance Theory (ART-1)

This program illustrates the binary ART-1 model. Construct a 16-by-4 ART-1 network as follows.

1. Run ART.
2. Use "Set Load Parameters" to set the F_1 x-dimension and y-dimension to 4 and the F_2 x-dimension to 4. Set "L" to 1.1. For large "L" values, ART-1 reduces to a competitive ABAM.
3. Select "Open Network" to create the network.
4. Set the Run Parameter "V" to 0.99. "V" is the "vigilance parameter" or minimal acceptable degree of match.
5. Enter an F_1 pattern by setting to 1 (white) desired nodes in the F_1 display.
6. Use "Step" to watch the network reach its final state. You may need to use "Step" repeatedly as the network serially searches its "grandmother cell."
7. Note the final state. Is this state "resonance," "full," "reset," or "ready"?
8. If the final state was "resonance," note the F_2 node chosen. If the final state was not "resonance," you made an error and need to begin again.

Question 1: Why must the first pattern presented to an ART-1 network always result in "resonance"?

Question 2: Does it always take the same number of steps to learn the first pattern?

9. Use "Edit Input" to turn off (darken) the input array's bottom two rows.

Question 3: With this new input, does the network choose the same F_2 grandmother cell as before?

10. Use "Step" to reach a stable state with the new input.
11. Use "Edit Input" to turn off (darken) the input array's top two rows. Set the bottom to 1 (white).

Question 4: With this new input, does the network choose a new F_2 grandmother cell? Explain.

12. Use "Step" to reach a stable state with the new input.
13. Use "Edit Input" to turn off (darken) the input array's leftmost two rows. Set the rightmost to 1 (white).

Question 5: With this new input, which cell does the network choose?

14. Use "Step" to reach a stable state with the new input.

15. Use "Edit Input" to turn off the input array's rightmost two rows. Set the leftmost to 1.

Question 6:   With this new input, which cell does the network choose?

Question 7:   Have you used all F_2 nodes for training? Which nodes (grandmother cells) represent which patterns?

16. Use "Edit Input" to turn off the input array's center four squares. Set the outermost to 1.

Question 8:   How do you predict the network will respond to this new input?

17. Use "Step" to reach a stable state with the new input.

Question 9:   What does the "FULL" network state mean?

Question 10:   Can the network detect the "FULL" state by itself? (You may need to consult the 1987 Carpenter and Grossberg paper.)

Now experiment with lower "vigilance parameter" values. The lower the vigilance parameter, the more mismatch the network tolerates.

18. Close the network opened above.

19. Set the Run Parameter to "V" to 0.01.

20. Create the ART-1 network as above.

21. Repeat the above encoding experiment.

Question 11:   Is it still possible to "fill up" the network? If so, give an example.

Question 12:   How does the network respond to noisy inputs?

PART TWO

ADAPTIVE
FUZZY SYSTEMS

FUZZINESS VERSUS PROBABILITY

*So far as the laws of mathematics refer to reality, they are not certain.
And so far as they are certain, they do not refer to reality.*

Albert Einstein
Geometrie und Erfahrung

FUZZY SETS AND SYSTEMS

We now explore fuzziness as an alternative to randomness for describing uncertainty. We develop the new *sets-as-points* geometric view of fuzzy sets. This view identifies a fuzzy set with a point in a unit hypercube, a nonfuzzy set with a vertex of the cube, and a fuzzy system as a mapping between hypercubes. Chapter 8 examines fuzzy systems.

Paradoxes of two-valued logic and set theory, such as Russell's paradox, correspond to the midpoint of the fuzzy cube. We geometrically answer the fundamental questions of fuzzy theory—How fuzzy is a fuzzy set? How much is one fuzzy set a subset of another?—with the fuzzy entropy theorem and the fuzzy subsethood theorem.

We develop a new geometric proof of the subsethood theorem. A corollary shows that the apparently probabilistic relative frequency n_A/N equals the deterministic subsethood $S(X, A)$, the degree to which the sample space X is contained in its subset A. So the frequency of successful trials equals the degree to which all trials are successful. We examine recent Bayesian polemics against fuzzy theory in light of the new sets-as-points theorems.

An element belongs to a multivalued or "fuzzy" set to some degree in $[0, 1]$. An element belongs to a nonfuzzy set all or none, 1 or 0. More fundamentally, one set contains another set to some degree. Sets fuzzily contain subsets as well as elements. Subsethood generalizes elementhood. We shall argue that subsethood generalizes probability as well.

FUZZINESS IN A PROBABILISTIC WORLD

Is uncertainty the same as randomness? If we are not sure about something, is it only up to chance? Do the notions of likelihood and probability exhaust our notions of uncertainty?

Many people, trained in probability and statistics, believe so. Some even say so, and say so loudly. These voices often arise from the Bayesian camp of statistics, where probabilists view probability not as a frequency or other objective testable quantity, but as a subjective *state of knowledge*.

Bayesian physicist E. T. Jaynes [1979] says that "any method of inference in which we represent degrees of plausibility by real numbers, is necessarily either equivalent to Laplace's [probability], or inconsistent." He claims physicist R. T. Cox [1946] has proven this as a theorem, a claim we examine below.

More recently, Bayesian statistician Dennis Lindley [1987] issued an explicit challenge: "probability is the only sensible description of uncertainty and is adequate for all problems involving uncertainty. All other methods are inadequate."

Lindley directs his challenge in large part at *fuzzy theory*, the theory that *all things admit degrees*, but admit them deterministically. We accept the probabilist's challenge from the fuzzy viewpoint. We will defend fuzziness with new geometric first principles and will question the reasonableness and the axiomatic status of randomness. The new view is the *sets-as-points view* [Kosko, 1987] of fuzzy sets: A fuzzy set defines a point in a unit-hypercube, and a nonfuzzy set defines a corner of the hypercube.

Randomness and fuzziness differ conceptually and theoretically. We can illustrate some differences with examples. Others we can prove with theorems, as we show below.

Randomness and fuzziness also share many similarities. Both systems describe uncertainty with numbers in the unit interval $[0, 1]$. This ultimately means that both systems describe uncertainty numerically. Both systems combine sets and propositions associatively, commutatively, and distributively. The key distinction concerns how the systems jointly treat a set A and its opposite A^c. Classical set theory demands $A \cap A^c = \emptyset$, and probability theory conforms: $P(A \cap A^c) = P(\emptyset) = 0$. So $A \cap A^c$ represents a probabilistically impossible event. But fuzziness begins when $A \cap A^c \neq \emptyset$.

Questions raise doubt, and doubt suggests room for change. So to commence the exposition, consider the following two questions, one fuzzy and the other probabilistic:

1. Is it always and everywhere true that $A \cap A^c = \emptyset$?

2. Do we *derive* or *assume* the conditional probability operator

$$P(B \mid A) = \frac{P(A \cap B)}{P(A)} \ ? \tag{7-1}$$

The second question may appear less fundamental than the first, which asks whether fuzziness exists. The entropy-subsethood theorem below shows that the first question reduces to the second: We measure the fuzziness of fuzzy set A when we measure how much the superset $A \cup A^c$ is a subset of its own subset $A \cap A^c$, a paradoxical relationship unique to fuzzy theory. In contrast, in probability theory the like relationship is impossible (has zero probability): $P(A \cap A^c \mid A \cup A^c) = P(\emptyset \mid X) = 0$, where X denotes the sample space or "sure event."

The conditioning or subsethood in the second question lies at the heart of Bayesian probabilistic systems. We may accept the absence of a first-principles derivation of $P(B \mid A)$. We may simply agree to take the ratio relationship as an axiom. But the new sets-as-points view of fuzzy sets *derives* its conditioning operator as a theorem from first principles. The history of science suggests that systems that hold theorems as axioms continue to evolve.

The first question asks whether we can logically or factually violate the law of noncontradiction—one of Aristotle's three "laws of thought" along with the laws of excluded middle, $A \cup A^c = X$, and identity, $A = A$. Set fuzziness occurs when, and only when, it is violated. Classical logic and set theory assume that we cannot violate the law of noncontradiction or, equivalently, the law of excluded middle. This makes the classical theory black or white. Fuzziness begins where Western logic ends—where contradictions begin.

RANDOMNESS VS. AMBIGUITY: WHETHER VS. HOW MUCH

Fuzziness describes *event ambiguity*. It measures the degree to which an event occurs, not whether it occurs. Randomness describes the uncertainty of *event occurrence*. An event occurs or not, and you can bet on it. The issue concerns the occurring event: Is it uncertain in any way? Can we unambiguously distinguish the event from its opposite?

Whether an event occurs is "random." To what degree it occurs is fuzzy. Whether an ambiguous event occurs—as when we say there is 20 percent chance of light rain tomorrow—involves compound uncertainties, the probability of a fuzzy event.

We regularly apply probabilities to fuzzy events: small errors, satisfied customers, A students, safe investments, developing countries, noisy signals, spiking neurons, dying cells, charged particles, nimbus clouds, planetary atmospheres, galactic clusters. We understand that, at least around the edges, some satisfied customers can be somewhat unsatisfied, some A students might equally be B+ students, some stars are as much in a galactic cluster as out of it. Events can transition more or less smoothly to their opposites, making classification hard near the midpoint of the transition. But in theory—in formal descriptions and in textbooks—the events and their opposites are black and white. A hill is a mountain if it is at least x meters tall, not a mountain if it is one micron less than x in height [Quine, 1981]. Every molecule in the universe either is or is not a pencil molecule, even those that hover about the pencil's surface.

Consider some further examples. The probability that this chapter gets published is one thing. The degree to which it gets published is another. The chapter may be edited in hundreds of ways. Or typographical errors may distort the chapter.

Question: Does quantum mechanics deal with the probability that an unambiguous electron occupies spacetime points? Or does it deal with the degree to which an electron, or an electron smear, occurs at spacetime points? Does $|\psi|^2 dV$ measure the probability that a random-point electron occurs in infinitesimal volume dV? Or [Kosko, 1991] does it measure the degree to which a deterministic electron cloud occurs in dV? Different interpretation, different universe. Perhaps even existence admits degrees at the quantum level.

Suppose there is 50% chance that there is an apple in the refrigerator (electron in a cell). That is one state of affairs, perhaps arrived at through frequency calculations or a Bayesian state of knowledge. Now suppose there is half an apple in the refrigerator. That is another state of affairs. Both states of affairs are superficially equivalent in terms of their numerical uncertainty. Yet physically, ontologically, they differ. One is "random," the other fuzzy.

Consider parking your car in a parking lot with painted parking spaces. You can park in any space with some probability. Your car will totally occupy one space and totally unoccupy all other spaces. The probability number reflects a frequency history or Bayesian brain state that summarizes which parking space your car will totally occupy. Alternatively, you can park in every space to some degree. Your car will partially, and deterministically, occupy every space. In practice your car will occupy most spaces to zero degree. Finally, we can use numbers in $[0, 1]$ to describe, for each parking space, the occurrence probability of each degree of partial occupancy—probabilities of fuzzy events.

If we *assume* events are unambiguous, as in balls-in-urns experiments, there is no set fuzziness. Only "randomness" remains. But when we discuss the physical universe, every assertion of event ambiguity or nonambiguity is an empirical *hypothesis*. We habitually overlook this when we apply probability theory. Years of such oversight have entrenched the sentiment that uncertainty is randomness, and randomness alone. We systematically assume away event ambiguity. We call the

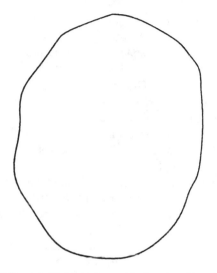

FIGURE 7.1 Inexact oval. Which statement better describes the situation: "It is probably an ellipse" or "It is a fuzzy ellipse"? *Source:* Kosko, B., "Fuzziness vs. Probability," *International Journal of General Systems*, vol. 17, nos. 2-3, Gordon and Breach Science Publishers, 1990.

partially empty glass empty and call the small number zero. This silent assumption of universal nonambiguity resembles the pre-relativistic assumption of an uncurved universe. $A \cap A^c = \emptyset$ is the "parallel postulate" of classical set theory and logic, indeed of Western thought.

If fuzziness is a genuine type of uncertainty, if fuzziness exists, the physical consequences are universal, and the sociological consequence is startling: Scientists, especially physicists, have overlooked an entire mode of reality.

Fuzziness is a type of deterministic uncertainty. Ambiguity is a property of physical phenomena. Unlike fuzziness, probability dissipates with increasing information. After the fact "randomness" looks like fiction. Yet many of the laws of science are time reversible, invariant if we replace time t with time $-t$. If we run the universe in reverse as if it were a video tape, where does the "randomness" go? There is as much ambiguity after a sample-space experiment as before. Increasing information specifies the degrees of occurrence. Even if science had run its course and all the facts were in, a platypus would remain only roughly a mammal, a large hill only roughly a mountain, an oval squiggle only roughly an ellipse. Fuzziness does not require that God play dice.

Consider the inexact oval in Figure 7.1. Does it make more sense to say that the oval is *probably* an ellipse, or that it *is* a fuzzy ellipse? There seems nothing random about the matter. The situation is deterministic: All the facts are in. Yet uncertainty remains. The uncertainty arises from the simultaneous occurrence of two properties: to some extent the inexact oval is an ellipse, and to some extent it is not an ellipse.

More formally, does $m_A(x)$, the degree to which element x belongs to fuzzy set A, equal the probability that x belongs to A? Is $m_A(x) = \text{Prob}\{x \in A\}$ true? Cardinality-wise, sample spaces cannot be too big. Else a positive measure cannot be both countably additive and finite, and thus in general cannot be a probability measure [Chung, 1974]. The space of all possible oval figures is too big, since there are more of these than real numbers. Almost all sets are too big for us to define probability measures on them, yet we can always define fuzzy sets on them.

Probability theory is a chapter in the book of finite measure theory. Many probabilists do not care for this classification, but they fall back upon it when defining terms [Kac, 1959]. How reasonable is it to believe that finite measure theory—ultimately, the summing of nonnegative numbers to unity—exhaustively describes the universe? Does it really describe *any* thing?

Surely from time to time every probabilist wonders whether probability describes anything real. From Democritus to Einstein, there has been the suspicion that, as David Hume [1748] put it, "Though there be no such thing as *chance* in the world, our ignorance of the real cause of any event has the same influence on the understanding and begets a like species of belief." When we model noisy processes by extending differential equations to stochastic differential equations, as in Chapters 4 through 6, we introduce the formalism only as a working approximation to several underlying unspecified processes, processes that presumably obey deterministic differential equations. In this sense conditional expectations and martingale techniques might seem reasonably applied, for example, to stock options or commodity futures phenomena, where the behavior involved consists of aggregates of aggregates of aggregates. The same techniques seem less reasonably applied to quarks, leptons, and void.

THE UNIVERSE AS A FUZZY SET

The world, as Wittgenstein [1922] observed, is everything that is the case. In this spirit we can summarize the ontological case for fuzziness: *The universe consists of all subsets of the universe.* The only subsets of the universe that are not in principle fuzzy are the constructs of classical mathematics. The integer 2 belongs to the even integers, and does not belong to the odd or negative integers. All other sets—sets of particles, cells, tissues, people, ideas, galaxies—in principle contain elements to different degrees. Their membership is partial, graded, inexact, ambiguous, or uncertain.

The same universal circumstance holds at the level of logic and truth. The only logically true or false statements—statements S with truth value $t(S)$ in $\{0, 1\}$—are tautologies, theorems, and contradictions. If statement S describes the universe, if S is an *empirical* statement, then $0 < t(S) < 1$ holds by the canons of scientific method and by the lack of a single demonstrated factual statement S with $t(S) = 1$ or

$t(S) = 0$. Philosopher Immanuel Kant [1787] wrote volumes in search of factually true logical statements and logically true factual statements.

Logical truth differs in kind from factual truth. "$2 = 1 + 1$" has truth value 1. "Grass is green" has truth value less than 1 but greater than 0. This produces the math/universe crisis Einstein laments in his quote at the beginning of this chapter. Scientists have imposed a two-valued mathematics, shot through with logical "paradoxes" or antinomies [Kline, 1980], on a multivalued universe. Last century John Stuart Mill [1843] argued that logical truths represent limiting cases of factual truths. This accurately summarized the truth-value distinction between $0 < t(S) < 1$ and $t(S) = 0$ or $t(S) = 1$ but, cast in linguistic form, seems not to have persuaded modern philosophers. The Heisenberg uncertainty principle, with its continuum of indeterminacy, forced multivaluedness on science, though few Western philosophers [Quine, 1981] have accepted multivaluedness. Lukasiewicz, Gödel, and Black [Rescher, 1969] did accept it and developed the first multivalued or "fuzzy" logic and set systems.

Fuzziness arises from the ambiguity or vagueness [Black, 1937] between a thing A and its opposite A^c. If we do not know A with certainty, we do not know A^c with certainty either. Else by double negation we would know A with certainty. This ambiguity produces nondegenerate *overlap*: $A \cap A^c \neq \emptyset$, which breaks the "law of noncontradiction." Equivalently, it also produces nondegenerate *underlap* [Kosko, 1986b]: $A \cup A^c \neq X$, which breaks the "law of excluded middle." Here X denotes the ground set or universe of discourse. (Probabilistic or stochastic logics [Gaines, 1983] do not break these laws: $P(A \text{ and not-}A) = 0$ and $P(A \text{ or not-}A) = 1$.) Formally, probability measures cannot take fuzzy sets as arguments. We must first quantize, round off, or defuzzify the fuzzy sets to the nearest nonfuzzy sets.

THE GEOMETRY OF FUZZY SETS: SETS AS POINTS

It helps to see the geometry of fuzzy sets when we discuss fuzziness. To date researchers have overlooked this geometry. Instead they have interpreted fuzzy sets as generalized indicator or membership functions [Zadeh, 1965], mappings m_A from domain X to range $[0, 1]$. But functions are hard to visualize. Fuzzy theorists [Klir, 1988] often picture membership functions as two-dimensional graphs, with the domain X represented as a one-dimensional axis. The geometry of fuzzy sets involves both the domain $X = \{x_1, \ldots, x_n\}$ and the range $[0, 1]$ of mappings $m_A \colon X \longrightarrow [0, 1]$. The geometry of fuzzy sets aids us when we describe fuzziness, define fuzzy concepts, and prove fuzzy theorems. Visualizing this geometry may by itself provide the most powerful argument for fuzziness.

An odd question reveals the geometry of fuzzy sets: What does the fuzzy power set $F(2^X)$, the set of all fuzzy subsets of X, look like? It looks like a cube. What does a fuzzy set look like? A point in a cube. The set of all fuzzy subsets

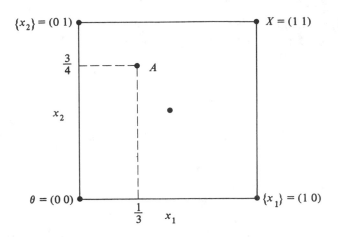

FIGURE 7.2 Sets as points. The fuzzy subset A is a point in the unit 2-cube with coordinates or fit values $(\frac{1}{3} \ \frac{3}{4})$. The first element x_1 fits in or belongs to A to degree $\frac{1}{3}$, the element x_2 to degree $\frac{3}{4}$. The cube consists of all possible fuzzy subsets of two elements $\{x_1, x_2\}$. The four corners represent the power set 2^X of $\{x_1, x_2\}$. Long diagonals connect nonfuzzy set complements. *Source:* Kosko, B., "Fuzziness vs. Probability," *International Journal of General Systems*, vol. 17, nos. 2-3, Gordon and Breach Science Publishers, 1990.

equals the unit hypercube $I^n = [0, 1]^n$. A fuzzy set is any point [Kosko, 1987] in the cube I^n. So (X, I^n) defines the fundamental measurable space of (finite) fuzzy theory. We can teach much of the theory of fuzzy sets—more accurately, the theory of multivalued or continuous sets—on a Rubik's cube.

Vertices of the cube I^n define nonfuzzy sets. So the ordinary power set 2^X, the set of all 2^n nonfuzzy subsets of X, equals the Boolean n-cube B^n: $2^X = B^n$. Fuzzy sets fill in the lattice B^n to produce the solid cube I^n: $F(2^X) = I^n$.

Consider the set of two elements $X = \{x_1, x_2\}$. The nonfuzzy power set 2^X contains four sets: $2^X = \{\emptyset, X, \{x_1\}, \{x_2\}\}$. These four sets correspond respectively to the four bit vectors $(0 \ 0)$, $(1 \ 1)$, $(1 \ 0)$, and $(0 \ 1)$. The 1s and 0s indicate the presence or absence of the ith element x_i in the subset. More abstractly, we can uniquely define each subset A as one of the two-valued membership functions $m_A : X \longrightarrow \{0, 1\}$.

Now consider the fuzzy subsets of X. We can view the fuzzy subset $A = (\frac{1}{3} \ \frac{3}{4})$ as one of the continuum-many continuous-valued membership functions $m_A: X \longrightarrow [0, 1]$. Indeed this corresponds to the classical Zadeh [1965] *sets-as-functions* definition of fuzzy sets. In this example element x_1 belongs to, or fits in, subset A a little bit—to degree $\frac{1}{3}$. Element x_2 has more membership than not at $\frac{3}{4}$. Analogous to the bit-vector representation of finite (countable) sets, we say that the *fit vector* $(\frac{1}{3} \ \frac{3}{4})$ represents A. The element $m_A(x_i)$ equals the ith *fit* [Kosko, 1986b] or *fuzzy unit* value. The sets-as-points view then geometrically represents the fuzzy subset A as a point in I^2, the unit square, as in Figure 7.2.

The midpoint of the cube I^n is maximally fuzzy. All its membership values equal $\frac{1}{2}$. The midpoint is unique in two respects. First, the midpoint is the only set A that not only equals its own opposite A^c but equals its own overlap and underlap as well:

$$A = A \cap A^c = A \cup A^c = A^c \qquad (7\text{-}2)$$

Second, the midpoint is the only point in the cube I^n equidistant to each of the 2^n vertices of the cube. The nearest corners are also the farthest. Figure 7.2 illustrates this metrical relationship.

We combine fuzzy sets pairwise with minimum, maximum, and order reversal, just as we combine nonfuzzy sets. So we combine set elements with the operators of Lukasiewicz continuous logic [Rescher, 1969]. We define fuzzy-set intersection fitwise by pairwise minimum (picking the smaller of the two elements), union by pairwise maximum, and complementation by order reversal:

$$m_{A \cap B} = \min(m_A, m_B) \qquad (7\text{-}3)$$

$$m_{A \cup B} = \max(m_A, m_B) \qquad (7\text{-}4)$$

$$m_{A^c} = 1 - m_A \qquad (7\text{-}5)$$

For example:

$$
\begin{aligned}
A &= (\ 1 \quad .8 \quad .4 \quad .5) \\
B &= (.9 \quad .4 \quad 0 \quad .7) \\
A \cap B &= (.9 \quad .4 \quad 0 \quad .5) \\
A \cup B &= (\ 1 \quad .8 \quad .4 \quad .7) \\
A^c &= (\ 0 \quad .2 \quad .6 \quad .5) \\
A \cap A^c &= (\ 0 \quad .2 \quad .4 \quad .5) \\
A \cup A^c &= (\ 1 \quad .8 \quad .6 \quad .5)
\end{aligned}
$$

The overlap fit vector $A \cap A^c$ in this example does not equal the vector of all zeroes, and the underlap fit vector $A \cup A^c$ does not equal the vector of all ones. This holds for all properly fuzzy sets, all points in I^n other than vertex points. Indeed the min-max definitions give at once the following fundamental characterization of fuzziness as nondegenerate overlap and nonexhaustive underlap.

> **Proposition** A is properly fuzzy   iff $A \cap A^c \neq \emptyset$
>
> iff $A \cup A^c \neq X$

The proposition says that Aristotle's laws of noncontradiction and excluded middle hold, but they hold only on a set of measure zero. They hold only at the 2^n vertices of I^n. In all other cases, and there are as many of these as there are real numbers, contradictions occur to some degree. In this sense contradictions in generalized set theory and logic represent the rule and not the exception. Fuzzy cubes box Aristotelian sets into corners.

Completing the fuzzy square illustrates this fundamental proposition. Consider again the two–dimensional fuzzy set A defined by the fit vector $(\frac{1}{3} \quad \frac{3}{4})$. We find the

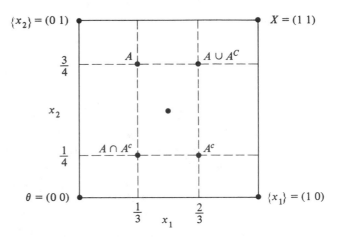

FIGURE 7.3 Completing the fuzzy square. The fuzzier A is, the closer A is to the midpoint of the fuzzy cube. As A approaches the midpoint, all four points— A, A^c, $A \cap A^c$, and $A \cup A^c$—contract to the midpoint. The less fuzzy A is, the closer A is to the nearest vertex. As A approaches the vertex, all four points spread out to the four vertices, and we recover the bivalent power set 2^X. In an n-dimensional fuzzy cube, the 2^n fuzzy sets with elements a_i or $1 - a_i$ similarly contract to the midpoint or expand to the 2^n vertices as A approaches total fuzziness or total bivalence. Interior diagonals connect fuzzy complements A and A^c, and $A \cap A^c$, and $A \cup A^c$. *Source:* Kosko, B., "Fuzziness vs. Probability," *International Journal of General Systems*, vol. 17, no. 2-3, Gordon and Breach Science Publishers, 1990.

corresponding overlap and underlap sets by first finding the complement set A^c and then combining the fit vectors pairwise with minimum and with maximum:

$$
\begin{aligned}
A &= (\tfrac{1}{3} \ \tfrac{3}{4}) \\
A^c &= (\tfrac{2}{3} \ \tfrac{1}{4}) \\
A \cap A^c &= (\tfrac{1}{3} \ \tfrac{1}{4}) \\
A \cup A^c &= (\tfrac{2}{3} \ \tfrac{3}{4})
\end{aligned}
$$

The sets-as-points view shows that these four points in the unit square hang together, and move together, in a very natural way. Consider the geometry of Figure 7.3.

In Figure 7.3 the four fuzzy sets involved in the fuzziness of set A—the sets A, A^c, $A \cap A^c$, and $A \cup A^c$—contract to the midpoint as A becomes maximally fuzzy and expand out to the Boolean corners of the cube as A becomes minimally fuzzy. The same contraction and expansion occurs in n dimensions for the 2^n fuzzy sets defined by all combinations of $m_A(x_1)$ and $m_{A^c}(x_1)$, ..., $m_A(x_n)$ and $m_{A^c}(x_n)$. The same contraction and expansion occurs in n dimensions for the 2^n fuzzy sets defined by all combinations of $m_A(x_1)$ and $m_{A^c}(x_1)$, ..., $m_A(x_n)$ and $m_{A^c}(x_n)$.

At the midpoint nothing is distinguishable. At the vertices everything is distinguishable. These extremes represent the two ends of the spectrum of logic and set theory. In this sense the midpoint represents the black hole of set theory.

Paradox at the Midpoint

The midpoint is full of paradox. Classical logic and set theory forbid the midpoint by the same axioms, noncontradiction and excluded middle, that generate the "paradoxes" or antinomies of bivalent systems. Where midpoint phenomena appear in Western thought, theorists have invariably labeled them "paradoxes" or denied them altogether. Midpoint phenomena include the half-empty and half-full cup, the Taoist Yin-Yang, the liar from Crete who said that all Cretans are liars, Bertrand Russell's set of all sets that are not members of themselves, and Russell's barber.

Russell's barber is a bewhiskered man who lives in a town and who shaves. His barber shop sign says that he shaves a man if and only if he does not shave himself. So who shaves the barber? If he shaves himself, then by definition he does not. But if he does not shave himself, then by definition he does. So he does and he does not—contradiction ("paradox"). Gaines [1983] observed that we can numerically interpret this paradoxical circumstance as follows.

Let S be the proposition that the barber shaves himself and not-S that he does not. Then since S implies not-S, and not-S implies S, the two propositions are logically equivalent: $S = $ not-S. Equivalent propositions have the same truth values:

$$t(S) \;=\; t(\text{not-}S) \tag{7-6}$$

$$=\; 1 - t(S) \tag{7-7}$$

Solving for $t(S)$ gives the midpoint point of the truth interval (the one-dimensional cube $[0, 1]$): $t(S) = \frac{1}{2}$. The midpoint is equidistant to the vertices 0 and 1. In the bivalent (two-valued) case, roundoff is impossible and paradox occurs. Equations (7-6) and (7-7) describe the logical *form* of the many paradoxes, though different paradoxes involve different descriptions [Quine, 1987].

In bivalent logic both statements S and not-S must have truth value zero or unity. The fuzzy resolution of the paradox uses only the fact that the truth values are equal. It does not constrain their range. The midpoint value $\frac{1}{2}$ emerges from the structure of the problem and the order-reversing effect of negation.

The paradoxes of classical set theory and logic illustrate the price we pay for an arbitrary insistence on bivalence [Quine, 1981]. Scientists often insist on bivalence in the name of science. But in the end this insistence reduces to a mere cultural preference, a reflection of an educational predilection that goes back at least to Aristotle. Fuzziness shows that there are limits to logical certainty. We can no longer assert the laws of noncontradiction and excluded middle *for sure*—and *for free*.

Fuzziness carries with it intellectual responsibility. We must explain how fuzziness fits in bivalent systems, or vice versa. The fuzzy theorist must explain why so many people have been in error for so long. We now have the machinery to offer an explanation: We round off. Rounding off, quantizing, simplifies life and often costs little. We agree to call empty the near empty cup, and call present the large pulse and absent the small pulse. We round off points inside the fuzzy cube to the nearest vertex. This round-off heuristic works well as a first approximation to describing the universe until we get near the midpoint of the cube. We find these phenomena harder to round off. In the logically extreme case, at the midpoint of the cube, the procedure breaks down completely, because every vertex is equally close. If we still insist on bivalence, we can only give up and declare paradox.

Faced with midpoint phenomena, the fuzzy skeptic resembles the flat-earther, who denies that the earth's surface is curved, when she stands at the north pole, looks at her compass, and wants to go south.

Counting with Fuzzy Sets

How big is a fuzzy set? The size or cardinality of A, $M(A)$, equals the sum of the fit values of A:

$$M(A) \;=\; \sum_{i=1}^{n} m_A(x_i) \tag{7-8}$$

The count of $A = (\frac{1}{3} \quad \frac{3}{4})$ equals $M(A) = \frac{1}{3} + \frac{3}{4} = \frac{13}{12}$. Some fuzzy theorists [Zadeh, 1983] call the cardinality measure M the *sigma-count*. The measure M generalizes [Kosko, 1986a] the classical counting measure of combinatorics and measure theory. (So (X, I^n, M) defines the fundamental measure space of fuzzy theory.) In general the measure M does not yield integer values.

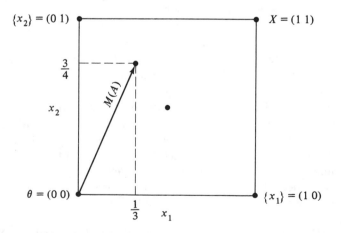

FIGURE 7.4 The count $M(A)$ of A equals the fuzzy Hamming norm (ℓ^1 norm) of the vector drawn from the origin to A. *Source:* Kosko, B., "Fuzziness vs. Probability," *International Journal of General Systems*, vol. 17, no. 2-3, Gordon and Breach Science Publishers, 1990.

The measure M has a natural geometric interpretation in the sets-as-points framework. $M(A)$ equals the magnitude of the vector drawn from the origin to the fuzzy set A, as Figure 7.4 illustrates.

Consider the ℓ^p distance between fuzzy sets A and B in ℓ^n:

$$\ell^p(A, B) = \sqrt[p]{\sum_{i=1}^{n} |m_A(x_i) - m_B(x_i)|^p} \qquad (7\text{-}9)$$

where $1 \leq p \leq \infty$. The ℓ^2 distance is the physical Euclidean distance actually illustrated in the figures. The simplest distance is the ℓ^1 or **fuzzy Hamming distance**, the sum of the absolute-fit differences. We shall use fuzzy Hamming distance throughout, though all results admit a general ℓ^p formulation. Using the fuzzy Hamming distance, we can rewrite the count M as the desired ℓ^1 norm:

$$M(A) = \sum_{i}^{n} m_A(x_i) \qquad (7\text{-}10)$$

$$= \sum_{i} |m_A(x_i) - 0| \qquad (7\text{-}11)$$

$$= \sum_{i} |m_A(x_i) - m_{\varnothing}(x_i)| \qquad (7\text{-}12)$$

$$= \ell^1(A, \varnothing) \qquad (7\text{-}13)$$

THE FUZZY ENTROPY THEOREM

How fuzzy is a fuzzy set? We measure fuzziness with a *fuzzy entropy* measure. Entropy is a generic notion. It need not be probabilistic. Entropy measures the uncertainty of a system or message. A fuzzy set describes a type of system or message. Its uncertainty equals its fuzziness.

The fuzzy entropy of A, $E(A)$, varies from 0 to 1 on the unit hypercube I^n. Only the cube vertices have zero entropy, since nonfuzzy sets are unambiguous. The cube midpoint uniquely has unity or maximum entropy. Fuzzy entropy smoothly increases as a set point moves from any vertex to the midpoint. Klir [1988] discusses the algebraic requirements for fuzzy entropy measures.

Simple geometric considerations lead to a ratio form for the fuzzy entropy [Kosko, 1986b]. The closer the fuzzy set A is to the nearest vertex A_{near}, the farther A is from the farthest vertex A_{far}. The farthest vertex A_{far} resides opposite the long diagonal from the nearest vertex A_{near}. Let a denote the distance $\ell^1(A, A_{\text{near}})$ to the nearest vertex, and let b denote the distance $\ell^1(A, A_{\text{far}})$ to the farthest vertex. Then the fuzzy entropy equals the ratio of a to b:

$$E(A) = \frac{a}{b} = \frac{\ell^1(A, A_{\text{near}})}{\ell^1(A, A_{\text{far}})} \qquad (7\text{-}14)$$

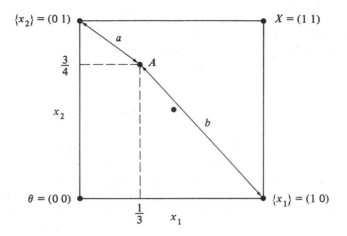

FIGURE 7.5 Fuzzy entropy, $E(A) = \frac{a}{b}$, balances distance to nearest vertex with distance to farthest vertex. *Source:* Kosko, B., "Fuzziness vs. Probability," *International Journal of General Systems*, vol. 17, no. 2-3, Gordon and Breach Science Publishers, 1990.

Figure 7.5 shows the sets-as-points interpretation of the fuzzy entropy measure, where $A = (\frac{1}{3} \ \frac{3}{4})$, $A_{\text{near}} = (0 \ 1)$, and $A_{\text{far}} = (1 \ 0)$. So $a = \frac{1}{3} + \frac{1}{4} = \frac{7}{12}$ and $b = \frac{2}{3} + \frac{3}{4} = \frac{17}{12}$. So $E(A) = \frac{7}{17}$.

Alternatively, if you read this in a room, you can imagine the room as the unit cube I^3 and your head as a fuzzy set in it. Once you locate the nearest corner of the room, the farthest corner resides opposite the long diagonal emanating from the nearest corner. If you put your head in a corner, then $a = 0$, and so $E(A) = 0$. If you put your head in the metrical center of the room, every corner is nearest and farthest. So $a = b$, and $E(A) = 1$.

Overlap and underlap characterize within-set fuzziness. So we can expect them to affect the measure of fuzziness. Figure 7.6 shows the connection. By symmetry, each of the four points A, A^c, $A \cap A^c$, and $A \cup A^c$ is equally close to its nearest vertex. The common distance equals a. Similarly, each point is equally far from its farthest vertex. The common distance equals b. One of the first four distances is the count $M(A \cap A^c)$. One of the second four distances is the count $M(A \cup A^c)$. This gives a geometric proof of the fuzzy entropy theorem [Kosko, 1986b, 1987], which states that fuzziness consists of counted violations of the law of noncontradiction balanced with counted violations of the law of excluded middle.

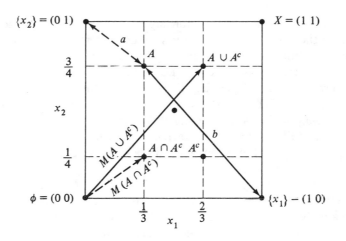

FIGURE 7.6 Geometry of the fuzzy entropy theorem. By symmetry each of the four points on the completed fuzzy square is equally close to its nearest vertex and equally far from its farthest vertex. *Source:* Kosko, B., "Fuzziness vs. Probability," *International Journal of General Systems*, vol. 17, no. 2-3, Gordon and Breach Science Publishers, 1990.

Fuzzy entropy theorem

$$E(A) \;=\; \frac{M(A \cap A^c)}{M(A \cup A^c)} \tag{7-15}$$

Proof. An algebraic proof is straightforward. The completed fuzzy square in Figure 7.6 contains a geometric proof (in this special case).

The fuzzy entropy theorem explains why set fuzziness begins where Western logic ends. When sets (or propositions) obey the laws of noncontradiction and excluded middle, overlap is empty and underlap is exhaustive. So $M(A \cap A^c) = 0$ and $M(A \cup A^c) = n$, and thus $E(A) = 0$.

The fuzzy entropy theorem also provides a first-principles derivation of the basic fuzzy-set operations of minimum (intersection), maximum (union), and order reversal (complementation) proposed in 1965 by Zadeh at the inception of fuzzy theory. (Lukasiewicz first proposed these operations for continuous or fuzzy logics in the 1920s [Rescher, 1969].)

For the fuzzy theorist, this result also shows that triangular norms or T-norms [Klir, 1988], which generalize conjunction or intersection, and the dual triangular co-norms C, which generalize disjunction or union, do not have the first-principles status of min and max. For, the triangular-norm inequalities,

$$T(x, y) \;\leq\; \min(x, y) \;\leq\; \max(x, y) \;\leq\; C(x, y) \tag{7-16}$$

show that replacing min with any T in the numerator term $M(A \cap A^c)$ can only make the numerator smaller. Replacing max with any C in the term $M(A \cup A^c)$

can only make the denominator larger. So any T or C not identically min or max makes the ratio smaller, strictly smaller if A is fuzzy. Then the entropy theorem does not hold, and the resulting pseudoentropy measure does not equal unity at the midpoint, though it continues to be maximized there. We can see this with the product T-norm [Prade, 1985] $T(x, y) = xy$ and its DeMorgan dual co-norm $C(x, y) = 1 - T(1-x, 1-y) = x+y-xy$, or with the bounded sum T-norm $T(x, y) = \max(0, x+y-1)$ and DeMorgan dual $C(x, y) = \min(1, x+y)$. The entropy theorem similarly fails in general if we replace the negation or complementation operator $N(x) = 1 - x$ with a parameterized operator $N_a(x) = (1-x)/(1+ax)$ for nonzero $a > -1$.

All probability distributions, all sets A in I^n with $M(A) = 1$, form an $n-1$ dimensional simplex $\mathbf{S}^n$. In the unit square the probability simplex equals the negatively sloped diagonal line. In the unit 3-cube it equals a solid triangle. In the unit 4-cube it equals a tetrahedron, and so on up.

If no probabilistic fit value p_i satisfies $p_i > \frac{1}{2}$, then the fuzzy entropy theorem implies [Kosko, 1987] that the the distribution P has fuzzy entropy $E(P) = 1/n - 1$. Else $E(P) < 1/(n-1)$. So the probability simplex $\mathbf{S}^n$ is entropically degenerate for large dimensions n. This result also shows that the uniform distribution $(1/n, \ldots, 1/n)$ maximizes fuzzy entropy on $\mathbf{S}^n$ but not uniquely. This in turn shows that fuzzy entropy differs from the average-information measure of probabilistic entropy, which the uniform distribution maximizes uniquely.

The fuzzy entropy theorem implies that, analogous to $\log 1/p$, a unit of fuzzy information equals $f/(1-f)$ or $(1-f)/f$, depending on whether the fit value f obeys $f \leq \frac{1}{2}$ or $f \geq \frac{1}{2}$.

The event x can be ambiguous or clear. It is ambiguous if f equals approximately $\frac{1}{2}$ and clear if f equals approximately 1 or 0. If an ambiguous event occurs, is observed, is disambiguated, etc., then it is maximally informative: $E(f) = E(\frac{1}{2}) = 1$. If a clear event occurs, is observed, etc., it is minimally informative: $E(f) = E(0) = E(1) = 0$. This agrees with the information interpretation of the probabilistic entropy measure $\log 1/p$, where the occurrence of a sure event $(p = 1)$ is minimally informative (zero entropy) and the occurrence of an impossible event $(p = 0)$ is maximally informative (infinite entropy).

THE SUBSETHOOD THEOREM

Sets contain subsets. A is a *subset* of B, denoted $A \subset B$, if and only if every element in A is an element of B. The *power set* 2^B contains all of B's subsets. So, alternatively [Bandler-Kohout, 1980], A is a subset of B iff A belongs to B's power set:

$$A \subset B \quad \text{if and only if} \quad A \in 2^B \qquad (7\text{-}17)$$

The subset relation corresponds to the implication relation in logic. In classical logic *truth* maps the set of statements $\{S\}$ to two truth values: $t: \{S\} \longrightarrow \{0, 1\}$.

Consider the truth-table definition of implication for bivalent propositions P and Q:

P	Q	$P \longrightarrow Q$
0	0	1
0	1	1
1	0	0
1	1	1

The implication is false if and only if the antecedent P is true and the consequent Q is false—when "truth implies falsehood."

The same holds for subsets. Representing sets as bivalent functions or "indicator" functions $m_A: X \longrightarrow \{0, 1\}$, A is a subset of B iff there is no element x that belongs to A but not to B, or $m_A(x) = 1$ but $m_B(x) = 0$. We can rewrite this membership-function definition as

$$A \subset B \quad \text{if and only if} \quad m_A(x) \leq m_B(x) \text{ for all } x \qquad (7\text{-}18)$$

Zadeh [1965] proposed the same relation for fuzzy-set containment. We refer to this as the *dominated membership function relationship*. If $A = (.3 \ 0 \ .7)$ and $B = (.4 \ .7 \ .9)$, then A is a fuzzy subset of B, but B is not a fuzzy subset of A. Either fuzzy set A is, or is not, a fuzzy subset of B. So Zadeh's relation of fuzzy subsethood is *not* fuzzy. It is black or white.

The sets-as-points view asks a geometric question: What do all fuzzy subsets of B look like? What does the *fuzzy power set* of B—$F(2^B)$, the set of all fuzzy subsets of B—look like? The dominated membership function relationship implies that $F(2^B)$ defines the hyperrectangle snug against the origin in a unit hypercube with side lengths equal to the fit values $m_A(x_i)$. Figure 7.7 displays the fuzzy power set of the set $B = (\frac{1}{3} \ \frac{2}{3})$. $F(2^B)$ has infinite cardinality if B is not empty. For finite-dimensional sets, we can measure the size of $F(2^B)$ [Kosko, 1987] as the Lebesgue measure or volume $V(B)$, the product of the fit values:

$$V(B) \quad = \quad \prod_{i=1}^{n} m_B(x_i) \qquad (7\text{-}19)$$

Figure 7.7 illustrates that $F(2^B)$ is not a fuzzy set. Either cube point A is or is not in the hyperrectangle $F(2^B)$. Different points A outside the hyperrectangle $F(2^B)$ resemble subsets of B to different degrees. The bivalent definition of subsethood ignores this.

The natural generalization defines fuzzy subsets on $F(2^B)$: Some sets A belong to $F(2^B)$ to different degrees. Then the abstract membership function $m_{F(2^B)}(A)$ can equal any number in $[0, 1]$. This defines degrees of subsethood.

Let $S(A, B)$ denote the degree to which A is a subset of B:

$$S(A, B) \quad = \quad \text{Degree}(A \subset B) \qquad (7\text{-}20)$$

$$= \quad m_{F(2^B)}(A) \qquad (7\text{-}21)$$

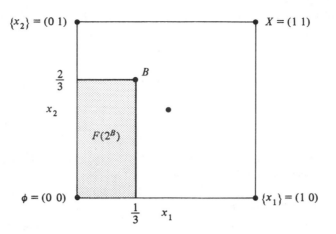

FIGURE 7.7 Fuzzy power set $F(2^B)$ as a hyperrectangle in the fuzzy cube. Side lengths equal the fit values $m_B(x_i)$. The size or volume of $F(2^B)$ equals the product of the fit values. *Source:* Kosko, B., "Fuzziness vs. Probability," *International Journal of General Systems*, vol. 17, no. 2-3, Gordon and Breach Science Publishers, 1990.

$S(., .)$ denotes the multivalued *subsethood measure*. $S(., .)$ takes values in $[0, 1]$. We will see that it provides the fundamental, unifying structure in fuzzy theory.

We want to measure $S(A, B)$. We will first present an earlier [Kosko, 1986b, 1987] algebraic derivation of the subsethood measure $S(A, B)$. We will then present a new, more fundamental, geometric derivation.

We call the algebraic derivation the *fit-violation strategy*. Intuitively we study a law by breaking it. Consider again the dominated membership function relationship: $A \subset B$ if and only if $m_A(x) \le m_B(x)$ for all x in X.

Suppose element x_v violates the dominated membership function relationship: $m_A(x_v) > m_B(x_v)$. Then A is not a subset of B, at least not totally. Suppose further that the dominated membership inequality holds for all other elements x. Only element x_v violates the relationship. For instance, X may consist of one hundred values: $X = \{x_1, \ldots, x_{100}\}$. The violation might occur, say, with the first element: $x_1 = x_v$. Then intuitively A is largely a subset of B. Suppose that X contains a thousand elements, or a trillion elements, and only the first element violates (7-18). Then it seems A is overwhelmingly a subset of B; perhaps $S(A, B) = 0.999999999999$.

This example suggests we should count fit violations in magnitude and frequency. The greater the violations in magnitude, $m_A(x_v) - m_B(x_v)$, and the greater the number of violations relative to the size $M(A)$ of A, the less A is a subset of B or, equivalently, the more A is a *superset* of B. For, both intuitively and by (7-18), supersethood and subsethood relate as additive but conserved opposites:

$$\text{SUPERSETHOOD}(A, B) \;=\; 1 - S(A, B) \qquad (7\text{-}22)$$

We count violations by adding them. If we sum over all x, the summand should equal $m_A(x_v) - m_B(x_v)$ when this difference is positive, and equal zero when it is nonpositive. So the summand equals $\max(0, m_A(x) - m_B(x))$. So the unnormalized count equals the sum of these maxima:

$$\sum_{x \in X} \max(0, m_A(x) - m_B(x)) \tag{7-23}$$

The count $M(A)$ provides a simple, and appropriate, normalization factor. Below we formally arrive at $M(A)$ by examining boundary cases in the geometric approach to subsethood. We can assume $M(A) > 0$, since $M(A) = 0$ if and only if A is empty. The empty set trivially satisfies the dominated membership function relationship (7-18). So it is a subset of every set. Normalization gives the minimal measure of nonsubsethood, of supersethood:

$$\text{SUPERSETHOOD}(A, B) = \frac{\sum_x \max(0, m_A(x) - m_B(x))}{M(A)} \tag{7-24}$$

Then subsethood is the negation of this ratio. This gives the minimal fit-violation measure of subsethood:

$$S(A, B) = 1 - \frac{\sum_x \max(0, m_A(x) - m_B(x))}{M(A)} \tag{7-25}$$

The subsethood measure may appear ungraceful at first, but it behaves as it should. $S(A, B) = 1$ if and only if (7-18) holds. For if (7-18) holds, (7-23) sums zero violations. Then $S(A, B) = 1 - 0 = 1$. If $S(A, B) = 1$, every numerator summand equals zero. So no violation occurs. At the other extreme, $S(A, B) = 0$ if and only if B is the empty set. The empty set uniquely contains no proper subsets, fuzzy or nonfuzzy. Degrees of subsethood occur between these extremes: $0 < S(A, B) < 1$.

The subsethood measure relates to logical implication. Viewed at the one-dimensional level of fuzzy logic, and so ignoring the normalizing count $(M(A) = 1)$, the subsethood measure reduces to the Lukasiewicz implication operator discussed in the homework problems in Chapter 1:

$$S(A, B) = 1 - \max(0, m_A - m_B) \tag{7-26}$$

$$= 1 - [1 - \min(1 - 0, 1 - (m_A - m_B))] \tag{7-27}$$

$$= \min(1, 1 - m_A + m_B) \tag{7-28}$$

$$= t_L(A \longrightarrow B) \tag{7-29}$$

The $\min(.)$ operator in (7-28) clearly generalizes the above truth-tabular definition of bivalent implication.

Consider the fit vectors $A = (.2\ 0\ .4\ .5)$ and $B = (.7\ .6\ .3\ .7)$. Neither set is a proper subset of the other. A is almost a subset of B but not quite, since

$m_A(x_3) - m_B(x_3) = .4 - .3 = .1 > 0$. Hence $S(A, B) = 1 - \frac{.1}{1.1} = \frac{10}{11}$. Similarly $S(B, A) = 1 - \frac{1.3}{2.3} = \frac{10}{23}$.

Subsethood applies to nonfuzzy sets, indeed here probability arises. Consider the sets $C = \{x_1, x_2, x_3, x_5, x_7, x_9, x_{10}, x_{12}, x_{14}\}$ and $D = \{x_2, x_3, x_4, x_5, x_6, x_7, x_8, x_9, x_{10}, x_{12}, x_{13}, x_{14}\}$ with corresponding bit vectors

$$C = (1\ 1\ 1\ 0\ 1\ 0\ 1\ 0\ 1\ 1\ 0\ 1\ 0\ 1)$$

$$D = (0\ 1\ 1\ 1\ 1\ 1\ 1\ 1\ 1\ 1\ 0\ 1\ 1\ 1)$$

C and D are not subsets of each other. But C should very nearly be a subset of D, since only x_1 violates (7-18). We find that $S(C, D) = 1 - \frac{1}{9} = \frac{8}{9}$ while $S(D, C) = 1 - \frac{4}{12} = \frac{2}{3}$. So D is more a subset of C than it is not. This holds because the two sets are largely equivalent. They have much overlap: $M(C \cap D) = 8$. This observation anticipates the fuzzy subsethood theorem presented below.

We now turn to a new [Kosko, 1990] and purely geometric derivation of the subsethood operator $S(A, B)$. Consider the sets-as-points geometry of subsethood in Figure 7.7. Set A either lies in the hyperrectangle $F(2^B)$ or not in it. Intuitively $S(A, B)$ should approach unity as A approaches the fuzzy power set $F(2^B)$. $S(A, B)$ should decrease, and the supersethood measure $1 - S(A, B)$ should increase, as A moves away from $F(2^B)$.

So the key idea is metrical: *How close is A to $F(2^B)$?* Let $d(A, F(2^B))$ denote this ℓ^p distance defined in (7-9). $d(A, B')$ denotes the distance between A and point B' in the hyperrectangle, and $B' \subset B$. Distance $d(A, F(2^B))$ equals the smallest such distance. Since the hyperrectangle $F(2^B)$ is closed and bounded (compact) and convex, some subset B^* of B achieves this minimum distance. So the infimum, the greatest lower bound, equals the distance $d(A, B^*)$:

$$d(A, F(2^B)) = \inf\{d(A, B') : B' \in F(2^B)\} \tag{7-30}$$

$$= d(A, B^*) \tag{7-31}$$

We can easily locate the closest set B^* in the hypercube geometry. If A is a subset of B—if A is in the hyperrectangle $F(2^B)$—then A equals the closest subset: $A = B^*$. So suppose A is not a proper subset of B. Then A lies outside the hyperrectangle $F(2^B)$.

We can slice the unit cube I^n into 2^n hyperrectangles by extending the sides of $F(2^B)$ to hyperplanes. The hyperplanes intersect perpendicularly (orthogonally), at least in the Euclidean case. $F(2^B)$ defines one of the hyperrectangles. The hyperrectangle interiors correspond to the 2^n cases whether $m_A(x_i) < m_B(x_i)$ or $m_A(x_i) > m_B(x_i)$ for fixed B and arbitrary A. The edges correspond to the loci of points where some $m_A(x_i) = m_B(x_i)$.

The 2^n hyperrectangles classify as *mixed* or *pure* membership domination. In the pure case either $m_A < m_B$, or $m_A > m_B$, holds in the hyperrectangle interior for all x and all interior points A. In the mixed case $m_A(x_i) < m_B(x_i)$ holds for some of the coordinates x_i, and $m_A(x_j) > m_B(x_j)$ holds for the remaining coordinates x_j in the interior for all interior A. So there are only two pure membership-domination

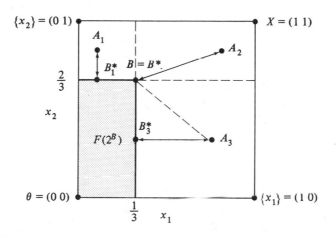

FIGURE 7.8 Partition of hypercube I^n into 2^n hyperrectangles by linearly extending the edges of $F(2^B)$. We find the nearest points B_1^* and B_3^* to points A_1 and A_3 in the northwest and southeast quadrants by the normals from $F(2^B)$ to A_1 and A_3. The nearest point B^* to point A_2 in the northeast quadrant is B itself. This "orthogonal" optimality condition gives $d(A, B)$, by the general Pythagorean theorem, as the hypotenuse in an ℓ^p "right" triangle. *Source:* Kosko, B., "Fuzziness vs. Probability," *International Journal of General Systems*, vol. 17, no. 2-3, Gordon and Breach Science Publishers, 1990.

hyperrectangles, the set of proper subsets $F(2^B)$, which includes the empty set, and the set of proper supersets, which includes X.

Figure 7.8 illustrates how the fuzzy power set $F(2^B)$ of $B = (\frac{1}{3} \ \frac{2}{3})$ linearly extends to partition the unit square into 2^2 rectangles. The nonsubsets A_1, A_2, and A_3 reside in distinct quadrants. The northwest and southeast quadrants define the mixed membership-domination rectangles. The southwest and the northeast quadrants define the pure rectangles.

B is the nearest set B^* to A in the pure superset hyperrectangle. To find the nearest set B^* in the mixed case we draw a perpendicular (orthogonal) line segment from A to $F(2^B)$. Convexity of $F(2^B)$ is ultimately responsible. In Figure 7.8 the perpendicular lines from A_1 and A_3 intersect line edges (one-dimensional linear subspaces) of the rectangle $F(2^B)$. The line from A_2 to B, the corner of $F(2^B)$, is degenerately perpendicular, since B is a zero-dimensional linear subspace.

These "orthogonality" conditions also hold in three dimensions. Let your room again be the unit 3-cube. Consider a large dictionary fitted snugly against the floor corner corresponding to the origin. Point B equals the dictionary corner farthest from the origin. Extending the three exposed faces of the dictionary partitions the room into eight octants. The dictionary occupies one octant. We connect points in the other seven octants to the nearest points on the dictionary by drawing lines, or tying strings, that perpendicularly intersect one of the three exposed faces, or one of the three exposed edges, or the corner B.

The "orthogonality" condition invokes a new [Kosko, 1990] ℓ^p-version of the

Pythagorean theorem. For our ℓ^1 purposes:

$$d(A, B) \ = \ d(A, B^*) + d(B, B^*) \tag{7-32}$$

The more familiar ℓ^2-version, actually pictured in Figure 7.8, requires squaring these distances. For the general ℓ^p case:

$$\|A - B\|^p \ = \ \|A - B^*\|^p + \|B^* - B\|^p \tag{7-33}$$

or equivalently,

$$\sum_{i=1}^{n} |a_i - b_i|^p \ = \ \sum_{i=1}^{n} |a_i - b_i^*|^p + \sum_{i=1}^{n} |b_i^* - b_i|^p \tag{7-34}$$

Equality holds for all $p \geq 1$ since, as we can see in Figure 7.8 and, in general, from the algebraic argument below, either $b_i^* = a_i$ or $b_i^* = b_i$.

This Pythagorean equality is surprising. We have come to think of the Pythagorean theorem (and orthogonality) as an ℓ^2 or Hilbert-space property. Yet here it holds in *every* ℓ^p space—if B^* is the set in $F(2^B)$ closest to A in ℓ^p distance. Of course for other sets strict inequality holds in general if $p \neq 2$. This suggests a special status for the closest set B^*. We shall see below that the subsethood theorem confirms this suggestion. We shall use the term "orthogonality" loosely to refer to this ℓ^p Pythagorean relationship, while remembering its customary restriction to ℓ^2 spaces and inner products.

A natural interpretation defines *super*sethood as the distance $d(A, F(2^B)) = d(A, B^*)$. Supersethood increases with this distance; subsethood decreases with it. To keep supersethood, and thus subsethood, unit-interval valued, we must suitably normalize the distance.

A constant provides the simplest normalization term for $d(A, B^*)$. That constant equals the maximum unit-cube distance, $n^{1/p}$ in the general ℓ^p case and n in our ℓ^1 case. This gives the candidate subsethood measure

$$S(A, B) \ = \ 1 - \frac{d(A, B^*)}{n} \tag{7-35}$$

This candidate subsethood measure fails in the boundary case when B is the empty set. For then $d(A, B^*) = d(A, B) = M(A)$. So the measure in (7-35) gives

$$S(A, \varnothing) \ = \ 1 - \frac{M(A)}{n} > 0$$

Equality holds exactly when $A = X$. But the empty set has no subsets. Only normalization factor $M(A)$ satisfies this boundary condition. Of course $M(A) = n$ when $A = X$. Explicitly we require $S(A, \varnothing) = 0$, as well as $S(\varnothing, A) = 1$.

Normalizing by n also treats all equidistant points the same. Consider points A_1 and A_2 in Figure 7.9. Both points are equidistant to their nearest $F(2^B)$ point: $d(A_1, B_1^*) = d(A_2, B_2^*)$. But A_1 is closer to B than A_2 is. In particular A_1 is closer to the horizontal line defined by the fit value $m_B(x_2) = \frac{2}{3}$. The count $M(A)$ reflects this: $M(A_1) > M(A_2)$. The count gap $M(A_1) - M(A_2)$ arises from the fit gap

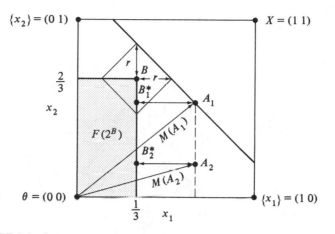

FIGURE 7.9 Dependence of subsethood on the count $M(A)$. A_1 and A_2 are equidistant to $F(2^B)$ but A_1 is closer to B than A_2 is; correspondingly, $M(A_1) > M(A_2)$. Loci of points A of constant count $M(A)$ define line segments parallel to the negatively sloping long diagonal. ℓ^1 spheres centered at B are diamond shaped. *Source:* Kosko, B., "Fuzziness vs. Probability," *International Journal of General Systems*, vol. 17, no. 2-3, Gordon and Breach Science Publishers, 1990.

involving x_1, and reflects $d(A_1, B) < d(A_2, B)$. In general the count $M(A)$ relates to this distance, as we can see by checking extreme cases of closeness of A to B (and drawing some diamond-shaped ℓ^1 spheres centered at B). Indeed if $m_A > m_B$ everywhere, $d(A, B) = M(A) - M(B)$.

Since $F(2^B)$ fits snugly against the origin, the count $M(A)$ in any of the other $2^n - 1$ hyperrectangles can be only larger than the count $M(B^*)$ of the nearest $F(2^B)$ points. The normalization choice of n leaves the candidate subsethood measure indifferent to which of the $2^n - 1$ hyperrectangles contains A and to where A resides in the hyperrectangle. Each point in each hyperrectangle involves a different combination of fit violations and satisfactions. The normalization choice of $M(A)$ reflects this fit-violation structure and uniquely satisfies the boundary conditions.

The normalization choice $M(A)$ leads to the subsethood measure

$$S(A, B) = 1 - \frac{d(A, B^*)}{M(A)} \tag{7-36}$$

We now show that this measure equals the subsethood measure (7-25) derived algebraically above.

Let B' be any subset of B. Then by definition the nearest subset B^* obeys the inequality:

$$\sqrt[p]{\sum_{i=1}^{n} |a_i - b_i^*|^p} \leq \sqrt[p]{\sum_{i=1}^{n} |a_i - b_i'|^p} \tag{7-37}$$

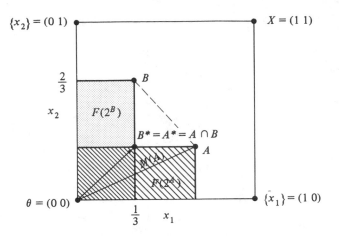

FIGURE 7.10 B^* as both the subset of B nearest A and the subset A^* of A nearest $B : B^* = A^* = A \cap B$. The distance $d(A, B^*) = M(A) - M(A \cap B)$ illustrates the subsethood theorem. *Source:* Kosko, B., "Fuzziness vs. Probability," *International Journal of General Systems*, vol. 17, no. 2-3, Gordon and Breach Science Publishers, 1990.

where for convenience $a_i = m_A(x_i)$, and $b_i = m_B(x_i)$. We will assume $p = 1$ but the following characterization of b_i^* also holds for any $p > 1$.

"Orthogonality" implies $a_i \geq b_i^*$. So first suppose $a_i = b_i^*$. This equality holds if and only if no violation occurs: $a_i \leq b_i$. (If this condition holds for all i, then $A = B^*$.) So $\max(0, a_i - b_i) = 0$. Next suppose $a_i > b_i^*$. This inequality holds if and only if a violation occurs: $a_i > b_i$. (If this holds for all i, then $B = B^*$.) So $b_i^* = b_i$ since B^* is the subset of B nearest to A. Equivalently, $a_i > b_i$ holds if and only if $\max(0, a_i - b_i) = a_i - b_i$. The two cases together prove that $\max(0, a_i - b_i) = |a_i - b_i^*|$. Summing over all x_i gives

$$d(A, B^*) = \sum_{i=1}^{n} \max(0, m_A(x_i) - m_B(x_i)) \qquad (7\text{-}38)$$

So the two subsethood measures (7-25) and (7-36) are equivalent.

This proof also proves a deeper characterization of the optimal subset B^*:

$$B^* = A \cap B \qquad (7\text{-}39)$$

For if a violation occurs, then $a_i > b_i$, and $b_i = b_i^*$. So $\min(a_i, b_i) = b_i^*$. Otherwise $a_i = b_i^*$, and so $\min(a_i, b_i) = b_i^*$. So $B^* = A \cap B$.

This in turn proves that B^* is a point of double optimality. B^* is both the subset of B nearest A, and A^*, the subset of A nearest to B:

$$d(B, F(2^A)) = d(B, A^*) = d(B, B^*) \qquad (7\text{-}40)$$

Figure 7.10 illustrates that $B^* = A \cap B = A^*$ identifies the set within both the hyperrectangle $F(2^A)$ and the hyperrectangle $F(2^B)$ that has maximal count $M(A \cap B)$.

Figure 7.10 also shows that the distance $d(A, B^*)$ equals a vector magnitude difference: $d(A, B^*) = M(A) - M(A \cap B)$. Dividing both sides of this equality by $M(A)$ and rearranging proves a still deeper structural characterization of subsethood, the subsethood theorem.

Subsethood theorem

$$S(A, B) = \frac{M(A \cap B)}{M(A)} \qquad (7\text{-}41)$$

The subsethood theorem immediately implies a Bayes theorem:

$$S(A, B) = \frac{M(B)S(B, A)}{M(A)} \qquad (7\text{-}42)$$

since (7-41) implies $M(A \cap B) = M(B)S(B, A)$.

The ratio form of the subsethood measure $S(A, B)$ has the same ratio form as the conditional probability $P(B \mid A)$ has in (7-1). We *derived* the ratio form for the subsethood measure $S(A, B)$ but *assumed* it for the conditional probability $P(B \mid A)$. Since every probability is a conditional probability, $P(A) = P(A \mid X)$, this suggests we can reduce probability to subsethood. We shall argue that this reduction holds for both frequentist or "objective" probability and axiomatic or Bayesian or "subjective" probability.

Consider the physical interpretation of randomness as the relative frequency n_A/n. n_A denotes the number of successes that occur in n trials. Historically probabilists have called the success ratio (or its limit) n_A/n the "probability of success" or $P(A)$. We can now derive the relative-frequency definition of probability as $S(X, A)$, the degree to which a bivalent superset X, the sample space, is a subset of its own subset A. The concept of "randomness" never enters the deterministic set-theoretic framework. This holds equally for flipping coins, drawing balls from urns, or computing Einstein-Bose statistics.

Suppose A is a nonfuzzy subset of X. Then (7-41) gives

$$S(X, A) = \frac{M(A \cap X)}{M(X)} \qquad (7\text{-}43)$$

$$= \frac{M(A)}{M(X)} \qquad (7\text{-}44)$$

$$= \frac{n_A}{n} \qquad (7\text{-}45)$$

The n elements of X constitute the *de facto* universe of discourse of the "experiment." (We can take the limit of the ratio $S(X, A)$ if it mathematically makes sense to do so [Kac, 1959].) The "probability" n_A/n has reduced to a degree of subsethood, a purely multivalued set-theoretical relationship. Perhaps if, centuries ago, scientists had developed set theory before they formalized gambling, the undefined notion of "randomness" might never have culturally prevailed, if even survived, in the age of modern science.

The measure of overlap $M(A \cap X)$ provides the key component of relative frequency. This count does not involve "randomness." $M(A \cap X)$ counts which elements are identical or similar. The phenomena themselves are deterministic and black or white. The same situation gives the same number. We may use the number to place bets or to switch a phone line, but it remains part of the description of a specific state of affairs. We need not invoke an undefined "randomness" to further describe the situation.

Subsethood subsumes elementhood. We can interpret the membership degree $m_A(x_i)$ as the subsethood degree $S(\{x_i\}, A)$, where $\{x_i\}$ denotes a singleton subset or "element" x_i of X. $\{x_i\}$ corresponds to a bit vector with a 1 in the ith slot and 0s elsewhere: $\{x_i\} = (0, \ldots, 0, 1, 0, \ldots, 0)$. If we view A as the fit vector $(a_1, \ldots, a_i, \ldots, a_n)$, then $\{x_i\} \cap A = (0, \ldots, 0, a_i, 0, \ldots, 0)$, the ith coordinate projection. Since the count $M(\{x_i\})$ equals one, the subsethood theorem gives

$$S(\{x_i\}, A) = \frac{M(\{x_i\} \cap A)}{M(\{x_i\})} \qquad (7\text{-}46)$$

$$= M((0, \ldots, a_i, \ldots, 0)) \qquad (7\text{-}47)$$

$$= a_i \qquad (7\text{-}48)$$

$$= m_A(x_i) \qquad (7\text{-}49)$$

$$= \text{Degree } (x_i \in A) \qquad (7\text{-}50)$$

So subsethood reduces to elementhood if antecedent sets are bivalent singleton sets.

The subsethood orthogonality conditions project A onto the facing side of the hyperrectangle $F(2^B)$. This projection gives the "normal equations" of least-squares parameter estimation [Sorenson, 1980], an LMS version of which we saw in Chapter 5. In general for two R^n vectors $\mathbf{x}$ and $\mathbf{y}$, we project $\mathbf{x}$ onto $\mathbf{y}$ to give the projection vector $\mathbf{p} = c\mathbf{y}$. The difference $\mathbf{x} - \mathbf{p}$ is orthogonal to $\mathbf{y}$: $(\mathbf{x} - \mathbf{p}) \perp \mathbf{y}$. So

$$0 = (\mathbf{x} - \mathbf{p})\mathbf{y}^T \qquad (7\text{-}51)$$

$$= (\mathbf{x} - c\mathbf{y})\mathbf{y}^T \qquad (7\text{-}52)$$

$$= \mathbf{x}\mathbf{y}^T - c\mathbf{y}\mathbf{y}^T \qquad (7\text{-}53)$$

where column vector $\mathbf{y}^T$ denotes the transpose of row vector $\mathbf{y}$. Equation (7-53) gives the projection coefficient c as the familiar normal equations:

$$c = \frac{\mathbf{x}\mathbf{y}^T}{\mathbf{y}\mathbf{y}^T} \qquad (7\text{-}54)$$

$$= \frac{\sum_{i=1}^{n} x_i y_i}{\sum_{i=1}^{n} y_i^2} \qquad (7\text{-}55)$$

Consider the unit square in Figure 7.10 with the same A, say $A = (\frac{3}{4} \ \frac{1}{2})$. But suppose we shift B directly to the left to $B = (0 \ \frac{2}{3})$. This contracts the rectangle $F(2^B)$ to the line segment $[0 \ \frac{2}{3}]$ along the vertical axis. These assumptions simplify the correlation mathematics yet still preserve the least-squares structure. We expect that $B^* = cB$, or $cB = A \cap B$, when we project A onto $F(2^B)$ or, equivalently in this special case, when we project A onto B. The intersection $A \cap B$ equals the minimum fit vector $(0 \ \frac{1}{2})$, $AB^T = 0 + \frac{2}{6} = \frac{1}{3}$, and $BB^T = 0 + (\frac{2}{3})^2 = \frac{4}{9}$. Then

$$c = \frac{AB^T}{BB^T}$$

$$= \frac{\frac{1}{3}}{\frac{4}{9}} = \frac{3}{4}$$

and

$$B^* = cB$$

$$= \frac{3}{4}(0 \ \frac{2}{3})$$

$$= (0 \ \frac{1}{2})$$

$$= A \cap B$$

as expected. More generally, if $B = (b_1 b_2)$, $b_1 = 0$, $b_2 > 0$, and $a_2 \le b_2$, then

$$c = \frac{AB^T}{BB^T} = \frac{a_1 b_1 + a_2 b_2}{b_1^2 + b_2^2} \qquad (7\text{-}56)$$

$$= \frac{a_2 b_2}{b_2^2} \qquad (7\text{-}57)$$

$$= \frac{a_2}{b_2} \qquad (7\text{-}58)$$

Then $cB = (0 \ a_2 b_2 / b_2) = (0 \ a_2) = A \cap B$, since $a_2 \le b_2$.

Subsethood has extended the Pythagorean theorem, relative frequency, and elementhood, and involves the normal equations of least-square estimation. We shall now see how subsethood relates to axiomatic or Bayesian probability and to fuzzy entropy.

Bayesian Polemics

Bayesian probabilists interpret probability as a subjective state of knowledge. In practice they use relative frequencies (subsethood degrees) but only to approximate these "states of knowledge."

Bayesianism is often a polemical doctrine. Some Bayesians claim that they, and only they, use all and only the available uncertainty information in the description of uncertain phenomena. This stems from the Bayes-theorem expansion of the "a posteriori" conditional probability $P(H_i \,|\, E)$, the probability that H_i, the ith of k-many disjoint hypotheses $\{H_j\}$, is true given observed evidence E:

$$P(H_i \,|\, E) \;=\; \frac{P(E \cap H_i)}{P(E)} \tag{7-59}$$

$$=\; \frac{P(E \,|\, H_i)P(H_i)}{P(E)} \tag{7-60}$$

$$=\; \frac{P(E \,|\, H_i)P(H_i)}{\displaystyle\sum_{j=1}^{k} P(E \,|\, H_j)P(H_j)} \tag{7-61}$$

since the hypotheses partition the sample space X: $H_1 \cup H_2 \cup \ldots \cup H_k = X$ and $H_i \cap H_j = \emptyset$ if $i \neq j$.

The Bayesian approach uses all available information in computing the posterior distribution $P(H_i \,|\, E)$ by using the "a priori" or prior distribution $P(H_i)$ of the hypotheses. The Bayesian approach stems from the ratio form of the conditional probability measure.

The subsethood theorem trivially implies Bayes theorem when the hypotheses $\{H_i\}$ and evidence E are nonfuzzy subsets. More important, the subsethood theorem implies the fuzzy Bayes Theorem in the more interesting case when the observed data E is fuzzy:

$$S(E,\, H_i) \;=\; \frac{S(H_i,\, E)M(H_i)}{\displaystyle\sum_{j=1}^{k} S(H_j,\, E)M(H_j)} \tag{7-62}$$

$$=\; \frac{S(H_i,\, E)f_i}{\displaystyle\sum_{j=1}^{k} S(H_j,\, E)f_j} \tag{7-63}$$

where $f_i = M(H_i)/M(X) = M(H_i)/n = S(X,\, H_i)$ gives the "relative frequency" of H_i, the degree to which all the hypotheses are H_i.

The subsethood theorem implies inequality when the partitioning hypotheses are fuzzy. For instance, if $k = 2$, H^c is the complement of an arbitrary fuzzy set H, and evidence E is fuzzy, then [Kosko, 1986b] the occurrence of nondegenerate hypothesis overlap and underlap gives a lower bound on the posterior subsethood:

$$S(E,\, H) \;\geq\; \frac{S(H,\, E)f_H}{S(H,\, E)f_H + S(H^c,\, E)f_{H^c}} \tag{7-64}$$

where $f_H = S(X, H)$. The lower bound increases with $M(H)$ and decreases with $M(H^c)$. Since a like lower bound holds for $S(E, H^c)$, adding the two posterior subsethoods gives the additive inequality

$$S(E, H) + S(E, H^c) \geq 1 \qquad (7\text{-}65)$$

an inequality Zadeh [1983] arrived at independently by directly defining a "relative sigma-count" as the subsethood measure given by the subsethood theorem. If H is nonfuzzy, equality holds as in the additive law of conditional probability:

$$P(H \mid E) + P(H^c \mid E) = 1 \qquad (7\text{-}66)$$

The subsethood theorem implies a deeper Bayes theorem for arbitrary fuzzy sets, the

Odds-form fuzzy Bayes theorem

$$\frac{S(A_1 \cap H, A_2)}{S(A_1 \cap H, A_2^c)} = \frac{S(A_2 \cap H, A_1)}{S(A_2^c \cap H, A_1)} \frac{S(H, A_2)}{S(H, A_2^c)} \qquad (7\text{-}67)$$

We prove this theorem directly by replacing the subsethood terms on the right-hand side with their equivalent ratios of counts, canceling like terms three times, multiplying by $M(A_1 \cap H)/M(A_1 \cap H)$, rearranging, and applying the subsethood theorem a second time.

We have now developed enough fuzzy theory to examine critically the recent anti-fuzzy polemics of Lindley [1987] and Jaynes [1979] (and thus Cheeseman [1985], who uses Jaynes' arguments). To begin we observe four more corollaries of the subsethood theorem:

(i) $\quad 0 \leq S(H, A) \leq 1$ $\qquad (7\text{-}68)$

(ii) $\quad S(H, A) = 1 \quad$ if $H \subset A$ $\qquad (7\text{-}69)$

(iii) $\quad S(H, A_1 \cup A_2) = S(H, A_1) + S(H, A_2) - S(H, A_1 \cap A_2)$ $\quad (7\text{-}70)$

(iv) $\quad S(H, A_1 \cap A_2) = S(H, A_1)S(A_1 \cap H, A_2)$ $\qquad (7\text{-}71)$

Each relationship follows from the ratio form of $S(A, B)$. The third relationship (7-70) uses the additivity of the count $M(A)$, which follows from $\min(x, y) + \max(x, y) = x + y$.

Suppose we make the notational identification $S(H, A) = P(A \mid H)$. We then obtain the defining relationships of conditional probability Lindley proposed:

Convexity: $\quad 0 \leq P(A \mid H) \leq 1$ and $P(A \mid H) = 1$ if H implies A $\quad (7\text{-}72)$

Addition: $\quad P(A_1 \cup A_2 \mid H)$

$$= P(A_1 \mid H) + P(A_2 \mid H) - P(A_1 \cap A_2 \mid H) \qquad (7\text{-}73)$$

Multiplication: $\quad P(A_1 \cap A_2 \mid H) = P(A_1 \mid H)P(A_2 \mid A_1 \cap H)$ $\qquad (7\text{-}74)$

"From these three rules," Lindley tells us, "all of the many, rich and wonderful results of the probability calculus follow. They may be described as the axioms of probability." Lindley takes these as "unassailable" axioms: "We really have no choice about the rules governing our measurement of uncertainty: they are dictated to us by the inexorable laws of logic." Lindley proceeds to build a "coherence" argument around the odds-form Bayes theorem, which he correctly deduces from the axioms as the equality

$$\frac{P(A_2 \mid A_1 \cap H)}{P(A_2^c \mid A_1 \cap H)} \quad = \quad \frac{P(A_1 \mid A_2 \cap H)}{P(A_1 \mid A_2^c \cap H)} \frac{P(A_2 \mid H)}{P(A_2^c \mid H)} \tag{7-75}$$

where here we interpret A^c as not-A. "Any other procedure," Lindley claims, "is incoherent." This polemic evaporates in the face of the above four subsethood corollaries and the odds-form fuzzy Bayes theorem. Ironically, rather than establish the primacy of axiomatic probability, Lindley seems to argue that it is fuzziness in disguise.

Maximum-entropy estimation provides another source of Bayesian probability polemic [Cheeseman, 1985]. Here the axiomatic argument rests on the so-called Cox's theorem [1946].

According to physicist E. T. Jaynes [1979]: "Cox proved that any method of inference in which we represent degrees of plausibility by real numbers, is necessarily either equivalent to Laplace's, or inconsistent," where Jaynes cites Laplace as an early Bayesian probabilist. In fact Cox used *bivalent* logic (Boolean algebra) and other assumptions to show that, again according to Jaynes, the "conditions of consistency can be be stated in the form of functional equations," namely the probabilistic product and sum rules:

$$P(A \cap B \mid C) \quad = \quad P(A \mid B \cap C)P(B \mid C) \tag{7-76}$$

$$P(B \mid A) + P(B^c \mid A) \quad = \quad 1 \tag{7-77}$$

The subsethood theorem implies

$$S(C, A \cap B) \quad = \quad S(B \cap C, A)S(C, B) \tag{7-78}$$

$$S(A, B) + S(A, B^c) \quad \geq \quad 1 \tag{7-79}$$

with, as we have seen, equality holding for the second subsethood relationship when B is nonfuzzy, which holds in the Cox-Jaynes setting.

In the probabilistic case overlap and underlap are degenerate. So $P(A \cap A^c \mid B) = P(\emptyset \mid B) = P(\emptyset)/P(B) = 0$, and $P(B \mid A \cap A^c) = P(B \mid \emptyset)$ is undefined. Yet in general $S(B, A \cap A^c) > 0$, and we can define $S(A \cap A^c, B)$ when A is fuzzy and B is fuzzy or nonfuzzy.

Jaynes' claim either is false or concedes that probability is a special case of fuzziness. For strictly speaking, since the subsethood measure $S(A, B)$ satisfies the multiplicative and additive laws specified by Cox and yet differs from the conditional probability $P(B \mid A)$, Jaynes' claim is false.

Presumably Jaynes was unaware of fuzzy sets. He suggests that the frequency theory of probability provides the only alternative uncertainty theory, and we have reduced relative frequency to the subsethood measure $S(X, A)$. So if we restrict consideration to nonfuzzy sets A and B, equality holds in the above subsethood relations, and Jaynes argues correctly: probability and fuzziness coincide. But fuzziness exists, indeed abounds, outside this restriction and classical probability theory does not. So fuzzy theory extends probability theory. Equivalently, probability represents a special case of fuzziness.

When we examine Cox's actual arguments, we find that Cox assumes that the uncertainty combination operators are continuously *twice differentiable*. Min and max are not twice differentiable. Technically, Cox's theorem does not apply.

THE ENTROPY-SUBSETHOOD THEOREM

We independently derived the fuzzy entropy theorem and the subsethood theorem from first principles, from sets-as-points unit-cube geometry. Both theorems involve ratios of cardinalities. So we can suspect a connection.

The entropy-subsethood theorem shows that the connection involves overlap $A \cap A^c$ and underlap $A \cup A^c$. The theorem eliminates fuzzy entropy in favor of subsethood. So subsethood emerges as the fundamental, characterizing quantity of fuzziness—and, arguably, of probability as well.

Entropy-subsethood theorem

$$E(A) \;=\; S(A \cup A^c, A \cap A^c) \tag{7-80}$$

Proof. The theorem follows if we replace B and A in the Subsethood Theorem with respectively overlap $A \cap A^c$ and underlap $A \cup A^c$. Since overlap is a subset of underlap, since $S(A \cap A^c, A \cup A^c) = 1$, the intersection of the two sets equals the overlap. **Q.E.D.**

The entropy-subsethood theorem describes a peculiar relationship. It gives fuzziness or ambiguity as the degree to which the superset $A \cup A^c$ is a subset of its own subset $A \cap A^c$, the degree to which the whole is a part of one of its own parts, a relationship Western logic forbids.

This relationship violates our ingrained Venn-diagram intuitions of unambiguous set inclusion. Only the midpoint of I^n yields total containment of underlap in overlap. The cube vertices yield zero containment. This parallels in the extreme the relative frequency relationship $S(X, A) = n_A/N$, where nonfuzzy subset A contains to some degree its nonfuzzy superset X.

Figure 7.11 illustrates the entropy-subsethood theorem. It shows that d^*, the shortest distance from underlap $A \cup A^c$ to the hyperrectangle that defines the fuzzy

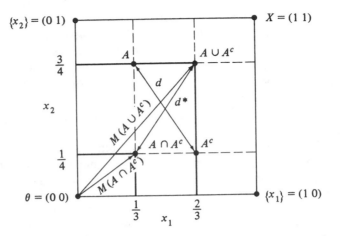

FIGURE 7.11 Entropy-subsethood theorem in two dimensions. Just as the long diagonals have equal length, $d(A, A^c) = d(A \cup A^c, A \cap A^c) = d^* = M(A \cup A^c) - M(A \cap A^c)$, the shortest distance from $A \cup A^c$ to the fuzzy power set of $A \cap A^c$. *Source:* Kosko, B., "Fuzziness vs. Probability," *International Journal of General Systems*, vol. 17, no. 2-3, Gordon and Breach Science Publishers, 1990.

power set of overlap $A \cap A^c$, equals the distance $d(A \cup A^c, A \cap A^c) = d(A, A^c)$ and equals a difference of vector magnitudes: $d^* = M(A \cup A^c) - M(A \cap A^c)$.

The entropy-subsethood theorem implies that no probability measure measures fuzziness. For the moment, suppose not. Suppose fuzzy entropy measures nothing new; fuzziness is simply disguised probability. Suppose, as Lindley [1987] claims, that probability theory "is adequate for all problems involving uncertainty." Then there exists some probability measure P such that $P = E$. P cannot equal zero everywhere because $P(X) = 1$. Then there is some A such that $P(A) = E(A) > 0$. But in a probability space overlap or underlap are degenerate: $A \cap A^c = \emptyset$, and $A \cup A^c = X$.

The entropy-subsethood theorem then implies that $0 < P(A) = E(A) = S(A \cup A^c, A \cap A^c) = S(X, \emptyset)$. X can be a subset to nonzero degree of the empty set only if X itself is empty, and hence only if A is empty: $X = A = \emptyset$. Then the sure event X is impossible: $P(X) = P(\emptyset) = 0$. Or the impossible event is sure: $P(\emptyset) = 1$. Either outcome gives a bivalent contradiction, impervious to normalization. So there exists no probability measure P that measures fuzziness. Fuzziness exists.

REFERENCES

Bandler, W., and Kohout, L., "Fuzzy Power Sets and Fuzzy Implication Operators," *Fuzzy Sets and Systems*, vol. 4, 13-30, 1980.

Black, M., "Vagueness: An Exercise in Logical Analysis," *Philosophy of Science*, vol. 4, 427-455, 1937.

Cheeseman, P., "In Defense of Probability," *Proceedings of the IJCAI-85*, 1002-1009, August 1985.

Chung, K. L., *A Course in Probability Theory*, Academic Press, Orlando, FL, 1974.

Cox, R. T., "Probability, Frequency, and Reasonable Expectations," *American Journal of Physics*, vol. 14, no. 1, 1-13, January/February 1946.

Gaines, B. R., "Precise Past, Fuzzy Future," *International Journal of Man-Machine Studies*, vol. 19, 117-134, 1983.

Hume, D., *An Inquiry Concerning Human Understanding*, 1748.

Jaynes, E. T., "Where Do We Stand on Maximum Entropy?" in *The Maximum Entropy Formalism*, Levine and Tribus (eds.), M.I.T. Press, Cambridge, MA, 1979.

Kac, M., *Probability and Related Topics in Physical Sciences, Lectures in Applied Mathematics*, vol. I, Interscience, New York, 1959.

Kant, I., *Critique of Pure Reason*, 2nd ed., 1787.

Kline, M., *Mathematics: The Loss of Certainty*, Oxford University Press, New York, 1980.

Klir, G. J., and Folger, T. A., *Fuzzy Sets, Uncertainty, and Information*, Prentice Hall, Englewood Cliffs, NJ, 1988.

Kosko, B., "Counting With Fuzzy Sets," *IEEE Transactions on Pattern Analysis and Machine Intelligence*, vol. PAMI-8, 556-557, July 1986(a).

Kosko, B., "Fuzzy Entropy and Conditioning," *Information Sciences*, vol. 40, 165-174, 1986(b).

Kosko, B., *Foundations of Fuzzy Estimation Theory*, Ph.D. dissertation, June 1987, Department of Electrical Engineering, University of California at Irvine, Order Number 8801936, University Microfilms International, 300 N. Zeeb Road, Ann Arbor, MI, 48106.

Kosko, B., "Fuzzy Quantum States," in preparation, 1991.

Lindley, D. V., "The Probability Approach to the Treatment of Uncertainty in Artificial Intelligence and Expert Systems," *Statistical Science*, vol. 2, no. 1, 17-24, February 1987.

Mill, J. S., *A System of Logic*, 1843.

Prade, H., "A Computational Approach to Approximate and Plausible Reasoning with Applications to Expert Systems," *IEEE Transactions on Pattern Analysis and Machine Intelligence*, vol. PAMI-7, 260-283, 1985.

Quine, W. V. O., "What Price Bivalence?" *Journal of Philosophy*, vol. 78, no. 2, 90-95, February 1981.

Quine, W. V. O., *Quiddities*, Harvard University Press, Cambridge, MA, 1987.

Rescher, N., *Many-valued Logic*, McGraw-Hill, New York, 1969.

Sorenson, H. W., *Parameter Estimation*, Marcel Dekker, New York, 1980.

Wittgenstein, L., *Tractatus Logico-Philosophicus*, Routledge & Kegan Paul Ltd., London, 1922.

Zadeh, L. A., "Fuzzy Sets," *Information and Control*, vol. 8, 338-353, 1965.

Zadeh, L. A., "A Computational Approach to Fuzzy Quantifiers in Natural Languages," *Computers and Mathematics*, vol. 9, no. 1, 149-184, 1983.

PROBLEMS

Problems 7.1–7.6 refer to the following identifications:

$$
\begin{array}{llccccc}
 & & x_1 & x_2 & x_3 & x_4 & x_5 \\
X & = (& 1 & 1 & 1 & 1 & 1 &) \\
\emptyset & = (& 0 & 0 & 0 & 0 & 0 &) \\
A & = (& .2 & .6 & .7 & .9 & 0 &) \\
B & = (& .3 & .5 & .2 & .8 & .1 &) \\
C & = (& .5 & .8 & .7 & 1 & .5 &) \\
D & = (& 1 & 0 & 1 & 0 & 1 &) \\
F & = (& 1 & 1 & 1 & 0 & 0 &) \\
P & = (& .5 & .5 & .5 & .5 & .5 &)
\end{array}
$$

7.1. Find $(A \cap B) \cup (A \cap B^c)$ and $[(A^c \cap B^c) \cup C]^c$.

7.2. Find $\ell^1(A, A_{near})$, $\ell^1(A, A_{far})$, and $\ell^{55}(F, P)$.

7.3. Find $E(A)$, $E(D)$, and $E(P)$.

7.4. Use the fit-violation definition of subsethood to find the following:

    (a) $S(A, B)$

    (b) $S(B, A)$

    (c) $S(A, C)$

    (d) $S(B, C)$

    (e) $S(D, F)$

7.5. Repeat Problem 7.4 using $S(A, B) = 1 - d(A, B^*)/M(A)$.

7.6. Repeat Problem 7.4 using the subsethood theorem.

7.7. Prove the fuzzy DeMorgan laws:

    (a) $A \cap B = (A^c \cup B^c)^c$

    (b) $A \cup B = (A^c \cap B^c)^c$

7.8. Prove:

$$
0 \ \leq \ \ell^p(A, A_{near}) \ \leq \ \frac{n^{1/p}}{2} \ \leq \ \ell^p(A, A_{far}) \ \leq \ n^{1/p}, \quad p \geq 1
$$

7.9. Prove the ℓ^1-version of the fuzzy entropy theorem:

$$
E(A) \ = \ \frac{\ell^1(A, A_{near})}{\ell^1(A, A_{far})} \ = \ \frac{M(A \cap A^c)}{M(A \cup A^c)}
$$

7.10. Prove: $M(A) + M(B) = M(A \cap B) + M(A \cup B)$.

7.11. Prove:

$$
\frac{1}{n} M(A \cap A^c) + \frac{1}{n} M(A \cup A^c) \ = \ 1
$$

7.12. Prove:

(a) $E(P) = \dfrac{1}{n-1}$ if $M(P) = 1$ and all $p_i \le 1/2$.

(b) $E(P) < \dfrac{1}{n-1}$ if $M(P) = 1$ and some $p_j > 1/2$.

7.13. Prove the fit-violation version of the subsethood theorem:

$$S(A, B) \;=\; 1 - \frac{\sum\limits_{x} \max(0,\, m_A - m_B(x))}{M(A)} \;=\; \frac{M(A \cap B)}{M(A)}$$

7.14. Prove:

$$S(E, H_i) \;=\; \frac{S(H_i, E)\, f_i}{\sum\limits_{j=1}^{K} S(H_j, E)\, f_j}$$

where $f_i = S(X, H_i)$, the nonfuzzy sets $H_1, \ldots, H_K$ partition X, and E is fuzzy.

7.15. Prove:

$$S(E, H) \;\ge\; \frac{S(H, E) f_H}{S(H, E) f_H + S(H^c, E) f_{H^c}}$$

where $f_H = S(X, H)$ and E and H are arbitrary fuzzy sets.

7.16. Prove the odds-form Bayes theorem:

$$\frac{S(A_1 \cap H, A_2)}{S(A_1 \cap H, A_2^c)} \;=\; \frac{S(A_2 \cap H, A_1)}{S(A_2^c \cap H, A_1)} \frac{S(H, A_2)}{S(H, A_2^c)}$$

for arbitrary fuzzy sets A_1, A_2, and H.

7.17. Prove directly the additive inequality: $S(A, B) + S(A, B^c) \ge 1$.

7.18. Prove:

(a) $0 \le S(H, A) \le 1$.

(b) $S(H, A) = 1$ if $H \subset A$.

(c) $S(H, A_1 \cup A_2) = S(H, A_1) + S(H, A_2) - S(H, A_1 \cap A_2)$.

(d) $S(H, A_1 \cap A_2) = S(H, A_1) S(A_1 \cap H, A_2)$.

7.19. Show that $N_a(N_a(x)) = x$ for the generalized negation operator

$$N_a(x) \;=\; \frac{1-x}{1+ax}, \qquad a > -1, \quad 0 \le x \le 1$$

7.20. If we define intersection $\bigcap_T$ pointwise by the Yager t-norm

$$T(x, y) = 1 - \min(1, [(1-x)^p + (1-y)^p]^{1/p}), \qquad p > 0$$

how should we define the corresponding DeMorgan dual union $\bigcup_S$?

7.21. What DeMorgan dual union operator corresponds to the intersection operator

$$\max(0, x + y - 1) \ ?$$

7.22. Zadeh's consequent conjunction syllogism schematizes as

$$Q_1 \ \text{As are } Bs$$
$$Q_2 \ \text{As are } Cs$$

Therefore: Q As are Bs and Cs

Show that if $Q_1 = S(A, B)$ and $Q_2 = S(A, C)$, then the fuzzy quantifier Q obeys

$$\max(0, \ Q_1 + Q_2 - 1) \ \leq \ Q \ \leq \ \min(Q_1, Q_2)$$

7.23. Define the *volume subsethood* measure $V(A, B)$ as

$$V(A, B) \ = \ \frac{v(A \cap B)}{v(A)}$$

for fit vectors $A = (a_1, \ldots, a_n)$ and $B = (b_1, \ldots, b_n)$ such that $a_i > 0$. $v(A)$ is the Lebesgue or volume measure of A:

$$v(A) \ = \ \prod_{i=1}^{n} a_i$$

The volume subsethood measure $V(A, B)$ measures the ratio of the volume of the overlap hyperrectangle $F(2^{A \cap B})$ to the volume of A's fuzzy power set $F(2^A)$. Prove that the volume subsethood measure $V(A, B)$ underestimates the subsethood measure $S(A, B)$:

$$V(A, B) \ \leq \ S(A, B)$$

7.24. Prove:

$$E(A) \ = \ E(A^c) \ = \ E(A \cap A^c) \ = \ E(A \cup A^c)$$

8

FUZZY ASSOCIATIVE MEMORIES

FUZZY SYSTEMS AS BETWEEN-CUBE MAPPINGS

Chapter 7 introduced multivalued or fuzzy sets as points in the unit hypercube $I^n = [0, 1]^n$. Within the cube we were interested in the distance between points. This led to measures of the size and fuzziness of a fuzzy set and, more fundamentally, to a measure of how much one fuzzy set is a subset of another fuzzy set. This *within-cube* theory directly extends to the continuous case where the space X is a subset of R^n or, in general, where X is a subset of products of real or complex spaces.

The next step considers mappings *between* fuzzy cubes. This level of abstraction provides a surprising and fruitful alternative to the propositional and predicate-calculus reasoning techniques used in artificial-intelligence (AI) expert systems. It allows us to reason with sets instead of propositions.

The fuzzy-set framework is numerical and multidimensional. The AI framework is symbolic and one-dimensional, with usually only bivalent expert "rules" or propositions allowed. Both frameworks can encode structured knowledge in linguistic form. But the fuzzy approach translates the structured knowledge into a flexible *numerical* framework and processes it in a manner that resembles neural-network processing. The numerical framework also allows us to adaptively infer and modify fuzzy systems, perhaps with neural or statistical techniques, directly from problem-domain sample data.

Between-cube theory is fuzzy-systems theory. A fuzzy set defines a point in a cube. A fuzzy system defines a mapping between cubes. A fuzzy system S maps fuzzy sets to fuzzy sets. Thus a **fuzzy system** S is a transformation $S: I^n \rightarrow I^p$. The n-dimensional unit hypercube I^n houses all the fuzzy subsets of the domain space, or input *universe of discourse*, $X = \{x_1, \ldots, x_n\}$. I^p houses all the fuzzy subsets of the range space, or output universe of discourse, $Y = \{y_1, \ldots, y_p\}$. X and Y can also denote subsets of R^n and R^p. Then the fuzzy power sets $F(2^X)$ and $F(2^Y)$ replace I^n and I^p.

In general a fuzzy system S maps families of fuzzy sets to families of fuzzy sets, thus $S: I^{n_1} \times \cdots \times I^{n_r} \rightarrow I^{p_1} \times \cdots \times I^{p_s}$. Here too we can extend the definition of a fuzzy system to allow arbitrary products of arbitrary mathematical spaces to serve as the domain or range spaces of the fuzzy sets.

(A technical comment is in order for sake of historical clarification. A tenet, perhaps the defining tenet, of the classical theory [Dubois, 1980] of fuzzy sets as functions concerns the fuzzy extension of any mathematical function. This tenet holds that any function $f: X \rightarrow Y$ that maps points in X to points in Y extends to map the fuzzy subsets of X to the fuzzy subsets of Y. The so-called *extension principle* defines the set-function $f: F(2^X) \rightarrow F(2^Y)$, where $F(2^X)$ denotes the fuzzy power set of X, the set of all fuzzy subsets of X. The formal definition of the extension principle is complicated. The key idea is a supremum of pairwise minima. Unfortunately, the extension principle achieves generality at the price of triviality. In general [Kosko, 1986a, 1987] the extension principle extends functions to fuzzy sets by stripping the fuzzy sets of their fuzziness, mapping the fuzzy sets into bit vectors of nearly all 1s. This shortcoming, combined with the tendency of the extension-principle framework to push fuzzy theory into largely inaccessible regions of abstract mathematics, led in part to the development of the alternative sets-as-points geometric framework of fuzzy theory.)

We shall focus on fuzzy systems $S: I^n \rightarrow I^p$ that map *balls* of fuzzy sets in I^n to balls of fuzzy sets in I^p. These continuous fuzzy systems behave as associative memories. They map close inputs to close outputs. We shall refer to them as **fuzzy associative memories**, or FAMs.

The simplest FAM encodes the **FAM rule** or association (A_i, B_i), which associates the p-dimensional fuzzy set B_i with the n-dimensional fuzzy set A_i. These minimal FAMs essentially map one ball in I^n to one ball in I^p. They are comparable to simple neural networks. But we need not adaptively train the minimal FAMs. As discussed below, we can directly encode structured knowledge of the form "If traffic is heavy in this direction, then keep the stop light green longer" in a Hebbian-style FAM correlation matrix. In practice we sidestep this large numerical matrix with a virtual representation scheme. In place of the matrix the user encodes the fuzzy-set association (HEAVY, LONGER) as a single linguistic entry in a FAM-bank linguistic matrix.

In general a **FAM system** $F: I^n \to I^p$ encodes and processes in parallel a **FAM bank** of m FAM rules $(A_1, B_1), \ldots, (A_m, B_m)$. Each input A to the FAM system activates each stored FAM rule to different degree. The minimal FAM that stores (A_i, B_i) maps input A to B_i', a partially activated version of B_i. The more A resembles A_i, the more B_i' resembles B_i. The corresponding output fuzzy set B combines these partially activated fuzzy sets $B_1', \ldots, B_m'$. B equals a weighted average of the partially activated sets:

$$B = w_1 B_1' + \cdots + w_m B_m'$$

where w_i reflects the credibility, frequency, or strength of the fuzzy association (A_i, B_i). In practice we usually "defuzzify" the output waveform B to a single numerical value y_j in Y by computing the fuzzy centroid of B with respect to the output universe of discourse Y.

More general still, a FAM system encodes a bank of compound FAM rules that associate multiple output or consequent fuzzy sets $B_i^1, \ldots, B_i^s$ with multiple input or antecedent fuzzy sets $A_i^1, \ldots, A_i^r$. We can treat compound FAM rules as compound linguistic conditionals. This allows us to naturally, and in many cases easily, obtain structural knowledge. We combine antecedent and consequent sets with logical conjunction, disjunction, or negation. For instance, we would interpret the compound association $(A^1, A^2; B)$ linguistically as the compound conditional "IF X^1 is A^1 AND X^2 is A^2, THEN Y is B" if the comma in the fuzzy association $(A^1, A^2; B)$ denotes conjunction instead of, say, disjunction.

We specify in advance the numerical universes of discourse for fuzzy variables X^1, X^2, and Y. For each universe of discourse or fuzzy variable X, we specify an appropriate *library* of fuzzy-set values, $A_1^r, \ldots, A_k^r$. Contiguous fuzzy sets in a library overlap. In principle a neural network can estimate these libraries of fuzzy sets. In practice this is usually unnecessary. The library sets represent a weighted, though overlapping, quantization of the input space X. They represent the fuzzy-set *values* assumed by a fuzzy *variable*. A different library of fuzzy sets similarly quantizes the output space Y. Once we define the library of fuzzy sets, we construct the FAM by choosing appropriate combinations of input and output fuzzy sets. Adaptive techniques can make, assist, or modify these choices.

An **adaptive FAM** (AFAM) is a *time-varying* FAM system. System parameters gradually change as the FAM system samples and processes data. Below we discuss how neural network algorithms can adaptively infer FAM rules from training data. In principle, learning can modify other FAM system components, such as the libraries of fuzzy sets or the FAM-rule weights w_i.

Below we propose and illustrate an unsupervised adaptive clustering scheme, based on competitive learning, to "blindly" generate and refine the bank of FAM rules. In some cases we can use supervised learning techniques if we have additional information to accurately generate error estimates.

FUZZY AND NEURAL FUNCTION ESTIMATORS

Neural and fuzzy systems estimate sampled functions and behave as associative memories. They share a key advantage over traditional statistical-estimation and adaptive-control approaches to function estimation. They are *model-free* estimators. Neural and fuzzy systems estimate a function without requiring a mathematical description of how the output functionally depends on the input. They "learn from example." More precisely, they learn from samples.

Both approaches are numerical, can be partially described with theorems, and admit an algorithmic characterization that favors silicon and optical implementation. These properties distinguish neural and fuzzy approaches from the symbolic processing approaches of artificial intelligence.

Neural and fuzzy systems differ in how they estimate sampled functions. They differ in the kind of samples used, how they represent and store those samples, and how they associatively "inference" or map inputs to outputs.

These differences appear during system construction. The neural approach requires the specification of a nonlinear dynamical system, usually feedforward, the acquisition of a sufficiently representative set of numerical training samples, and the encoding of those training samples in the dynamical system by repeated learning cycles. The fuzzy system requires only that we partially fill in a linguistic "rule matrix." This task is markedly simpler than designing and training a neural network. Once we construct the systems, we can present the same numerical inputs to either system. The outputs will reside in the same numerical space of alternatives. So both systems define a surface or manifold in the input-output product space $X \times Y$. We present examples of these surfaces in Chapters 9, 10, and 11.

Which system, neural or fuzzy, is more appropriate for a particular problem depends on the nature of the problem and the availability of numerical and structured data. To date engineers have applied fuzzy techniques largely to control problems. These problems often permit comparison with standard control-theoretic and expert-system approaches. Neural networks so far seem best applied to ill-defined two-class pattern-recognition problems (defective or nondefective, bomb or not, etc.).

Fuzzy systems estimate functions with *fuzzy-set* samples (A_i, B_i). Neural systems use *numerical-point* samples (x_i, y_i). Both kinds of samples reside in the input-output product space $X \times Y$.

Figure 8.1 illustrates the geometry of fuzzy-set and numerical-point samples taken from the function $f: X \to Y$.

Engineers sometimes call the fuzzy-set association (A_i, B_i) a "rule." This is misleading, since reasoning with sets is not the same as reasoning with propositions. Reasoning with sets is harder. Sets are multidimensional, and matrices, not proportional conditionals, house associations. We must take care how we define each term and operation. We shall refer to the antecedent term A_i in the fuzzy association (A_i, B_i) as the **input associant** and the consequent term B_i as the **output associant**.

The fuzzy-set sample (A_i, B_i) encodes *structure*. It represents a mapping, a minimal *fuzzy association* of part of the output space with part of the input space.

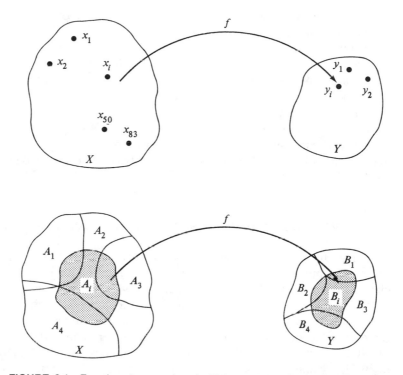

FIGURE 8.1 Function f maps domain X to range Y. In the first illustration we use several numerical-point samples (x_i, y_i) to estimate $f\colon X \longrightarrow Y$. In the second case we use only a few fuzzy subsets A_i of X and B_i of Y. The fuzzy association (A_i, B_i) represents system structure, as an adaptive clustering algorithm might infer or as an expert might articulate. In practice there are usually fewer different output associants or "rule" consequents B_i than input associants or antecedents A_i.

In practice this resembles a meta-rule—IF A_i, THEN B_i—the type of structured linguistic rule an expert might articulate to build an expert-system "knowledge base." The association might also represent the result of an adaptive clustering algorithm.

Consider a fuzzy association for the intelligent control of a traffic light: "If the traffic is heavy in this direction, then keep the light green longer." The fuzzy association is (HEAVY, LONGER). The input fuzzy variable *traffic density* assumes the fuzzy-set value HEAVY. The output fuzzy variable *green light duration* assumes the fuzzy-set value LONGER. Another fuzzy association might be (LIGHT, SHORTER). The fuzzy system encodes each linguistic association or "rule" in a numerical *fuzzy associative memory* (FAM) mapping. The FAM then numerically processes numerical input data. A measured description of traffic density (e.g., 150 cars per unit road surface area) then corresponds to a unique numerical output (e.g., 3 seconds), the "recalled" output.

The degree to which a particular measurement of traffic density is heavy depends on how we define the fuzzy set of heavy traffic. The definition may arise from statistical or neural clustering of historical data or from pooling the responses of experts. In practice the fuzzy engineer and the problem-domain expert agree on one of many possible libraries of fuzzy-set values for the fuzzy variables.

The degree to which the traffic light stays green longer depends on the degree to which the measured traffic density is heavy. In the simplest case the two degrees are the same. In general the two degrees differ. In actual fuzzy systems the output-control variables—in this case the single variable green-light duration—depend on many FAM-rule antecedents or associants activated to different degrees by incoming data.

Neural vs. Fuzzy Representation of Structured Knowledge

The distinction between fuzzy and neural systems begins with how they represent structured knowledge. How would a neural network encode the same associative information? How would a neural network encode the structured knowledge "If the traffic is heavy in this direction, then keep the light green longer"?

The simplest method encodes two associated numerical vectors. One vector represents the input associant HEAVY. The other represents the output associant LONGER. But this is too simple. For the neural network's fault tolerance now works to its disadvantage. The network tends to reconstruct partial inputs to complete sample inputs. It erases the desired partial degrees of activation. If an input is close to A_i, the output will tend to be B_i. If the output is distant from A_i, the output will tend to be some other sampled output vector or a spurious output altogether.

A better neural approach encodes a mapping from the heavy-traffic subspace to the longer-time subspace. Then the neural network needs a representative sample set to capture this structure. Statistical networks, such as adaptive vector quantizers, may need thousands of statistically representative samples. Feedforward multilayer neural networks trained with the backpropagation algorithm in Chapter 5 may need hundreds of representative numerical input-output pairs and may need to recycle these samples tens of thousands of times in the learning process.

The neural approach suffers a deeper problem than just the computational burden of training. *What* does it encode? How do we know the network encodes the original structure? What does it recall? There is no natural inferential audit trail. System nonlinearities wash it away. Unlike an expert system, we do not know which inferential paths the network uses to reach a given output or even which inferential paths exist. There is only a large system of synchronous or asynchronous nonlinear functions. Unlike, say, the adaptive Kalman filter, we cannot appeal to a postulated mathematical model of how the output state depends on the input state. Model-free estimation is, after all, the central computational advantage of neural networks. The cost is system inscrutability.

We are left with an unstructured computational black box. We do not know what the neural network encoded during training or what it will encode or forget in further training. (For competitive adaptive vector quantizers we do know that synaptic vectors asymptotically estimate sample-space centroids and perhaps higher-order moments.) We can characterize the neural network's behavior only by exhaustively passing all inputs through the black box and recording the recalled outputs. The characterization may use a summary scalar like mean-squared error.

This black-box characterization of the network's behavior involves a computational *dilemma*. On the one hand, for most problems the number of input-output cases we need to check is computationally prohibitive. On the other, when the number of input-output cases is tractable, we may as well store these pairs and appeal to them directly, and without error, as a look-up table. In the first case the neural network is unreliable. In the second case it is unnecessary.

A further problem is sample generation. Where did the original numerical point samples come from? Did we ask an expert to give numbers? How reliable are such numerical vectors, especially when the expert feels most comfortable giving the original linguistic data? This procedure seems at most as reliable as the expert-system method of asking an expert to give condition-action rules with numerical uncertainty weights.

Statistical neural estimators require a "statistically representative" sample set. We may need to randomly "create" these samples from an initial small sample set by bootstrap techniques or by random-number generation of points clustered near the original samples. Both sample-augmentation procedures assume that the initial sample set sufficiently represents the underlying probability distribution. The problem of where the original sample set comes from remains. The fuzziness of the notion "statistically representative" compounds the problem. In general we do not know in advance how well a given sample set reflects an unknown underlying distribution of points. Indeed when the network adapts on-line, we know only past samples. The remainder of the sample set resides in the unsampled future.

In contrast, fuzzy systems directly encode the linguistic sample (HEAVY, LONGER) in a dedicated numerical matrix, perhaps of infinite dimensions. The default encoding technique is the fuzzy Hebb procedure discussed below. For practical problems we need not store this large, perhaps infinite, numerical matrix. Instead we use a virtual representation scheme. Numerical point inputs permit this simplification. Mathematically we implicitly pass large unit bit vectors, or delta pulses in the continuous case, through the FAM-rule matrix. In general we describe inputs by an uncertainty distribution, probabilistic or fuzzy. Then we must use the entire matrix or reduce the input to a scalar by averaging.

For instance, if the *heavy traffic* input is 150 cars, we can omit the FAM matrix. Below we refer to these systems as binary input-output FAMs, or BIOFAMs. But if the input is a Gaussian curve with mean 150, then in principle we must process the vector input with a FAM matrix. (In practice we might use only the mean.) The dimensions of the *linguistic* FAM-bank matrix are usually small. The dimensions reflect the quantization levels of the input and output spaces, the number of fuzzy-set values assumed by the fuzzy variables.

The fuzzy approach combines the purely numerical approaches of neural networks and mathematical modeling with the symbolic, structure-rich approaches of artificial intelligence. We acquire knowledge symbolically—or numerically if we use adaptive techniques—but represent it numerically. We also process data numerically. Adaptive FAM rules correspond to common-sense, often nonarticulated, behavioral rules that improve with experience.

This approach does not abandon neural-network techniques. Instead, it limits them to *unstructured* parameter and state estimation, pattern recognition, and cluster formation. The system *architecture* remains fuzzy.

FAMs as Mappings

Fuzzy associative memories (FAMs) are transformations. *FAMs map fuzzy sets to fuzzy sets.* They map unit cubes to unit cubes, as in Figure 8.1. In the simplest case the FAM system consists of a single association, such as (HEAVY, LONGER). In general the FAM system consists of a bank of different FAM associations. Each association corresponds to a different numerical FAM matrix, or a different entry in a linguistic FAM-bank matrix. We do not combine these matrices as we combine or superimpose neural-network associative-memory (outer-product) matrices. (An exception is the *fuzzy cognitive map* [Kosko, 1988; Taber, 1987, 1991].) We store the matrices separately and access them in parallel. This avoids crosstalk. Since we use a virtual (BIOFAM) representation scheme, the computational burden of the parallel access is light.

We begin with single-association FAMs. For concreteness let the fuzzy-set pair (A, B) encode the traffic-control association (HEAVY, LIGHT). We quantize the domain of traffic density to the n numerical variables $x_1, x_2, \ldots, x_n$. We quantize the range of green-light duration to the p variables $y_1, y_2, \ldots, y_p$. The elements x_i and y_j belong respectively to the ground sets $X = \{x_1, \ldots, x_n\}$ and $Y = \{y_1, \ldots, y_p\}$. x_1 might represent zero traffic density. y_p might represent 10 seconds.

The fuzzy sets A and B are multivalued or fuzzy subsets of X and Y. So A defines a point in the n-dimensional unit hypercube $I^n = [0, 1]^n$, and B defines a point in the p-dimensional fuzzy cube I^p. Equivalently, A and B define the membership functions m_A and m_B that map the elements x_i of X and y_j of Y to degrees of membership in $[0, 1]$. The membership values, or *fit* (fuzzy unit) values, indicate how much x_i belongs to or fits in subset A, and how much y_j belongs to B. We describe this with the abstract functions $m_A: X \longrightarrow [0, 1]$ and $m_B: Y \longrightarrow [0, 1]$. We shall freely view sets both as functions and as points in fuzzy power sets.

The geometric *sets-as-points* interpretation of finite fuzzy sets A and B as points in unit cubes allows a natural vector representation. We represent A and B by the numerical *fit vectors* $A = (a_1, \ldots, a_n)$ and $B = (b_1, \ldots, b_p)$, where $a_i = m_A(x_i)$, and $b_j = m_B(y_j)$. We can interpret the identifications A = HEAVY and B = LONGER to suit the problem at hand. Intuitively the a_i values should increase

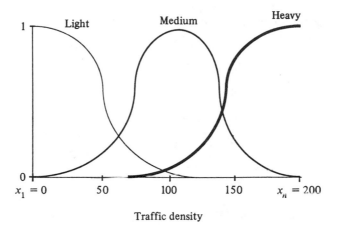

FIGURE 8.2 Three possible fuzzy subsets of traffic-density space X. Each fuzzy sample corresponds to such a subset. We draw the fuzzy sets as continuous membership functions. In practice membership values are sampled or quantized. So the sets are points in some unit hypercube I^n.

as the index i increases, perhaps approximating a sigmoid membership function. Figure 8.2 illustrates three possible fuzzy subsets of the universe of discourse X.

Fuzzy Vector-Matrix Multiplication: Max-Min Composition

Fuzzy vector-matrix multiplication resembles classical vector-matrix multiplication. We replace pairwise multiplications with pairwise minima. We replace column (row) sums with column (row) maxima. We denote this **fuzzy vector-matrix composition** relation, or the **max-min composition** relation [Klir, 1988], by the composition operator "∘". For row fit vectors A and B and fuzzy n-by-p matrix M (a point in $I^{n \times p}$):

$$A \circ M = B \qquad (8\text{-}1)$$

where we compute the "recalled" component b_j by taking the fuzzy inner product of fit vector A with the jth column of M:

$$b_j = \max_{1 \le i \le n} \min(a_i, m_{ij}) \qquad (8\text{-}2)$$

Suppose we compose the fit vector $A = (.3 \ .4 \ .8 \ 1)$ with the fuzzy matrix M given by

$$M = \begin{pmatrix} .2 & .8 & .7 \\ .7 & .6 & .6 \\ .8 & .1 & .5 \\ 0 & .2 & .3 \end{pmatrix}$$

Then we compute the "recalled" fit vector $B = A \circ M$ componentwise as

$$b_1 = \max\{\min(.3, .2), \min(.4, .7), \min(.8, .8), \min(1, 0)\}$$

$$= \max(.2, .4, .8, 0)$$

$$= .8$$

$$b_2 = \max(.3, .4, .1, .2)$$

$$= .4$$

$$b_3 = \max(.3, .4, .5, .3)$$

$$= .5$$

So $B = (.8 \ .4 \ .5)$. If we somehow encoded (A, B) in the FAM matrix M, we would say that the FAM system exhibits *perfect recall* in the forward direction: $A \circ M = B$.

The neural interpretation of max-min composition is that each neuron in field F_Y (or field F_B) generates its signal/activation value by fuzzy linear composition. Passing information back through M^T allows us to interpret the fuzzy system as a bidirectional associative memory (BAM). The bidirectional FAM theorems below characterize successful BAM recall for fuzzy correlation or Hebbian learning.

For completeness we also mention the **max-product composition** operator, which replaces minimum with product in (8-2):

$$b_j = \max_{1 \leq i \leq n} a_i m_{ij}$$

Fuzzy literature often confuses this composition operator with the fuzzy correlation encoding scheme discussed below. Max-product composition is a method for "multiplying" fuzzy matrices or vectors. Fuzzy correlation, which also uses pairwise products of fit values, constructs fuzzy matrices. In practice, and in the following discussion, we use only max-min composition.

FUZZY HEBB FAMS

Most fuzzy systems found in applications are fuzzy Hebb FAMs [Kosko, 1986b]. They are fuzzy systems $S: I^n \to I^p$ constructed in a simple neurallike manner. As discussed in Chapter 4, in neural-network theory we interpret the classical Hebbian hypothesis of correlation synaptic learning [Hebb, 1949] as unsupervised learning with the signal product $S_i S_j$:

$$\dot{m}_{ij} = -m_{ij} + S_i(x_i) S_j(y_j) \tag{8-3}$$

For a given pair of bipolar row vectors (X, Y), the neural interpretation gives the *outer-product* correlation matrix

$$M = X^T Y \tag{8-4}$$

We define pointwise the **fuzzy Hebb matrix** by the minimum of the "signals" a_i and b_j, an encoding scheme we shall call **correlation-minimum encoding**:

$$m_{ij} \ = \ \min(a_i, b_j) \qquad (8\text{-}5)$$

given in matrix notation as the *fuzzy outer-product*

$$M \ = \ A^T \circ B \qquad (8\text{-}6)$$

Mamdani [1977] and Togai [1986] independently arrived at the fuzzy Hebbian prescription (8-5) as a multivalued logical-implication operator: truth $(a_i \rightarrow b_i) = \min(a_i, b_j)$. The min operator, though, is a symmetric truth operator. So it does not properly generalize the classical implication $P \rightarrow Q$, which is false if and only if the antecedent P is true and the consequent Q is false, $t(P) = 1$ and $t(Q) = 0$. In contrast, a like desire to define a "conditional-possibility" matrix pointwise with continuous implication values led Zadeh [1983] to choose the Lukasiewicz implication operator: $m_{ij} = \text{truth } (a_i \rightarrow b_j) = \min(1, 1 - a_i + b_j)$. Unfortunately the Lukasiewicz operator usually equals or approximates unity, for $\min(1, 1 - a_i + b_j) < 1$ iff $a_i > b_j$. Most entries of the resulting matrix M are unity or near unity. This ignores the information in the association (A, B). So $A' \circ M$ tends to equal the largest fit value a'_k for any system input A'.

We construct an *autoassociative* fuzzy Hebb FAM matrix by encoding the redundant pair (A, A) in (8-6) as the fuzzy autocorrelation matrix:

$$M = A^T \circ A \qquad (8\text{-}7)$$

In the previous example the matrix M was such that the input $A = (.3 \ .4 \ .8 \ 1)$ recalled fit vector $B = (.8 \ .4 \ .5)$ upon max-min composition: $A \circ M = B$. Will A still recall B if we replace the original matrix M with the fuzzy Hebb matrix found with (8-6)? Substituting A and B in (8-6) gives

$$M \ = \ A^T \circ B \ = \ \begin{pmatrix} .3 \\ .4 \\ .8 \\ 1 \end{pmatrix} \circ (.8 \ .4 \ .5) \ = \ \begin{pmatrix} .3 & .3 & .3 \\ .4 & .4 & .4 \\ .8 & .4 & .5 \\ .8 & .4 & .5 \end{pmatrix}$$

This fuzzy Hebb matrix M illustrates two key properties. First, the ith row of M equals the pairwise minimum of a_i and the output associant B. Symmetrically, the jth column of M equals the pairwise minimum of b_j and the input associant A:

$$M \ = \ \begin{bmatrix} a_1 \wedge B \\ \vdots \\ a_n \wedge B \end{bmatrix} \qquad (8\text{-}8)$$

$$= \ [b_1 \wedge A^T | \dots | b_m \wedge A^T] \qquad (8\text{-}9)$$

where the cap operator denotes pairwise minimum: $a_i \wedge b_j = \min(a_i, b_j)$. The term $a_i \wedge B$ indicates componentwise minimum:

$$a_i \wedge B \ = \ (a_i \wedge b_1, \dots, a_i \wedge b_n) \qquad (8\text{-}10)$$

Hence if some $a_k = 1$, then the kth row of M equals B. If some $b_l = 1$, the lth column of M equals A. More generally, if some a_k is at least as large as every b_j, then the kth row of the fuzzy Hebb matrix M equals B.

Second, the third and fourth rows of M equal the fit vector B. Yet no column equals A. This allows perfect recall in the forward direction, $A \circ M = B$, but not in the backward direction, $B \circ M^T \neq A$:

$$A \circ M \quad = \quad (.8 \ .4 \ .5) \quad = \quad B$$

$$B \circ M^T \quad = \quad (.3 \ .4 \ .8 \ .8) \quad = \quad A' \subset A$$

A' is a proper subset of A: $A' \neq A$ and $S(A', A) = 1$, where S measures the degree of subsethood of A' in A, as discussed in Chapter 7. In other words, $a'_i \leq a_i$ for each i and $a'_k < a_k$ for at least one k. The bidirectional FAM theorems below show that this holds in general: If $B' = A \circ M$ differs from B, then B' is a proper subset of B. Hence fuzzy subsets map to fuzzy subsets.

The Bidirectional FAM Theorem for Correlation-Minimum Encoding

Analysis of FAM recall uses the traditional [Klir, 1988] fuzzy-set notions of the *height* and the *normality* of fuzzy sets. The **height** $H(A)$ of fuzzy set A is the maximum fit value of A:

$$H(A) \quad = \quad \max_{1 \leq i \leq n} a_i$$

A fuzzy set is **normal** if $H(A) = 1$, if at least one fit value a_k is maximal: $a_k = 1$. In practice fuzzy sets are usually normal. We can extend a nonnormal fuzzy set to a normal fuzzy set by adding a dummy dimension with corresponding fit value $a_{n+1} = 1$.

Recall accuracy in fuzzy Hebb FAMs constructed with correlation-minimum encoding depends on the heights $H(A)$ and $H(B)$. Normal fuzzy sets exhibit perfect recall. Indeed (A, B) is a bidirectional fixed point—$A \circ M = B$, and $B \circ M^T = A$— if and only if $H(A) = H(B)$, which always holds if A and B are normal. This is a corollary of the bidirectional FAM theorem [Kosko, 1986a] for correlation-minimum encoding. Below we present a similar theorem for correlation-product encoding.

Correlation-minimum bidirectional FAM theorem.   If $M = A^T \circ B$, then

$$(i) \quad A \circ M = B \quad \text{iff } H(A) \geq H(B)$$
$$(ii) \quad B \circ M^T = A \quad \text{iff } H(B) \geq H(A)$$
$$(iii) \quad A' \circ M \subset B \quad \text{for any } A'$$
$$(iv) \quad B' \circ M^T \subset A \quad \text{for any } B'$$

Proof. Observe that the height $H(A)$ equals the *fuzzy norm* of A:

$$A \circ A^T \quad = \quad \max_i a_i \wedge a_i \quad = \quad \max_i a_i \quad = \quad H(A)$$

Then

$$
\begin{aligned}
A \circ M \quad &= \quad A \circ (A^T \circ B) \\
&= \quad (A \circ A^T) \circ B \\
&= \quad H(A) \circ B \\
&= \quad H(A) \wedge B
\end{aligned}
$$

So $H(A) \wedge B = B$ iff $H(A) \geq H(B)$, establishing (i). Now suppose A' is an arbitrary fit vector in I^n. Then

$$
\begin{aligned}
A' \circ M \quad &= \quad (A' \circ A^T) \circ B \\
&= \quad (A' \circ A^T) \wedge B
\end{aligned}
$$

which establishes (iii) since $A' \circ A^T \leq H(A)$. A similar argument using $M^T = B^T \circ A$ establishes (ii) and (iv). **Q.E.D.**

The equality $A \circ A^T = H(A)$ implies an immediate corollary of the bidirectional FAM theorem. Supersets $A' \supset A$ behave the same as the encoded input associant A: $A' \circ M = B$ if $A \circ M = B$. Fuzzy Hebb FAMs ignore the information in the difference $A' - A$, when $A \subset A'$.

Correlation-Product Encoding

Correlation-product encoding provides an alternative fuzzy Hebbian encoding scheme. The standard mathematical outer product of the fit vectors A and B forms the FAM matrix M. Then

$$m_{ij} \quad = \quad a_i b_j \tag{8-11}$$

and in matrix notation,

$$M \quad = \quad A^T B \tag{8-12}$$

So the ith row of M equals the fit-scaled fuzzy set $a_i B$, and the jth column of M equals $b_j A^T$:

$$
M \quad = \quad
\begin{bmatrix}
a_1 B \\
\hline
\vdots \\
\hline
a_n B
\end{bmatrix}
\tag{8-13}
$$

$$
\quad = \quad [b_1 A^T | \ldots | b_m A^T] \tag{8-14}
$$

If $A = (.3\ .4\ .8\ 1)$ and $B = (.8\ .4\ .5)$ as above, we encode the FAM rule (A, B) with correlation product in the following matrix M:

$$M = \begin{pmatrix} .24 & .12 & .15 \\ .32 & .16 & .2 \\ .64 & .32 & .4 \\ .8 & .4 & .5 \end{pmatrix}$$

Note that if $A' = (0\ 0\ 0\ 1)$, then $A' \circ M = B$. The FAM system recalls output associant B to maximal degree. If $A' = (1\ 0\ 0\ 0)$, then $A' \circ M = (.24\ .12\ .15)$. The FAM system recalls output B only to degree .3.

Correlation-minimum encoding produces a matrix of clipped B sets, while correlation-product encoding produces a matrix of scaled B sets. In membership-function plots, the scaled fuzzy sets $a_i B$ all have the same shape as B. The clipped fuzzy sets $a_i \wedge B$ are flat at or above the a_i value. In this sense correlation-product encoding preserves more information than correlation-minimum encoding, an important point in fuzzy applications when we add output fuzzy sets together as in Equation (8-17) below. In the fuzzy-applications literature this often leads to the selection of correlation-product encoding.

Unfortunately, the fuzzy literature invariably confuses the correlation-product *encoding* scheme with the max-product composition method of recall or *inference*, as mentioned above. This widespread confusion warrants formal clarification.

In practice, and in the fuzzy applications developed in the next chapters, the input fuzzy set A' is a binary vector with one 1 and all other elements 0—a row of the n-by-n identity matrix (or a delta pulse in the continuous case). A' represents the occurrence of the crisp measurement datum x_i, such as a traffic density value of 30. When applied to the encoded FAM rule (A, B), the measurement value x_i activates A to degree a_i. This is part of the max-min composition recall process, for $A' \circ M = (A' \circ A^T) \circ B = a_i \wedge B$ or $a_i B$ depending on whether we encoded (A, B) in M with correlation-minimum or correlation-product encoding. We activate or "fire" the output associant B of the "rule" to degree a_i.

Since the values a_i' are binary, $a_i' m_{ij} = a_i' \wedge m_{ij}$. So the max-min and max-product composition operators coincide. We avoid this confusion by referring to both the recall process and the correlation encoding scheme as **correlation-minimum inference** when we combine correlation-minimum encoding with max-min composition, and as **correlation-product inference** when we combine correlation-product encoding with max-min composition.

We now prove the correlation-product version of the bidirectional FAM theorem.

Correlation-product bidirectional FAM theorem. If $M = A^T B$ and A and B are nonnull fit vectors, then

$$\begin{aligned}
\text{(i)} \quad & A \circ M = B && \text{iff } H(A) = 1 \\
\text{(ii)} \quad & B \circ M^T = A && \text{iff } H(B) = 1 \\
\text{(iii)} \quad & A' \circ M \subset B && \text{for any } A' \\
\text{(iv)} \quad & B' \circ M^T \subset A && \text{for any } B'
\end{aligned}$$

Proof.

$$\begin{aligned}
A \circ M &= A \circ (A^T B) \\
&= (A \circ A^T) B \\
&= H(A) B
\end{aligned}$$

Since B is not the empty set, $H(A)B = B$ iff $H(A) = 1$, establishing (i). ($A \circ M = B$ holds trivially if B is the empty set.) For an arbitrary fit vector A' in I^n:

$$\begin{aligned}
A' \circ M &= (A' \circ A^T) B \\
&\subset H(A) B \\
&\subset B
\end{aligned}$$

since $A' \circ A \leq H(A)$, establishing (iii). (ii) and (iv) follow similarly using $M^T = B^T A$. **Q.E.D.**

Superimposing FAM Rules

Now suppose we have m FAM rules or associations $(A_1, B_1), \ldots, (A_m, B_m)$. The fuzzy Hebb encoding scheme (8-6) leads to m FAM matrices $M_1, \ldots, M_m$ to encode the associations. The natural neural-network temptation is to add, or in this case maximum, the m matrices pointwise to distributively encode the associations in a single matrix M:

$$M = \max_{1 \leq k \leq m} M_k \tag{8-15}$$

This superimposition scheme fails for fuzzy Hebbian encoding. The superimposed result tends to be the matrix $A^T \circ B$, where A and B denote the pointwise maximum of the respective m fit vectors A_k and B_k. We can see this from the pointwise inequality

$$\max_{1 \leq k \leq m} \min(a_i^k, b_j^k) \leq \min(\max_{1 \leq k \leq m} a_i^k, \max_{1 \leq k \leq m} b_j^k) \tag{8-16}$$

Inequality (8-16) tends to hold with equality as m increases, since all maximum terms approach unity [Kosko, 1986a]. We lose the information in the m associations (A_k, B_k).

The fuzzy approach to the superimposition problem *additively superimposes the* m *recalled vectors* B_k' instead of the fuzzy Hebb matrices M_k. B_k' and M_k correspond to

$$A \circ M_k = A \circ (A_k^T \circ B_k)$$
$$= B_k'$$

for any fit-vector input A applied in parallel to the bank of FAM rules (A_k, B_k). This requires separately storing the m associations (A_k, B_k), as if each association in the FAM bank represents a separate feedforward neural network.

Separate storage of FAM associations consumes space but provides an "audit trail" of the FAM inference procedure and avoids crosstalk. The user can directly determine which FAM rules contributed how much membership activation to a "concluded" output. Separate storage also provides knowledge-base modularity. The user can add or delete FAM-structured knowledge without disturbing stored knowledge. Both of these benefits are advantages over a pure neural-network architecture for encoding the same associations (A_k, B_k). Of course we can use neural networks exogenously to estimate, or even individually house, the associations (A_k, B_k).

Separate storage of FAM rules brings out another distinction between FAM systems and neural networks. A fit-vector input A activates all the FAM rules (A_k, B_k) in parallel but to different degrees. If A only partially "satisfies" the antecedent associant A_k, the consequent associant B_k only partially activates. If A does not satisfy A_k at all, B_k does not activate at all. B_k' equals the null vector.

Neural networks behave differently. They try to reconstruct the entire association (A_k, B_k) when stimulated with A. If A and A_k mismatch severely, a neural network will tend to emit a nonnull output B_k', perhaps the result of the network dynamical system falling into a "spurious" attractor in the state space. We may desire this for metrical classification problems, but not for inferential problems and, arguably, for associative-memory problems. When we ask an expert a question outside his field of knowledge, it may be more prudent if he gives no response than if he gives an educated guess.

Recalled Outputs and "Defuzzification"

The recalled fit-vector output B equals a weighted sum of the individual recalled vectors B_k':

$$B = \sum_{k=1}^{m} w_k B_k' \tag{8-17}$$

where the nonnegative weight w_k summarizes the credibility or strength of the kth FAM rule (A_k, B_k). The credibility weights w_k are immediate candidates for adaptive modification. In practice we choose $w_1 = \ldots = w_m = 1$ as a default.

In principle, though not in practice, the recalled fit-vector output equals a normalized sum of the B'_k fit vectors. This keeps the components of B unit-interval valued. We do not use normalization in practice because we invariably "defuzzify" the output distribution B to produce a single numerical output, a single value in the output universe of discourse $Y = \{y_1, \ldots, y_p\}$. The information in the output waveform B resides largely in the relative values of the membership degrees.

The simplest defuzzification scheme chooses that element $y_{\max}$ that has maximal membership in the output fuzzy set B:

$$m_B(y_{\max}) \quad = \quad \max_{1 \le j \le k} m_B(y_j) \tag{8-18}$$

The popular probabilistic methods of maximum-likelihood and maximum-a-posteriori parameter estimation motivate this **maximum-membership defuzzification** scheme.

The maximum-membership defuzzification scheme has two fundamental problems. First, the mode of the B distribution is not unique. This problem affects correlation-minimum encoding, as the representation (8-8) shows, more than it affects correlation-product encoding. Since the minimum operator clips off the top of the B_k fit vectors, the additively combined output fit vector B tends to be flat over many regions of universe of discourse Y. For continuous membership functions this leads to infinitely many modes. Even for quantized fuzzy sets, there may be many modes.

In practice we can average multiple modes. For large FAM banks of "independent" FAM rules, some form of the central limit theorem (whose proof ultimately depends on Fourier transformability, not probability) tends to hold. The waveform B tends to resemble a Gaussian membership function. So a unique mode tends to emerge. It tends to emerge with fewer samples if we use correlation-product encoding.

Second, the maximum-membership scheme ignores the information in much of the waveform B. Again correlation-minimum encoding compounds the problem. In practice B is often highly asymmetric, even if it is unimodal. Infinitely many output distributions can share the same mode.

The natural alternative is the **fuzzy centroid defuzzification** scheme. We directly compute the real-valued output as a (normalized) convex combination of fit values, the *fuzzy centroid* $\bar{B}$ of fit-vector B with respect to output space Y:

$$\bar{B} \quad = \quad \frac{\displaystyle\sum_{j=1}^{p} y_j m_B(y_j)}{\displaystyle\sum_{j=1}^{p} m_B(y_j)} \tag{8-19}$$

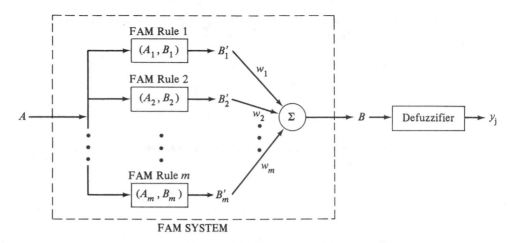

FAM SYSTEM

FIGURE 8.3 FAM system architecture. The FAM system F maps fuzzy sets in the unit cube I^n to fuzzy sets in the unit cube I^p. Binary input sets model exact input data. In general only an uncertainty estimate of the system state confronts the FAM system. So A is a proper fuzzy set. The user can defuzzify output fuzzy set B to yield exact output data, reducing the FAM system to a mapping between Boolean cubes.

The fuzzy centroid is unique and uses all the information in the output distribution B. For symmetric unimodal distributions the mode and fuzzy centroid coincide. In many cases we must replace the discrete sums in (8-19) with integrals over continuously infinite spaces. We show in Chapter 11, though, that for libraries of trapezoidal fuzzy-set values we can replace such a ratio of integrals with a ratio of simple discrete sums.

Computing the centroid (8-19) is the only step in the FAM inference procedure that requires division. All other operations are inner products, pairwise minima, and additions. This promises realization in a fuzzy optical processor. Already some form of this FAM-inference scheme has led to digital [Togai, 1986] and analog [Yamakawa, 1987, 1988] VLSI circuitry.

FAM System Architecture

Figure 8.3 schematizes the architecture of the nonlinear FAM system F. Note that F maps fuzzy sets to fuzzy sets: $F(A) = B$. So F defines a fuzzy-system transformation $F: I^n \to I^p$. In practice A equals a bit vector with one unity value, $a_i = 1$, and all other fit values zero, $a_j = 0$, or a delta pulse.

We defuzzify the output fuzzy set B with the centroid technique to produce an exact element y_j in the output universe of discourse Y. In effect defuzzification produces an output binary vector O, again with one element 1 and the rest 0s. At this level the FAM system F maps sets to sets, reducing the fuzzy system F to a mapping between Boolean cubes, $F: \{0, 1\}^n \rightarrow \{0, 1\}^p$. In many applications we model X and Y as continuous universes of discourse. So n and p are quite large. We shall call such systems binary input-output FAMs.

Binary Input-Output FAMs: Inverted-Pendulum Example

Binary input-output FAMs (BIOFAMs) are the most popular fuzzy systems for applications. BIOFAMs map system state-variable data to control, classification, or other output data. In the case of traffic control, a BIOFAM maps traffic densities to green (and red) light durations.

BIOFAMs easily extend to multiple FAM-rule antecedents, to mappings from product cubes to product cubes. There has been little theoretical justification for this extension, aside from Mamdani's [1977] original suggestion to multiply relational matrices. In the next section we present a general method for dealing with multiantecedent FAM rules. First, though, we present the BIOFAM algorithm by illustrating it, and the FAM construction procedure, on a standard control problem.

Consider an inverted pendulum. We wish to adjust a motor to balance an inverted pendulum in two dimensions. The inverted pendulum is a classical control problem and admits a math-model control solution. This provides a formal benchmark for BIOFAM pendulum controllers.

There are two state fuzzy variables and one control fuzzy variable. The first state fuzzy variable is the *angle* θ that the pendulum shaft makes with the vertical. Zero angle corresponds to the vertical position. Positive angles are to the right of the vertical, negative angles to the left.

The second state fuzzy variable is the *angular velocity* $\Delta\theta$. In practice we approximate the instantaneous angular velocity $\Delta\theta$ as the difference between the present angle measurement θ_t and the previous angle measurement θ_{t-1}:

$$\Delta\theta_t = \theta_t - \theta_{t-1}$$

The control fuzzy variable is the *motor current* or angular velocity v_t. The velocity can be positive or negative. We expect that if the pendulum falls to the right, the motor velocity should be negative to compensate. If the pendulum falls to the left, the motor velocity should be positive. If the pendulum successfully balances at the vertical, the motor velocity should be zero.

The real line R is the universe of discourse of the three fuzzy variables. In practice we restrict each universe of discourse to a comparatively small interval, such as $[-90, 90]$ for the pendulum angle, centered about zero.

We can quantize each universe of discourse into five overlapping fuzzy-set values. We know that the fuzzy variables can be positive, zero, or negative. We can quantize the magnitudes of the fuzzy variables finely or coarsely. Suppose we quantize the magnitudes as small, medium, and large. This leads to seven fuzzy-set values:

NL: Negative Large
NM: Negative Medium
NS: Negative Small
ZE: Zero
PS: Positive Small
PM: Positive Medium
PL: Positive Large

For example, θ is a fuzzy *variable* that takes NL as a fuzzy-set *value*. Different fuzzy quantizations of the angle universe of discourse allow the fuzzy variable θ to assume different fuzzy-set values. The expressive power of the FAM approach stems from these fuzzy-set quantizations. In one stroke we reduce system dimensions, and we describe a nonlinear numerical process with linguistic commonsense terms.

We are not concerned with the exact shape of the fuzzy sets defined on each of the three universes of discourse. In practice the quantizing fuzzy sets are usually symmetric triangles or trapezoids centered about representative values. (We can think of such sets as *fuzzy numbers*.) The set ZE may define a Gaussian curve for the pendulum angle θ, a triangle for the angular velocity $\Delta\theta$, and a trapezoid for the motor current v. But all the ZE fuzzy sets center about the numerical value zero, which will have maximum membership in the set of zero values.

How much should contiguous fuzzy sets overlap? This design issue depends on the problem at hand. Too much overlap blurs the distinction between the fuzzy-set values. Too little overlap tends to resemble bivalent control, producing excessive overshoot and undershoot. In Chapter 11 we determine experimentally the following default heuristic for ideal overlap: *Contiguous fuzzy sets in a library should overlap approximately 25 percent.*

Inverted-pendulum FAM rules are triples, such as (NM, ZE; PM). They describe how to modify the control variable for observed values of the pendulum state variables. A FAM rule associates a motor-velocity fuzzy-set value with a pendulum-angle fuzzy-set value and an angular-velocity fuzzy-set value. So we can interpret the triple (NM, ZE; PM) as the set-level implication

IF the pendulum angle θ is negative but medium
AND the angular velocity $\Delta\theta$ is about zero,
THEN the motor velocity should be positive but medium

These commonsensical FAM rules are comparatively easy to articulate in natural language. Consider a terser linguistic version of the same two-antecedent FAM rule:

IF $\theta = \text{NM}$ AND $\Delta\theta = \text{ZE}$
THEN $v = \text{PM}$

Even this mild level of formalism may inhibit the knowledge-acquisition process.

On the other hand, the still terser FAM triple (NM, ZE; PM) allows knowledge to be acquired simply by filling in a few entries in a linguistic FAM-bank matrix. In practice this often allows us to develop a working system in minutes.

We specify the pendulum FAM system when we choose a *FAM bank* of two-antecedent FAM rules. Perhaps the first FAM rule to choose is the *steady-state FAM rule*:(ZE, ZE; ZE). The steady-state FAM rule describes what to do in equilibrium. For the inverted pendulum we should do nothing.

Many control problems require nulling a scalar error measure. We can control many multivariable problems by nulling the norms of the system error vector and error-velocity vectors, or, better, by directly nulling the individual scalar variables. (Chapter 11 shows how error nulling can control a real-time target tracking system.) Adaptive error-nulling extends the FAM methodology to nonlinear estimation, control, and decision problems of high dimension.

The pendulum FAM bank is a 7-by-7 matrix with linguistic fuzzy-set entries. We index the columns by the seven fuzzy sets that quantize the angle θ universe of discourse. We index the rows by the seven fuzzy sets that quantize the angular velocity $\Delta\theta$ universe of discourse.

Each matrix entry can equal one of seven motor-current fuzzy-set values or equal no fuzzy set at all. Since a FAM rule is a mapping or function, there is exactly one output motor-current value for every pair of angle and angular-velocity values. So the 49 entries in the FAM bank matrix represent a subset of the 343 (7^3) possible two-antecedent FAM rules. In practice most of the entries are blank. In the adaptive FAM case discussed below, we adaptively generate the entries from process sample data.

Common sense and engineering judgment dictate the entries in the pendulum FAM-bank matrix. Suppose the pendulum does not move. So $\Delta\theta =$ ZE. If the pendulum tilts to the right of vertical, the motor velocity should be negative to compensate. The farther the pendulum tilts to the right, the larger the negative motor velocity should be. The motor velocity should be positive if the pendulum tilts to the left. So the fourth row of the FAM bank matrix, which corresponds to $\Delta\theta =$ ZE, should equal the ordinal inverse of the θ row values. This assignment includes the steady-state FAM rule (ZE, ZE; ZE).

Now suppose the angle θ is zero but the pendulum moves. If the angular velocity is negative, the pendulum will overshoot to the left. So the motor velocity should be positive to compensate. If the angular velocity is positive, the motor velocity should be negative. The greater the angular velocity is in magnitude, the greater the motor velocity should be in magnitude. So the fourth column of the FAM-bank matrix, which corresponds to $\theta =$ ZE, should equal the ordinal inverse of the $\Delta\theta$ column values. This assignment also includes the steady-state FAM rule.

Positive θ values with negative $\Delta\theta$ values should produce negative motor-current values, since the pendulum heads toward the vertical. So (PS, NS; NS) is a candidate FAM rule. Symmetrically, negative θ values with positive $\Delta\theta$ values should produce positive motor-current values. So (NS, PS; PS) is another candidate FAM rule.

This gives 15 FAM rules altogether. In practice these rules can successfully balance an inverted pendulum. Different, and smaller, subsets of FAM rules can also balance the pendulum. The software problems at the end of the chapter explore these cases.

We can represent the bank of 15 FAM rules as the 7-by-7 linguistic matrix

θ

$\Delta\theta$ \ θ	NL	NM	NS	ZE	PS	PM	PL
NL				PL			
NM				PM			
NS				PS	NS		
$\Delta\theta$ ZE	PL	PM	PS	ZE	NS	NM	NL
PS			PS	NS			
PM				NM			
PL				NL			

The BIOFAM system F admits a geometric interpretation. The set of all possible input-outpairs $(\theta, \Delta\theta; F(\theta, \Delta\theta))$ defines a *FAM surface* in the input-output product space, in this case in R^3. We plot examples of these control surfaces in Chapters 9, 10 and 11.

The BIOFAM *inference procedure* activates in parallel the antecedents of all 15 FAM rules. The binary or pulse nature of inputs picks off single fit values from the quantizing fuzzy-set values of the fuzzy variables. We can use either the correlation-minimum or correlation-product inferencing technique. For simplicity we shall illustrate the procedure with correlation-minimum inferencing.

Suppose the current pendulum angle θ equals 15 degrees and the angular velocity $\Delta\theta$ equals -10. This amounts to passing two bit vectors of one 1 and all else 0 through the BIOFAM system. What is the corresponding motor-current value $v = F(15, -10)$?

Consider first how the input data pair $(15, -10)$ activates the steady-state FAM rule (ZE, ZE; ZE). Suppose we define the antecedent and consequent fuzzy sets for ZE with the triangular fuzzy-set membership functions in Figure 8.4. Then the angle datum 15 defines a zero angle value to degree .2: $m_{ZE}^{\theta}(15) = .2$. The angular-velocity datum -10 defines a zero angular-velocity value to degree .5 : $m_{ZE}^{\Delta\theta}(-10) = .5$.

We combine the antecedent fit values with minimum or maximum depending on whether we combine the antecedent fuzzy sets with the conjunctive AND or the disjunctive OR. Intuitively, it should be at least as difficult to satisfy both antecedent conditions as to satisfy either one separately.

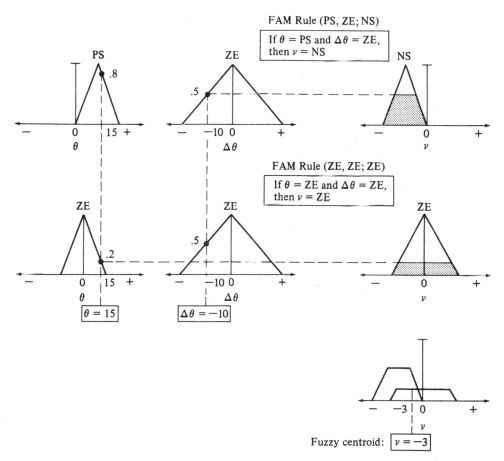

FIGURE 8.4 FAM correlation-minimum inference procedure. The FAM system consists of the two two-antecedent FAM rules (PS, ZE; NS) and (ZE, ZE; ZE). The input angle datum equals 15 and is more a small but positive angle value than a zero angle value. The input angular-velocity datum equals −10, and is a zero angular-velocity value only to degree .5. The system combines antecedent flt values with minimum, since the conjunction AND combines the antecedent terms. The combined fit value then scales the consequent fuzzy set with pairwise minimum. The system adds the minimum-scaled output fuzzy sets and computes the fuzzy centroid of this output waveform. This yields the system output-current value −3.

The FAM-rule notation (ZE, ZE; ZE) implicitly assumes that we combine antecedent fuzzy sets conjunctively with AND. So the data satisfy the compound antecedent of the FAM rule $(ZE, ZE; ZE)$ to degree

$$\min(m^{\theta}_{ZE}(15), \; m^{\Delta\theta}_{ZE}(-10)) \; = \; \min(.2, .5)$$

$$= \; .2$$

This methodology extends to any number of antecedent terms connected with arbitrary logical (set-theoretical) connectives.

The system should now activate the consequent fuzzy set of zero-motor-current values to degree .2. This differs from activating the ZE motor-current fuzzy set 100 percent with probability .2, and certainly differs from Prob $\{v = 0\} = .2$. Instead a deterministic 20 percent of ZE should result and, according to the additive combination formula (8-17), we should add this truncated fuzzy set to the final output fuzzy set.

The correlation-minimum inference procedure activates the angular-velocity fuzzy set ZE to degree .2 by taking the pairwise minimum of .2 and the ZE fuzzy set m^v_{ZE}:

$$\min(m^\theta_{ZE}(15), m^{\Delta\theta}_{ZE}(-10)) \wedge m^v_{ZE}(v) = .2 \wedge m^v_{ZE}(v)$$

for all velocity values v. The correlation-product inference procedure would multiply the zero-angular-velocity fuzzy set by .2: $.2m^v_{ZE}(v)$ for all v.

The data similarly activate the FAM rule (PS, ZE; NS) depicted in Figure 8.4. The angle datum 15 is a small but positive angle value to degree .8. The angular-velocity datum -10 is a zero-angular-velocity value to degree .5. So we scale the output motor-velocity fuzzy set of small but negative motor-velocity values by .5, the lesser of the two antecedent fit values:

$$\min(m^\theta_{PS}(15), m^{\Delta\theta}_{ZE}(-10)) \wedge m^v_{NS}(v) = .5 \wedge m^v_{NS}(v)$$

for all velocity values v. So the data activate the FAM rule (PS, ZE; NS) to greater degree than it activates the steady-state FAM rule (ZE, ZE; ZE) since in this example an angle value of 15 degrees is more a small but positive angle value than a zero angle value.

The data similarly activate the other 13 FAM rules. We combine the resulting minimum-scaled consequent fuzzy sets according to (8-17) by summing pointwise. We can then compute the fuzzy centroid with Equation (8-19), with perhaps integrals replacing the discrete sums, to determine the specific output motor velocity v. In Chapter 11 we show that, for symmetric fuzzy-set values of fuzzy variables, we can always compute the centroid exactly with simple discrete sums even if the fuzzy sets are continuous. In many real-time applications we must repeat this entire FAM inference procedure hundreds, perhaps thousands, of times per second. This may require fuzzy VLSI or optical processors.

Figure 8.4 illustrates the equal-weight additive combination procedure for just the FAM rules (ZE, ZE; ZE) and (PS, ZE; NS). In this case the fuzzy-centroidal motor-velocity value equals -3.

Multiantecedent FAM Rules: Decompositional Inference

BIOFAM inference treats antecedent fuzzy sets as propositions with fuzzy truth values. This holds because fuzzy logic corresponds to one-dimensional fuzzy-set theory and because we use binary (or delta-pulse) or exact inputs. We now formally develop the connection between BIOFAMs and the FAM theory presented earlier.

Consider the compound FAM rule "IF X is A AND Y is B, THEN C is Z," or $(A, B; C)$ for short. Let the universes of discourse X, Y, and Z have dimensions n, p, and q: $X = \{x_1, \ldots, x_n\}$, $Y = \{y_1, \ldots, y_p\}$, and $Z = \{z_1, \ldots, z_q\}$. We can directly extend this framework to multiple antecedent and consequent terms. Continuously infinite universes of discourse require a delta-pulse formulation.

In our notation X, Y, and Z double as universes of discourse and fuzzy variables. The fuzzy *variable* X can assume the fuzzy-set *values* A_1, A_2, ..., and similarly for the fuzzy variables Y and Z. When controlling an inverted pendulum, the identification "X is A" might represent the natural-language description "The pendulum angle is positive but small."

What matrix represents the FAM rule $(A, B; C)$? The question is nontrivial, since A, B, and C are fuzzy subsets of different universes of discourse, points in different unit cubes (fuzzy power sets). Their dimensions and interpretations differ. Mamdani [1977] and others have suggested representing such rules as fuzzy multidimensional relations or arrays. Then the FAM rule $(A, B; C)$ would define a fuzzy subset of the product space $X \times Y \times Z$. Engineers do not use this representation in practice, since actual systems present only exact inputs or measurements to FAM systems, and the BIOFAM procedure applies. If we presented the system with a genuine fuzzy-set input, we could preprocess the fuzzy set with a centroidal or maximum-fit-value technique, so we could still apply the BIOFAM inference procedure.

We present an alternative representation that decomposes, then recomposes, the FAM rule $(A, B; C)$ in accord with the FAM inference procedure. This representation allows neural networks to adaptively estimate, store, and modify the decomposed FAM rules. The representation requires far less storage than the multidimensional-array representation.

Let the fuzzy Hebb matrices M_{AC} and M_{BC} store the simple FAM associations (A, C) and (B, C):

$$M_{AC} = A^T \circ C \qquad (8\text{-}20)$$

$$M_{BC} = B^T \circ C \qquad (8\text{-}21)$$

The fuzzy Hebb matrices M_{AC} and M_{BC} *split* the compound FAM rule $(A, B; C)$. We can construct the *splitting matrices* with correlation-product encoding.

Let $I_X^i = (0 \ \ldots \ 0 \ 1 \ 0 \ \ldots \ 0)$ denote an n-dimensional bit vector with ith element 1 and all other elements 0. (In the continuous case I_x^i denotes a delta pulse in a convolution integral.) I_X^i equals the ith row of the n-by-n identity matrix. Similarly, I_Y^j and I_Z^k equal the respective jth and kth rows of the p-by-p and q-by-q identity matrices. The bit vector I_X^i represents the occurrence of the exact input x_i.

We will call the proposed FAM representation scheme **FAM decompositional inference**, in the spirit of the max-min compositional inference scheme discussed above. FAM decompositional inference *decomposes* the compound FAM rule $(A, B; C)$ into the component rules (A, C) and (B, C). The BIOFAM processes the simpler component rules in parallel. New fuzzy-set inputs A' and B' pass through the FAM matrices M_{AC} and M_{BC}. Max-min composition then gives the recalled fuzzy sets $C_{A'}$ and $C_{B'}$:

$$C_{A'} = A' \circ M_{AC} \tag{8-22}$$

$$C_{B'} = B' \circ M_{BC} \tag{8-23}$$

The trick is to *recompose* the fuzzy sets $C_{A'}$ and $C_{B'}$ with intersection or union depending on whether we combine the antecedent terms "X is A" and "Y is B" with AND or OR. The negated antecedent term "X is NOT A" requires forming the set complement $C_{A'}^c$ for input fuzzy set A'.

Suppose we present the new inputs A' and B' to the single-FAM-rule system F that stores the FAM rule $(A, B; C)$. Then the recalled output fuzzy set C' equals the intersection of $C_{A'}$ and $C_{B'}$:

$$\begin{aligned} F(A', B') &= [A' \circ M_{AC}] \cap [B' \circ M_{BC}] \\ &= C_{A'} \cap C_{B'} \\ &= C' \end{aligned} \tag{8-24}$$

We can then defuzzify C' to yield the exact output I_Z^k.

Logical connectives apply to antecedent terms of different dimension and meaning. Decompositional inference applies the set-theoretic analogues of the logical connectives to subsets of Z. All subsets C' of Z have the same dimension and meaning.

We now prove that decompositional inference generalizes BIOFAM inference. This generalization is not simply formal. It opens an immediate path to adaptation with arbitrary neural-network techniques.

Suppose we present the exact inputs x_i and y_j to the single-FAM-rule system F that stores $(A, B; C)$. So we present the unit bit vectors I_X^i and I_Y^j to F as nonfuzzy set inputs. Then

$$\begin{aligned} F(x_i, y_j) = F(I_X^i, I_Y^j) &= [I_X^i \circ M_{AC}] \cap [I_Y^j \circ M_{BC}] \\ &= a_i \wedge C \cap b_j \wedge C \tag{8-25} \\ &= \min(a_i, b_j) \wedge C \tag{8-26} \end{aligned}$$

Equation (8-25) follows from (8-8). Representing C with its membership function m_C, (8-26) corresponds to the BIOFAM prescription

$$\min(a_i, b_j) \wedge m_C(z) \tag{8-27}$$

for all z in Z.

If we encode the simple FAM rules (A, C) and (B, C) with correlation-product encoding, decompositional inference gives the BIOFAM version of correlation-product inference:

$$
\begin{aligned}
F(I_X^i, I_Y^j) &= [I_X^i \circ A^T C] \cap [I_Y^j \circ B^T C] \\
&= a_i C \cap b_j C & (8\text{-}28) \\
&= \min(a_i, b_j)C & (8\text{-}29) \\
&= \min(a_i, b_j)m_C(z) & (8\text{-}30)
\end{aligned}
$$

for all z in Z. Equation (8-13) implies (8-28). $\min(a_i c_k, b_j c_k) = \min(a_i, b_j)c_k$ implies (8-29).

Decompositional inference allows arbitrary fuzzy sets, waveforms, or distributions A' and B' to pass through a FAM system. The FAM system can house an arbitrary FAM bank of compound FAM rules. If we use the FAM system to control a process, the input fuzzy sets A' and B' can equal the output of an independent state-*estimation* system, such as a Kalman filter. A' and B' might then represent probability distributions on the exact input spaces X and Y. The filter-controller cascade is a common engineering architecture.

We can split compound consequents as desired. We can split the compound FAM rule "IF X is A AND Y is B, THEN Z is C OR W is D," or $(A, B; C, D)$, into the FAM rules $(A, B; C)$ and $(A, B; D)$. We can use the same split for the consequent logical-connective AND.

We can give a propositional-calculus justification for the decompositional inference technique. Let A, B, and C denote bivalent *propositions* with truth values $t(A)$, $t(B)$, and $t(C)$ in $\{0, 1\}$. Then truth tables prove the two consequent-splitting tautologies used in decompositional inference:

$$
[A \longrightarrow (B \text{ OR } C)] \quad \longrightarrow \quad [(A \longrightarrow B) \text{ OR } (A \longrightarrow C)] \qquad (8\text{-}31)
$$

$$
[A \longrightarrow (B \text{ AND } C)] \quad \longrightarrow \quad [(A \longrightarrow B) \text{ AND } (A \longrightarrow C)] \qquad (8\text{-}32)
$$

where the arrow denotes logical implication.

In bivalent logic, the implication $A \to B$ is false iff the antecedent A is true and the consequent B is false. Equivalently, $t(A \to B) = 1$ iff $t(A) = 1$ and $t(B) = 0$. This allows a "brief" truth-table check for validity. We implicitly chose truth values for the terms in the consequent of the overall implication (8-31) or (8-32) to make the consequent false. Given those restrictions, if we cannot find truth values to make the antecedent true, the statement is a tautology. In Equation (8-31), if $t((A \to B) \text{ OR } (A \to C)) = 0$, then $t(A) = 1$ and $t(B) = t(C) = 0$, since a disjunction is false iff both disjuncts are false. This forces the antecedent $A \to (B \text{ OR } C)$ to be false. So Equation (8-31) is a **tautology**: a statement is true in all cases.

A propositional tautology also justifies splitting the compound FAM rule "IF X is A OR Y is B, THEN Z is C" into the disjunction (union) of the two simple FAM rules "IF X is A, THEN Z is C" and "IF Y is B, THEN Z is C":

$$[(A \text{ OR } B) \longrightarrow C] \longrightarrow [(A \longrightarrow C) \text{ OR } (B \longrightarrow C)] \qquad (8\text{-}33)$$

Now consider splitting the original compound FAM rule "IF X is A AND Y is B, THEN Z is C" into the conjunction (intersection) of the two simple FAM rules "IF X is A, THEN Z is C" and "IF Y is B, THEN Z is C." A problem arises in the truth table of the corresponding proposition

$$[(A \text{ AND } B) \longrightarrow C] \longrightarrow [(A \longrightarrow C) \text{ AND } (B \longrightarrow C)] \qquad (8\text{-}34)$$

Proposition (8-34) is not always true, and hence not a tautology. The implication is false if A is true and if B and C are false, or if A and C are false and if B is true. But the implication (8-34) is valid if *both* antecedent terms A and B are true. So if $t(A) = t(B) = 1$, the compound conditional $(A \text{ AND } B) \to C$ implies both $A \to C$ and $B \to C$.

The simultaneous occurrence of the data values x_i and y_j satisfies this condition. Recall that logic is one-dimensional set theory. The condition $t(A) = t(B) = 1$ arises from the 1 in I_X^i and the 1 in I_X^j. We can interpret the unit bit vectors I_X^i and I_Y^j as the (true) bivalent propositions "X is x_i" and "Y is y_j." Propositional logic applies coordinatewise. A similar argument holds for the converse of (8-33).

For general fuzzy-set inputs A' and B' the argument still holds in the sense of continuous-valued logic. But the truth values of the logical implications may be less than unity while greater than zero. If A' is a null vector and B' is not, or vice versa, the implication (8-34) is false coordinatewise, at least if one coordinate of the nonnull vector equals unity. But in this case the decompositional inference scheme yields an output null vector C'. In effect the FAM system indicates the propositional falsehood.

Adaptive Decompositional Inference

The decompositional-inference scheme allows arbitrary splitting matrices M_{AC} and M_{BC}. Indeed it allows us to eliminate them altogether.

Let $N_X\colon I^n \to I^q$ define an arbitrary *neural-network* system that maps fuzzy subsets A' of X to fuzzy subsets C' of Z. $N_Y\colon I^p \to I^q$ can define a different neural network. In general N_X and N_Y chage with time.

The adaptive decompositional inference (ADI) scheme allows neural networks to adaptively split, store, and modify compound FAM rules. We can split the compound FAM rule "IF X is A AND Y is B, THEN Z is C," or $(A, B; C)$, with N_X and N_Y. N_X can house the simple FAM association (A, C). N_Y can house (B, C). Then for arbitrary fuzzy-set inputs A' and B', ADI proceeds as before for an adaptive FAM system $F: I^n \times I^p \to I^q$ that houses the FAM rule $(A, B; C)$ or a bank of such FAM rules:

$$
\begin{aligned}
F(A', B') &= N_X(A') \cap N_Y(B') \qquad\qquad\qquad (8\text{-}35) \\
&= C_{A'} \cap C_{B'} \\
&= C'
\end{aligned}
$$

Any neural network can define the system operation. The backpropagation algorithm, discussed in Chapter 5, provides a reasonable candidate for many unstructured problems. The primary concerns are memory space and training time. We can often train several small neural networks in parallel faster, and more accurately, than we can train a single large neural network.

The ADI approach illustrates one way neural algorithms embed in a FAM architecture. Below we introduce a more practical way that uses unsupervised clustering algorithms.

ADAPTIVE FAMS: PRODUCT-SPACE CLUSTERING IN FAM CELLS

An **adaptive FAM (AFAM)** is a time-varying mapping between fuzzy cubes. In principle the adaptive decompositional-inference technique generates AFAMs. But we shall reserve the label AFAM for systems that generate FAM rules from training data but that do not require splitting and recombining FAM data.

We propose a geometric AFAM procedure. The procedure adaptively clusters training samples in the FAM system *input-output product space*. FAM mappings define balls or clusters in the input-output product space. These clusters correspond to fuzzy Hebb matrices, which define fuzzy Cartesian products. The procedure "blindly" generates weighted FAM rules from training data. Further training modifies the weighted set of FAM rules. We call this unsupervised procedure **product-space clustering**.

Consider first a discrete one-dimensional FAM system $S: I^n \to I^p$. Then a FAM rule has the form "IF X is A_i, THEN Y is B_i" or (A_i, B_i). The input-output product space is $I^n \times I^p$.

What does the FAM rule (A_i, B_i) look like in the product space $I^n \times I^p$? It looks like a cluster of points centered at the numerical point (A_i, B_i). The FAM system maps points A near A_i to points B near B_i. The closer A is to A_i, the closer the point (A, B) is to the point (A_i, B_i) in the product space $I^n \times I^p$. In this sense FAMs map balls in I^n to balls in I^p. The notation is ambiguous, since (A_i, B_i) stands for both the FAM-rule mapping, or fuzzy subset (Cartesian product) of $I^n \times I^p$, and the numerical fit-vector point in $I^n \times I^p$.

Adaptive clustering algorithms can estimate the unknown FAM rule (A_i, B_i) from training samples of the form (A, B). In general there are m unknown FAM rules $(A_1, B_1), \ldots, (A_m, B_m)$, and we do not know m. The user may select m arbitrarily in many applications.

Competitive *adaptive vector-quantization* (AVQ) algorithms can adaptively estimate both the unknown FAM rules (A_i, B_i) and the unknown number m of FAM rules from FAM system input-output data. The AVQ algorithms do not require fuzzy-set data. Scalar BIOFAM data suffices, as we illustrate below for adaptive estimation of inverted-pendulum control FAM rules.

Suppose the r fuzzy sets $A_1, \ldots, A_r$ quantize the input universe of discourse X. The s fuzzy sets $B_1, \ldots, B_s$ quantize the output universe of discourse Y. If we double up on notation and view X and Y as fuzzy variables as well, then fuzzy variable X assumes fuzzy-set values A_i, and fuzzy variable Y assumes fuzzy-set values B_j. In general r and s do not relate to each other or to the number m of FAM rules (A_i, B_i). The user must specify r and s and the shape of the fuzzy sets A_i and B_i. In practice this is not difficult. Quantizing fuzzy sets are usually trapezoidal or triangular, and r and s are less than 10.

The quantizing collections $\{A_i\}$ and $\{B_j\}$ define rs **FAM cells** F_{ij} in the input-output product space $I^n \times I^p$. The FAM cells F_{ij} overlap, since contiguous quantizing fuzzy sets A_i and A_{i+1}, and B_j and B_{j+1}, overlap. So the FAM cell collection $\{F_{ij}\}$ does not partition the product space $I^n \times I^p$. The union of all FAM cells also does not equal $I^n \times I^p$, since the patches F_{ij} are fuzzy subsets of $I^n \times I^p$. The union provides only a fuzzy "cover" for $I^n \times I^p$.

The *fuzzy Cartesian product* $A_i \times B_i$ defines the FAM cell F_{ij}. $A_i \times B_i$ equals the fuzzy outer product $A_i^T \circ B_i$ in (8-6) or the correlation product $A_i^T B_i$ in (8-12). A FAM cell F_{ij} corresponds to the fuzzy correlation-minimum or correlation-product matrix $M_{ij} : F_{ij} = M_{ij}$. Thus we connect fuzzy geometry with fuzzy algebra and arrive at the following equalities: product-space cluster = FAM cell = $A_i \times B_j$ = M_{ij} = FAM rule (A_i, B_j).

Adaptive FAM-Rule Generation

Let $\mathbf{m}_1, \ldots, \mathbf{m}_k$ denote k quantization vectors in the input-output product space $I^n \times I^p$ or, equivalently, in I^{n+p}. $\mathbf{m}_j$ defines the jth column of the synaptic connection matrix $\mathbf{M}$. $\mathbf{M}$ has $n + p$ rows and k columns.

Suppose, for instance, $\mathbf{m}_j$ changes in time according to the differential competitive learning (DCL) AVQ algorithm discussed in Chapters 4 and 6, and in Chapter 1 of the companion volume [Kosko, 1991]. The competitive system samples concatenated fuzzy-set samples of the form $[A \,|\, B]$. The augmented fuzzy set $[A \,|\, B]$ is a point in the unit hypercube I^{n+p}.

The synaptic vectors $\mathbf{m}_j$ converge to FAM-matrix centroids in $I^n \times I^p$. More generally they estimate the density or distribution of the FAM rules in $I^n \times I^p$. The quantizing synaptic vectors naturally weight an estimated FAM rule. The more synaptic vectors clustered about a centroidal FAM rule, the greater its weight w_i in (8-17).

Suppose there are 15 FAM-rule centroids in $I^n \times I^p$ and $k > 15$. Suppose k_i synaptic vectors $\mathbf{m}_j$ cluster around the ith centroid. So $k_1 + \cdots + k_{15} = k$. Suppose the *cluster counts* k_i obey

$$k_1 \geq k_2 \geq \cdots \geq k_{15} \tag{8-36}$$

The first centroidal FAM rule is at least as frequent as the second centroidal FAM rule, and so on. This gives the adaptive FAM-rule weighting scheme

$$w_i = \frac{k_i}{k} \tag{8-37}$$

The FAM rule weights w_i evolve in time as the FAM system samples new augmented fuzzy sets $[A \mid B]$. In practice we may want only the 15 most-frequent FAM rules or only the FAM rules with at least some minimum frequency $w_{\min}$. Then (8-37) provides a quantitative solution.

We count the number k_{ij} of quantizing vectors in each FAM cell F_{ij}. We can define FAM-cell boundaries in advance or even designate some FAM cells as always sufficiently occupied. High-count FAM cells outrank low-count FAM cells. Most FAM cells contain zero or few synaptic vectors.

Product-space clustering extends to compound FAM rules and product spaces. The FAM rule "IF X is A AND Y is B, THEN Z is C," or $(A, B; C)$, defines a point in $I^n \times I^p \times I^q$. The t fuzzy sets $C_1, \ldots, C_t$ quantize the output space Z. There are rst FAM cells F_{ijk}. Equations (8-36) and (8-37) extend similarly and X, Y, and Z can be continuous. The adaptive clustering procedure extends to any number of FAM-rule antecedent terms.

Adaptive BIOFAM Clustering

BIOFAM data clusters more efficiently than fuzzy-set FAM data. We can more easily obtain and process paired numbers than we can obtain and process paired fit vectors. This allows system input-output data to directly generate FAM systems.

In control applications, human or automatic controllers generate streams of "well-controlled" system input-output data. Adaptive BIOFAM clustering converts this data to weighted FAM rules. The adaptive system transduces behavioral data to behavioral rules. The fuzzy system learns causal patterns. It learns which control inputs cause which control outputs. The system approximates these causal patterns when it acts as the controller.

Adaptive BIOFAMs cluster in the input-output product space $X \times Y$. The product space $X \times Y$ is vastly smaller than the power-set product space $I^n \times I^p$ used above. The adaptive synaptic vectors $\mathbf{m}_j$ are now two-dimensional instead of $(n + p)$-dimensional. On the other hand, competitive BIOFAM clustering requires many more input-output data pairs $(x_i, y_i) \in R^2$ than augmented fuzzy-set samples $[A \mid B] \in I^{n+p}$.

Again we double up on notation. We now use x_i as the numerical sample from X at sample time i. Earlier x_i denoted the ith ordered element in the finite nonfuzzy set $X = \{x_1, \ldots, x_n\}$. So now X can denote a continuous set, say R^n, usually R.

BIOFAM clustering counts synaptic quantization vectors in FAM cells. The system samples the nonfuzzy input-output stream $(x_1, y_1), (x_2, y_2), \ldots$. Unsupervised competitive learning distributes the k synaptic quantization vectors $\mathbf{m}_1, \ldots, \mathbf{m}_k$ in $X \times Y$. Learning distributes them to different FAM cells F_{ij}. The FAM cells F_{ij} overlap but are nonfuzzy subcubes of $X \times Y$. The BIOFAM FAM cells F_{ij} cover $X \times Y$. The key idea is *cluster equals rule*.

F_{ij} contains k_{ij} quantization vectors at each sample time. The cell counts k_{ij} define a frequency *histogram*, since all k_{ij} sum to k. So $w_{ij} = k_{ij}/k$ weights the FAM rule "IF X is A_i, THEN Y is B_j."

Suppose the overlapping fuzzy sets NL, NM, NS, ZE, PS, PM, PL quantize the input space X. Suppose seven similar fuzzy sets quantize the output space Y. We can define the fuzzy sets arbitrarily. In practice they are symmetric trapezoids or triangles. (The boundary fuzzy sets NL and PL are ramp functions or clipped trapezoids.) X and Y may each equal the real line. A typical FAM rule is "IF X is NL, THEN Y is PS," or (NL; PS).

Input datum x_i is nonfuzzy. When $X = x_i$ holds, the relations $X = $ NL, $\ldots$, $X = $ PL hold to different degrees. Most hold to zero degree. $X = $ NM holds to degree $m_{\text{NM}}(x_i)$. Input datum x_i partially activates the FAM rule "IF X is NM, THEN Y is ZE" or, equivalently, (NM; ZE). Since the FAM rules have single antecedents, x_i activates the consequent fuzzy set ZE to degree $m_{\text{NM}}(x_i)$ as well. Multiantecedent FAM rules activate output consequent sets according to a logic-based function of antecedent term membership values, as discussed above on BIOFAM inference.

Suppose that Figure 8.5 represents the input-output data stream (x_1, y_1), (x_2, y_2), $\ldots$ in the planar product space $X \times Y$, and that the sample data in Figure 8.5 trains a DCL system. Also suppose competitive learning distributes ten two-dimensional synaptic vectors $\mathbf{m}_1, \ldots, \mathbf{m}_{10}$ as in Figure 8.6. (We can use other types of learning as well.)

FAM cells do not overlap in Figure 8.5 and Figure 8.6 for convenience's sake. The corresponding quantizing fuzzy sets touch but do not overlap. The FAM cells partition the input-output product space.

Figure 8.5 reveals six sample-data clusters. The six quantization-vector clusters in Figure 8.6 estimate the six sample-data clusters. The single synaptic vector in FAM cell (PM; NS) indicates a smaller cluster. Since $k = 10$, the number of quantization vectors in each FAM cell measures the percentage or frequency weight w_{ij} of each possible FAM rule.

In general the additive combination rule (8-17) does not require normalizing the quantization-vector count k_{ij}. $w_{ij} = k_{ij}$ is acceptable. This holds for both maximum-membership defuzzification (8-18) and fuzzy-centroid defuzzification (8-19).

The ten quantization vectors in Figure 8.6 estimate at most six FAM rules.

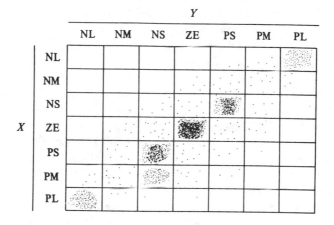

FIGURE 8.5   Distribution of input-output data (x_i, y_i) in the input-output product space $X \times Y$. Data clusters reflect FAM rules, such as the steady-state FAM rule "IF X is ZE, THEN Y is ZE."

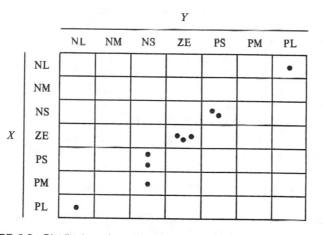

FIGURE 8.6   Distribution of ten two-dimensional synaptic quantization vectors $\mathbf{m}_1, \ldots, \mathbf{m}_{10}$ in the input-output product space $X \times Y$. As the FAM system samples nonfuzzy data (x_i, y_i), competitive learning distributes the synaptic vectors in $X \times Y$. The synaptic vectors estimate the frequency distribution of the sampled input-output data, and thus estimate FAM rules. Clusters define rules.

From most to least frequent or "important," the FAM rules are (ZE; ZE), (PS; NS), (NS; PS), (PM; NS), (PL; NL), and (NL; PL). These FAM rules suggest that fuzzy variable X behaves as an error variable or an error-velocity variable, since the steady-state FAM rule (ZE; ZE) is most important. If we sample a system only in steady-state equilibrium, we will estimate only the steady-state FAM rule. We can accurately estimate the FAM system's global behavior only if we representatively sample the system's input-output behavior. Experts must exhibit their expertise in problem cases.

The "corner" FAM rules (PL; NL) and (NL; PL) may be more important than their frequencies suggest. The boundary sets Negative Large (NL) and Positive Large (PL) usually define ramp functions as negatively and positively sloped lines. NL and PL alone cover the important end-point regions of the universe of discourse X. They give $m_{NL}(x) = m_{PL}(x) = 1$ only if x falls at or near the end point of X, since NL and PL are ramp functions, not trapezoids. NL and PL cover these end-point regions "briefly." Their corresponding FAM cells tend to be smaller than the other FAM cells. The end-point regions must be covered in most control problems, especially error-nulling problems like stabilizing an inverted pendulum. The user can weight these FAM-cell counts more highly, for instance $w_{ij} = ck_{ij}$ for scaling constant $c > 0$. Or the user can simply include these end-point FAM rules in every operative FAM bank.

Most FAM cells do not generate FAM rules, for we estimate every possible FAM rule but usually with zero or near-zero frequency weight w_{ij}. For large numbers of multiple FAM-rule antecedents, system input-output data streams through comparatively few FAM cells. Structured trajectories in $X \times Y$ may be few.

A FAM-rule's mapping structure also limits the number of estimated FAM rules. A FAM rule maps fuzzy sets in I^n or $F(2^X)$ to fuzzy sets in I^p or $F(2^Y)$. A fuzzy associative memory maps every domain fuzzy set A to a unique range fuzzy set B. Fuzzy set A cannot map to multiple fuzzy sets B, B', B'', and so on. We write the FAM rule as $(A; B)$ not $(A; B$ or B' or B'' or ...$)$. So we estimate *at most* one rule per FAM-cell row in Figure 8.6.

If two FAM cells in a row are equally and highly frequent, we can pick arbitrarily either FAM rule to include in the FAM bank. This occurs infrequently but can occur. In principle we could estimate the FAM rule as a compound FAM rule with a disjunctive consequent. The simplest strategy picks only the highest-frequency FAM cell per row.

The user can estimate FAM rules without counting the quantization vectors in each FAM cell. There may be too many FAM cells to search at each estimation iteration. The user never need examine FAM cells. Instead the user checks the synaptic-vector components m_{ij}. The user defines in advance fuzzy-set intervals, such as $[l_{NL}, u_{NL}]$ for NL. If $l_{NL} \le m_{ij} \le u_{NL}$, then the FAM-antecedent reads "IF X is NL."

Suppose the input and output spaces X and Y are the same, the real interval $[-35, 35]$. Suppose we partition X and Y into the same seven disjoint fuzzy sets:

$$
\begin{aligned}
\text{NL} &= [-35, -25] \\
\text{NM} &= [-25, -15] \\
\text{NS} &= [-15, -5] \\
\text{ZE} &= [-5, 5] \\
\text{PS} &= [5, 15] \\
\text{PM} &= [15, 25] \\
\text{PL} &= [25, 35]
\end{aligned}
$$

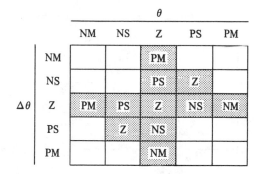

FIGURE 8.7 Inverted-pendulum FAM bank used in simulation. This BIOFAM generated 1000 sample vectors of the form $(\theta, \Delta\theta, v)$.

Then the observed synaptic vector $\mathbf{m}_j = [9, -10]$ increases the count of FAM cell PS × NS and increases the weight of FAM rule " IF X is PS, THEN Y is NS."

This amounts to nearest-neighbor classification of synaptic quantization vectors. We assign quantization vector $\mathbf{m}_k$ to FAM cell F_{ij} iff $\mathbf{m}_k$ is closer to the centroid of F_{ij} than to all other FAM-cell centroids. We break ties arbitrarily. Centroid classification allows the FAM cells to overlap.

Adaptive BIOFAM Example: Inverted Pendulum

We used DCL to train an AFAM to control the inverted pendulum discussed above. We used the accompanying C-software to generate 1000 pendulum trajectory data. These product-space training vectors $(\theta, \Delta\theta, v)$ were points in R^3. Pendulum angle θ data ranged between -90 and 90. Pendulum angular-velocity $\Delta\theta$ data ranged from -150 to 150.

We defined FAM cells by uniformly partitioning the product-space "cube" in R^3. Fuzzy variables could assume only the five fuzzy-set values NM, NS, ZE, PS, and PM. So there were 125 possible FAM rules. For instance, the steady-state FAM rule took the form (ZE, ZE; ZE) or, more completely, "IF $\theta =$ ZE AND $\Delta\theta =$ ZE, THEN $v =$ ZE."

A BIOFAM controlled the inverted pendulum. The BIOFAM restored the pendulum to equilibrium as we knocked it over to the right and to the left. (Function keys F9 and F10 knock the pendulum over to the left and to the right. Input-output sample data reads automatically to a training data file.) Eleven FAM rules described the BIOFAM controller. Figure 8.7 displays this FAM bank. The zero (ZE) row and column are ordinal inverses of the respective row and column indices.

We trained 125 three-dimensional synaptic quantization vectors with differential competitive learning, as discussed in Chapters 4 and 6, and in Chapter 1 of the companion volume [Kosko, 1991]. In principle the 125 synaptic vectors could describe a uniform distribution of product-space trajectory data. Then the 125 FAM

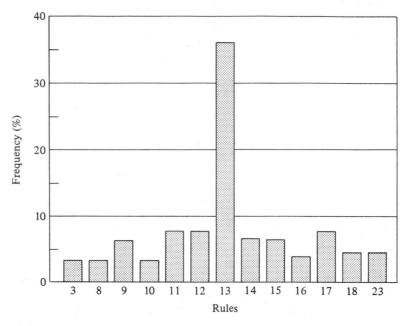

FIGURE 8.8 Synaptic-vector histogram. Differential competitive learning allocated 125 three-dimensional synaptic vectors to the 125 FAM cells. Here the adaptive system has sampled 1000 representative pendulum-control data. DCL allocates the synaptic vectors to only 13 FAM cells. The steady-state FAM cell (ZE, ZE; ZE) is most frequent.

cells would each contain one synaptic vector. Alternatively, if we used a vertically stabilized pendulum to generate the 1000 training vectors, all 125 synaptic vectors would concentrate in the (ZE, ZE; ZE) FAM cell. This would still be true if we perturbed the pendulum only mildly from vertical equilibrium.

DCL distributed the 125 synaptic vectors to 13 FAM cells. So we estimated 13 FAM rules. Some FAM cells contained more synaptic vectors than others. Figure 8.8 displays the synaptic-vector histogram after the DCL system samples the 1000 samples. Actually Figure 8.8 displays a truncated histogram. The horizontal axis should list all 125 FAM cells, all 125 FAM-rule weights w_k in (8-17). The missing 112 entries have zero synaptic-vector frequency.

Figure 8.8 gives a snapshot of the adaptive process. Successive data gradually modify the histogram. "Good" training samples should include a significant number of equilibrium samples. In Figure 8.8 the steady-state FAM cell (ZE, ZE; ZE) is clearly the most frequent.

Figure 8.9 displays the DCL-estimated FAM bank. The product-space clustering method rapidly recovered the 11 original FAM rules. It also estimated the two additional FAM rules (PS, NM; ZE) and (NS, PM; ZE), which did not affect the BIOFAM system's performance. The estimated FAM bank defined a BIOFAM,

θ

		NM	NS	Z	PS	PM
	NM			PM	Z	
	NS			PS	Z	
$\Delta\theta$	Z	PM	PS	Z	NS	NM
	PS		Z	NS		
	PM		Z	NM		

FIGURE 8.9 DCL-estimated FAM bank. Product-space clustering recovered the original 11 FAM rules and estimated two new FAM rules. The new and original BIOFAM systems controlled the inverted pendulum equally well.

with all 13 FAM-rule weights set w_k equal to unity, that controlled the pendulum as well as the original BIOFAM controlled it.

In non-real-time applications we can in principle omit the adaptive step altogether. We can directly compute the FAM-cell histogram if we count all sampled data. Then the (growing) number of synaptic vectors equals the number of training samples. This procedure equally weights all samples, and so tends not to "track" an evolving process. Competitive learning weights more recent samples more heavily. Competitive learning's metrical-classification step also helps filter noise from the stream of sample data.

REFERENCES

Dubois, D., and Prade, H., *Fuzzy Sets and Systems: Theory and Applications*, Academic Press, Orlando, FL, 1980.

Hebb, D., *The Organization of Behavior*, Wiley, New York, 1949.

Klir, G. J., and Foger, T. A., *Fuzzy Sets, Uncertainty, and Information*, Prentice Hall, Englewood Cliffs, NJ, 1988.

Kosko, B., "Fuzzy Knowledge Combination," *International Journal of Intelligent Systems*, vol. 1, 293-320, 1986.

Kosko, B., "Fuzzy Associative Memory Systems," *Fuzzy Expert Systems*, A. Kandel (Ed.), Addison-Wesley, Reading, MA, in press, December 1986.

Kosko, B., "Fuzzy Entropy and Conditioning," *Information Sciences*, vol. 40, 165-174, 1986.

Kosko, B., *Foundations of Fuzzy Estimation Theory*, Ph.D. dissertation, Department of Electrical Engineering, University of California at Irvine, June 1987; Order Number 8801936, University Microfilms International, 300 N. Zeeb Road, Ann Arbor, MI, 48106.

Kosko, B., "Hidden Patterns in Combined and Adaptive Knowledge Networks," *International Journal of Approximate Reasoning*, vol. 2, no. 4, 377-393, October 1988.

Kosko, B., *Neural Networks for Signal Processing*, Prentice Hall, Englewood Cliffs, NJ, 1991.

Mamdani, E. H., "Application of Fuzzy Logic to Approximate Reasoning Using Linguistic Synthesis," *IEEE Transactions on Computers*, vol. C-26, no. 12, 1182-1191, December 1977.

Taber, W. R., and Siegel, M. A., "Estimation of Expert Weights Using Fuzzy Cognitive Maps," *Proceedings of the IEEE 1st International Conference on Neural Networks (ICNN-87)*, vol. II, 319-325, June 1987.

Taber, W. R., "Knowledge Processing with Fuzzy Cognitive Maps," *Expert Systems with Applications*, vol. 2, no. 1, 82-87, February 1991.

Togai, M., and Watanabe, H., "Expert System on a Chip: An Engine for Realtime Approximate Reasoning, *IEEE Expert*, vol. 1, no. 3, 1986.

Yamakawa, T., "A Simple Fuzzy Computer Hardware System Employing MIN & MAX Operations," *Proceedings of the Second International Fuzzy Systems Association (IFSA)*, Tokyo, 827-830, July 1987.

Yamakawa, T., "Fuzzy Microprocessors–Rule Chip and Defuzzification Chip," *Proceedings of the International Workshop on Fuzzy Systems Applications*, Iizuka-88, Kyushu Institute of Technology, 51-52, August 1988.

Zadeh, L. A., "A Computational Approach to Fuzzy Quantifiers in Natural Languages," *Computers and Mathematics*, vol. 9, no. 1, 149-184, 1983.

PROBLEMS

8.1. Use correlation-minimum encoding to construct the FAM matrix M from the fit-vector pair (A, B) if $A = (.6\ 1\ .2\ .9)$ and $B = (.8\ .3\ 1)$. Is (A, B) a bidirectional fixed point? Pass $A' = (.2\ .9\ .3\ .2)$ through M, and $B' = (.9\ .5\ 1)$ through M^T. Do the recalled fuzzy sets differ from B and A?

8.2. Repeat Problem 8.1 using correlation-product encoding.

8.3. Compute the fuzzy entropy $E(M)$ of M in Problems 8.1 and 8.2.

8.4. If $M = A^T \circ B$ in Problem 8.1, find a different FAM matrix M' with greater fuzzy entropy, $E(M') > E(M)$, but that still gives perfect recall: $A \circ M' = B$. Find the *maximum-entropy fuzzy associative memory* (MEFAM) matrix $\mathbf{M}^*$ such that $A \circ \mathbf{M}^* = B$.

8.5. Prove: If $M = A^T \circ B$ or $M = A^T B$, $A \circ M = B$, and $A \subset A'$, then $A' \circ M = B$.

8.6. Prove: $\displaystyle\max_{1 \le k \le m} \min(a_k, b_k) \le \min(\max_{1 \le k \le m} a_k,\ \max_{1 \le k \le m} b_k)$.

8.7. Use truth tables to prove the two-valued propositional tautologies:

    (a) $[A \longrightarrow (B\ OR\ C)] \longrightarrow [(A \longrightarrow B)\ OR\ (A \longrightarrow C)]$.

    (b) $[A \longrightarrow (B\ AND\ C)] \longrightarrow [(A \longrightarrow B)\ AND\ (A \longrightarrow C)]$.

    (c) $[(A\ OR\ B) \longrightarrow C] \longrightarrow [(A \longrightarrow C)\ OR\ (B \longrightarrow C)]$.

(d) $[(A \longrightarrow C) \text{ AND } (B \longrightarrow C)] \longrightarrow [(A \text{ AND } B) \longrightarrow C]$.

Is the converse of (c) a tautology? Explain whether this affects BIOFAM inference.

8.8. *(BIOFAM inference.)* Suppose the input spaces X and Y both equal $[-10, 10]$, and the output space Z equals $[-100, 100]$. Define five trapezoidal fuzzy sets— NL, NS, ZE, PS, PL—on X, Y, and Z. Suppose the underlying (unknown) system transfer function is $z = x^2 - y^2$. State at least five FAM rules that accurately describe the system's behavior. Use $z = x^2 - y^2$ to generate streams of sample data. Use BIOFAM inference and fuzzy-centroid defuzzification to map input pairs (x, y) to output data z. Plot the BIOFAM outputs and the desired outputs z. What is the arithmetic average of the squared errors $(F(x, y) - x^2 + y^2)^2$? Divide the product space $X \times Y \times Z$ into 125 overlapping FAM cells. Estimate FAM rules from clustered system data (x, y, z). Use these FAM rules to control the system. Evaluate the performance.

SOFTWARE PROBLEMS

The following problems use the accompanying FAM software for controlling an inverted pendulum.

1. Explain why the pendulum stabilizes in the diagonal position if the pendulum bob mass increases to maximum and the motor current decreases slightly. The pendulum stabilizes in the vertical position if you remove which FAM rules?

2. Oscillation results if you remove which FAM rules? The pendulum sticks in a horizontal equilibrium if you remove which FAM rules?

3. Use DCL to train a new FAM system. Use the F3 and F4 function keys to gradually generate interesting control trajectories. Try first to recover the original FAM rules. Try next to recover only half of the original FAM rules. Does the FAM system still stabilize the inverted pendulum?

COMPARISON OF FUZZY AND NEURAL TRUCK BACKER-UPPER CONTROL SYSTEMS

Seong-Gon Kong and Bart Kosko

FUZZY AND NEURAL CONTROL SYSTEMS

In this chapter we develop fuzzy and neural systems to back up a simulated truck, and truck-and-trailer, to a loading dock in a planar parking lot. We use differential competitive learning and the product-space clustering technique, discussed in Chapter 8, to adaptively generate fuzzy-associative-memory (FAM) rules from training data taken from the fuzzy and neural simulations.

We developed the neural truck systems on the design recently proposed by Nguyen and Widrow [1989]. We trained the neural truck systems with the back-propagation learning algorithm, discussed in Chapter 5. In principle product-space clustering can convert any neural black-box system into a representative set of FAM rules.

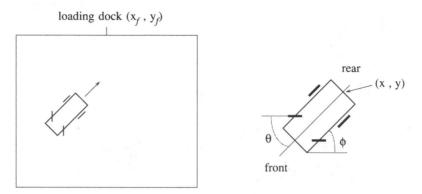

FIGURE 9.1   Diagram of simulated truck and loading zone.

BACKING UP A TRUCK

Figure 9.1 shows the simulated truck and loading zone. The truck corresponds to the cab part of the neural truck in the Nguyen-Widrow neural truck backer-upper system. The three state variables ϕ, x, and y exactly determine the truck position. ϕ specifies the angle of the truck with the horizontal. The coordinate pair (x, y) specifies the position of the rear center of the truck in the plane.

The goal was to make the truck arrive at the loading dock at a right angle ($\phi_f = 90°$) and to align the position (x, y) of the truck with the desired loading dock (x_f, y_f). We considered only backing up. The truck moved backward by some fixed distance at every stage. The loading zone corresponded to the plane $[0, 100] \times [0, 100]$, and (x_f, y_f) equaled $(50, 100)$.

At every stage the fuzzy and neural controllers should produce the steering angle θ that backs up the truck to the loading dock from any initial position and from any angle in the loading zone.

Fuzzy Truck Backer-Upper System

We first specified each controller's input and output variables. The input variables were the truck angle ϕ and the x-position coordinate x. The output variable was the steering-angle signal θ. We assumed enough clearance between the truck and the loading dock so we could ignore the y-position coordinate. The variable ranges were as follows:

$$0 \leq x \leq 100$$

$$-90 \leq \phi \leq 270$$

$$-30 \leq \theta \leq 30$$

Positive values of θ represented clockwise rotations of the steering wheel. Negative values represented counterclockwise rotations. We discretized all values to reduce computation. The resolution of ϕ and θ was one degree each. The resolution of x was 0.1.

Next we specified the fuzzy-set values of the input and output fuzzy variables. The fuzzy sets numerically represented linguistic terms, the sort of linguistic terms an expert might use to describe the control system's behavior. We chose the fuzzy-set values of the fuzzy variables as follows:

Angle ϕ		x-position x		Steering-angle signal θ	
RB:	Right Below	LE:	Left	NB:	Negative Big
RU:	Right Upper	LC:	Left Center	NM:	Negative Medium
RV:	Right Vertical	CE:	Center	NS:	Negative Small
VE:	Vertical	RC:	Right Center	ZE:	Zero
LV:	Left Vertical	RI:	Right	PS:	Positive Small
LU:	Left Upper			PM:	Positive Medium
LB:	Left Below			PB:	Positive Big

Fuzzy subsets contain elements with degrees of membership. A fuzzy membership function $m_A: Z \longrightarrow [0, 1]$ assigns a real number between 0 and 1 to every element z in the universe of discourse Z. This number $m_A(z)$ indicates the degree to which the object or data z belongs to the fuzzy set A. Equivalently, $m_A(z)$ defines the *fit* (fuzzy unit) value [Kosko, 1986] of element z in A.

Fuzzy membership functions can have different shapes depending on the designer's preference or experience. In practice fuzzy engineers have found triangular and trapezoidal shapes help capture the modeler's sense of fuzzy numbers and simplify computation. Figure 9.2 shows membership-function graphs of the fuzzy subsets above. In the third graph, for example, $\theta = 20°$ is Positive Medium to degree 0.5, but only Positive Big to degree 0.3.

In Figure 9.2 the fuzzy sets CE, VE, and ZE are narrower than the other fuzzy sets. These narrow fuzzy sets permit fine control near the loading dock. We used wider fuzzy sets to describe the endpoints of the range of the fuzzy variables ϕ, x, and θ. The wider fuzzy sets permitted rough control far from the loading dock.

Next we specified the fuzzy "rulebase" or bank of *fuzzy associative memory* (FAM) rules. Fuzzy associations or "rules" (A, B) associate output fuzzy sets B of control values with input fuzzy sets A of input-variable values. We can write fuzzy associations as antecedent-consequent pairs or IF-THEN statements.

In the truck backer-upper case, the FAM bank contained the 35 FAM rules in Figure 9.3. For example, the FAM rule of the left upper block (FAM rule 1) corresponds to the following fuzzy association:

$$\text{IF } x = \text{LE AND } \phi = \text{RB}, \text{ THEN } \theta = \text{PS}$$

FAM rule 18 indicates that if the truck is in near the equilibrium position, then the controller should not produce a positive or negative steering-angle signal. The FAM rules in the FAM-bank matrix reflect the symmetry of the controlled system.

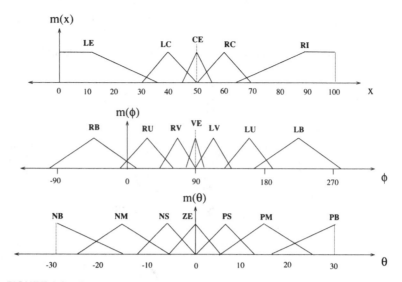

FIGURE 9.2 Fuzzy membership functions for each linguistic fuzzy-set value. To allow finer control, the fuzzy sets that correspond to near the loading dock are narrower than the fuzzy sets that correspond to far from the loading dock.

X

		LE	LC	CE	RC	RI
ϕ	RB	1 PS	2 PM	3 PM	4 PB	5 PB
	RU	6 NS	7 PS	PM	PB	PB
	RV	NM	NS	PS	PM	PB
	VE	NM	NM	18 ZE	PM	PM
	LV	NB	NM	NS	PS	PM
	LU	NB	NB	NM	NS	PS
	LB	NB	NB	NM	NM	35 NS

FIGURE 9.3 FAM-bank matrix for the fuzzy truck backer-upper controller.

For the initial condition $x = 50$ and $\phi = 270$, the fuzzy truck did not perform well. The symmetry of the FAM rules and the fuzzy sets cancelled the fuzzy controller output in a rare saddle point. For this initial condition, the neural controller (and truck-and-trailer below) also performed poorly. Any perturbation breaks the symmetry. For example, the rule (IF $x = 50$ AND $\phi = 270$, THEN $\theta = 5$) corrected the problem.

The three-dimensional control surfaces in Figure 9.4 show steering-angle signal outputs θ that correspond to all combinations of values of the two input state

FAM rule 2 (LC, RB; PM)

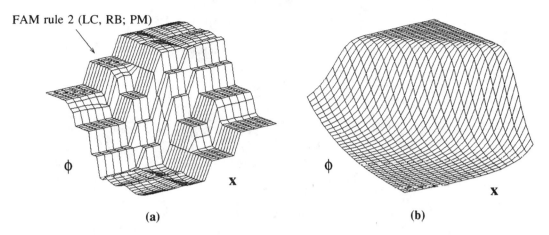

(a) (b)

FIGURE 9.4 (a) Control surface of the fuzzy controller. Fuzzy-set values determined the input and output combination corresponding to FAM rule 2 (IF $x = $ LC AND $\phi = $ RB, THEN $\theta = $ PM). (b) Corresponding control surface of the neural controller for constant value $y = 20$.

variables ϕ and x. The control surface defines the fuzzy controller. In this simulation the correlation-minimum FAM inference procedure, discussed in Chapter 8, determined the fuzzy control surface. If the control surface changes with sampled variable values, the system behaves as an *adaptive* fuzzy controller. Below we demonstrate unsupervised adaptive control of the truck and the truck-and-trailer systems.

Finally, we determined the output action given the input conditions. We used the correlation-minimum inference method illustrated in Figure 9.5. Each FAM rule produced the output fuzzy set clipped at the degree of membership determined by the input conditions and the FAM rule. Alternatively, correlation-product inference would combine FAM rules multiplicatively. Each FAM rule emitted a fit-weighted output fuzzy set O_i at each iteration. The total output O added these weighted outputs:

$$O \;=\; \sum_i O_i \tag{9-1}$$

$$\;=\; \sum_i \min(f_i,\, S_i) \tag{9-2}$$

where f_i denotes the antecedent fit value and S_i represents the consequent fuzzy set of steering-angle values in the ith FAM rule. Earlier fuzzy systems combined the output sets O_i with pairwise maxima. But this tends to produce a uniform output set O as the number of FAM rules increases. Adding the output sets O_i invokes the fuzzy version of the central limit theorem. This tends to produce a symmetric, unimodal output fuzzy set O of steering-angle values.

Fuzzy systems map fuzzy sets to fuzzy sets. The fuzzy control system's

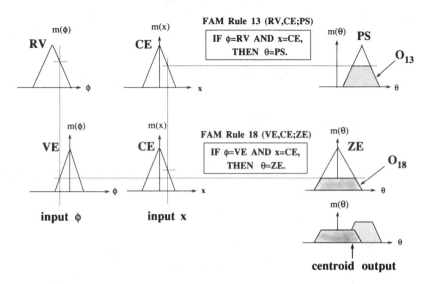

FIGURE 9.5 Correlation-minimum inference with centroid defuzzification method. Then FAM-rule antecedents combined with AND use the *minimum* fit value to activate consequents. Those combined with OR would use the *maximum* fit value.

output defines the fuzzy set O of steering-angle values at each iteration. We must "defuzzify" the fuzzy set O to produce a numerical (point-estimate) steering-angle output value θ.

As discussed in Chapter 8, the simplest defuzzification scheme selects the value corresponding to the *maximum fit* value in the fuzzy set. This mode-selection approach ignores most of the information in the output fuzzy set and requires an additional decision algorithm when multiple modes occur.

Centroid defuzzification provides a more effective procedure. This method uses the *fuzzy centroid* $\bar{\theta}$ as output:

$$\bar{\theta} = \frac{\displaystyle\sum_{j=1}^{p} \theta_j m_O(\theta_j)}{\displaystyle\sum_{j=1}^{p} m_O(\theta_j)} \tag{9-3}$$

where O defines a fuzzy subset of the steering-angle universe of discourse $\Theta = \{\theta_1, \ldots, \theta_p\}$. The central-limit-theorem effect produced by adding output fuzzy set O_i benefits both max-mode and centroid defuzzification. Figure 9.5 shows the correlation-minimum inference and centroid defuzzification applied to FAM rules 13 and 18. We used centroid defuzzification in all simulations.

With 35 FAM rules, the fuzzy truck controller produced successful truck backing-up trajectories starting from any initial position. Figure 9.6 shows typical

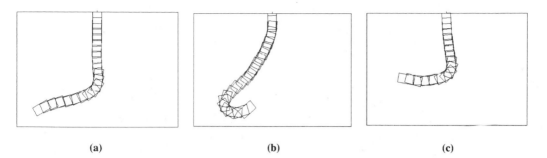

(a)   (b)   (c)

FIGURE 9.6 Sample truck trajectories of the fuzzy controller for initial positions (x, y, ϕ): (a) $(20, 20, 30)$, (b) $(30, 10, 220)$, and (c) $(30, 40, -10)$.

examples of the fuzzy-controlled truck trajectories from different initial positions. The fuzzy control system did not use ("fire") all FAM rules at each iteration. Equivalently most output consequent sets are empty. In most cases the system used only one or two FAM rules at each iteration. The system used at most 4 FAM rules at once.

Neural Truck Backer-Upper System

The neural truck backer-upper of Nguyen and Widrow [1989] consisted of multilayer feedforward neural networks trained with the backpropagation gradient-descent (stochastic-approximation) algorithm. The *neural control system* consisted of two neural networks: the controller network and the truck emulator network. The *controller network* produced an appropriate steering-angle signal output given any parking-lot coordinates (x, y), and the angle ϕ. The *emulator network* computed the next position of the truck. The emulator network took as input the previous truck position and the current steering-angle output computed by the controller network.

We did not train the emulator network since we could not obtain "universal" synaptic connection weights for the truck-emulator network. The backpropagation learning algorithm did not converge for some sets of training samples. The number of training samples for the emulator network might exceed 3000. For example, the combinations of training samples of a given angle ϕ, x-position, y-position, and steering-angle signal θ might correspond to 3150 ($18 \times 5 \times 5 \times 7$) samples, depending on the division of the input-output product space. Moreover, the training samples were numerically similar, since the neuronal signals assumed scaled values in $[0, 1]$ or $[-1, 1]$. For example, we treated close values, such as 0.40 and 0.41, as distinct sample values.

Simple kinematic equations replaced the truck-emulator network. If the truck moved backward from (x, y) to (x', y') at an iteration, then

$$x' = x + r\cos(\phi') \tag{9-4}$$

$$y' = y + r\sin(\phi') \tag{9-5}$$

$$\phi' = \phi + \theta \tag{9-6}$$

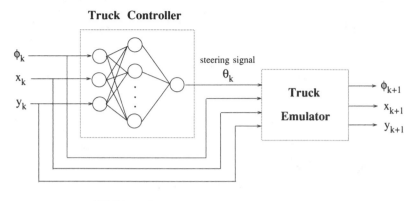

FIGURE 9.7   Topology of our neural control system.

r denotes the fixed driving distance of the truck for all backing movements. We used Equations (9-4)–(9-6) instead of the emulator network. This did not affect the posttraining performance of the neural truck backer-upper, since the truck emulator network backpropagated only errors.

We trained only the controller network with backpropagation. The controller network used 24 "hidden" neurons with logistic sigmoid functions. In the training of the truck controller, we estimated the ideal steering-angle signal at each stage before we trained the controller network. In the simulation, we used the arc-shaped truck trajectory produced by the fuzzy controller as the ideal trajectory. The fuzzy controller generated each training sample (x, y, ϕ, θ) at each iteration of the backing-up process. We used 35 training-sample vectors and needed more than 100,000 iterations to train the controller network.

Figure 9.4(b) shows the resulting neural control surface for $y = 20$. The neural control surface shows less structure than the corresponding fuzzy control surface. This reflects the unstructured nature of black-box supervised learning. Figure 9.7 shows the network connection topology for our neural truck backer-upper control system.

Figure 9.8 shows typical examples of the neural-controlled truck trajectories from several initial positions. Even though we trained the neural network to follow the smooth arc-shaped path, some learned truck trajectories were nonoptimal.

Comparison of Fuzzy and Neural Systems

As shown in Figures 9.6 and 9.8, the fuzzy controller always smoothly backed up the truck, but the neural controller did not. The neural-controlled truck sometimes followed an irregular path.

Training the neural control system was time-consuming. The backpropagation

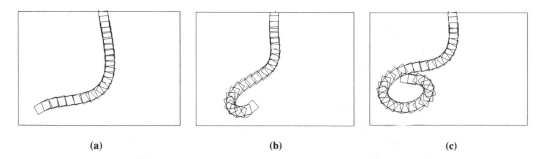

    (a)   (b)   (c)

FIGURE 9.8 Sample truck trajectories of the neural controller for initial positions (x, y, ϕ): (a) (20, 20, 30), (b) (30, 10, 220), and (c) (30, 40, −10).

algorithm required thousands of back ups to train the controller network. In some cases, the learning algorithm did not converge.

    We "trained" the fuzzy controller by encoding our own common-sense FAM rules. Once we develop the FAM-rule bank, we can compute control outputs from the resulting FAM-bank matrix or control surface. The fuzzy controller did not need a truck emulator and did not require a math model of how outputs depended on inputs.

    The fuzzy controller was computationally lighter than the neural controller. Most computation operations in the neural controller involved the multiplication, addition, or logarithm of two real numbers. In the fuzzy controller, most computational operations involved comparing and adding two real numbers.

Sensitivity Analysis

    We studied the sensitivity of the fuzzy controller in two ways. We replaced the FAM rules with destructive or "sabotage" FAM rules, and we randomly removed FAM rules. We deliberately chose sabotage FAM rules to confound the system. Figure 9.9 shows the trajectory when two sabotage FAM rules replaced the important steady-state FAM rule—FAM rule 18: the fuzzy controller should produce zero output when the truck is nearly in the correct parking position. Figure 9.10 shows the truck trajectory after we removed four randomly chosen FAM rules (7, 13, 18, and 23). These perturbations did not significantly affect the fuzzy controller's performance.

    We studied robustness of each controller by examining failure rates. For the fuzzy controller we removed fixed percentages of randomly selected FAM rules from the system. For the neural controller we removed training data. Figure 9.11 shows performance errors averaged over ten typical back-ups with missing FAM rules for the fuzzy controller and missing training data for the neural controller. The missing FAM rules and training data ranged from 0 to 100 percent of the total. In Figure 9.11(a), the docking error equaled the Euclidean distance from the actual

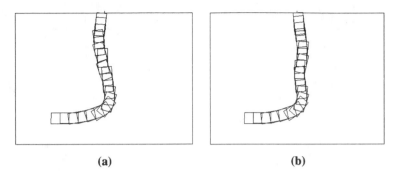

(a)   (b)

FIGURE 9.9 The fuzzy truck trajectory after we replaced the key steady-state FAM rule 18 by the two worst rules: (a) IF $x = $ CE AND $\phi = $ VE, THEN $\theta = $ PB; (b) IF $x = $ CE AND $\phi = $ VE, THEN $\theta = $ NB.

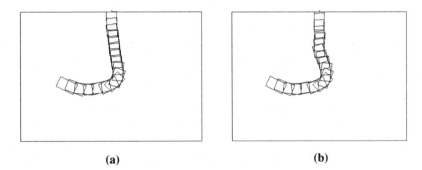

(a)   (b)

FIGURE 9.10 Fuzzy truck trajectory when (a) no FAM rules are removed and (b) FAM rules 7, 13, 18, and 23 are removed.

final position (ϕ, x, y) to the desired final position (ϕ_f, x_f, y_f):

$$\text{Docking error} = \sqrt{(\phi_f - \phi)^2 + (x_f - x)^2 + (y_f - y)^2} \qquad (9\text{-}7)$$

In Figure 9.11(b), the trajectory error equaled the ratio of the actual trajectory length of the truck divided by the straight-line distance to the loading dock:

$$\text{Trajectory error} = \frac{\text{length of truck trajectory}}{\text{distance(initial position, desired final position)}} \qquad (9\text{-}8)$$

Adaptive Fuzzy Truck Backer-Upper

Adaptive FAM (AFAM) systems generate FAM rules directly from training data. A one-dimensional FAM system, $S: I^n \longrightarrow I^p$, defines a FAM rule, a single association of the form (A_i, B_i). In this case the input-output product space equals $I^n \times I^p$. As discussed in Chapter 8, a FAM rule (A_i, B_i) defines a cluster or ball

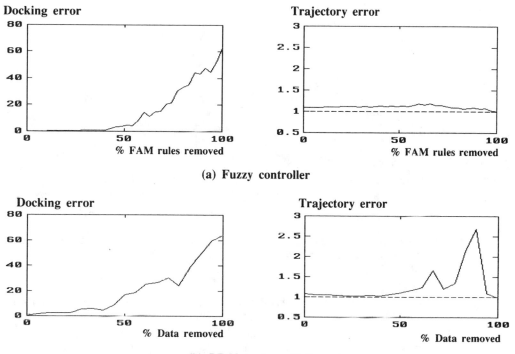

(a) **Fuzzy controller**

(b) **BP-Neural controller**

FIGURE 9.11   Comparison of robustness of the controllers: (a) Docking and trajectory error of the fuzzy controller; (b) Docking and trajectory error of the neural controller.

of points in the product-space cube $I^n \times I^p$ centered at the point (A_i, B_i). Adaptive clustering algorithms can estimate the unknown FAM rule (A_i, B_i) from training samples in R^2. We used differential competitive learning (DCL) to recover the bank of FAM rules that generated the truck-training data.

We generated 2230 truck samples from seven different initial positions and varying angles. We chose the initial positions (20, 20), (30, 20), (45, 20), (50, 20), (55, 20), (70, 20), and (80, 20). We changed the angle from $-60°$ to $240°$ at each initial position. At each step, the fuzzy controller produced output steering angle θ. The training vectors (x, ϕ, θ) defined points in a three-dimensional product-space. x had five fuzzy-set values: LE, LC, CE, RC, and RI. ϕ had seven fuzzy-set values: RB, RU, RV, VE, LV, LU, and LB. θ had seven fuzzy-set values: NB, NM, NS, ZE, PS, PM, and PB. So there were 245 ($5 \times 7 \times 7$) possible FAM cells.

We defined FAM cells by partitioning the effective product space as follows. We divided the space $0 \le x \le 100$ into five nonuniform intervals [0, 32.5], [32.5, 47.5], [47.5, 52.5], [52.5, 67.5], and [67.5, 100]. Each interval represented the five fuzzy-set values LE, LC, CE, RC, and RI. This choice corresponded to the nonoverlapping intervals of the fuzzy membership function graphs $m(x)$

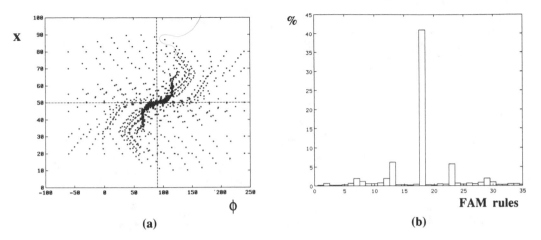

FIGURE 9.12 (a) Input data distribution. (b) Synaptic-vector histogram. Differential competitive learning allocated synaptic quantization vectors to FAM cells. The steady-state FAM cell (CE, VE; ZE) contained the most synaptic vectors.

in Figure 9.2. Similarly, we divided the space $-90 \leq \phi \leq 270$ into seven nonuniform intervals $[-90, 0]$, $[0, 66.5]$, $[66.5, 86]$, $[86, 94]$, $[94, 113.5]$, $[113.5, 182.5]$, and $[182.5, 270]$, which corresponded respectively to RB, RU, RV, VE, LV, LU, and LB. We divided the space $-30 \leq \theta \leq 30$ into seven nonuniform intervals $[-30, -20]$, $[-20, -7.5]$, $[-7.5, -2.5]$, $[-2.5, 2.5]$, $[2.5, 7.5]$, $[7.5, 20]$, and $[20, 30]$, which corresponded to NB, NM, NS, ZE, PS, PM, and PB. FAM cells near the center were smaller than outer FAM cells because we chose narrow membership functions near the steady-state FAM cell. Uniform partitions of the product space produced poor estimates of the original FAM rules. As in Figure 9.2, this reflected the need to judiciously define the fuzzy-set values of the system fuzzy variables.

We performed product-space clustering with the version of DCL discussed in Chapter 1 of the companion volume [Kosko, 1991]. If a FAM cell contained at least one of the 245 synaptic quantization vectors, we entered the corresponding FAM rule in the FAM matrix. In case of ties we chose the FAM cell with the most densely clustered data.

Figure 9.12(a) shows the input sample distribution of (x, ϕ). We did not include the variable θ in the figure. Training data clustered near the steady-state position ($x = 50$ and $\phi = 90°$). Figure 9.12(b) displays the synaptic-vector histogram after DCL classified 2230 training vectors for 35 FAM rules. Since successful FAM system generated the training samples, most training samples, and thus most synaptic vectors, clustered in the steady-state FAM cell.

DCL product-space clustering estimated 35 new FAM rules. Figure 9.13 shows the DCL-estimated FAM bank and the corresponding control surface. The DCL-estimated control surface visually resembles the underlying unknown control surface in Figure 9.4(a). The two systems produce nearly equivalent truck-backing behavior. This suggests adaptive product-space clustering can estimate the FAM

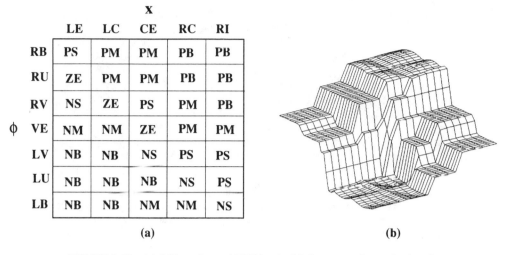

X

	LE	LC	CE	RC	RI
RB	PS	PM	PM	PB	PB
RU	ZE	PM	PM	PB	PB
RV	NS	ZE	PS	PM	PB
φ VE	NM	NM	ZE	PM	PM
LV	NB	NB	NS	PS	PS
LU	NB	NB	NB	NS	PS
LB	NB	NB	NM	NM	NS

(a)   (b)

FIGURE 9.13 (a) DCL-estimated FAM bank. (b) Corresponding control surface.

X

	LE	LC	CE	RC	RI
RB	PS	PB	PB	PB	PB
RU	NM	ZE	PM	PB	PB
RV	NM	NM	NS	PS	PB
φ VE	NM	NM	NM	ZE	PB
LV	NM	NM	NM	NS	PB
LU	NM	NM	NM	NM	PM
LB	NM	NM	NM	NM	NM

(a)   (b)

FIGURE 9.14 (a) FAM bank generated by the neural control surface in Figure 9.4b. (b) Control surface of the new BP-AFAM system in (a).

rules underlying expert behavior in many cases, even when the expert or fuzzy engineer cannot articulate the FAM rules.

We also used the neural control surface in Figure 9.4(b) to estimate FAM rules. We divided the rectangle $[0, 100] \times [-90, 270]$ into 35 nonuniform squares with the same divisions as in the fuzzy control case. Then we added and averaged the control-surface values in the square. We added a FAM rule to the FAM bank if the averaged value corresponded to one of the seven FAM cells. Figure 9.14 shows the resulting FAM bank and corresponding control surface generated by the neural

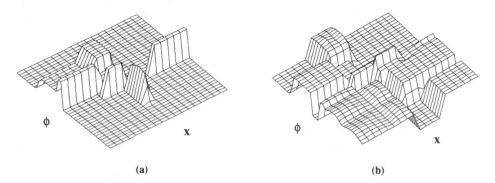

(a)   (b)

FIGURE 9.15   (a) Absolute difference of the FAM surface in Figure 9.4(a) and the DCL-estimated FAM surface in Figure 9.13(b). (b) Absolute difference of the FAM surface in Figure 9.4(a) and the neural-estimated FAM surface in Figure 9.14(b).

control surface in Figure 9.4(b). This new control surface resembles the original fuzzy control surface in Figure 9.4(a) more than it resembles the neural control surface in Figure 9.4(b). Note the absence of a steady-state FAM rule in the FAM matrix in Figure 9.14(a).

Figure 9.15 compares the DCL-AFAM and BP-AFAM control surfaces with the fuzzy control surface in Figure 9.4(a). Figure 9.15 shows the absolute difference of the control surfaces. As expected, the DCL-AFAM system produced less absolute error than the BP-AFAM system produced.

Figure 9.16 shows the docking and trajectory errors of the two AFAM control systems. The DCL-AFAM system produced less docking error than the BP-AFAM system produced for 100 arbitrary backing-up trials. The two AFAM systems generated similar backing-up trajectories. This suggests that black-box neural estimators can define the front end of FAM-structured systems. In principle we can use this technique to generate structured FAM rules for *any* neural application. We can then inspect and refine these rules and perhaps replace the original neural system with the tuned FAM system.

Fuzzy Truck-and-Trailer Controller

We added a trailer to the truck system, as in the original Nguyen-Widrow model. Figure 9.17 shows the simulated truck-and-trailer system. We added one more variable (cab angle, ϕ_c) to the three state variables of the trailerless truck. In this case a FAM rule takes the form

$$\text{IF } x = \text{LE AND } \phi_t = \text{RB AND } \phi_c = \text{PO, THEN } \beta = \text{NS}$$

The four state variables x, y, ϕ_t, and ϕ_c determined the position of the truck-and-trailer system in the plane. Fuzzy variable ϕ_t corresponded to ϕ for the trailerless truck. Fuzzy variable ϕ_c specified the relative cab angle with respect to the center

(a) Docking Error

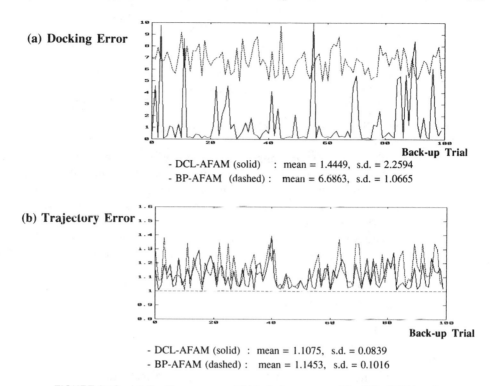

- DCL-AFAM (solid) : mean = 1.4449, s.d. = 2.2594
- BP-AFAM (dashed) :   mean = 6.6863, s.d. = 1.0665

(b) Trajectory Error

- DCL-AFAM (solid) : mean = 1.1075, s.d. = 0.0839
- BP-AFAM (dashed) :   mean = 1.1453, s.d. = 0.1016

FIGURE 9.16   (a) Docking errors and (b) trajectory errors of the DCL-AFAM and BP-AFAM control systems.

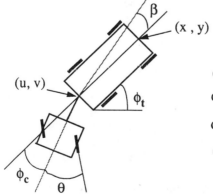

(x , y) : Cartesian coordinate of the rear end, [0,100].

(u , v) : Cartesian coordinate of the joint.

ϕ_t : Angle of the trailer with horizontal, [-90,270].

ϕ_c : Relative angle of the cab with trailer, [-90,90].

θ : Steering angle, [-30,30].

β : Angle of the trailer updated at each step, [-30,30].

FIGURE 9.17   Diagram of the simulated truck-and-trailer system.

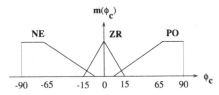

FIGURE 9.18 Membership graphs of the three fuzzy-set values of fuzzy variable ϕ_c.

line along the trailer. ϕ_c ranged from $-90°$ to $90°$. The extreme cab angles $90°$ and $-90°$ corresponded to two "jackknife" positions of the cab with respect to the trailer. Positive ϕ_c value indicated that the cab resided on the left-hand side of the trailer. Negative value indicated that it resided on the right-hand side. Figure 9.17 shows a positive angle value of ϕ_c.

Fuzzy variables x, ϕ_t, and ϕ_c defined the input variables. Fuzzy variable β defined the output variable. β measured the angle that we needed to update the trailer at each iteration. We computed the steering-angle output θ with the following geometric relationship. With the output β value computed, the trailer position (x, y) moved to the new position (x', y'):

$$x' \;=\; x + r\cos(\phi_t + \beta) \tag{9-9}$$

$$y' \;=\; y + r\sin(\phi_t + \beta) \tag{9-10}$$

where r denotes a fixed backing distance. Then the joint of the cab and the trailer (u, v) moved to the new position (u', v'):

$$u' \;=\; x' - l\cos(\phi_t + \beta) \tag{9-11}$$

$$v' \;=\; y' - l\sin(\phi_t + \beta) \tag{9-12}$$

where l denotes the trailer length. We updated the directional vector $(\mathrm{dir}U, \mathrm{dir}V)$, which defined the cab angle, by

$$\mathrm{dir}U' \;=\; \mathrm{dir}U + \Delta u \tag{9-13}$$

$$\mathrm{dir}V' \;=\; \mathrm{dir}V + \Delta v \tag{9-14}$$

where $\Delta u = u' - u$, and $\Delta v = v' - v$. The new directional vector $(\mathrm{dir}U', \mathrm{dir}V')$ defines the new cab angle ϕ_c'. Then we obtain the steering-angle value as $\theta = \phi_{c,h}' - \phi_{c,h}$, where $\phi_{c,h}$ denotes the cab angle with the horizontal. We chose the same fuzzy-set values and membership functions for β as we chose for θ. β ranged from $-30°$ to $30°$. We chose the fuzzy-set values of ϕ_c as NE, ZR and PO as in Figure 9.18.

Figure 9.19 displays the five FAM-rule matrices in the FAM bank of the fuzzy truck-and-trailer system. In Figure 9.19 we fixed the fuzzy variable x as LE, LC, CE, RC, and RI. There were 735 ($7 \times 5 \times 7 \times 3$) possible FAM rules and only 105 actual FAM rules.

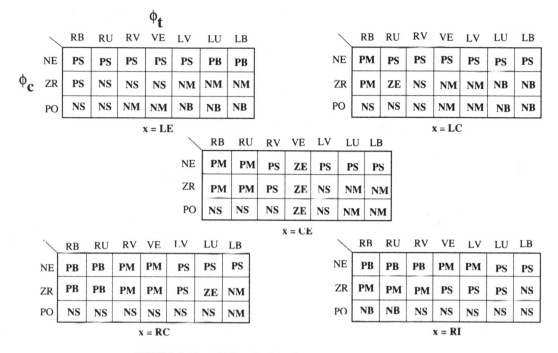

ϕ_t

ϕ_c	RB	RU	RV	VE	LV	LU	LB
NE	PS	PS	PS	PS	PS	PB	PB
ZR	PS	NS	NS	NS	NM	NM	NM
PO	NS	NS	NM	NM	NB	NB	NB

x = LE

	RB	RU	RV	VE	LV	LU	LB
NE	PM	PS	PS	PS	PS	PS	PS
ZR	PM	ZE	NS	NM	NM	NB	NB
PO	NS	NS	NS	NM	NM	NB	NB

x = LC

	RB	RU	RV	VE	LV	LU	LB
NE	PM	PM	PS	ZE	PS	PS	PS
ZR	PM	PM	PS	ZE	NS	NM	NM
PO	NS	NS	NS	ZE	NS	NM	NM

x = CE

	RB	RU	RV	VE	LV	LU	LB
NE	PB	PB	PM	PM	PS	PS	PS
ZR	PB	PB	PM	PM	PS	ZE	NM
PO	NS	NS	NS	NS	NS	NS	NM

x = RC

	RB	RU	RV	VE	LV	LU	LB
NE	PB	PB	PB	PM	PM	PS	PS
ZR	PM	PM	PM	PS	PS	PS	NS
PO	NB	NB	NS	NS	NS	NS	NS

x = RI

FIGURE 9.19 FAM bank of the fuzzy truck-and-trailer control system.

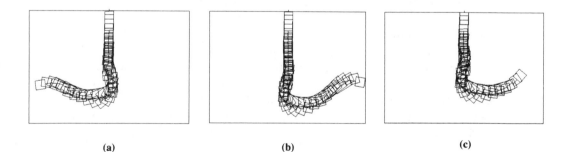

(a) (b) (c)

FIGURE 9.20 Sample truck-and-trailer trajectories from the fuzzy controller for initial positions (x, y, ϕ_t, ϕ_c): (a) $(25, 30, -20, 30)$, (b) $(80, 30, 210, -40)$, and (c) $(70, 30, 200, 30)$.

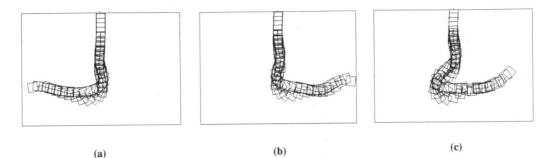

(a) (b) (c)

FIGURE 9.21 Sample truck-and-trailer trajectories of the BP-trained controller for initial positions (x, y, ϕ_t, ϕ_c): (a) $(25, 30, -20, 30)$, (b) $(80, 30, 210, -40)$, and (c) $(70, 30, 200, 30)$.

Figure 9.20 shows typical backing-up trajectories of the fuzzy truck-and-trailer control system from different initial positions. The truck-and-trailer backed up in different directions, depending on the relative position of the cab with respect to the trailer. The fuzzy control systems successfully controlled the truck-and-trailer in jackknife positions.

BP Truck-and-Trailer Control Systems

We added the cab-angle variable ϕ_c to the backpropagation-trained neural truck controller as an input. The controller network contained 24 hidden neurons with output variable β. The training samples consisted of five-dimensional space of the form $(x, y, \phi_t, \phi_c, \beta)$. We trained the controller network with 52 training samples from the fuzzy controller: 26 samples for the left half of the plane, 26 samples for the right half of the plane. We used equations (9-9)–(9-14) instead of the emulator network. Training required more than 200,000 iterations. Some training sequences did not converge. The BP-trained controller performed well except in a few cases. Figure 9.21 shows typical backing-up trajectories of the BP truck-and-trailer control system from the same initial positions used in Figure 9.20.

We performed the same robustness tests for the fuzzy and BP-trained truck-and-trailer controllers as in the trailerless truck case. Figure 9.22 shows performance errors averaged over ten typical back-ups from ten different initial positions. These performance graphs resemble closely the performance graphs for the trailerless truck systems in Figure 9.11.

AFAM Truck-and-Trailer Control Systems

We generated 6250 truck-and-trailer data using the original FAM system in Figure 9.19. We backed up the truck-and-trailer from the same initial positions as in the trailerless-truck case. The trailer angle ϕ_t ranged from $-60°$ to $240°$, and the cab angle ϕ_c assumed only the three values $-45°$, $0°$, and $45°$. The training vectors

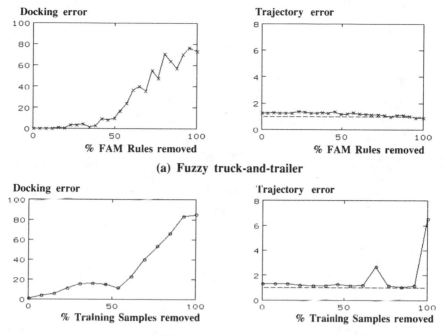

FIGURE 9.22 Comparison of robustness of the two truck-and-trailer controllers. (a) Docking and trajectory error of the fuzzy controller. (b) Docking and trajectory error of the BP controller.

$(x, \phi_t, \phi_c, \beta)$ defined points in the four-dimensional input-output product space. We nonuniformly partitioned the product space into FAM cells to allow narrower fuzzy-set values near the steady-state FAM cell.

We used DCL to train the AFAM truck-and-trailer controller. The total number of FAM cells equaled 735 (7 × 5 × 7 × 3). We used 735 synaptic quantization vectors. The DCL algorithm classified the 6250 data into 105 FAM cells. We treated the trailerless-truck fuzzy variables as before. For the cab angle ϕ_c, we divided the space $-90 \le \phi_c \le 90$ into three intervals $[-90, -12.5]$, $[-12.5, 12.5]$, and $[12.5, 90]$, which corresponded to NE, ZR, and PO. There were 735 FAM cells, and 735 possible FAM rules, of the form $(x, \phi_t, \phi_c; \beta)$. Figure 9.23 shows the synaptic-vector histogram corresponding to the 105 FAM rules. Figure 9.24 shows the estimated FAM bank by the DCL algorithm. Figure 9.25 shows the original and DCL-estimated control surfaces for the fuzzy truck-and-trailer systems.

Figure 9.26 shows the trajectories of the original FAM and the DCL-estimated AFAM truck-and-trailer controllers. Figures 9.26(a) and (b) show the two trajectories from the initial position $(x, y, \phi_t, \phi_c) = (30, 30, 10, 45)$. Figures 9.26(c) and (d) show the trajectories from initial position $(60, 30, 210, -60)$. The original FAM and DCL-estimated AFAM systems exhibited comparable truck-and-trailer control

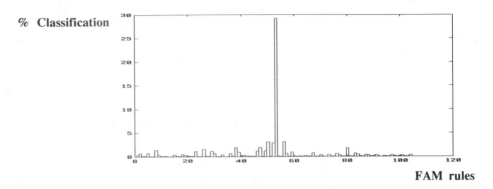

FIGURE 9.23 Synaptic-vector histogram for the AFAM truck-and-trailer system.

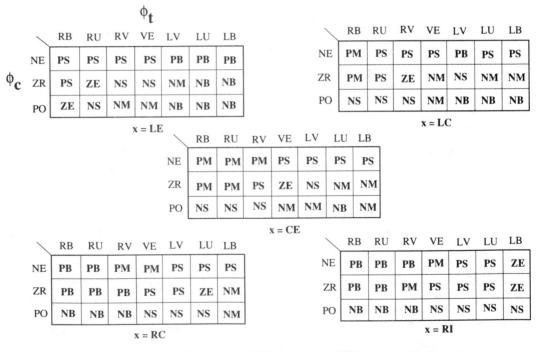

ϕ_t

$x = LE$

ϕ_c	RB	RU	RV	VE	LV	LU	LB
NE	PS	PS	PS	PS	PB	PB	PB
ZR	PS	ZE	NS	NS	NM	NB	NB
PO	ZE	NS	NM	NM	NB	NB	NB

$x = LC$

	RB	RU	RV	VE	LV	LU	LB
NE	PM	PS	PS	PS	PB	PS	PS
ZR	PM	PS	ZE	NM	NS	NM	NM
PO	NS	NS	NS	NM	NB	NB	NB

$x = CE$

	RB	RU	RV	VE	LV	LU	LB
NE	PM	PM	PM	PS	PS	PS	PS
ZR	PM	PM	PS	ZE	NS	NM	NM
PO	NS	NS	NS	NM	NM	NB	NM

$x = RC$

	RB	RU	RV	VE	LV	LU	LB
NE	PB	PB	PM	PM	PS	PS	PS
ZR	PB	PB	PB	PS	PS	ZE	NM
PO	NB	NB	NB	NS	NS	NS	NM

$x = RI$

	RB	RU	RV	VE	LV	LU	LB
NE	PB	PB	PB	PM	PS	PS	ZE
ZR	PB	PB	PM	PS	PS	PS	ZE
PO	NB	NB	NB	NS	NS	NS	NS

FIGURE 9.24 DCL-estimated FAM bank for the AFAM truck-and-trailer system.

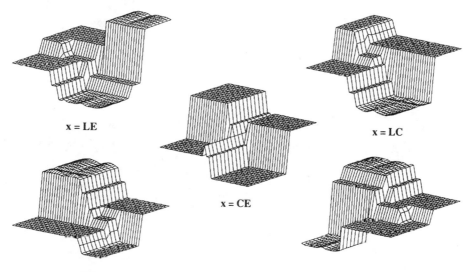

x = LE

x = CE

x = LC

x = RC

x = RI

(a)

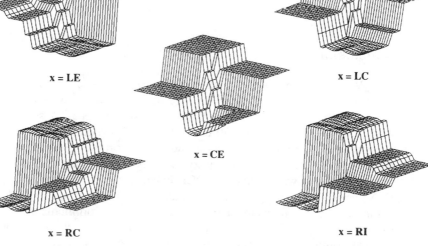

x = LE

x = CE

x = LC

x = RC

x = RI

(b)

FIGURE 9.25   (a) Original control surface. (b) DCL-estimated control surface.

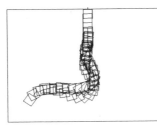

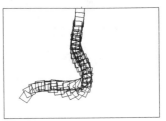

(a) Original FAM   **(b) DCL-estimated FAM**

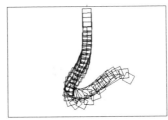

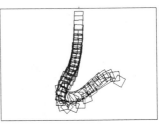

(c) Original FAM   **(d) DCL-estimated FAM**

FIGURE 9.26 Sample truck-and-trailer trajectories from the original and the DCL-estimated FAM systems starting at initial positions $(x, \ y, \ \phi_t, \ \phi_c) = (30, 30, 10, 45)$ and $(60, 30, 210, -60)$.

performance except in a few cases, where the DCL-estimated AFAM trajectories were irregular.

Conclusion

We quickly engineered fuzzy systems to successfully back up a truck and truck-and-trailer system in a parking lot. We used only common sense and error-nulling intuitions to generate sufficient banks of FAM rules. These systems performed well until we removed over 50 percent of the FAM rules. This extreme robustness suggests that, for many estimation and control problems, different fuzzy engineers can rapidly develop prototype fuzzy systems that perform similarly and well.

The speed with which the DCL clustering technique recovers the underlying FAM bank further suggests that we can likewise construct fuzzy systems for more complex, higher-dimensional problems. For these problems we may have access to only incomplete numerical input-output data. Pure neural-network or statistical-process-control approaches may generate systems with comparable performance. But these systems will involve far greater computational effort, will be more difficult to modify, and will not provide a structured representation of the system's throughput.

Our neural experiments suggests that whenever we model a system with a neural network, for little extra computational cost we can generate a set of structured FAM rules that approximate the neural system's behavior. We can then tune the fuzzy system by refining the FAM-rule bank with fuzzy-engineering rules of thumb and with further training data.

REFERENCES

Kosko, B., "Fuzzy Entropy and Conditioning," *Information Sciences*, vol. 40, 165-174, 1986.

Kosko, B., *Neural Networks for Signal Processing*, Prentice Hall, Englewood Cliffs, NJ, 1991.

Nguyen, D., and Widrow, B., "The Truck Backer-Upper: An Example of Self-Learning in Neural Networks," *Proceedings of International Joint Conference on Neural Networks (IICNN 89)*, vol. II, 357-363, June 1989.

10

FUZZY IMAGE TRANSFORM CODING

Seong-Gon Kong and Bart Kosko

TRANSFORM IMAGE CODING WITH ADAPTIVE FUZZY SYSTEMS

Transmission of digital image data increases communication accuracy but requires increased bandwidth. Limited channel capacity favors image-compression techniques. These techniques attempt to minimize the number of bits needed to represent an image and to reconstruct it with little visible distortion.

Pixel gray levels correlate highly between pixels in natural images. Image coding represents an image with *uncorrelated* data. The uncorrelated data has no redundancy and contains the essential information in the original image.

Image coding often takes the form of predictive coding or transform coding. In predictive coding, we directly exploit redundancy in the data to predict similar behavior in similar image contexts. In transform coding, a transformation, perhaps an energy-preserving transform such as the discrete cosine transform, converts an image to uncorrelated data. We keep the transform coefficients with high energy and discard the coefficients with low energy, and thus compress the image data. Jain [1981] and Netravali [1980] review several image-compression techniques. We shall apply the adaptive fuzzy methodology to a simple form of transform coding.

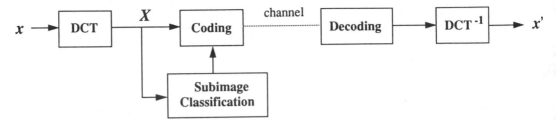

FIGURE 10.1   Block diagram of adaptive cosine transform coding.

A neural system estimates a fuzzy system, which in turn approximates a function that classifies subimages.

High-definition television (HDTV) systems [Schreiber, 1989] have reinvigorated the image-coding field. In principle we can extend fuzzy image-coding to three-dimensional TV image coding. TV images correlate more highly in the time domain than in the spatial domain. Such time correlation permits even higher compression than we can usually achieve with still image coding.

Adaptive cosine transform coding [Chen, 1977] produces high-quality compressed images at the less than 1-bit/pixel rate. The Chen system classifies subimages into four classes according to their AC energy level and encodes each class with different bit maps. The system assigns more bits to a subimage if the subimage contains much detail (large AC energy), and less bits if it contains less detail (small AC energy). DC energy refers to the constant background intensity in an image and behaves as an average. AC energy measures intensity deviations about the background DC average. So the AC energy behaves as a sample-variance statistic.

Fuzzy transform image coding uses common-sense fuzzy rules for subimage classification. Fuzzy associative memory (FAM) rules encode structured knowledge as fuzzy associations. The fuzzy association (A_i, B_i) represents the linguistic rule "IF X is A_i, THEN Y is B_i." In fuzzy transform image coding, A_i represents the AC energy distribution of a subimage, and B_i denotes its class membership. Figure 10.1 shows a simplified block diagram of adaptive cosine transform coding systems. This chapter addresses only subimage classification, the key step in the image-compression process.

Product-space clustering, discussed in Chapter 8, estimates FAM rules from training data generated by the Chen system. The resulting FAM system estimates the nonlinear subimage classification function $f: E \longrightarrow m$, where E denotes the AC energy distribution of a subimage, and m denotes the class membership of a subimage.

An integer value n in a **bit map** defines 2^n quantization levels in a quantization function. There are eight quantization functions in the bit maps of Figure 10.10 for 1-bit-per-pixel encoding. We discard transform coefficients corresponding to 0 values in the bit maps. We used fixed bit maps for multiple images to reduce side information and used FAM-subimage classification to achieve high-resolution compressed-image quality. Computing bit maps for each image introduces system complexity and increases side information.

Adaptive Cosine Transform Coding of Images

We define a two-dimensional discrete cosine transform (DCT) pair of $N \times N$ image data $x(m, n)$ as

$$X(u, v) = \frac{4c(u)c(v)}{N^2} \sum_{m=0}^{N-1} \sum_{n=0}^{N-1} x(m, n) \cos \frac{(2m+1)u\pi}{2N} \cos \frac{(2n+1)v\pi}{2N} \quad (10\text{-}1)$$

$$x(m, n) = \sum_{u=0}^{N-1} \sum_{v=0}^{N-1} c(u)c(v)X(u, v) \cos \frac{(2m+1)u\pi}{2N} \cos \frac{(2n+1)v\pi}{2N} \quad (10\text{-}2)$$

where $c(0) = 1/\sqrt{2}$, and $c(k) = 1$ for $k = 1, \ldots, N-1$. $\{x(m, n)\}$ denotes a random field of image intensities or gray scales. DCT's energy-compression capability approaches that of the optimal Karhunen-Loeve transform (KLT) [Rosenfeld, 1982]. The DCT provides fast implementation, real-valued transform coefficients, and small boundary effects. We divided an original 256×256 image into 256 16×16 subimages and transformed each subimage with the two-dimensional DCT.

Chen's adaptive transform coding system sorts subimages according to their AC energy and divides them into, say, the four classes 1, 2, 3, and 4. Class 1 contains the highest-activity subimages. Class 4 contains the least-activity subimages. The AC energy within each subimage measures the activity level of that subimage. Each class contains 64 subimages for a total of 256 subimages.

We define the "DC" and "AC" energies of a subimage as

$$\text{DC energy} = X^2(0, 0) \quad (10\text{-}3)$$

$$\text{AC energy} = \sum_u \sum_v X^2(u, v) - X^2(0, 0) \quad (10\text{-}4)$$

where $X(u, v)$ denotes the two-dimensional DCT coefficients of an image $x(m, n)$. $X(0, 0)$ defines the background intensity level, the "DC term." The other $X(u, v)$ terms contribute to the AC energy as part of an unnormalized sample variance. For subimages with uniform background intensity, most energy concentrated near $X(0, 0)$. For subimages with gray levels that varied widely, the energy spread widely.

The Chen system uses the variance of each picture element to compute four bit maps for each subimage class. We assign more bits for pixels with large variance, less bits for pixels with small variance, and normalize the transform coefficients with the standard deviations before quantization. In decoding, we multiply the quantized transform coefficients by the normalization coefficients. "Side information" refers to the information needed to decode the compressed image data at the receiver. In the Chen system side information for each image consists of one classification map, normalization coefficients, and four bit maps.

We used the signal-to-noise ratio (SNR) to evaluate coded image quality:

$$\text{SNR} = 10 \log_{10} \left(\frac{255^2}{\sigma_r^2} \right) \text{ dB} \tag{10-5}$$

where σ_r^2 denotes the sample variance of the random reconstruction error $r(m, n)$, $r(m, n) = x(m, n) - \hat{x}(m, n)$, if $r(m, n)$ is zero-mean, and 255 indicates the difference between the maximum (255) and minimum (0) gray levels:

$$\sigma_r^2 = \frac{1}{N^2} \sum_{m=0}^{N-1} \sum_{n=0}^{N-1} r^2(m, n) \tag{10-6}$$

We define the average code bits-per-pixel R as

$$R = \frac{B_c}{N^2} \text{ bits/pixel} \tag{10-7}$$

where B_c denotes the total number of bits used to encode the image. N denotes the image size ($N = 256$). For example, if we encode images with $R = 0.5$, the compression ratio equals 16:1, since we have encoded the original images with $R = 8$.

ADAPTIVE FAM SYSTEMS FOR TRANSFORM CODING

An adaptive FAM (AFAM) system generates FAM rules from training data. We classified subimages into four *fuzzy* classes B represented by the four fuzzy-set values HI (high), MH (medium high), ML (medium low), and LO (low). We encoded the HI subimage with more bits and the LO subimage with less bits. The four fuzzy sets BG (big), MD (medium), SL (small), and VS (very small) quantized the total AC power T of a subimage. So the fuzzy variable T assumed only the four fuzzy-set values BG, MD, SL, and VS, as shown in Figure 10.2. To help make fuzzy decisions, we introduced the fuzzy variable L as the low-frequency AC power. L assumed only the two fuzzy-set values SM (small) and LG (large).

We computed the total AC power T and the low-frequency AC power L of a subimage in terms of the DCT coefficients $\{X(u, v)\}$ as

$$T = \sum_{u=0}^{m-1} \sum_{v=0}^{m-1} |X(u, v)| - |X(0, 0)| \tag{10-8}$$

$$L = \sum_{u=0}^{m/2-1} \sum_{v=0}^{m/2-1} |X(u, v)| - |X(0, 0)| \tag{10-9}$$

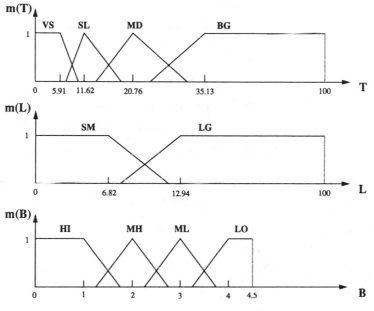

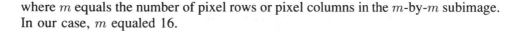

FIGURE 10.2   Fuzzy-set values of fuzzy variables T, L, and B.

where m equals the number of pixel rows or pixel columns in the m-by-m subimage. In our case, m equaled 16.

Selection of Quantizing Fuzzy-Set Values

We estimated FAM rules of the form $(T_i, L_i; B_i)$ from the training data generated by the Chen system. Index i ranged from 1 to 256, the number of subimages. T_i and L_i denote the total and low-frequency AC power of the ith subimage. B_i denotes the class membership (1, 2, 3, or 4) of the ith subimage.

We used percentage-scaled values of T_i and L_i scaled by the maximum possible AC power value. We computed the maximum AC power $T_{\max}$ from the DCT coefficients of the subimage filled with random numbers from 0 to 255. $T_{\max}$ equaled 16,129.46.

We calculated the arithmetic *average* AC powers $\bar{T}_j$ and $\bar{L}_j$ for each class $j = 1, \ldots, 4$:

$$\bar{T}_j \;=\; \frac{1}{64}\sum_{i=1}^{64} T_j^i \tag{10-10}$$

$$\bar{L}_j \;=\; \frac{1}{64}\sum_{i=1}^{64} L_j^i \tag{10-11}$$

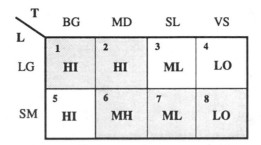

FIGURE 10.3 FAM-rule bank for the fuzzy subimage classification.

where T_j^i and L_j^i denote the ith T and L values in the class j. For the Lenna image in Figure 10.5, $\bar{T}_1 = 35.13$, $\bar{T}_2 = 20.76$, $\bar{T}_3 = 11.62$, and $\bar{T}_4 = 5.91$; $\bar{L}_1 = 19.89$, $\bar{L}_2 = 12.94$, $\bar{L}_3 = 6.82$, and $\bar{L}_4 = 2.66$. Triangular and trapezoidal fuzzy sets described BG, MD, SL, and VS for fuzzy variable T and peaked at the values $\bar{T}_1$, $\bar{T}_2$, $\bar{T}_3$, and $\bar{T}_4$. The fuzzy sets LG and SM peaked at $\bar{L}_2$ and $\bar{L}_3$. Figure 10.2 shows the fuzzy-set values for the three fuzzy variables, T, L, and B. The four fuzzy sets HI, MH, ML, and LO of the subimage class B peaked respectively at 1, 2, 3, and 4.

Product-Space Clustering to Estimate FAM Rules

Product-space clustering with competitive learning adaptively quantizes pattern clusters in the input-output product-space R^n. Stochastic competitive learning systems are neural adaptive vector quantization (AVQ) systems. p neurons compete for the activation induced by randomly sampled input-output patterns. The corresponding synaptic fan-in vectors $\mathbf{m}_j$ adaptively quantize the pattern space R^n. The p synaptic vectors $\mathbf{m}_j$ define the p columns of a synaptic connection matrix M.

Fuzzy rules $(T_i, L_i; B_i)$ define clusters or FAM cells in the input-output product-space R^3. We defined FAM-cell edges with the nonoverlapping intervals of the fuzzy-set values in Figure 10.2. The interval $0 \leq T \leq 100$ had four nonuniform subintervals [0, 8.76], [8.76, 16.19], [16.19, 27.94], and [27.94, 100]. These four subintervals represented the four fuzzy-set values VS, SL, MD, and BG assumed by the fuzzy variable T. The interval $0 \leq L \leq 100$ had two nonuniform subintervals [0, 9.88] and [9.88, 100] that represented SM and LG. The interval $0 \leq B \leq 4.5$ had four subintervals [0, 1.5], [1.5, 2.5], [2.5, 3.5], and [3.5, 4.5] that represented HI, MH, ML, and LO. There were total 32 $(4 \times 2 \times 4)$ possible FAM cells and thus 32 possible FAM rules. We used 32 synaptic vectors of quantization to estimate the FAM rules.

Differential competitive learning (DCL) classified each of the 256 input-output

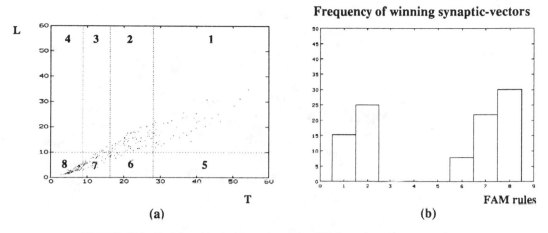

FIGURE 10.4 Training data in (a) produced the DCL-based product-space clustering of synaptic vectors in (b). (a) shows two-dimensional projections of the 32 $(4 \times 2 \times 4)$ possible FAM cells.

data vectors generated from the Chen system into one of the 32 FAM cells. We added a FAM rule to the FAM system if a DCL-trained synaptic vector fell in the FAM cell. In Figure 10.3, DCL-based product-space clustering estimated the five FAM rules (1, 2, 6, 7, and 8). We added three common-sense FAM rules (3, 4, and 5) to cover the whole input space.

Figure 10.4(a) plots the two-dimensional training data (T_i, L_i). The synaptic-vector histogram shown in Figure 10.4(b) illustrates the five FAM rules generated from the Chen system data. For example, FAM rule 1 (BG, LG; HI) represents the association:

IF the total AC power T is BG AND the low-frequency AC power L is LG,

THEN encode the subimage with the class B corresponding to HI.

Figure 10.6 shows the subimage classification maps generated by the FAM method and the Chen method. The Chen system sorts subimages according to their AC-energy content to produce the subimage-classification mapping. Sorting real numbers requires comparatively heavy computation. The FAM system does not sort subimages. Once we have trained the FAM system, the FAM system classifies subimages with almost no computation. The FAM system only adds and multiplies comparatively few real numbers. Figure 10.7 shows the subimage-classification surfaces for the FAM system and the Chen system.

(a)

FIGURE 10.5 Original and compressed Lenna images by FAM and Chen systems for approximately 0.5 bits/pixel. (a) Original Lenna; (b) FAM encoding (15.9:1); (c) Chen encoding (15.1:1).

Differential Competitive Learning

As discussed in Chapter 4, differential competitive synapses learn only if the competing F_Y neuron changes its competitive status:

$$\dot{\mathbf{m}}_j = \dot{S}_j(y_j)[\mathbf{S}(\mathbf{x}) - \mathbf{m}_j] \tag{10-12}$$

where $\mathbf{S}(\mathbf{x}) = (S_1(x_1), \ldots, S_n(x_n))$ and $\mathbf{m}_j = (m_{1j}, \ldots, m_{nj})$. For discrete implementation, we used the DCL algorithm as a stochastic difference equation:

$$\mathbf{m}_j(t+1) = \mathbf{m}_j(t) + c_t \,\Delta y_j(t)[\mathbf{x}(t) - \mathbf{m}_j(t)] \quad \text{if the } j\text{th neuron wins} \tag{10-13}$$

$$\mathbf{m}_i(t+1) = \mathbf{m}_i(t) \qquad\qquad\qquad\qquad\qquad \text{if the } i\text{th neuron loses} \tag{10-14}$$

FIGURE 10.5 (contd.)

(b)

(c)

```
4 3 4 4 4 4 4 4 4 4 3 4 3 1 4 1      4 3 4 4 4 4 4 4 4 4 3 4 3 2 4 2      · · · · · · · · · · · · x · x
3 3 4 4 4 3 3 2 2 4 4 4 4 1 1 4      3 3 4 4 4 3 2 2 2 4 4 4 4 2 1 4      · · · · · · x · · · · · x · ·
2 4 4 4 3 3 3 2 2 1 4 3 4 1 2 1      3 3 3 4 3 3 3 2 2 1 4 3 4 1 2 2      x x x · · · · · · · · · · · x
3 4 4 2 2 3 2 2 3 2 1 1 1 1 2 3      3 4 4 2 2 3 2 2 3 2 1 1 1 1 2 3      · · · · · · · · · · · · · · ·
4 4 4 1 2 2 2 3 3 3 2 1 2 2 1 4      4 3 4 1 2 2 2 3 3 3 2 1 2 1 2 4      · x · · · · · · · · · x x ·
4 4 4 1 2 3 2 1 1 1 4 1 1 1 4 4      4 4 4 1 2 3 2 1 1 1 3 1 1 2 4 4      · · · · · · · · · · x · · x ·
4 4 4 2 1 2 2 1 1 3 1 1 2 2 4 4      4 3 4 2 1 2 2 1 1 3 1 1 2 2 4 4      · x · · · · · · · · · · · · ·
4 4 4 3 1 1 2 1 1 3 1 1 3 3 4 4      4 3 4 3 1 1 2 1 1 3 1 1 3 3 4 4      · x · · · · · · · · · · · · ·
3 4 2 1 1 1 1 1 1 2 1 2 2 4 4 3      3 4 2 1 1 1 1 1 1 2 1 2 2 4 4 3      · · · · · · · · · · · · · · ·
3 4 2 1 1 1 1 3 3 2 3 1 2 4 3 4      3 4 2 1 1 1 1 3 3 2 3 1 2 4 3 4      · · · · · · · · · · · · · · ·
3 4 2 1 1 1 1 4 3 1 2 2 3 4 3 4      3 4 2 1 1 1 1 4 3 1 2 2 3 4 3 4      · · · · · · · · · · · · · · ·
3 3 1 1 1 1 3 3 3 1 2 2 4 4 3 3      3 3 1 1 1 1 3 3 3 1 2 2 4 4 3 3      · · · · · · · · · · · · · · ·
1 3 1 1 1 1 4 2 3 1 1 2 3 3 1 1      2 3 1 1 1 1 4 2 3 2 1 2 3 3 2 1      x · · · · · · · x · · · · x ·
1 3 1 2 1 1 2 3 4 4 4 1 4 1 1 4      2 3 1 2 1 1 2 3 4 4 4 1 4 2 2 3      x · · · · · · · · · · · · x x x
1 3 2 3 1 1 3 2 4 4 4 1 3 2 1 3      2 3 2 3 1 1 2 2 4 4 4 1 3 2 2 3      x · · · · · · x · · · · · x ·
1 3 3 2 1 2 2 4 4 4 4 1 3 1 3 3      2 3 3 2 1 2 2 4 4 4 4 1 3 2 3 3      x · · · · · · · · · · · · x ·
         (a)                                 (b)                                 (c)
```

FIGURE 10.6 Subimage classification maps of the Lenna image in Figure 10.5. (a) FAM method and (b) Chen method. 1 indicates the highest-activity subimages; 4 indicates the least-activity subimages. (c) Difference between the two classification maps. × denotes a differently classified subimage.

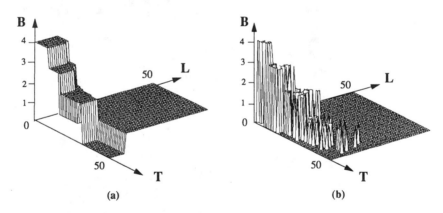

(a)　　　　　　　　　　　　　　　　　(b)

FIGURE 10.7 Classification surfaces for (a) the Chen-trained FAM system and (b) the Chen-system training data.

$\Delta y_j(t)$ denotes the time change of the jth neuron's competition signal y_j:

$$\Delta y_j(t) \quad = \quad \text{sgn}[y_j(t+1) - y_j(t)] \tag{10-15}$$

$\{c_t\}$ denotes a slowly decreasing sequence of learning coefficients, such as $c_t = 0.1(1 - t/M)$ for M training samples. m_{ij} denotes the synaptic weight between x_i and y_j. $\dot{m}_j$ and $\dot{S}_j(y_j)$ denote the time derivatives of m_j and $S_j(y_j)$, synaptic and signal velocities.

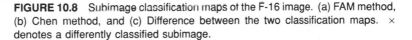

FIGURE 10.8 Subimage classification maps of the F-16 image. (a) FAM method, (b) Chen method, and (c) Difference between the two classification maps. × denotes a differently classified subimage.

The neuronal activations y_j update according to the additive model

$$y_j(t+1) = y_j(t) + \sum_i^n x_i(t)m_{ij}(t) + \sum_k^p y_k(t)w_{kj} \qquad (10\text{-}16)$$

Positive learning, $\Delta m_{ij} > 0$, tends to occur when $\mathbf{x}$ is close to the jth synaptic vector $\mathbf{m}_j$. The $p \times p$ matrix W contains the F_Y within-field synaptic connection strengths. W has positive diagonal entries and negative off-diagonal entries. Winning neurons excite themselves and inhibit all other neurons.

Simulation

The FAM system encoded the Lenna image both at approximately 1 bit/pixel and at 0.5 bits/pixel. The FAM system required fewer bits per sample than did the Chen system to reproduce a high-quality image. Figure 10.5 shows the original Lenna and the compressed Lenna at approximately 0.5 bits/pixel or 16-to-1 compression.

We then used the FAM system developed from the training data of the Lenna image to encode an F-16 jetfighter image. The FAM system also performed well for this image. Figure 10.8 shows the subimage classification maps of the F-16 image. Figure 10.9 shows the original F-16 image and the compressed versions at approximately 0.5 bits/pixel.

When we encode multiple images with fixed bit maps, we cannot optimize or tune the bit maps to a specific image. We used the four bit maps in [Chen, 1977] shown in Figure 10.10 for a 1-bit/pixel compression and used the bit maps shown in

(a)

FIGURE 10.9 Original and compressed F-16 images. FAM and Chen systems encoded the image at approximately 0.5 bits/pixel. (a) Original F-16; (b) FAM encoding (18.4:1); (c) Chen encoding (15.1:1).

Figure 10.11 for 0.5-bits/pixel compression. Table 10.1 shows that FAM encoding performed slightly better (had a larger signal-to-noise ratio) than did Chen encoding and maintained a slightly higher compression ratio (fewer bits/pixel).

Conclusion

The adaptive FAM system required fewer bits per pixel than did the Chen system to maintain comparable compressed-image quality. The FAM system also reduced side information and used only eight FAM rules to achieve 16-to-1 image compression. The FAM system approximated the input-output behavior of the Chen system. By tuning the FAM rules, we improved upon the original FAM system and hence upon the Chen system. In principle we can apply this unsupervised technique

FIGURE 10.9 (contd.)

(b)

(c)

```
8 7 6 5 4 3 3 2 2 1 1 1 0 0 0 0
7 6 6 5 4 3 3 2 2 1 1 1 0 0 0 0
6 5 5 4 4 3 3 2 2 1 1 0 0 0 0 0
5 5 4 4 4 3 3 3 2 2 1 1 1 0 0 0
4 4 4 3 3 3 3 2 2 2 1 1 1 0 0 0
3 3 3 3 2 2 2 2 2 1 1 1 1 0 0
3 3 3 3 3 2 2 2 2 1 1 1 1 1 0 0
2 2 2 3 2 2 2 1 1 1 1 1 1 0 0 0
2 2 2 3 2 2 1 1 1 1 1 1 1 0 0
2 2 2 2 2 2 1 1 1 1 1 1 1 0 0 0
1 1 1 1 1 1 1 1 1 1 1 0 0 0 0 0
1 1 1 1 1 1 1 1 1 1 1 0 0 0 0 0
1 1 1 1 1 1 1 1 1 1 1 0 0 0 0 0
1 1 1 1 1 1 1 1 1 1 0 0 0 0 0
1 1 1 1 1 1 1 1 1 1 0 0 0 0 0
```
Class 1

```
8 6 5 4 4 3 2 2 2 1 1 1 0 0 0 0
6 5 5 4 4 3 2 2 2 1 1 1 0 0 0 0
5 5 4 4 3 3 2 2 1 1 1 0 0 0 0 0
4 4 4 3 3 3 3 2 2 2 1 1 0 0 0 0
3 3 3 3 3 3 3 2 2 2 1 1 1 0 0 0
3 3 3 3 2 2 2 2 2 1 1 1 1 0 0
2 2 2 2 2 2 2 2 1 1 1 1 0 0 0
2 2 2 3 3 2 1 1 1 1 1 1 1 1 0 0
2 2 2 3 4 3 2 1 1 1 1 1 0 0 0 0
1 2 1 2 3 2 1 1 1 1 1 0 1 0 0 0
1 1 1 1 1 1 1 1 1 1 0 0 0 0 0
1 1 1 1 1 1 1 1 1 1 0 0 0 0 0
1 1 1 1 1 1 1 1 1 1 0 0 0 0 0
1 1 1 1 1 1 1 1 1 1 0 0 0 0 0
1 1 1 1 1 1 1 2 2 1 0 0 0 0 0
```
Class 2

```
8 5 4 3 3 2 2 1 1 1 1 0 0 0 0 0
5 5 4 4 3 3 2 2 1 1 0 0 0 0 0
4 4 3 3 3 3 2 2 1 1 1 0 0 0 0 0
3 3 3 3 3 2 2 1 1 1 0 0 0 0 0
3 3 3 2 2 2 2 1 1 1 1 0 0 0 0 0
2 2 2 2 2 2 1 1 1 1 0 0 0 0 0
2 2 2 2 1 1 1 1 1 1 0 0 0 0 0
2 1 1 1 2 1 1 1 0 0 0 0 0 0 0
1 1 1 2 2 2 1 1 1 0 0 0 0 0 0
1 1 1 1 1 1 1 1 0 0 0 0 0 0 0
1 1 1 1 1 1 1 0 0 0 0 0 0 0 0
1 1 1 0 1 1 1 1 0 0 0 0 0 0 0
1 0 0 0 0 0 1 0 0 0 0 0 0 0 0
0 0 0 0 0 0 1 1 0 0 0 0 0 0
1 0 0 0 0 0 0 1 1 0 0 0 0 0 0
```
Class 3

```
8 3 3 2 1 1 0 0 0 0 0 0 0 0 0 0
3 3 2 2 1 1 1 1 0 0 0 0 0 0 0 0
2 2 2 1 1 1 1 1 0 0 0 0 0 0 0 0
1 1 1 1 1 1 1 0 0 0 0 0 0 0 0 0
1 1 1 1 0 0 0 0 0 0 0 0 0 0 0 0
1 1 1 0 0 0 0 0 0 0 0 0 0 0 0 0
1 0 0 0 0 0 0 0 0 0 0 0 0 0 0 0
0 0 0 0 0 0 0 0 0 0 0 0 0 0 0 0
0 0 0 0 0 0 0 0 0 0 0 0 0 0 0 0
0 0 0 0 0 0 0 0 0 0 0 0 0 0 0 0
0 0 0 0 0 0 0 0 0 0 0 0 0 0 0 0
0 0 0 0 0 0 0 0 0 0 0 0 0 0 0 0
0 0 0 0 0 0 0 0 0 0 0 0 0 0 0 0
0 0 0 0 0 0 0 0 0 0 0 0 0 0 0 0
0 0 0 0 0 0 0 0 0 0 0 0 0 0 0 0
```
Class 4

FIGURE 10.10 Example of Chen-system bit maps at approximately 1 bit/pixel.

Table 10.1 Summary of FAM and Chen System Performance.

		SNR (dB)	R	Comp. ratio	SNR (dB)	R	Comp. ratio
Lenna	FAM	28.24	0.963	8.3:1	25.72	0.504	15.9:1
	Chen	28.10	0.976	8.2:1	25.68	0.528	15.1:1
F-16	FAM	26.35	0.898	8.9:1	24.56	0.435	18.4:1
	Chen	26.02	0.976	8.2:1	24.41	0.528	15.1:1

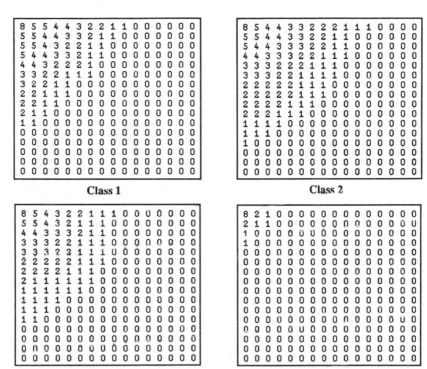

FIGURE 10.11 Example of bit maps for approximately 0.5 bits/pixel. Bit-map values determined by DCT energy intensities.

for rule generation to arbitrary image and signal-processing problems. If a system leaves numerical input-output footprints in the data, an AFAM system can leave similar footprints in similar contexts. Judicious fuzzy engineering can then refine the system and sharpen the footprints.

REFERENCES

Chen, W. H., and Smith, C. H., "Adaptive Coding of Monochrome and Color Images," *IEEE Transactions on Communications*, vol. COM-25, no. 11, 1285-1292, November 1977.

Jain, A. K. "Image Data Compression: A Review," *Proceedings of the IEEE*, vol. 69, no. 3, 349-389, March 1981.

Netravali, A. N., and Limb, J. O., "Picture Coding: A Review," *Proceedings of the IEEE*, vol. 68, no. 3, 366-406, March 1980.

Rosenfeld, A., and Kak, A. C., *Digital Picture Processing*, 2nd ed., vol. 1, Academic Press, Orlando, FL, 1982.

Schreiber, W. F., Adelson, E. H., Lippman, A. B., Rongshu, G., Monta, P., Popat, A., Sallic, H., Shen, P., Tom, A., Zangi, K., and Netravali, A. N., "A Compatible High-Definition Television System Using the Noise-Margin Method of Hiding Enhancement Information," *Society of Motion Picture and Television Engineers Journal*, 873-879, December 1989.

PROBLEMS

10.1. Show that the average squared reconstruction error of the real-valued sequence $x(0), \ldots, x(M-1)$ equals the average squared reconstruction error of the transformed values:

$$\frac{1}{M}\sum_{m=0}^{M-1}[x(m)-\hat{x}(m)]^2 \;=\; \frac{1}{M}\sum_{u=0}^{M-1}|X(u)-\hat{X}(u)|^2$$

if $[X(0), \ldots, X(M-1)] = [x(0), \ldots, x(M-1)]\,\mathbf{U}$. $\mathbf{U}$ denotes a unitary matrix and $\mathbf{U}^*$ denotes its complex conjugate transpose. So $\mathbf{U}^*\mathbf{U} = \mathbf{I}$, where $\mathbf{I}$ denotes the M-by-M identity matrix.

10.2. Show that the discrete cosine transform of $[x(0), x(1)]$ corresponds to the 2-by-2 Walsh-Hadamard transform:

$$[X(0), X(1)] \;=\; [x(0), x(1)]\begin{bmatrix} \frac{1}{\sqrt{2}} & \frac{1}{\sqrt{2}} \\ \frac{1}{\sqrt{2}} & -\frac{1}{\sqrt{2}} \end{bmatrix}$$

This property need not hold in higher dimensions.

10.3. Define the one-dimensional discrete cosine transform of real-valued sequence $x(0), \ldots, x(M-1)$ as

$$X(u) \;=\; \frac{2c(u)}{M}\sum_{m=0}^{M-1}x(m)\cos\frac{(2m+1)u\pi}{2M}$$

where $c(0) = 1/\sqrt{2}$, and $c(u) = 1$ for $u = 1, \ldots, M-1$. Then derive the discrete cosine transform in terms of the discrete Fourier transform:

$$X(u) \;=\; \frac{2c(u)}{M}\mathrm{Re}\left\{\exp\left(\frac{iu\pi}{2M}\right)\sum_{m=0}^{2M-1}x'(m)\exp\left(\frac{i2\pi mu}{2M}\right)\right\}$$

where $i = \sqrt{-1}$, and

$$x'(m) \;=\; \begin{cases} x(m) & \text{if } 0 \le m \le M-1 \\ 0 & \text{if } M \le m \le 2M-1 \end{cases}$$

COMPARISON OF FUZZY AND KALMAN-FILTER TARGET-TRACKING CONTROL SYSTEMS

Peter J. Pacini and Bart Kosko

In Chapter 9, we compared fuzzy and neural systems for the comparatively simple control problem of backing up a truck to a fixed loading dock in an empty parking lot. In this chapter, we compare a fuzzy system with a Kalman filter system for real-time target tracking. The Kalman filter is an optimal stochastic linear adaptive filter, or controller, and requires an explicit mathematical model of how control outputs depend on control inputs. In this sense the Kalman filter is a paragon of math-model controllers—and a challenging benchmark for alternative control systems.

FUZZY AND MATH-MODEL CONTROLLERS

Fuzzy controllers differ from classical math-model controllers. Fuzzy controllers do not require a mathematical model of how control outputs functionally depend on control inputs. Fuzzy controllers also differ in the type of uncertainty they represent and how they represent it. The fuzzy approach represents ambiguous or fuzzy-system behavior as partial implications or approximate "rules of thumb"—as fuzzy associations (A_i, B_i).

Fuzzy controllers are fuzzy systems. A finite *fuzzy set A* is a *point* [Kosko, 1987] in a unit hypercube $I^n = [0, 1]^n$. A *fuzzy system F*: $I^n \rightarrow I^p$ is a *mapping* between unit hypercubes. I^n contains all fuzzy subsets of the domain space $X = \{x_1, \ldots, x_n\}$. I^n is the *fuzzy power set* $F(2^x)$ of X. I^p contains all the fuzzy subsets of the range space $Y = \{y_1, \ldots, y_p\}$. Element $x_i \in X$ belongs to fuzzy set A to degree $m_A(x_i)$. The 2^n nonfuzzy subsets of X correspond to the 2^n corners of the fuzzy cube I^n. The fuzzy system F maps fuzzy subsets of X to fuzzy subsets of Y. In general, X and Y are continuous, not discrete, sets.

Math-model controllers usually represent system uncertainty with probability distributions. Probability models describe system behavior with first-order and second-order statistics—with conditional means and covariances. They usually describe unmodeled effects and measurement imperfections with additive "noise" processes.

Mathematical models of the system state and measurement processes facilitate a mean-squared-error analysis of system behavior. In general we cannot accurately articulate such mathematical models. This greatly restricts the range of real-world applications. In practice we often use linear or quasi-linear (Markov) mathematical models.

Mathematical state and measurement models also make it difficult to add nonmathematical knowledge to the system. Experts may articulate such knowledge, or neural networks may adaptively infer it from sample data. In practice, once we have articulated the math model, we use human expertise only to estimate the initial state and covariance conditions.

Fuzzy controllers consist of a bank of *fuzzy associative memory* (FAM) "rules" or associations (A_i, B_i) operating in parallel, and operating to different degrees. Each FAM rule is a set-level implication. It represents ambiguous expert knowledge or learned input-output transformations. A FAM rule can also summarize the behavior of a specific mathematical model. The system nonlinearly transforms exact or fuzzy state inputs to a fuzzy-set output. This output fuzzy set is usually "defuzzified" with a centroid operation to generate an exact numerical output. In principle the system can use the entire fuzzy distribution as the output. We can easily construct, process, and modify the FAM bank of FAM rules in software or in digital VLSI circuitry.

Fuzzy controllers require that we articulate or estimate the FAM rules. The fuzzy-set framework provides more expressiveness than, say, traditional expert-system approaches, which encode bivalent propositional associations. But the fuzzy framework does not eliminate the burden of knowledge acquisition. We can use neural-network systems to estimate the FAM rules. But neural systems also require an accurate (statistically representative) set of articulated input-output numerical samples. Below we use unsupervised competitive learning to adaptively generate target-tracking FAM rules.

Experts can hedge their system descriptions with fuzzy concepts. Although fuzzy controllers are numerical systems, experts can contribute their knowledge in natural language. This is especially important in complex problem domains, such as economics, medicine, and history, where we may not know how to mathematically model system behavior.

Below we compare a fuzzy controller with a Kalman-filter controller for real-time target tracking. This problem admits a simple and reasonably accurate mathematical description of its state and measurement processes. We chose the Kalman filter as a benchmark because of its many optimal linear-systems properties. We wanted to see whether this "optimal" controller remains optimal when compared with a computationally lighter fuzzy controller in different uncertainty environments.

We indirectly compared the sensitivity of the two controllers by varying their system uncertainties. We randomly removed FAM rules from the fuzzy controller. We also added "sabotage" FAM rules to the controller. Both techniques modeled less-stuctured control environments. For the Kalman filter, we varied the noise variance of the unmodeled-effects noise process.

Both systems performed well for mildly uncertain target environments. They degraded differently as the system uncertainty increased. The fuzzy controller's performance degraded when we removed more than half the FAM rules. The Kalman-filter controller's performance quickly degraded when the additive state noise process increased in variance.

REAL-TIME TARGET TRACKING

A target-tracking system maps azimuth-elevation inputs to motor-control outputs. The nominal target moves through azimuth-elevation space. Two motors adjust the position of a platform to continuously point at the target.

The platform can be any directional device that accurately points at the target. The device may be a laser, video camera, or high-gain antenna. We assume we have available a radar or other device that can detect the direction from the platform to the target.

The radar sends azimuth and elevation coordinates to the tracking system at the end of each time interval. We calculate the current *error* e_k in platform position and *change in error* $\dot{e}_k$. Then a fuzzy or Kalman-filter controller determines the control outputs for the motors, one each for azimuth and elevation. The control outputs reposition the platform.

We can independently control movement along azimuth and elevation if we apply the same algorithm twice. This reduces the problem to matching the target's position and velocity in only one dimension.

Figure 11.1 shows a block diagram of the target-tracking system. The controller's output v_k gives the estimated change in angle required during the next time interval. In principle a hardware system must transduce the angular velocity v_k into a voltage or current.

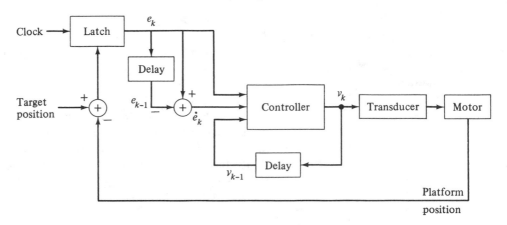

FIGURE 11.1 Target-tracking system.

FUZZY CONTROLLER

We restrict the output angular velocity v_k of the fuzzy controller to the interval $[-6, 6]$. So we must insert a gain element before the voltage transduction. This gain must equal one-sixth the maximum angle through which the platform can turn in one time interval. Similarly, the position error e_k must be scaled so that 6 equals the maximum error. The product of this scale factor and the output gain provides a design parameter—the "gain" of the fuzzy controller.

The fuzzy controller uses heuristic control set-level "rules" or *fuzzy associative memory* (FAM) associations based on quantized values of e_k, $\dot{e}_k$, and v_{k-1}. We define seven fuzzy levels by the following library of fuzzy-set values of the fuzzy variables e_k, $\dot{e}_k$, and v_{k-1}:

LN: Large Negative
MN: Medium Negative
SN: Small Negative
ZE: Zero
SP: Small Positive
MP: Medium Positive
LP: Large Positive

We do not quantize inputs in the classical sense that we assign each input to exactly one output level. Instead, each linguistic value equals a fuzzy set that overlaps with adjacent fuzzy sets. The fuzzy controller uses trapezoidal fuzzy-set values, as Figure 11.2 shows. The lengths of the upper and lower bases provide design parameters that we must calibrate for satisfactory performance. A good rule of thumb is *adjacent fuzzy-set values should overlap approximately 25 percent*. Below we discuss examples of calibrated and uncalibrated systems. The fuzzy controller attained its best performance with upper and lower bases of 1.2 and

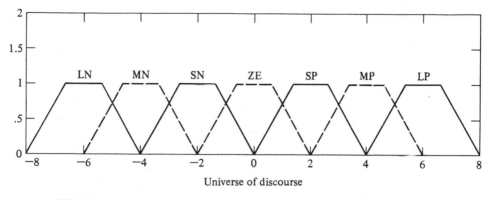

FIGURE 11.2 Library of overlapping fuzzy-set values defined on a universe of discourse.

3.9—26.2 percent overlap. Different target scenarios may require more or less overlap.

We assign each system input to a fit vector of length 7, where the ith *fit*, or *fuzzy unit* [Kosko, 1986], equals the value of the ith fuzzy set at the input value. In other words, the ith fit measures the degree to which the input belongs to the ith fuzzy-set value. For instance, we apply the input values 1, -4, and 3.8 to the seven fuzzy sets in the library to obtain the fit vectors

$$1 \longrightarrow (0 \quad 0 \quad 0 \quad .7 \quad .7 \quad 0 \quad 0)$$
$$-4 \longrightarrow (0 \quad 1 \quad 0 \quad 0 \quad 0 \quad 0 \quad 0)$$
$$3.8 \longrightarrow (0 \quad 0 \quad 0 \quad 0 \quad .1 \quad 1 \quad 0)$$

We determine these fit values above by convolving a Dirac delta function centered at the input value with each of the seven fuzzy sets:

$$m_{SP}(3.8) \quad = \quad \delta(y - 3.8) * m_{SP}(y) \quad = \quad 0.1 \qquad (11\text{-}1)$$

If we use a discretized universe of discourse, then we use a Kronecker delta function instead. Equivalently, for the discrete case n-dimensional universe of discourse $X = \{x_1, \ldots, x_n\}$, a control input corresponds to a *bit* (binary unit) vector B of length n. A single 1 element in the ith slot represents the "crisp" input value x_i. Similarly, we represent the kth library fuzzy set by an n-dimensional fit vector A_k that contains samples of the fuzzy set at the n discrete points within the universe of discourse X. The degree to which the crisp input x_i activates each fuzzy set equals the inner product $B \cdot A_k$ of the bit vector B and the corresponding fit vector A_k.

We formulate control FAM rules by associating output fuzzy sets with input fuzzy sets. The antecedent of each FAM rule conjoins e_k, $\dot{e}_k$, and v_{k-1} fuzzy-set values. For example,

IF $e_k = $ MP AND $\dot{e}_k = $ SN AND $v_{k-1} = $ ZE

THEN $v_k = $ SP

We abbreviate this as (MP, SN, ZE; SP).

The scalar activation value w_i of the ith FAM rule's consequent equals the *minimum* of the three antecedent conjuncts' values. If alternatively we combine the antecedents disjunctively with OR, the activation degree of the consequent equals the *maximum* of the three antecedent disjuncts' values. In the following example, $m_A(e_k)$ denotes the degree to which e_k belongs to the fuzzy set A:

				LN	MN	SN	ZE	SP	MP	LP
e_k	=	2.6	$\longrightarrow$	(0	0	0	0	1	.4	0)
$\dot{e}_k$	=	-2.0	$\longrightarrow$	(0	0	1	0	0	0	0)
v_{k-1}	=	1.8	$\longrightarrow$	(0	0	0	.1	1	0	0)

$$m_{\text{MP}}(e_k) = 0.4$$
$$m_{\text{SN}}(\dot{e}_k) = 1$$
$$m_{\text{ZE}}(v_{k-1}) = 0.1$$
$$w_i = \min(0.4, 1, 0.1) = 0.1$$

So the system activates the consequent fuzzy set SP to degree $w_i = .1$.

The output fuzzy set's shape depends on the FAM-rule encoding scheme used, as discussed in Chapter 8. With *correlation-minimum* encoding, we clip the consequent fuzzy set L_i in the library of fuzzy-set values to degree w_i with pointwise minimum:

$$m_{O_i}(y) = \min(w_i, m_{L_i}(y)) \tag{11-2}$$

With *correlation-product* encoding, we multiply L_i by w_i:

$$m_{O_i}(y) = w_i m_{L_i}(y) \tag{11-3}$$

or equivalently,

$$O_i = w_i L_i \tag{11-4}$$

Figure 11.3 illustrates how both inference procedures transform L_i to scaled output O_i. For the example above, correlation-product inference gives output fuzzy set $O_i = 0.1$ SP, where $L_i = $ SP denotes the fuzzy set of small but positive angular-velocity values.

The fuzzy system activates each FAM-rule consequent set to a different degree. For the ith FAM rule this yields the output fuzzy set O_i. The system then sums the O_i to form the combined output fuzzy set O:

$$O = \sum_{i=1}^{N} O_i \tag{11-5}$$

or equivalently,

$$m_O(y) = \sum_{i=1}^{N} m_{O_i}(y) \tag{11-6}$$

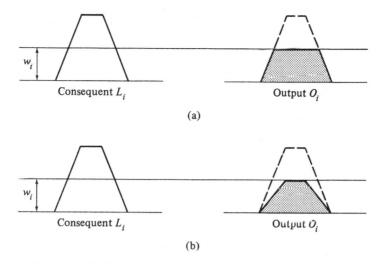

(a)

(b)

FIGURE 11.3 FAM inference procedure depends on FAM-rule encoding procedure: (a) correlation-minimum encoding, (b) correlation-product encoding.

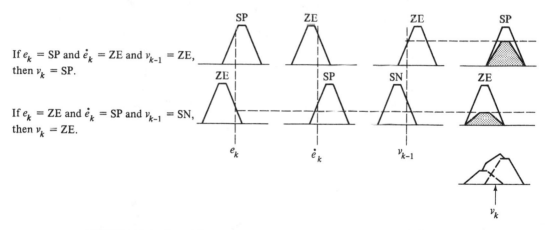

If e_k = SP and $\dot{e}_k$ = ZE and v_{k-1} = ZE, then v_k = SP.

If e_k = ZE and $\dot{e}_k$ = SP and v_{k-1} = SN, then v_k = ZE.

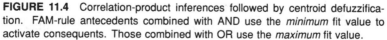

FIGURE 11.4 Correlation-product inferences followed by centroid defuzzification. FAM-rule antecedents combined with AND use the *minimum* fit value to activate consequents. Those combined with OR use the *maximum* fit value.

The control output v_k equals the *fuzzy centroid* of O:

$$v_k = \frac{\int y m_O(y)\, dy}{\int m_O(y)\, dy} \tag{11-7}$$

where the limits of integration correspond to the entire universe of discourse Y of angular velocity values. Figure 11.4 shows an example of correlation-product inference for two FAM rules followed by centroid defuzzification of the combined output fuzzy set.

To reduce computations, we can discretize the output universe of discourse Y to p values, $Y = \{y_1, \ldots, y_p\}$, which gives the discrete fuzzy centroid

$$v_k = \frac{\displaystyle\sum_{j=1}^{p} y_j m_O(y_j)}{\displaystyle\sum_{j=1}^{p} m_O(y_j)} \tag{11-8}$$

as used in Chapters 8 and 9.

Fuzzy-Centroid Computation

We now develop two discrete methods for computing the fuzzy centroid (11-7). Theorem 11.1 states that we can compute the global centroid v_k from local FAM-rule centroids. Theorem 11.2 states that v_k can be computed from only seven sample points if all the fuzzy sets are symmetric and unimodal (in the broad sense of a trapezoid peak), though otherwise arbitrary. Both results reduce computation and favor digital implementation.

Theorem 11.1.   If correlation-product inference determines the output fuzzy sets, then we can compute the global centroid v_k from local FAM-rule centroids:

$$v_k = \frac{\displaystyle\sum_{i=1}^{N} w_i c_i I_i}{\displaystyle\sum_{i=1}^{N} w_i I_i} \tag{11-9}$$

Proof. The consequent fuzzy set of each FAM rule equals one of the fuzzy-set values shown in Figure 11.2. We assume each fuzzy set includes at least one unity value, $m_A(x) = 1$. Define I_i and c_i as the respective area and centroid of the ith FAM rule's consequent set L_i:

$$I_i = \int m_{L_i}(y) \, dy \tag{11-10}$$

$$c_i = \frac{\displaystyle\int y m_{L_i}(y) \, dy}{\displaystyle\int m_{L_i}(y) \, dy}$$

$$= \frac{\int y m_{L_i}(y)\,dy}{I_i}$$

substituting from (11-10). Hence

$$\int y m_{L_i}(y)\,dy \;=\; c_i\,I_i \qquad\qquad (11\text{-}11)$$

Using (11-3), the result of correlation-product inference, we get

$$\int y m_{O_i}(y)\,dy \;=\; \int y w_i m_{L_i}(y)\,dy$$

$$- \; w_i \int y m_{L_i}(y)\,dy$$

$$= \; w_i c_i I_i \qquad\qquad (11\text{-}12)$$

substituting from (11-11). Similarly,

$$\int m_{O_i}(y)\,dy \;=\; \int w_i m_{L_i}(y)\,dy$$

$$= \; w_i I_i \qquad\qquad (11\text{-}13)$$

substituting from (11-10).

We can use (11-12) and (11-13) to derive a discrete expression equivalent to (11-7):

$$\int y m_O(y)\,dy \;=\; \int y \Big[\sum_{i=1}^{N} m_{O_i}(y)\Big]\,dy \qquad \text{substituting from (11-6)}$$

$$= \; \sum_{i} \int y m_{O_i}(y)\,dy$$

$$= \; \sum_{i} w_i c_i I_i \qquad\qquad (11\text{-}14)$$

from (11-12). Similarly,

$$\int m_O(y)\,dy \;=\; \int \sum_{i=1}^{N} m_{O_i}(y)\,dy$$

$$= \; \sum_{i} \int m_{O_i}(y)\,dy$$

$$= \; \sum_{i} w_i I_i \qquad\qquad (11\text{-}15)$$

from (11-13). Substituting (11-14) and (11-15) into (11-7), we derive a new form for the centroid:

$$v_k \;=\; \frac{\displaystyle\sum_{i=1}^{N} w_i c_i I_i}{\displaystyle\sum_{i=1}^{N} w_i I_i} \tag{11-16}$$

which is equivalent to (11-9). Each summand in each summation of (11-16) depends on only a single FAM rule. So we can compute the global output centroid from local FAM-rule centroids. **Q.E.D.**

Theorem 11.2. If the 7 library fuzzy sets are symmetric and unimodal (in the trapezoidal sense) and we use correlation-product inference, then we can compute the centroid v_k from only 7 samples of the combined output fuzzy set O:

$$v_k \;=\; \frac{\displaystyle\sum_{j=1}^{7} m_O(y_j) y_j J_j}{\displaystyle\sum_{j=1}^{7} m_O(y_j) J_j} \tag{11-17}$$

The 7 sample points are the centroids of the output fuzzy-set values.

Proof. Define $\bar{O}_i$ as a fit vector of length 7, where the fit value corresponding to the ith consequent set has the value w_i, and the other entries equal zero. If all the fuzzy sets are symmetric and unimodal, then the jth fit value of $\bar{O}_i$ is a sample of m_{O_i} at the centroid of the jth fuzzy set. The combined output fit vector is

$$\bar{O} \;=\; \sum_{i=1}^{N} \bar{O}_i \tag{11-18}$$

Since

$$m_O(y) \;=\; \sum_{i=1}^{N} m_{O_i}(y)$$

the jth fit value of $\bar{O}$ is a sample of m_O at the centroid of the jth fuzzy set. Equivalently, the jth fit value of $\bar{O}$ equals the sum of the output activations w_i from the FAM rules with consequent fuzzy sets equal to the jth library fuzzy-set value.

Define the reduced universe of discourse as $Y = \{y_1, \ldots, y_7\}$ such that y_j equals the centroid of the jth output fuzzy set. In vector form

$$Y \;=\; (y_1, \ldots, y_7)$$
$$=\; (-6, -4, -2, 0, 2, 4, 6)$$

for the library of fuzzy sets in Figure 11.2. Also define the diagonal matrix

$$J \;=\; \mathrm{diag}\,(J_1, \,\ldots,\, J_7) \tag{11-19}$$

where J_j denotes the area of the jth fuzzy-set value. If the ith FAM rule's consequent fuzzy set equals the jth fuzzy-set value, then the jth fit value of $\bar{O}$ increases by w_i, $c_i = y_j$, and $I_i = J_j$. So

$$\bar{O}JY^T \;=\; \sum_{j=1}^{7} m_O(y_j)y_j J_j \;=\; \sum_{i=1}^{N} w_i c_i I_i \tag{11-20}$$

Also,

$$\bar{O}J\mathbf{1}^T \;=\; \sum_{j=1}^{7} m_O(y_j) J_j \;=\; \sum_{i=1}^{N} w_i I_i \tag{11-21}$$

where $\mathbf{1} = (1, \,\ldots,\, 1)$. Substituting (11-20) and (11-21) into (11-16) gives

$$v_k \;=\; \frac{\displaystyle\sum_{j=1}^{7} m_O(y_j)y_j J_j}{\displaystyle\sum_{j=1}^{7} m_O(y_j) J_j} \tag{11-22}$$

which is equivalent to (11-17). Therefore, (11-22) gives a simpler, but equivalent form of the centroid (11-7) if all the fuzzy sets are symmetric and unimodal, and if we use correlation-product inference to form the output fuzzy sets O_i. **Q.E.D.**

Consider a fuzzy controller with the fuzzy sets defined in Figure 11.2, and 7 FAM rules with the following outputs:

i	w_i	Consequent
1	0.0	MP
2	0.2	SP
3	1.0	ZE
4	0.4	SN
5	0.1	SP
6	0.8	ZE
7	0.6	SN

Figure 11.5 shows the combined output fuzzy set O, with the SN, ZE, and SP components displayed with dotted lines. Using (11-7) we get a velocity output of -0.452. Alternatively, the combined output fit vector $\bar{O}$ equals $(0, 0, 1.0, 1.8, 0.3, 0, 0)$. From (11-22) we get

$$v_k \;=\; \frac{-2 \times 1 + 0 \times 1.8 + 2 \times 0.3}{1 + 1.8 + 0.3} \;=\; -0.452$$

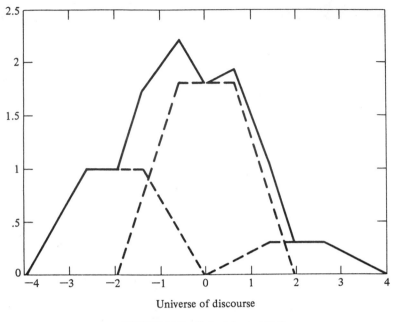

Universe of discourse

FIGURE 11.5   Output fuzzy set O.

Fuzzy-Controller Implementation

A FAM bank or "rulebase" of FAM rules defines the fuzzy controller. Each FAM rule associates one consequent fuzzy set with three antecedent fuzzy-set conjuncts.

Suppose the ith FAM rule is (MP, SN, ZE; SP). Suppose the inputs at time k are $e_k = 2.6$, $\dot{e}_k = -2.0$, and $v_{k-1} = 1.8$. Then

$$
\begin{aligned}
w_i &= \min(m_{\text{MP}}(e_k),\ m_{\text{SN}}(\dot{e}_k),\ m_{\text{ZE}}(v_{k-1})) \\
&= \min(0.4, 1, 0.1) \\
&= 0.1
\end{aligned}
$$

If all the fuzzy sets have the same shape, then they correspond to shifted versions of a single fuzzy set ZE:

$$
m_{\text{SP}}(y) = m_{\text{ZE}}(y - 2)
$$

Define e^i, $\dot{e}^i$, and v^i as the centroids of the corresponding antecedent fuzzy sets in the example above. So $e^i = 4$, $\dot{e}^i = -2$, and $v^i = 0$. Then the output activation equals

$$
\begin{aligned}
w_i &= \min(m_{\text{ZE}}(e_k - e^i),\ m_{\text{ZE}}(\dot{e}_k - \dot{e}^i),\ m_{\text{ZE}}(v_{k-1} - v^i)) \\
&= \min(m_{\text{ZE}}(-1.4),\ m_{\text{ZE}}(0),\ m_{\text{ZE}}(1.8))
\end{aligned}
$$

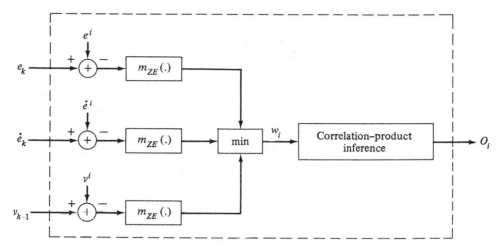

FIGURE 11.6 Algorithmic structure of a FAM rule for the special case of identically shaped fuzzy sets and correlation-product inference.

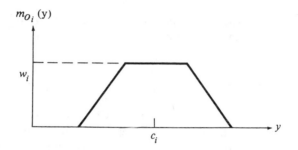

FIGURE 11.7 Trapezoidal output fuzzy set O_i.

$$= \quad \min(0.4, 1, 0.1)$$

$$= \quad 0.1$$

as computed above. Figure 11.6 schematizes such a FAM rule when presented with crisp inputs.

The output fuzzy set O_i in Figure 11.6 equals the fuzzy set ZE scaled by w_i and shifted by c_i:

$$m_{O_i}(y) \quad = \quad w_i m_{ZE}(y - c_i) \tag{11-23}$$

Figure 11.7 illustrates O_i.

The fuzzy control system activates a bank of FAM rules operated in parallel, as shown in Figure 11.8. The system sums the output fuzzy sets to form the total output set O, which the system converts to a "defuzzified" scalar output by computing its fuzzy centroid.

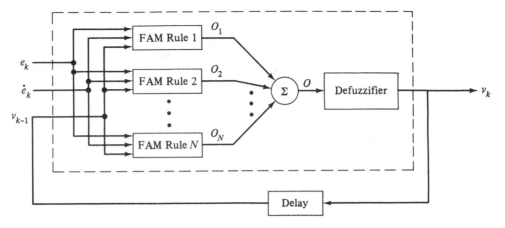

FIGURE 11.8 Fuzzy control system as a parallel FAM bank with centroidal output.

KALMAN-FILTER CONTROLLER

We designed a one-dimensional **Kalman filter** to act as an alternative controller. The *state* and *measurement* equations take the general form

$$
\begin{aligned}
x_{k+1} &= \Phi_{k+1,k}x_k + \Gamma_{k+1,k}w_k + \Psi_{k+1,k}u_k \\
z_k &= H_k x_k + V_k
\end{aligned}
\tag{11-24}
$$

where V_k denotes Gaussian white noise with covariance matrix R_k. If V_k is colored noise or if $R_k = 0$, then the filtering-error covariance matrix $P_{k|k}$ becomes singular. The state x_k and the measurements z_k are jointly Gaussian. Mendel [1987] gives details of this model.

Assume the following one-dimensional model:

$$
\begin{aligned}
\Phi_{k+1,k} &= \Gamma_{k+1,k} = \Psi_{k+1,k} = H_k = 1 \qquad \text{for all } k \\
u_k &= e_k + \dot{e}_k
\end{aligned}
\tag{11-25}
$$

Let x_{k+1} denote the output velocity required at time k to exactly lock onto the target at time $k + 1$. So the controller output at time k equals the "predictive" estimate $\hat{x}_{k+1|k} = v_k$. Note that

$$
\begin{aligned}
e_k &= x_k - \hat{x}_{k|k-1} \\
&= \tilde{x}_{k|k-1} \\
\dot{e}_k &= e_k - e_{k-1}
\end{aligned}
$$

Substituting (11-25) into (11-24), we get the new state equation

$$
x_{k+1} = x_k + e_k + \dot{e}_k + w_k
\tag{11-26}
$$

where w_k denotes white noise that models target acceleration or other unmodeled effects. The new measurement equation is

$$
\begin{aligned}
z_k &= x_k + V_k \\
&= \hat{x}_{k|k-1} + \tilde{x}_{k|k-1} + V_k \\
&= \hat{x}_{k|k-1} + V_k'
\end{aligned} \tag{11-27}
$$

Since we assume $\hat{x}_{k|k-1}$ and V_k are uncorrelated, the variance of V_k' is

$$
\begin{aligned}
R_k' &= E[V_k'^2] \\
&= E[\tilde{x}_{k|k-1}^2] + E[V_k^2] \\
&= P_{k|k-1} + R_k
\end{aligned} \tag{11-28}
$$

The recursive Kalman-filter equations take the general form

$$
\begin{aligned}
\hat{x}_{k|k} &= \hat{x}_{k|k-1} + K_k[z_k - H_k\hat{x}_{k|k-1}] \\
K_k &= P_{k|k-1}H_k^T[H_kP_{k|k-1}H_k^T + R_k]^{-1} \\
\hat{x}_{k+1|k} &= \Phi_{k+1,k}\hat{x}_{k|k} + \Psi_{k+1,k}u_k \\
P_{k|k-1} &= \Phi_{k,k-1}P_{k-1|k-1}\Phi_{k,k-1}^T + \Gamma_{k,k-1}Q_{k-1}\Gamma_{k,k-1}^T \\
P_{k|k} &= [I - K_kH_k]P_{k|k-1}
\end{aligned} \tag{11-29}
$$

where $Q_k = \text{Var}(w_k) = E[w_kw_k^T]$. Substituting (11-25), (11-27), (11-28) and the definition of v_k into (11-29), we get the following one-dimensional Kalman filter:

$$
\begin{aligned}
\hat{x}_{k|k} &= v_{k-1} + K_kV_k' \\
K_k &= \frac{P_{k|k-1}}{R_k'} \\
v_k &= \hat{x}_{k|k} + e_k + \dot{e}_k \\
P_{k|k-1} &= P_{k-1|k-1} + Q_{k-1} \\
P_{k|k} &= [1 - K_k]P_{k|k-1}
\end{aligned} \tag{11-30}
$$

Unlike the fuzzy controller, this Kalman filter does not automatically restrict the output v_k to a usable range. We must apply a threshold immediately after the controller. To remain consistent with the fuzzy controller, we set the following thresholds:

$$
|v_k| \leq 9 \text{ degrees azimuth}
$$

$$
|v_k| \leq 4.5 \text{ degrees elevation}
$$

$$v_{k-1}$$

		LN	MN	SN	ZE	SP	MP	LP
	LN	LN	LN	LN	LN	MN	SN	ZE
	MN	LN	LN	LN	MN	SN	ZE	SP
	SN	LN	LN	MN	SN	ZE	SP	MP
$\dot{e}_k$	ZE	LN	MN	SN	ZE	SP	MP	LP
	SP	MN	SN	ZE	SP	MP	LP	LP
	MP	SN	ZE	SP	MP	LP	LP	LP
	LP	ZE	SP	MP	LP	LP	LP	LP

FIGURE 11.9 e_k = ZE cross section of the fuzzy control system's FAM bank. Each entry represents one FAM rule with e_k = ZE as the first antecedent term. The shaded FAM rule is "IF e_k = ZE AND $\dot{e}_k$ = SP AND v_{k-1} = SN, THEN v_k = ZE," abbreviated as (ZE, SP, SN; ZE). Note the ordinal antisymmetry of this FAM-bank matrix. The six other cross-section FAM-bank matrices are similar. We can eliminate many FAM-rule entries without greatly perturbing the fuzzy controller's behavior.

Fuzzy and Kalman-Filter Control Surfaces

Each control system maps inputs to outputs. Geometrically, these input-output transformations define *control surfaces*. The control surfaces are sheets in the input space (since the output velocity v_k is a scalar). Three inputs and one output give rise to a four-dimensional control surface, which we cannot plot. Instead, for each controller we can plot a family of three-dimensional control surfaces indexed by constant values of the fourth variable, the error e_k, for example. Then each control surface corresponds to a different value of the error e_k.

The fuzzy control surface characterizes the fuzzy system's fuzzy-set value definitions and its bank of FAM rules. Different sets of FAM rules yield different fuzzy controllers, and hence different control surfaces. Figure 11.9 shows a cross section of the FAM bank when e_k = ZE. Each entry in this linguistic matrix represents one FAM rule with e_k = ZE as the first antecedent term.

The entire FAM bank—including cross sections for e_k equal to each of the seven fuzzy-set values LN, MN, SN, ZE, SP, MP, and LP—determines how the system maps input fuzzy sets to output fuzzy sets. The fuzzy-set membership functions shown in Figure 11.2 determine the degree to which each crisp input value belongs to each fuzzy-set value. So both the fuzzy-set value definitions and the FAM bank determine the defuzzified output v_k for any set of crisp input values e_k, $\dot{e}_k$, and v_{k-1}.

Figure 11.10 shows the control surface of the fuzzy controller for $e_k = 0$. We plotted the control output v_k against $\dot{e}_k$ and v_{k-1}. Since we use the same algorithm

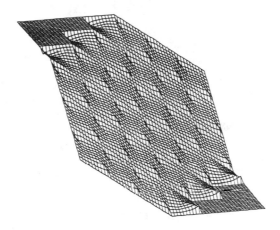

FIGURE 11.10 Control surface of the fuzzy controller for constant error $e_k = 0$. We plotted the control output v_k against $\dot{e}_k$ and v_{k-1} along the respective west and south borders.

for tracking in azimuth and elevation, the control surfaces for the two dimensions differ in scale only by a factor of two.

The Kalman filter has a random control surface that depends on a time-varying parameter. From (11-30) we see that

$$v_k = \hat{x}_{k|k} + e_k + \dot{e}_k$$

$$\hat{x}_{k|k} = v_{k-1} + K_k V_k'$$

where V_k' denotes white noise with variance given by (11-28). Combining these two equations gives the equation for the random control surface:

$$v_k = v_{k-1} + e_k + \dot{e}_k + K_k V_k' \tag{11-31}$$

At time k the noise term $K_k V_k'$ has variance

$$\sigma_k^2 = K_k^2 R_k' \tag{11-32}$$

$$= \frac{P_{k|k-1}^2}{R_k'} \quad \text{upon substituting from (11-30)}$$

$$= \frac{P_{k|k-1}^2}{P_{k|k-1} + R_k}$$

substituting from (11-28). Combining (11-31) and (11-32) gives a new control-surface equation:

$$v_k = v_{k-1} + e_k + \dot{e}_k + \sigma_k V_k'' \tag{11-33}$$

where V_k'' denotes unit-variance Gaussian noise. So the Kalman filter's control output equals the sum of the three input variables plus additive Gaussian noise with

Table 11.1 Convergence rates and steady-state values of σ_k for different values of the variance Var(w) of the white-noise, unmodeled-effects process w_k.

Var(w)	Steady-state value of σ_k	Number of iterations required for convergence
1.00	0.79	2
0.25	0.46	4
0.05	0.22	9

time-dependent variance σ_k^2. For constant error e_k, we can interpret (11-33) as a smooth control surface in R^3 defined by

$$v_k \;=\; v_{k-1} + e_k + \dot{e}_k$$

and perturbed at time k by Gaussian noise with variance σ_k^2.

In our simulations the standard deviation σ_k converged after only a few iterations. We used unity initial conditions: $P_{0|0} = R_k = 1$ for all k.

Table 11.1 lists the convergence rates and steady-state values of σ_k for three different values of the variance Var(w) of the white-noise, unmodeled-effects process w_k. For Var(w) = 0, σ_k decreases rapidly at first—$\sigma_8 = 0.10$, $\sigma_{17} = 0.05$—but does not attain its steady-state value of zero within 100 iterations.

Figure 11.11 shows four realizations of the Kalman filter's random control surface for $e_k = 0$, each at a time k when σ_k has converged to its steady-state value. For each plot, we used output thresholds and initial variances for the azimuth case: $|v_k| \leq 9.0$, $R_k = P_{0|0} = 1.0$. As with the fuzzy controller, elevation control surfaces equal scaled versions of the corresponding azimuth control surfaces.

SIMULATION RESULTS

Our target-tracking simulations model several real-world scenarios. Suppose we have mounted the target-tracking system on the side of a vehicle, aircraft, or ship. The system tracks a missile that cuts across the detection range on a straight flight path. The target maintains a constant speed of 1870 miles per hour and comes within 3.5 miles of the platform at closest approach. The platform can scan from 0 to 180 degrees in azimuth at a maximum rate of 36 degrees per second, and from 0 (vertical) to 90 degrees in elevation at a maximum rate of 18 degrees per second. The sampling interval is 1/4 of a second. The gain of the fuzzy controller equals 0.9. So the maximum error considered is 10 degrees azimuth and 5 degrees elevation. We threshold all error values above this level.

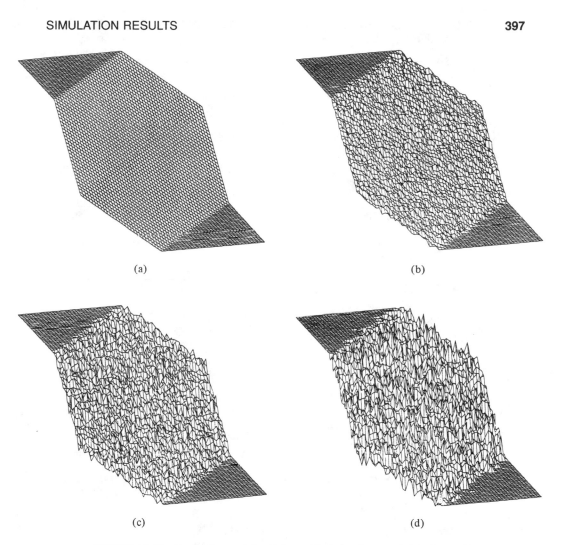

(a)   (b)

(c)   (d)

FIGURE 11.11 Realizations of the Kalman filter's random control surface with $e_k = 0$ for different values of the variance $\text{Var}(w)$ and steady-state values of the standard deviation σ_k. (a) $\text{Var}(w) = \sigma_k = 0$. (b) $\text{Var}(w) = 0.05$, $\sigma_k = 0.22$. (c) $\text{Var}(w) = 0.25$, $\sigma_k = 0.46$. (d) $\text{Var}(w) = 1.0$, $\sigma_k = 0.79$.

Figure 11.12 demonstrates the best performance of the fuzzy controller for a simulated scenario. The solid lines indicate target position. The dotted lines indicate platform position. To achieve this performance, we calibrated the three design parameters—upper and lower trapezoid bases and the gain. Figures 11.13 and 11.14 show examples of uncalibrated systems. Too much overlap causes excessive overshoot. Too little overlap causes lead or lag for several consecutive time intervals. A

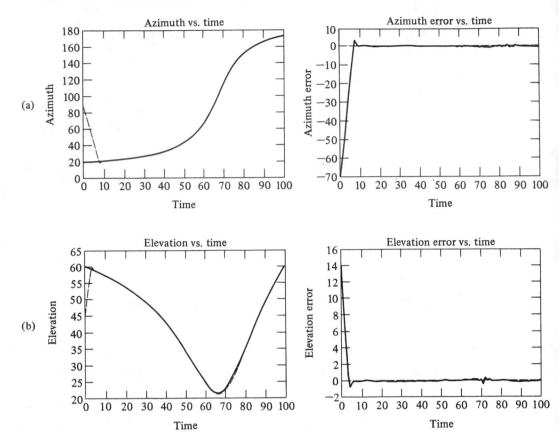

FIGURE 11.12 Best performance of the fuzzy controller. (a) Azimuth position and error. (b) Elevation position and error. Fuzzy-set overlap is 26.2 percent.

gain of 0.9 suffices for most scenarios. We can fine-tune the fuzzy control system by altering the percentage overlap between adjacent fuzzy sets.

Figure 11.15 demonstrates the best performance of the Kalman-filter controller for the same scenario used to test the fuzzy controller. For simplicity, $R_k = P_{0|0}$ for all values of k. For this study we chose the values 1.0 (unit variance) for azimuth and 0.25 for elevation. This 1/4 ratio reflects the difference in scanning range. We set Q_k to 0 for optimal performance. Figure 11.16 shows the Kalman-filter controller's performance when $Q_k = 1.0$ azimuth, 0.25 elevation.

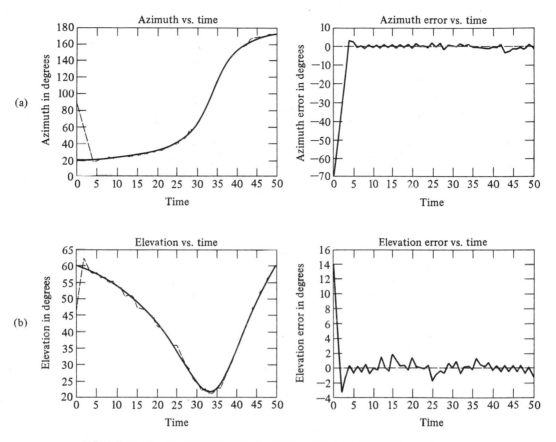

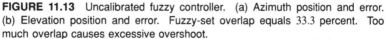

FIGURE 11.13 Uncalibrated fuzzy controller. (a) Azimuth position and error. (b) Elevation position and error. Fuzzy-set overlap equals 33.3 percent. Too much overlap causes excessive overshoot.

Sensitivity Analysis

We compared the uncertainty sensitivity of the fuzzy and Kalman-filter control systems. Under normal operating conditions, when the FAM bank contains all fuzzy control rules, and when the unmodeled-effects noise variance $\text{Var}(w)$ is small, the controllers perform almost identically. Under more uncertain conditions their performance differs. The Kalman filter's state equation (11-26) contains the noise term w_k whose variance we must assume. When $\text{Var}(w)$ increases, the state equation becomes more uncertain. The fuzzy-control FAM rules depend implicitly on this same equation, but without the noise term. Instead, the fuzziness of the FAM rules accounts for the system uncertainty. This suggests that we can increase the uncertainty of the implicit state equation by omitting randomly selected FAM rules.

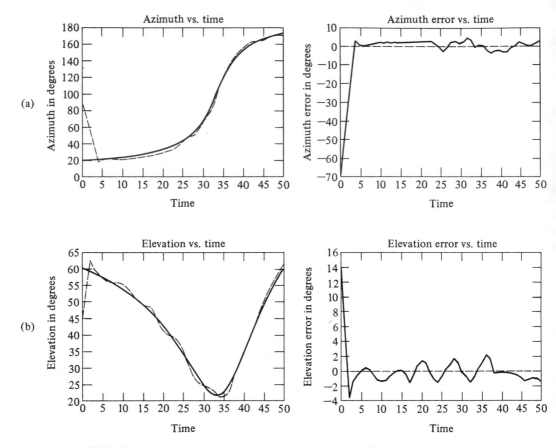

FIGURE 11.14 Uncalibrated fuzzy controller. (a) Azimuth position and error. (b) Elevation position and error. Fuzzy-set overlap equals 12.5 percent. Too little overlap causes lead or lag for several consecutive time intervals.

Figures 11.17 and 11.18 show the effect on the *root-mean-squared error* (RMSE) in degrees when we omit FAM rules and increase Var(w). Each data point averages ten runs.

The controllers behave differently as uncertainty increases. The RMSE of the fuzzy controller increases little until we omit nearly 60 percent of the FAM rules. The RMSE of the Kalman filter increases steeply for small values of Var(w), then gradually levels off.

We also tested the fuzzy controller's robustness by "sabotaging" the most vulnerable FAM rule. This could reflect lack of accurate expertise, or a highly

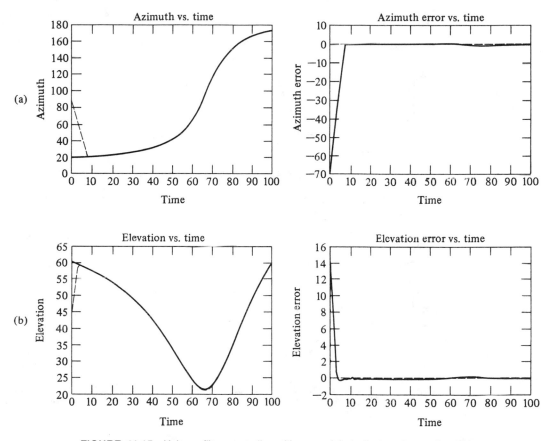

FIGURE 11.15 Kalman-filter controller with unmodeled-effects noise variance Var(w) = 0. (a) Azimuth position and error. (b) Elevation position and error.

unstructured problem. Changing the consequent of the steady-state FAM rule (ZE, ZE, ZE; ZE) to LP gives the following nonsensical FAM rule:

> IF the platform points directly at the target
>
>   AND both the target and the platform are stationary
>
> THEN turn in the positive direction with maximum velocity

Figure 11.19 shows the fuzzy system's performance when this sabotage FAM rule replaces the steady-state FAM rule. When the sabotage FAM rule activates, the system quickly adjusts to decrease the error again. The fuzzy system is piecewise stable.

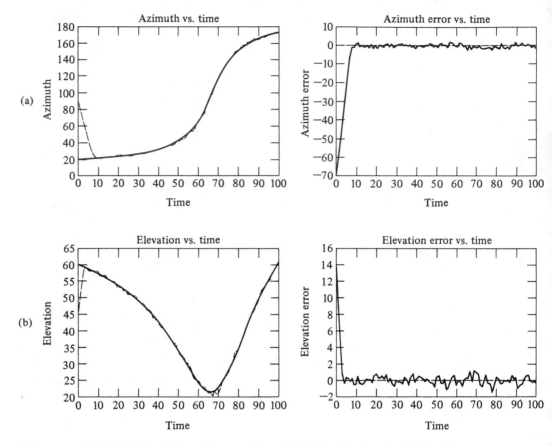

FIGURE 11.16   Kalman-filter controller with Var(w) = 1.0 azimuth, 0.25 elevation.
(a) Azimuth position and error. (b) Elevation position and error.

Adaptive FAM (AFAM)

We used unsupervised product-space clustering to train an adaptive FAM (AFAM) fuzzy controller. Differential competitive learning (DCL) adaptively clustered input-output pairs. A manually designed FAM bank and 80 random target trajectories generated 19,236 training vectors. Each product-space training vector $(e_k, \dot{e}_k, v_{k-1}, v_k)$ defined a point in R^4.

The FAM cells, as discussed in Chapter 8, uniformly partitioned the four-dimensional product space. Each FAM cell represented a single FAM rule. The

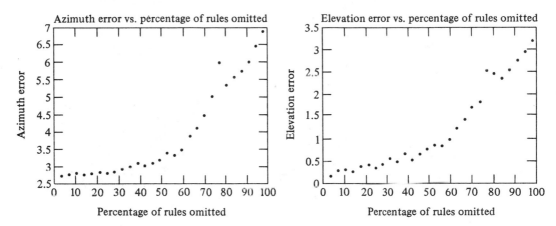

FIGURE 11.17   Root-mean-squared error of the fuzzy controller with randomly selected FAM rules omitted.

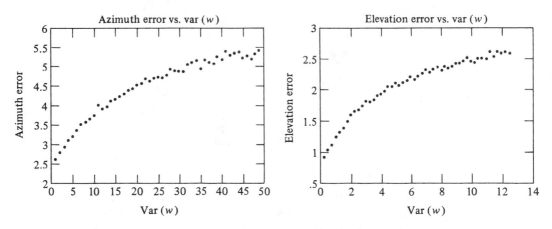

FIGURE 11.18   Root-mean-squared error of the Kalman-filter controller as Var(w) varies.

four fuzzy variables could assume only the seven fuzzy-set values LN, MN, SN, ZE, SP, MP, and LP. So the product space contained $7^4 = 2401$ FAM cells.

We trained 6000 synaptic quantization vectors with DCL. At the end of the training period, we defined a FAM cell as occupied only if it contained at least one synaptic vector. For some combinations of antecedent fuzzy sets, synaptic vectors occupied more than one FAM cell with different consequent fuzzy sets. In these cases we computed the centroid of the consequent fuzzy sets weighted by the

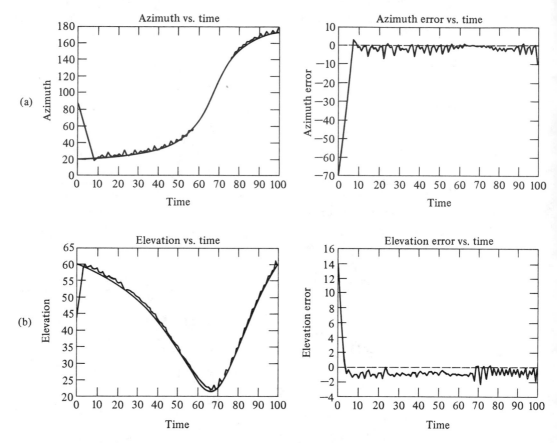

FIGURE 11.19   Fuzzy controller with a "sabotage" FAM rule. (a) Azimuth position and error. (b) Elevation position and error. The sabotage rule (ZE, ZE, ZE; LP) replaces the steady-state FAM rule (ZE, ZE, ZE; ZE). The system quickly adjusts each time the sabotage rule activates.

number of synaptic vectors in their FAM cells. We chose the consequent fuzzy set as that output fuzzy-set value with centroid nearest the weighted centroid value. We ignored other FAM rules with the same antecedents but different consequent fuzzy sets.

Figure 11.20(a) shows the $e_k = ZE$ cross section of the original FAM bank used to generate the training samples. Figure 11.20(b) shows the same cross section of the DCL-estimated FAM bank. Figure 11.21 shows the original and DCL-estimated control surfaces for constant error $e_k = 0$.

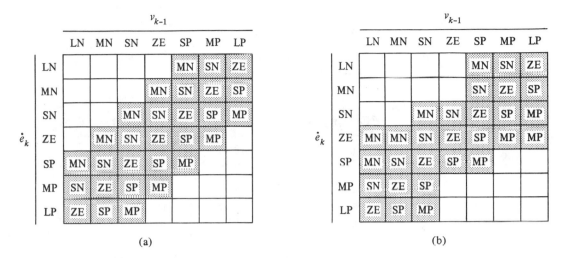

(a)   (b)

FIGURE 11.20 Cross sections of the original and DCL-estimated FAM banks when $e_k = $ ZE. (a) Original. (b) DCL-estimated.

(a)

	v_{k-1}						
$\dot{e}_k$	LN	MN	SN	ZE	SP	MP	LP
LN					MN	SN	ZE
MN				MN	SN	ZE	SP
SN			MN	SN	ZE	SP	MP
ZE		MN	SN	ZE	SP	MP	
SP	MN	SN	ZE	SP	MP		
MP	SN	ZE	SP	MP			
LP	ZE	SP	MP				

(b)

	v_{k-1}						
$\dot{e}_k$	LN	MN	SN	ZE	SP	MP	LP
LN					MN	SN	ZE
MN					SN	ZE	SP
SN			MN	SN	ZE	SP	MP
ZE	MN	MN	SN	ZE	SP	MP	MP
SP	MN	SN	ZE	SP	MP		
MP	SN	ZE	SP				
LP	ZE	SP	MP				

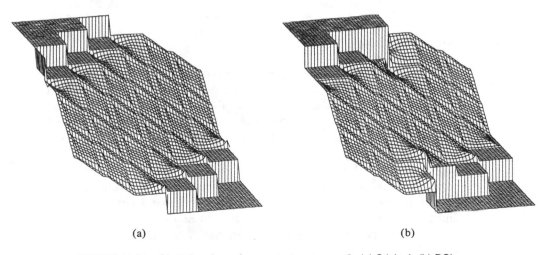

(a)   (b)

FIGURE 11.21 Control surfaces for constant error $e_k = 0$. (a) Original. (b) DCL-estimated.

The regions where the two control surfaces differ correspond to infrequent high-velocity situations. So the original and DCL-estimated control surfaces yield similar results. Table 11.2 compares the controllers' root-mean-squared errors for 10 randomly selected target trajectories.

Table 11.2 Root-mean-squared errors for 10 randomly selected target trajectories. The original and DCL-estimated FAM banks yielded similar results, since they differed only in regions corresponding to infrequent high-velocity situations.

Trajectory Number	Azimuth		Elevation	
	Original	Estimated	Original	Estimated
1	2.33	2.33	3.31	3.37
2	4.14	4.14	3.03	2.89
3	6.11	6.11	3.69	3.68
4	3.83	3.83	3.32	3.30
5	4.02	4.02	3.11	3.10
6	2.84	2.84	1.20	1.21
7	3.22	3.22	3.04	2.98
8	0.75	0.74	2.00	2.00
9	9.28	9.27	5.50	5.41
10	1.81	1.81	2.29	2.29
Average	3.83	3.83	3.05	3.02

CONCLUSION

We developed and compared a fuzzy control system and a Kalman-filter control system for real-time target tracking. The fuzzy system represented uncertainty with continuous or fuzzy sets, with the partial occurrence of multiple alternatives. The Kalman-filter system represented uncertainty with the random occurrence of an exact alternative. Accordingly, our simulations tested each system's response to a different family of uncertainty environments, one fuzzy and the other random. In general, representative training data can "blindly" generate the governing FAM rules.

These simulations suggest that in many cases fuzzy controllers may provide a robust, computationally effective alternative to linear Kalman-filter, indeed to nonlinear extended Kalman-filter, approaches to real-time system control—even when we can accurately articulate an input-output math model.

REFERENCES

Kosko, B., "Fuzzy Entropy and Conditioning," *Information Sciences*, vol. 40, 165-174, 1986.

Kosko, B., *Foundations of Fuzzy Estimation Theory*, Ph.D. dissertation, Department of Electrical Engineering, University of California at Irvine, June 1987; Order number 8801936, University Microfilms International, 300 N. Zeeb Road, Ann Arbor, MI, 48106.

Mendel, J. M., *Lessons in Digital Estimation Theory*, Prentice Hall, Englewood Cliffs, NJ, 1987.

APPENDIX

NEURAL AND FUZZY
SOFTWARE INSTRUCTIONS

Neural and fuzzy C software accompanies this book. Part of the fuzzy software includes source code.

The staff of Olmsted & Watkins developed the neural-network software and the software instructions below. The OWL neural-network models include ART-1, backpropagation applied to multilayer feedforward networks, discrete additive BAM, unsupervised competitive and differential competitive learning AVQ, and additive RABAM. The neural network models include menus for modifying parameters. The staff of Olmsted & Watkins also developed the software problems related to these neural-network models.

The staff of HyperLogic Corporation developed the fuzzy-associative-memory (FAM) demonstration software that illustrates the fuzzy truck backer-upper control system in Chapter 9. The user can delete or add FAM rules on-line to test system robustness or to explore alternative banks of FAM rules. The user can choose correlation-minimum or correlation-product inference.

The staff of Togai InfraLogic developed the FAM pendulum, adaptive FAM pendulum, and FAM-versus-Kalman filter demonstration software and related software instructions below. One FAM system controls a simulated inverted pendulum.

The user can delete or restore, but not add, FAM rules on-line to test system robustness. The user can modify the C source code to add arbitrary FAM rules to the FAM bank. The user can modify many other parameters as well. The staff of Togai InfraLogic developed the FAM software problems in Chapter 8.

The adaptive FAM demonstration uses differential competitive learning to generate pendulum-control FAM rules from pendulum trajectory training data. The target-tracking demonstration compares the FAM and Kalman-filter control systems discussed in Chapter 11. The AFAM-pendulum and target-tracking software programs do not include source code.

All three firms specialize in neural and fuzzy software, services, and products. Please direct questions about the demonstration software, or related technologies, to the appropriate firm:

> HyperLogic Corporation (and Olmsted & Watkins)
> 2411 East Valley Parkway
> Suite 294
> Escondido, California 92025
> (619) 746-2765 phone
> (619) 746-4089 fax

> Togai InfraLogic
> 30 Corporate Park
> Suite 107
> Irvine, California 92714
> (714) 975-8522 phone
> (714) 975-8524 fax

NEURAL NETWORK SOFTWARE INSTRUCTIONS: USING THE OWL DEMONSTRATION PROGRAMS (IBM-PC/AT)

General

This section describes the user options when running the Olmsted & Watkins OWL demonstration software included with this book. The software consists of five programs, which give users an opportunity to exercise various types of neural-network paradigms. The programs and the networks available through them are as follows:

ART:	ART-1 binary Adaptive Resonance Network
BAM:	BAM binary Bidirectional Associative Memory
BKP:	1-, 2-, or 3-layer Backpropagation
CL:	Three types of Competitive Learning Networks
RABAM:	RABAM Heteroassociator

Each program gives access to network parameters. You can safely manipulate the parameters as desired. Erroneous settings produce error messages.

The demonstration programs discussed below assume the following hardware:

IBM-PC/XT/AT

MS-DOS 3.0 or higher

512K memory minimum

EGA graphics adapter

EGA color monitor

1.2M floppy disk

Each program allows access to one or more neural-network simulations. Only one network can be active at any one time, but a given program may permit instantiation of multiple network types.

You can interact with the programs through Owlgraphics, a simple menu-based interface without mouse support. A menu is highlighted by horizontal movement through the menu bar, always visible at the top left of the screen. You move through menus with the arrow keys in the obvious way. If you press the ESCAPE key, the highlighted menu will toggle in or out of visibility.

You also move within a menu with the arrow keys. Once the desired choice is highlighted, activate it by pressing the RETURN key. The associated action will occur forthwith. Some actions complete immediately, while others may invoke a dialog box for further interaction.

A dialog box resembles a menu except that it usually occurs in midscreen instead of along the menu bar. Use the arrow keys to move within a box that has multiple options. To set a value in a dialog box, first select it via the arrow keys and then press RETURN. Owlgraphics will respond by placing an input box on the screen, usually on top of the dialog box. Some boxes require alphabetic or numeric input while others permit arrow selection only. Make the appropriate response, then punctuate it with the RETURN key. To terminate any box, press ESCAPE. Then the program accepts changes previously entered with the RETURN key.

ART

The adaptive-resonance-theory or ART program gives access to an ART-1 simulation. While a network is active, displays depict the input data, the F1 and F2 field arrays, the category representative, the current network state, and the current level of "vigilance." The program has two menus, "Network" and "Run."

A. "Network" Menu Selections

    1. Open Network. This option will instantiate a network according to the current instantiation parameter settings (see item 4, below) unless a network is already active.

2. **Restart Network.** This option will reinitialize the currently active network according to the current instantiation parameter settings (see below); if no network is active, an error message will result.

3. **Close Network.** This option will terminate the currently active network. If no network is active, an error message will result.

4. **Set Load Parameters.** This option allows access to five program variables that are settable only when no network is active. The F1X and F1Y values define the size of the network F1 field (two values are provided to facilitate two-dimensional representation). Maximum value for either F1X or F1Y is 20. The minimum in either case equals 1. The F2X value defines the size of the F2 field. F2 is presented in one dimension only, and the maximum size is 40. The program will trim overlarge specifications to the maximum size.

   You can also adjust the value of the network parameter L. This real-number value must always exceed 1.0.

   The pixel width adjustment allows some degree of control over the field display presentation format. Since bad choices can cause memory-control problems, the program may not honor the user specification. No particular warning is given in this case.

5. **Quit Program.** This option, if selected, produces an "Are you sure?" dialog box. If the user responds "Yes" (the default) by pressing the RETURN key, the program terminates immediately, whether or not a network is active. ESCAPE will abort the termination request.

B. "Run" Menu Selections

1. **Clear Input.** This option clears the input data array to all zeroes. The screen display will clear to black.

2. **Edit Input.** This option transfers control to the input editor so that you may specify a pattern for subsequent use. The Input Field will show a single highlighted "active" node. Pressing the SPACE bar will toggle it between red and blue. A red color indicates that the node is active (value 1), while blue indicates that it is off (value 0). Move between the nodes by means of the arrow keys, left, right, up, or down. When you establish the desired pattern, terminate the edit by pressing the ESCAPE key.

3. **Train.** This option trains the network (to equilibrium) using the current input pattern. Possible network-state results are READY, if the input was all zeroes; RESONANCE, if the pattern has been matched with an existing or newly established template; or FULL, if the network could neither match the pattern nor create a new category for it.

4. **Step.** This option takes the network through a single iteration of the training process. Any of the state results listed above for "train" are possible, and in addition the RESET state may occur if the vigilance

is sufficiently high and the pattern does not match any category exactly. Selecting this option repeatedly until RESET does not occur is equivalent to selecting the "train" option.

5. **Run.** This option is identical to "train," except that no learning takes place.

6. **Set Run Parameters.** This option allows adjustment of the vigilance parameter (only) via a dialog box. Acceptable values for this parameter lie in the range 0.0 to 1.0, inclusive.

BAM

The bidirectional-associative-memory or BAM program gives access to a BAM network simulation. While a network runs, displays depict the X and Y field arrays and the current "energy" of the system. The program has four menus: "Network," "Run," "Nodes," and "Weights."

A. "Network" Menu Selections

1. **Open Network.** This option will instantiate a network according to the current instantiation parameter settings (see item 3, below) unless a network is already active.

2. **Close Network.** This option will terminate the currently active network. If no network is active an error message will result.

3. **Set Load Parameters.** This option allows access to five program variables that are settable only when no network is active. The $X - x$ and $X - y$ values define the size of the network X field (two values are provided to facilitate two-dimensional representation). Maximum value for either $X - x$, $X - y$, $Y - x$, or $Y - y$ is 12; the minimum in all cases is 1. The program will trim overlarge specifications to the maximum size.

   The pixel-width adjustment allows some degree of control over the field display presentation format. Since bad choices can cause memory-control problems, the program may elect not to honor the user specification. No particular warning is given in this case.

4. **Quit.** This option, if selected, produces an "Are you sure?" dialog box. If the user responds "Yes" (the default) by pressing the RETURN key, the program terminates immediately, whether or not a network is active. ESCAPE will abort the termination request.

B. "Run" Menu Selections

1. **Step.** This option will cycle the BAM through one iteration. A network must be active lest an error message result.

2. Run. This option will cycle the BAM to equilibrium. A network must be active lest an error message result.

3. Set State Mode. This option selects the network node activation representation. Selection is by arrow key choice, either Binary or Bipolar. No network need be active.

C. "Nodes" Menu Selections

1. Clear X. This option will clear the X field of the active network. If no network is active, an error message will result.

2. Edit X. This option transfers control to the input editor so that a pattern may be specified for subsequent use. The X field will show a single highlighted "active" node. Pressing the SPACE bar will toggle it between red and blue. A red color indicates that the node is active (value 1), while blue indicates that it is off (value 0 or -1, depending on the state mode). Move between the nodes by means of the arrow keys, left, right, up, or down. When the desired pattern is established, terminate the edit by pressing the ESCAPE key.

3. Clear Y. This option is similar to "Clear X," except that it is the Y field which is involved.

4. Edit Y. This option is similar to "Edit X," except that it is the Y field which is involved.

5. Load Biases. This option will load threshold bias values from a previously opened bias file (see item 6, below). The bias file should contain at least as many bias values (in ASCII floating point representation), on per line. Either lack of open bias file or premature end of file will result in an error message. When a bias file is opened via this option, any previously opened bias file is closed.

6. Open Bias File. This option allows the user to select a bias file for subsequent use via option 5, above. Any previously opened bias file is closed.

D. "Weights" Menu Selections

1. Clear Weight Array. This option will clear the BAM weight array to all zeroes. A network must be active or an error message will result.

2. Add Pattern. This option will cause the outer product of the current X and Y field values to be added componentwise to the network weight matrix. The addition is appropriate for the current-state mode.

3. Delete Pattern. This option will cause the outer product of the current X and Y field values to be subtracted componentwise from the network weight matrix. The subtraction is appropriate for the current-state mode.

4. **Load Weights.** This option will load the weight matrix from values extracted from the currently open weights file. Format of the weights file is the similar to that of the bias file, except that the weight values must be integral.

5. **Open Weights File.** This option allows the user to select a weights file for later use via option 4, above. Any previously opened weights file is closed.

BKP

The BKP program simulates a backpropagation-trained network. While a network is active, displays depict the current iteration and either the weight arrays as two-dimensional areas, or else the network "anatomy," the current outputs and squared error, and a pairwise plot of weights. The program has four menus: "Network," "Run," "Data," and "Weights." In the discussion below, no network may have an input size larger than 20 units, no hidden layer may exceed 50 units in width, and the network output layer cannot exceed 10 units.

Any time training is in progress you may reset the display by pressing the SPACE bar.

A. **"Network" Menu Selections**

1. **Open 1-Layer Network.** This option will create a single-layer network (no "hidden" layer).

2. **Open 2-Layer Network.** This option will create a two-layer network (one "hidden" layer).

3. **Open 3-Layer Network.** This option will create a three-layer network (two "hidden" layers).

4. **Close Network.** This option will immediately close the currently active network, if there is one.

5. **Set Display Mode.** This option toggles the screen between either the color-coded weight arrays display or else the states, pairwise weights, and anatomy display.

6. **Quit.** This option, if selected, produces an "Are you sure?" dialog box. If the user responds "Yes" (the default) by pressing the RETURN key, the program terminates immediately, whether or not a network is active. ESCAPE will abort the termination request.

B. **"Run" Menu Selections**

1. **Run.** This selection will operate the network against the currently opened data file (see section C below). The file-handling options will be honored. No learning occurs.

2. **Train.** This selection will operate the network against the currently opened data file just as "Run" does, except that learning is enabled.

3. **Restart Network.** This option will randomize all the weight values in the active network.

4. **Signal Function.** This option permits choice of signal function for the network as a whole; you can choose "Logistic," "Bipolar Logistic," or "Gaussian."

5. **Run Parameters.** This option permits adjustment of parameters appropriate for the currently active network. Settings are available for the learning rates and momenta (by layer), whether to use momentum, the batch mode size (0 disables batching), the half-range for weight randomization, and a toggle to permit lower operation when running a small network on a fast machine.

C. "Data" Menu Selections

1. **Select Data File.** This option permits selection of a file to supply input data. The format is ASCII floating point, and you can place multiple items on a single line.

2. **Reset Data File.** This option performs an fseek to the beginning of the currently opened data file. The fseek function positions the file pointer, the index of the next file read.

3. **File Handling.** This option specifies the action upon end of run and end of file. If you select "Reset on run," an fseek replaces the file pointer after each run. If you select "Reset on EOF," the file is reset to the beginning when end-of-file is encountered. The third option, "Lesson Size," allows specification of the number of training examples in a single lesson.

D. "Weights" Menu Selections

1. **Select Weights File.** This option permits selection of a file to initialize the weight arrays. The format is ASCII floating point, and you can place multiple items on a single line.

2. **Load Weights.** This option will load the currently active network weight arrays from the currently active weights file.

CL

The CL program demonstrates three different competitive learning networks. You can select one of KOH ("standard" competitive learning), KCON (competitive learning with "conscience"), or DCL (differential competitive learning). The display illustrates the weight distribution as training proceeds. To simplify the presentation only one- or two-dimensional input spaces are permitted. The program draws the centroid(s) of the mode(s) of the selected distribution over the weights display so that you can visually monitor progress.

Any time training is in progress you can reset the display by pressing the SPACE bar. This helps when observing the weight motions.

A. "Network" Menu Selections

1. **Activate KOH.** This selection will bring up a load-time parameters dialog box and then activate a KOH network according to the specification.

2. **Activate KCON.** This selection will bring up a load-time parameters dialog box and then activate a KCON network according to the specification.

3. **Activate DCL.** This selection will bring up a load-time parameters dialog box and then activate a DCL network according to the specification.

4. **Close Network.** This option will close the currently active network.

5. **Quit.** This option, if selected, produces an "Are you sure?" dialog box. If the user responds "Yes" (the default) by pressing the RETURN key, the program terminates immediately, whether or not a network is active. ESCAPE will abort the termination request.

B. "Run" Menu Selections

1. **Train.** This option lets you train the network with random data generated according to the currently specified distribution (see below). The display plot is one-dimensional versus time, or else two-dimensional in space, according to the number of inputs chosen.

2. **Restart.** This option will randomize all the weight values in the active network according to the weight scatter radius and bias. The weights will be scattered uniformly over twice the range value and offset by the bias value. The bias applies to both dimensions if two-input networks are used.

3. **Run Parameters.** This option brings up a dialog box appropriate for the currently active network's run-time parameter list. Four values are set in common: the learning rate, the number of examples per training run, and the weight scatter radius and weight scatter bias (used when weights are initialized). KCON networks also permit setting the Bias Value and Bias Utilization parameters (see the literature for the interpretation of these settings). DCL networks allow setting of the differential and normal activation rates, the smoothing exponent, and the number of winners allowed per run, in addition to the basic four settings.

4. **Distribution.** This option sets the distribution of the random data to be shown to the network during training. The choices allowed give a fair selection of unimodal and multimodal distributions. The UNIFORM

choices are Uniform Distributions, while the TRIANGLE distributions are biased toward the center of each mode (via squaring). The PYRAMIDAL distribution has a heavy concentration of probability toward its center with a less likely "apron" around it. The PRODUCT distribution results (more or less) from multiplying two independently chosen Gaussian random variables.

5. **Magnification.** This option sets the visual magnification. One of $\times 1$, $\times 2$, or $\times 4$ is permitted.

6. **Mark Centroids.** This option toggles the centroid markers on or off. It is useful for those cases when the distributions are tight and the centroid markers obscure the data plots.

RABAM

The RABAM program demonstrates the RABAM heteroassociator. The display windows illustrate the "ideal" input data versus the actual input data, the "ideal" output equilibrium versus the current output values, the averaged state values of two nodes versus their ideal equilibrium values, and the current iteration number. A variety of noise distributions are permitted, and you can control separately the noise levels for the two network fields, the weight array, and the inputs.

A. **"Network" Menu Selections**

1. **Open Network.** This option will lead the user through a create-time and then a run-time parameter setting dialog box and then create a RABAM heteroassociator with the specified characteristics. The create-time options specify the number of neurons or nodes in each of the two layers of the network. Each layer must have at least one node, and there can be no more than 99 nodes total among the two layers.

   The run-time options are discussed below.

2. **Close Network.** This option will dismantle the currently active network.

3. **Set Colors.** This option permits setting the colors used for menus, boxes, windows, and other objects within Owlgraphics. Selections are entered via menu choice boxes and arrow keys in the usual way and take effect when next the affected object is drawn.

4. **Quit.** This option, if selected, produces an "Are you sure?" dialog box. If the user responds "Yes" (the default) by pressing the RETURN key, the program terminates immediately, whether or not a network is active. ESCAPE will abort the termination request.

B. **"Run" Menu Selections**

1. **Train.** This option trains the currently active network against random internally generated data.

2. **Restart Network.** This option randomizes the network weights and calculates a new "ideal equilibrium." The process may take some time on slow machines. The program erases the menus while this action is underway and restores them after it completes. This option should be invoked whenever the network signal function or learning law is changed.

3. **Set Run Parameters.** This option permits adjustment of the various parameters used to control the RABAM operation.

The signal-function submenu allows selection of one of six possible choices:

Logistic	Logistic Sigmoid
Blogistic	Bipolar Logistic Sigmoid
Slogistic	Logistic Sigmoid with inflection at 1
Unit Step	Step function with transition at lambda
Trunc Exp	Truncated Exponential
Unit Ramp	Truncated Linear

The *Signal Lambda* option sets the lambda value for the selected signal function. Different signal functions use lambda in different ways.

The *Learning Law* submenu allows selection of one of Signal Hebbian, Differential Hebbian, or Competitive learning laws; in addition, No Learn disables learning altogether.

The *Learning Rate* parameter sets the learning rate in the same way as other networks do.

Four *Noise Gain* parameters exist: X field noise, Y field noise, internal weight array noise, and input array noise. The values can be set arbitrarily as desired, but very large values can strain the internal digital representation used to simulate the network. All noise uses the selected distribution.

Display Node Number chooses two nodes for display in the average state plot. The nodes chosen are those with the index entered and the next subsequent node (wrapping around to the first node if necessary). For the purpose of numbering, the X nodes come first, then the Y nodes.

Weight Init Value sets the initial value for all the weights in the network. It is effective whenever a network is created.

Node Rate Factor is a rate control that helps maintain the digital simulation. When using large networks or large noise levels, this value should be small (close to zero but positive). Small networks with low noise levels can use higher values, but this parameter should never exceed 1.0. Ideally this value should be just large enough to render the squared-error plot as a curve.

Duty Cycle sets the number of iterations that the current input is held before a new input array is generated. This allows simulation of noise of different apparent frequencies.

FUZZY-ASSOCIATIVE-MEMORY SOFTWARE INSTRUCTIONS

The C software includes simulations of three fuzzy control systems. The first program simulates the fuzzy truck backer-upper control system in Chapter 9. The second program simulates the fuzzy-versus-Kalman-filter target-tracking systems in Chapter 11. The third program simulates an adaptive fuzzy controller that balances an inverted pendulum.

As the pendulum demo runs, a trace buffer stores the input-output data pairs. These sample vectors then train a synaptic adaptive vector quantizer (AVQ) system using differential competitive learning (DCL). Product-space clustering of the synaptic vectors approximates the underlying control surface that generated the data. By partitioning the product space into FAM cells, the system estimates a FAM-rule matrix. When training stops, the user can restart the pendulum demo with the default (manually designed) FAM-rule matrix, the FAM matrix from the previous run, or the new DCL-estimated FAM matrix.

All simulations are IBM PC compatible and require VGA graphics. The executable files are named TRUCK, TRACKER, and FUZZYDCL.

Fuzzy Truck Backer-Upper Control System

The program in file TRUCK.EXE simulates a planar truck as it backs up to a loading dock. A fuzzy controller, based on the system in Chapter 9, directs the truck's motion. The loading dock is located at the bottom center of the window labelled "Truck Yard."

The program uses a menu-driven graphical interface. Use the arrow keys to move among the menus or their suboptions. If you cannot see the menus, the ESCAPE key will return them to view. The RETURN key selects a menu item.

Some menu options produce a dialog box. Use the arrow keys to move among the options in the dialog box. Select options with the RETURN key. If the dialog box asks you for numeric data, end your numeric answers with the RETURN key. Press ESCAPE to end the dialog.

The FAM-bank matrix contains 35 FAM rules. You can add or delete rules to the FAM bank. First select the "Rules" option in the "Project" menu. This outlines a portion of the "Rule Activity" window. When you press ENTER, a selection box will appear to allow you to choose a FAM rule for the highlighted situation. Use the arrow keys to move among the selection options. Use the ENTER key to enter your choices. Use the ESCAPE key to end the edit process. The fuzzy variables

X and PHI assume fuzzy-set values along, respectively, the horizontal and vertical axes. Fuzzy variable X assumes, from left to right, the fuzzy sets LE, LC, CE, RC, and RI. Fuzzy variable PHI assumes, from top to bottom, RB, RU, RV, VE, LV, LU, and LB. See Chapter 9 for the fuzzy-set definitions of these acronyms.

FAM rules activate to different degrees as the simulation proceeds. The "Rules Active" window uses colors to indicate the relative activity of FAM rules. In order of increasing activity, these colors correspond to black, brown, red, yellow, green, blue, and white.

Use the "Situation" submenu, under the main "Run" menu, to select the truck's initial position and orientation in the parking lot. The parking lot has 100 units on each side with the origin at the lower left corner.

You can view the truck's path if you select "Yes" in the "Tracks" option in the "Situation" submenu. Otherwise, the truck moves in an animation across the parking lot. You can pause a running simulation at any time if you press any key. Selecting the "Run Truck" menu item resumes the simulation from the truck's current position and azimuth settings. The simulations stops automatically when the truck reaches the parking lot's edge.

If you set the "Auto Restart" to "Yes" in the "Run" menu, the truck simulation will start afresh, from randomly chosen initial conditions, each time the truck reaches the loading dock or hits the edge of the parking lot. Press any key to stop the process. To restart the truck from its last initial position, select the "Restart Truck" option in the "Run" menu. The "Run" option continues the simulation from the current position.

The simulation displays the steering-wheel position in the "Wheel" window. The steering ratio equals 2.75. The corresponding fuzzy variable THETA assumes the output fuzzy-set value displayed in the "Resultant" window. The red mark indicates the centroid of this output fuzzy set. THETA numerical values range from -30 to 30 along the horizontal axis.

You can select either correlation-minimum or correlation-product inference with the "Scale" option in the "Situation" option item in the "Run" menu. Use the "Speed" option in the "Situation" item of the "Run" menu to adjust the truck's speed from 1 to 5.

The "Helpd" option of the "Project" menu gives an on-line summary of this text.

To change the color scheme, select the "Set Colors" option from the "Project" menu. A dialog box will appear that allows you to select alternative features. Select the desired color from the appropriate dialog box.

Fuzzy Target-Tracking Demonstration

Design parameters and target scenario. Details of the fuzzy and Kalman-filter controllers appear in Chapter 11. The white Gaussian process noise has a variance $\text{Var}(w) = 0.08$ azimuth, 0.02 elevation. We set the three design parameters as

$$\text{Upper base} = 0.6$$
$$\text{Lower base} = 1.8$$
$$\text{Gain} = 1.0$$

All targets maintain constant velocities ranging from 1000 to 2000 miles per hour. The sampling interval is 1/10 second. The platform's maximum scan rate equals 90-degrees-per-second azimuth, 45-degrees-per-second elevation.

Display windows and user controls.   The demonstration screen consists of five windows. The system graphically displays only data for the fuzzy controller. An error display menu allows the user to compare error traces for the two controllers.

1. The lower left window shows the scene that an observer would see through a physical window. The target position is plotted in yellow, while the position where the platform is pointing is plotted in magenta. Target range in miles and velocity in thousands of miles per hour appear near the top of the window.

2. The upper left window shows a polar view of the same data plotted below. Azimuth sweeps from 0 degrees on the right to 180 degrees on the left. Elevation equals 0 degrees (vertical pointing) at the origin and 90 degrees (horizontal pointing) at the perimeter. Since the target trajectories are all constant-bearing, they appear linear in the visual scene, but nonlinear in azimuth-elevation coordinates.

3. The upper right window displays the tracking errors in degrees.

4. The middle right window shows the center of the error plot magnified by a factor of 50. The magnitude of chaotic movement about the origin can be decreased by fine-tuning the fuzzy sets and/or scaling the output by a factor slightly less than 1.

5. The lower right window is a numerical comparison of the two controllers' errors.

At the end of each trajectory, the screen is refreshed, and a new target appears. TARGET.DCT contains 25 trajectories with random velocity vectors and starting points. The demo continually cycles through all 25 trajectories.

SPACE BAR pauses the demo. Any key restarts it. ESC ends the demonstration. RETURN pauses the demo and calls up an error display menu with the following options:

1. Error trace for the current trajectory

2. Smoothed error trace for the current trajectory

3. Error trace for the previous trajectory

4. Smoothed error trace for the previous trajectory

5. Exit

Each option produces both azimuth and elevation plots, with data for both controllers plotted on the same axes. The system provides root-mean-square errors with plots of the previous trajectory. Smoothed errors are computed with a symmetric moving window of user-defined width. After the user selects an option, hitting any key returns to the menu. The Exit option restarts the demo where it left off.

Adaptive Fuzzy Control of Inverted Pendulum

This simulation shows a motor controlling the position of an inverted pendulum. If we perturb the pendulum from the vertical position, the motor drives it back to the upright position again. In a physical model, we would replace the motor with a sliding cart that moves beneath the pendulum to balance it.

Display windows and user controls.   The demonstration screen is divided into five display windows:

1. The center window shows the pendulum itself. The user can change both the bob mass and motor size. Color-coded triangles below the pendulum indicate the current settings of these quantities. Pressing F6 increases the bob's mass; shift-F6 decreases it. F5 similarly controls the motor size.

   The length of the pendulum's shaft can be varied sinusoidally to introduce further nonlinearities into the system. F7 toggles the oscillation on and off. When oscillation is turned off, a red dot appears next to the oscillation icon to the right of the pendulum.

   A "bopper" periodically strikes the pendulum from the left, disrupting the system temporarily. F8 toggles the bopper on and off. When the bopper is deactivated, a red dot appears next to the bopper icon to the left of the pendulum.

   To manually perturb the system the user can hit F3 and F4, which push the pendulum to the left and right, respectively. F9 and F10 start the pendulum to the extreme left and extreme right, respectively.

2. The upper right window shows the FAM-rule matrix that governs the fuzzy controller. Each element in the matrix represents one fuzzy associative memory or FAM rule. For example, the center element in the matrix defines the steady-state FAM rule and reads

$$\text{IF Theta is Z and dTheta is Z}$$
$$\text{THEN Current} = Z$$

When a rule in the FAM matrix fires to any nonzero degree, the corresponding matrix element is highlighted in gray.

Notice that the default FAM matrix contains only 11 of the possible 25 rules. The user can test the fuzzy controller's robustness by eliminating individual rules.

3. The lower right window consists of two main sections: the rule list and the currently selected rule display. The user can scroll through the rule list with the up and down arrow keys. Scrolling through the list dynamically sets the highlighted rule in the rule list to be the "current rule." The fuzzy inference procedure determining the output of the current rule is then graphically displayed to the right of the rule list. If the rule list contains more than 11 rules, the arrows cause scrolling. Otherwise, wraparound occurs.

Pressing F1 eliminates the current rule from the FAM-rule matrix. Pressing F1 again returns the rule to the FAM-rule matrix. A red dot next to a FAM rule indicates that the user has eliminated it.

4. The lower left window displays the system from a black-box perspective. The input-fuzzy-variable names, Theta and dTheta, appear at the top of the window. The output value fuzzy variable, motor current, appears at the bottom of the window.

The middle portion of the window shows the combined output fuzzy set determined by all the rules in the FAM matrix or they fire to different degrees. The nonfuzzy output value (for that cycle) equals the centroid of the combined output fuzzy set.

5. The upper left window is a trace buffer showing the motor current as it varies over time. The iteration number also appears in this window.

F2 freezes the demo. Any key restarts it. Hitting the question mark ("?") calls up a help menu, which lists all the function-key definitions. ESCape ends the pendulum demonstration and starts the DCL program. Since the AVQ requires 125 sample vectors to initialize its synaptic vectors, the program ignores ESC until the iteration counter reaches 125.

Fuzzy-set definitions.   Both input and output fuzzy variables assume the following fuzzy-set values:

NM   Negative Medium
NS   Negative Small
Z   Zero
PS   Positive Small
PM   Positive Medium

A positive position error, Theta, indicates a tilt to the right, and a positive output current drives the pendulum to the right.

FAM matrix design. We designed the FAM matrix by anticipating possible input combinations and applying common sense to determine the necessary motor current.

If the pendulum is vertical and stationary, then the motor current should be zero. Hence we write the steady-state FAM rule:

IF Theta = Z AND dTheta = Z, THEN Motor Current = Z

We write this as

IF Theta(Z) AND dTheta(Z), THEN Current(Z)

If the position error is positive small or medium and the velocity is near zero, the motor current should drive the error to zero. Hence,

IF Theta(PS) AND dTheta(Z), THEN Current(NS)
IF Theta(PM) AND dTheta(Z), THEN Current(NM)

By symmetry, we include the following FAM rules:

IF Theta(NS) AND dTheta(Z), THEN Current(PS)
IF Theta(NM) AND dTheta(Z), THEN Current(PM)

With these five rules, the controller can respond to a positional error.

As the pendulum approaches vertical, we wish to damp its velocity. Four FAM rules accomplish this task:

IF Theta(Z) AND dTheta(NS), THEN Current(PS)
IF Theta(Z) AND dTheta(NM), THEN Current(PM)
IF Theta(Z) AND dTheta(PS), THEN Current(NS)
IF Theta(Z) AND dTheta(PM), THEN Current(NM)

The following two FAM rules help stabilize the system:

IF Theta(PS) AND dTheta(NS), THEN Current(Z)
IF Theta(NS) AND dTheta(PS), THEN Current(Z)

Adding more rules may improve performance slightly. Removing rules may destabilize the system.

When the pendulum starts at the extreme right, the first FAM rule to fire is

IF Theta(PM) AND dTheta(Z), THEN Current(NM)

The FAM-rule firing order then moves counterclockwise around the FAM matrix, spiraling toward the origin as Theta and dTheta decrease simultaneously. Thus the motor current is initially negative medium (NM), then negative small (NS). Next it changes to positive medium (PM) to begin the damping cycle. The current decreases again to positive small (PS), and finally zero (Z). This cycle provides nearly critical damping.

If the bopper bops the pendulum, the FAM-rule firing pattern differs from the one described above. After dTheta goes to zero, the spiral pattern in the input plane occurs once again. But the pendulum recovers from roughly a 45-degree angle. So the FAM rules on the perimeter of the FAM matrix fire to a lesser degree.

FAM rule product-space clustering with DCL.   This program estimates the pendulum's FAM matrix from sample data vectors. Crisp values of Theta, dTheta, and Current are saved in a trace buffer while the pendulum demo runs. Each data vector represents a point in the three-dimensional product space. An adaptive vector quantization (AVQ) arises from differential competitive learning (DCL). DCL adaptively clusters these data vectors to estimate the underlying control surface that generated them. This partitions the product space into synaptic-vector-occupied FAM cells, each one representing a single FAM rule.

On the left side of the screen, two-dimensional projections of the data vectors appear as points in the input plane. The color of each point indicates which fuzzy-set value admits the highest degree of membership for that vector's crisp output value.

Cyan lines partitioning the input plane indicate FAM cell boundaries. If the number of data vectors in a cell exceeds a threshold, then the associated FAM rule is entered into the FAM matrix on the right side of the screen. Since points of more than one color often appear within a FAM cell, we sum the consequent fuzzy sets in each cell and employ centroid defuzzification. The consequent fuzzy set for each FAM rule equals the output fuzzy-set value with centroid nearest the defuzzified centroid for its cell. After all the samples are processed (or the user hits ESC to terminate training), the resulting FAM-rule matrix summarizes the underlying fuzzy control surface.

The DCL system's performance depends on the training data. Since the AVQ assumes no prior knowledge of the underlying control surface, FAM rules that are seldom used seldom appear in the final FAM matrix. So we automatically add the following rules to the estimated FAM matrix to cover boundary conditions:

IF Theta(NM) AND dTheta(Z),   THEN Current(PM)

IF Theta(PM) AND dTheta(Z),   THEN Current(NM)

When training stops, hitting any key will bring up a menu. The user may restart the pendulum demo with the default, previous, or new FAM-rule matrix.

INDEX

Note: Italicized page numbers locate references.

A

ABAM. *See* Adaptive bidirectional
    associative memories (ABAMs)
Acceleration term, 204
Action potentials, firing frequency of, 43
Action (signal pulse), 42
Activation(s)
    biological, 41–44
    input, 39
    operating range of, 94
Activation dynamics, 100–103
Activation models, 55–110
    additive. *See* Additive activation models
    Cohen-Grossberg, 99–103
    multiplicative, 96–97, 99–103
    shunting, 97, 98, 100–103
Activation "temperature" (annealing
    schedules), 254

Activation velocities, 41
Adaptation, 2, 22–24. *See also* Learning
Adaptation level index, 97
Adaptive bidirectional associative
    memories (ABAMs), 233–35. *See
    also* Random adaptive bidirectional
    associative memories (RABAMs)
    adaptive resonance, 240–41
    asymptotic stability of, 238–39
    Hebbian, 234, 236, 237–42
    Hopfield model, 92–94
    Lyapunov function, 245–46
    signal Hebbian, 237–38, 239–40
    structural stability of, 242–43
    theorem, 224, 236–39
        squared-velocity condition, 251
        stability-convergence dilemma and,
        236
Adaptive causal inference, 158–59

Adaptive clustering algorithms, 327–35, 348–52
Adaptive cosine transform coding, 364, 365–66
Adaptive decompositional inference (ADI), 326–27
Adaptive delta modulation, 168
Adaptive filter, 44
Adaptive fuzzy associative memories (AFAMs), 301, 327–35
  adaptive BIOFAM clustering, 329–35
  defined, 327
  rule generation, 328–29
  for target-tracking control systems, 402–5
  for transform image coding, 366–77
    differential competitive learning (DCL), 368–69, 370–73
    fuzzy set values, 367–68
    product-space clustering to estimate FAM rules, 368–69
    simulation, 373–74
Adaptive fuzzy systems, 18–19, 29–30
Adaptive resonance, 68
Adaptive resonance theory (ART), 166, 240–41
  binary (ART-1), 259–61
  capacity of, 91
Adaptive vector quantization (AVQ), 112, 223, 368
  algorithms, 225–32
    competitive, 227, 328
    differential competitive learning (DCL), 228
    stochastic differential equations, 225–27
    stochastic equilibrium and convergence, 228–32
    supervised competitive learning (SCL), 228
    unsupervised competitive learning (UCL), 227
  global stability of, 233
  theorems
    centroid, 228–31, 233
    convergence, 149, 230–32
    equilibrium, 148–49

Additive activation dynamics, 100
Additive activation models, 56–99
  BAM connection matrices, 79–94
    autoassociative (OLAM) filtering, 83–85
    BAM correlation encoding example, 85–91
    bipolar Hebbian or outer-product learning method, 79–81
    Hopfield model, 92–94, 100
    memory capacity, 91–92
    optimal linear associative memory (OLAM) matrices, 81–83
  bivalent, 63–79
    bidirectional stability, 68–69
    bivalent additive BAM, 63–68
    bivalent BAM theorem, 73–79
    Lyapunov functions, 69–73
  neuronal dynamics, 56–59
    additive external input, 58–59
    membrane resting potentials, 57–58
    membrane time constants, 57
    passive membrane decay, 56–57
  neuronal feedback, 59–61
    bidirectional and unidirectional connection topologies, 60–61
    synaptic connection matrices, 59–60
  noise in, 245
  noise-saturation dilemma and, 94–99
    saturation theorem, 95–99
Additive combination technique, 29, 31
Additive Grossberg model, 96
Adelson, E. H., 364, *378*
AI. *See* Artificial intelligence (AI)
Albert, A. E., 186, 189, *213*
Algorithm(s)
  adaptive clustering, 327–35, 348–52
  adaptive vector quantization (AVQ), 225–32
    competitive, 227, 328
    differential competitive learning (DCL), 228
    stochastic differential equations, 225–27
    stochastic equilibrium and convergence, 228–32

supervised competitive learning (SCL), 228
unsupervised competitive learning (UCL), 227
backpropagation, 181, 188, 196–212, 222, 327
  derivation, 203–10
  feedforward sigmoidal representation theorems, 199–201
  history of, 196–99
  multilayer feedforward network architectures, 201–3
  robust, 211–12
  as stochastic approximation, 210–11
differential competitive learning (DCL), 357–60
geometric supervised learning, 213
gradient-descent, 189
  estimated, 193–94
  ideal, 193
Greville, 81–82
Kalman-filter, 190
learning-automata, 181
least-mean-square (LMS), 181, 190–96, 210, 212
perceptron learning, 189
supervised, 113–14, 225
supervised learning, 180
two-class perceptron, 190
unsupervised, 113–14, 225
Almeida, L. B., 210, *213*
Amari, S., 69, 92, 93, *104*, *255*
Amari-Hopfield model, 92–94
Ambiguity, randomness vs., 265–68
Amplification function, 100
Anandan, P., 212, *213*
Anderson, B. D. O., 71, *103*
Anderson, C. W., 160, *170*
Anderson, J. A., 46, *52*, 82, 100, *103*, *104*, 187, 197, *213*, 222, *255*
Angular velocity, 317
Annealing, 253–55
  simulated, 253, 255
Antagonistic ionic pair, 43
Anti-Hebbian learning, 255
Aperiodic equilibrium attractors, 15
Approximation

associative-reward-penalty, 212
Newton's method, 193
stochastic, 114, 161–62, 181, 193, 198
  backpropagation algorithm as, 210–11
  backpropagation as, 210–11
  linear, 190–96
  supervised learning as, 185–96, 212
ART. *See* Adaptive resonance theory (ART)
Artificial intelligence (AI), 12, 24–25, 302. *See also* Expert systems
Artificial neural systems, 13
Associations, FAM. *See under* Fuzzy associative memories (FAMs)
Associative-reward-penalty algorithm, 212
Associativity, 115
Associators, parallel, 29–32
Asymptotic centroid estimation, 148–49
Asymptotic correlation encoding, 138–40
Asymptotic stability, 71–72
  of ABAM models, 238–39
Asynchronous recall, 89, 90
Asynchronous state-change patterns, 64
Asynchrony, 66–67
  simple, 64, 93–94
  subset, 64–66
Attentive processing, 3
Attractor basins, 114
Attractors
  equilibrium, 15
  spurious, 88
Augmented field, 46
Autoassociative matrix, 309
Autoassociative systems, 46, 60
  activation dynamics in, 100–103
Autoassociativity, 92
Autocorrelation, 93
Autocorrelation function, 128
Autocorrelation matrix, input, 191–92
Avalanche, 143
Axon hillock, 42–43

B

Backpropagation algorithm, 181, 188, 196–212, 222, 327
derivation, 203–10

Backpropagation algorithm *(cont.)*
  feedforward sigmoidal representation
    theorems, 199–201
  history of, 196–99
  multilayer feedforward network
    architectures, 201–3
  robust, 211–12
  as stochastic approximation, 210–11
Backpropagation control systems, 345–47
  sensitivity and robustness of, 347–48,
    349, 356, 357
Backward-chaining inference, 25
Backward projections, 60
Balls, stimulus and response, 20–21
BAM. *See* Bidirectional associative
    memories (BAMs)
Bandler, W., 278, *294*
Barto, A. C., 160, *170*, 181, 212, *213*
Bayesian polemics against fuzziness, 264,
    289–93
Bayes theorem, fuzzy, 290–93
  odds-form, 291–93
Between-set relationships, 11
Bidirectional associative memories
    (BAMs). *See also* Adaptive
    bidirectional associative memories
    (ABAMs); Random adaptive
    bidirectional associative memories
    (RABAMs)
  adaptive resonance (ABAMs), 240–41
  bivalent additive, 63–68
  bivalent theorem, 73–79
    database interpretation, 76–77
  connection matrices, 79–94
    autoassociative OLAM filtering,
      83–85
    BAM correlation encoding example,
      85–91
    bipolar Hebbian or outer-product
      learning method, 79–81
    Hopfield model, 92–94, 100
    memory capacity, 91–92
    optimal linear associative memory
      (OLAM) matrices, 81–83
  continuous, 101–2
    theorem, 102
  continuous additive, 62

convergence of trajectories, 78–79
  correlation encoding, 85–91
  Hebbian, dimensionality-capacity
    trade-off in, 141
  nonadaptive, global stability for, 238
  recall process, 89
  shunting, 103
Bidirectional equilibrium, 66–67
Bidirectional networks, 60–61
Bidirectional stability, 68–69, 74–76
  of additive bivalent activation model,
    68–69
Binary adaptive resonance theory (ART-1),
    software problems, 259–61
Binary input-output fuzzy associative
    memories (BIOFAMs), 305, 306,
    317–26
  adaptive, 333–35
  adaptive clustering, 329–35
  decompositional inference, 322–27
  defuzzification by, 30–31
  inference procedure, 320
Binary outer-product law, 79
Binary propositional rules, 200
Binary signal functions, 43–44
Binary signal vectors, 86–88
BIOFAM. *See* Binary input-output fuzzy
    associative memories (BIOFAMs)
Biological activations and signals, 41–44
Bipolar Hebbian (outer-product) learning
    method, 79–81
Bipolar outer-product law, 79
Bipolar signal functions, 43–44
Bipolar signal state vectors, 88–89
Birkhoff, G., 3, *34*
Bit map, 364
Bivalence, 11
Bivalent additive activation models. *See
    under* Additive activation models
Bivalent bidirectional associative memory
    (BAM) theorem, 73–79
  database interpretation, 76–77
Bivalent logic, 325
Bivalent paradoxes, 4–5, 273, 274
Bivalent signal functions, 64
Black, M., 6, *34*, 269, *294*
Blum, J. R., 186, *213*

Boltzmann machine learning, 255
Boolean outer-product law, 80
Borel measurable, 120
Borel field, 120, 121–22
Borel sct, 120
Borel sigma-algebra, 121
Brain(s)
    distributed encoding of, 14
    neurons in, 13
Brain-state-in-a-box (BSB) model, 46,
        100, 222–23
Brownian motion, 116, 118–19, 127–31,
        146
    centered, 134
Bucy, R., *214*
Budelli, R. W., 62, *105*

C

Capacitance, 57
Capacity-dimensionality trade-off, 14,
        91–92, 141
Cardinality measure, 274–75
Carpenter, G. A., 91, *104*, 146, 166, *170*,
        234, 240, 241, *255*
Cartesian product, fuzzy, 328
Cauchy noise, 247
Caulfield, H. J., 102, *105*
Causal inference, adaptive, 158–59
Cell membrane, 57–58
Cells, FAM, 28–29, 328
Central limit theorems, 127
    fuzzy version of, 31
Centroid(s), 149–50
    AVQ centroid theorem, 228–31, 233
    decision-class, 228
    defuzzification method, 315–16, 343,
        344
    estimation, 146, 148–49
    fuzzy, 315–16, 344, 385
        computation of, 386–90
    pattern-class, 228, 231–32
    synaptic convergence to, 225–32
Chain rule, 129
Change, learning as, 22–24, 111–15
Chaotic equilibrium attractors, 15
Charge, 57

Cheeseman, P., 291, 292, *295*
Chen, S., 152, *173*
Chen, W. H., 364, 365, 373, *377*
Chua, L. O., 71, *105*, 230–31, *256*
Chung, K. L., 121, *170*, 183, 184, 193,
        213, 268, *295*
Churchland, P. M., 3, *34*
Classes, decision, 23, 183–84
Classical conditioning, 181–82
Class membership, stochastic pattern
        learning with known, 183–85
Class-membership information, 114
Class probability, 125, 183–84
Closed synaptic loops (feedback), 17
Code instability, 166
Coefficients, correction, 141
Cognitive maps, fuzzy. *See* Fuzzy
        cognitive maps (FCMs)
Cohen, M. A., 99, 100, 102, *104*, 146,
        166, 233, 234, 236, 237, 240, 241,
        253, *255*
Cohen-Grossberg activation dynamics,
        99–103
Cohen-Grossberg model, 233–34
    global stability of, 100, 238
Cohen-Grossberg theorem, 236, 238
Colored noise, 131
"Combinatorial explosion," 200
Combinatorial optimization networks, 233
Competition, 185
Competitive adaptive bidirectional
        associative memories (CABAMs),
        234, 235, 236, 240–41
    differential, 235
    global stability for, 238
Competitive adaptive vector quantization
        (AVQ) algorithms, 227, 328
Competitive assumption, 238
Competitive connection topology, 61
Competitive learning, 145–51, 222, 225.
        See also Differential competitive
        learning (DCL); Supervised
        learning; Unsupervised learning
    asymptotic centroid estimation, 148–49
    competitive covariance estimation,
        149–51
    as correlation detection, 147–48

Competitive learning *(cont.)*
    feedforward. *See* Adaptive vector
        quantization (AVQ)
    as indication, 146
Competitive learning law, 115, 116–17,
    238
    differential, 115, 117–18, 137
    linear, 115, 117, 146
    supervised, 135, 185, 230
    unsupervised, 225
Competitive neuronal signals, 43–44
Competitive signal functions, 226
Competitive synaptic vectors, convergence
    to pattern-class centroids, 231–32
Component axes, principal, 195
Compression of images. *See* Transform
    image coding
Computational geometry, 188
Concomitant variation as statistical
    covariance, 161–63
Conditional expectation, 150
Conditional probability density function,
    125
Conditional probability measure, 290
Conditioning, 179, 181–82
Conductance, 43, 56
Conjugate gradient descent, 204
Conjunctive (correlation) learning laws,
    115
Connectionism, 197
Connection matrices
    BAM, 79–94
        autoassociative OLAM filtering,
            83–85
        BAM correlation encoding example,
            85–91
        bipolar Hebbian or outer-product
            learning method, 79–81
        Hopfield model, 92–94, 100
        memory capacity, 91–92
        optimal linear associative memory
            (OLAM) matrices, 81–83
    synaptic, 59–60
Connection topologies, 60–61
Conservation law (normalization rule), 97
Content-addressable-memory (CAM), 68,
    69, 232

Continuity assumption, 141–42
Continuous ("fuzzy") logics, 6
Control systems, 339. *See also*
        Target-tracking control systems;
        Truck backer-upper control systems
Control theory, 69
Convergence, 223–24, 228–32. *See also*
        Global stability
    almost everywhere, 126
    almost surely, 126
    of BAM trajectories, 78–79
    in distribution, 127
    everywhere, 126
    in globally stable neural network, 77, 78
    mean-squared, 127
    in probability, 126–27
    with probability one, 126, 127
    stochastic, 126–27
    of stochastic gradient systems, 253
    synaptic, to centroids, 225–32
    uniform, 126, 199
Convergence interval, LMS, 196
Convex hull, 188
Convex subsets, 188
Cooper, G. R., 147, 168, 169, *170*, 213
Cooper, L. N., 213, *214*
Correction coefficients, 141
Correction inequality, 142
Correlation, 124
    statistical, 168–69
Correlation (conjunctive) learning laws,
    115
Correlation decoding, Hebbian, 140–45
Correlation detection, competition as,
    147–48
Correlation encoding, 76
    asymptotic, 138–40
    BAM, 85–91
Correlation-minimum encoding, 309, 384,
    385
    bidirectional FAM theorem for, 310–11,
        312, 315
Correlation-minimum inference, 312, 321,
    343, 344
Correlation-product encoding, 311–13,
    384, 385

bidirectional FAM theorem for, 311–13, 315
Correlation-product inference, 312
Cosine law, 148
Countability, 121
Counterpropagation, 147
Covariance, 124
  statistical, concomitant variation as, 161–63
Covariance estimation, competitive, 149–51
Covariance learning law, 161–62
Covariance matrices, 124, 150
Cox, R. T., 264, 292–93, 295
Cox's theorem, 292–93
Creativity, 20–21
Credit-assignment problem, 203
Crick, F., 3, 34
Cross-correlation vector, 191–92
Cross covariance matrix, 124
Crosstalk, 14
Cruz, J. B., 63, 105
Cube. *See also* Hypercube
  fuzzy midpoints of, 270–74
  fuzzy sets as points in, 7–9, 269–71
Cumulative distribution function, 122–23
Cumulative error, 205
Cumulative probability function, 122–27
Current, 56
Current law, Kirchhoff's, 57
Curve, sigmoidal, 39
Cybernetics, 12

D

Decision-class centroids, 228
Decision classes, 23, 183–84
Decision surface, 189
Decision tree, deep vs. shallow, 32–33
Decoding, Hebbian correlation, 140–45
Decomposition
  signal-noise, 140–41
  spectral, 195
Decompositional inference, 322–26
  adaptive, 326–27
Definition, recognition without, 13–14
Defuzzification, 30–31, 314–16

centroid, 315–16, 343, 344
  maximum-membership, 315
Delta modulation, 118
  adaptive, 168
  differential competitive learning (DCL) as, 168–70
DeMorgan dual co-norm, 278
DeMorgan's law, 148
Density, Gaussian, 123
Density functions, 49, 122–27
  Gaussian, 128
  probability, 23, 123
  conditional, 125
Derivative, mean-squared, 129
DeSieno, D., 227, 256
Deterministic encoding, 91
Deterministic-signal Hebbian learning law, 165
Differential competitive ABAM, 235
Differential competitive learning (DCL), 166–70, 228, 349–52, 403–5
  algorithm, 357–60
  as delta modulation, 168–70
  law, 115, 117–18, 137, 230
  linear, 226–27
  pulse-coded, 167–68
  software problems, 176–77
  for transform image coding, 368–69, 370–73
Differential-competitive synaptic conjecture, 167
Differential equations, stochastic, 114, 225–27
Differential Hebbian ABAMs, 235, 241–42
Differential Hebbian learning, 152–65
  adaptive causal inference, 158–59
  by biological synapses, 182
  concomitant variation as statistical covariance, 161–63
  fuzzy cognitive maps (FCMs), 152–58
  Klopf's drive reinforcement model, 159–60
  pulse-coded, 163–65, 167–68
  law, 115, 117, 137, 167–68
  pulse-coded, 163–65, 167–68
Diffusion, Brownian-motion, 127–31

Dimensionality-capacity trade-off, 14, 91–92, 141
Dimension-independent property, 76
Dirac delta function, 130, 131
Discourse, universes of, 300, 301, 323
Discrete cosine transform (DCT), 365
Distributed encoding, 14
Distribution, convergence in, 127
Distribution functions, 49, 122–23
Dominated membership function relationship, 279
Drive, neuronal, 160
Drive-reinforcement model, 159–60
Dual triangular co-norms, 277–78
Dubois, D., 300, *335*
Dworkin, R. M., 33, *34*
Dynamical range, 94
Dynamical system(s)
  asymptotically stable, 71–72
  fixed point of, 15
  function estimators, 13–14
  fuzzy systems and applications, 18–19
  neuronal, 44–47, 55
    autonomous, 45
    neuronal activations as short-term memory, 47
    signal state spaces as hypercubes, 46–47
    state spaces of, 45–46
    time in, 45
  synaptic, 22
  trainable, 13, 14–17

E

Eigenvalues, 195
Eigenvectors, 195–96
Einstein, A., 263
Elbaum, C., 213, *214*
Elbert, T. F., *104*
Elementhood, 8, 288
Encoding
  correlation, 76, 85–91, 138–40
  correlation-minimum, 309, 384, 385
  correlation-product, 311–13, 384, 385
  deterministic, 91
  distributed, 14

learning as, 111–15
outer-product, 86–91
Energy, Lyapunov, 72, 78, 87, 88
Entropy, fuzzy, 7, 263, 275–78
Entropy-subsethood theorem, 265, 293–94
Equilibrium attractors, 15
Equilibrium/equilibria, 223–24. *See also* Adaptive bidirectional associative memories (ABAMs); Stability
  bidirectional, 66–67
  fixed-point, 67–68
  global, 223–24
  "invisible hand," 89–91
  stochastic, 133–37, 228–32
Equilibrium synaptic vector, 229
Equinorm property of synaptic vectors, 147–48
Ergodic process, 134
Error
  cumulative, 205
  defined, 180
  hidden-neuron, 202–3
  instantaneous, 191
  local error minimum, 198
  mean-squared (MSE), 181, 191, 203, 232
  root-mean-squared (RMSE), 400, 403, 404, 406
  squared, 204–5, 207
  summed absolute, 211
Error conditional covariance matrix, 150
Error information, 114
Error suppressor function, 211–12
Esperanto, 24
Estimated gradient descent, 188, 193–94
Estimation. *See also* Approximation
  centroid, 146, 148–49
  competitive covariance, 149–51
  least-square, 288–89
  maximum-entropy, 292
  polynomial, 199
  supervised function, 180–81
Estimation theory, fundamental theorem of, 150
Estimators
  function, 13–14, 302–8
  model-free, 13, 25–26

structured numerical, 26–28
unbiased, 165
Euclidean norm, matrix, 82
Events, 120
Evolution, punctuated, 21
Excitatory ions, 43
Excitatory synapses, 60
Excluded middle, law of, 9, 265, 269, 271, 272
Exclusive-OR (XOR) software problem, 218–19
Expectation, mathematical, 124
Expert systems, 25–26, 200. *See also* Artificial intelligence (AI)
    chaining through rules in, 32–34
    knowledge as rule trees in, 24–25
    reasoning techniques used in, 299
Exponential-distribution signal functions, 49
Exponential fading memory, 81
Exponents, Lyapunov, 17
Extended Kalman filters, 213
Extension principle, 31, 300

F

Factual truth, 269
Fading-memory (recency) effect, 80–81, 138
Fahlman, S. E., 255, *256*
FAM. *See* Fuzzy associative memories
Fan-out (outstar) neuron, 182
Feedback, 17, 59–61
Feedback loops in FCMs, 152
Feedback neural networks, 7, 13, 221
    fixed-point stability in, 15, 16
    global stability of, 62, 232–42
Feedforward neural networks, 221
    multilayer, 201–3
    noise-saturation dilemma in, 94
    supervised, 17, 223
    unsupervised, 91–92
Feedforward sigmoidal representation theorems, 199–201
Field(s)
    augmented, 46
    Borel, 120, 121–22

of neurons, 44
random, 119
receptive, 41
Filter(s)
    adaptive, 44
    Kalman, 151, 190, 379
        extended, 213
        for target-tracking, 392–96
    linear, 83
    novelty, 85
    OLAM, 83–85
Finite measure theory, 268
Firing frequency of action potentials, 43
Fit (fuzzy unit) value, 8, 47, 270, 383
Fit vector, 270
Fit-violation strategy, 280–82
Fixed learning rates, 194–95
Fixed-point equilibrium, 67–68
Fixed point of dynamical system, 15
Fleming, W. H., 185, *214*
FLIPS (fuzzy logical inferences per second), 18
Foger, T. A., 31, *35*, 269, 275, 277, *295*, 307, 310, *335*
Folk psychology, 3
Forgetting law, 138
Forgetting term, 145
Forward-chaining inference, 25
Forward projections, 60
Franklin, J. N., 195, *214*
Freeman, W. J., 15, *35*
Frequency, relative, 11, 288
Fu, K. S., 198, *214*
Fukushima, K., 212, *214*
Function(s), 19. *See also* Lyapunov functions; Signal function(s)
    amplification, 100
    autocorrelation, 128
    continuous, 20
    density, 23, 49, 122–27
    Dirac delta, 130, 131
    distribution, 49
    error suppressor, 211–12
    estimation, supervised, 180–81
    fuzzy sets as, 300
    indicator, 6, 125, 184
    linear discriminant, 188

Function(s) *(cont.)*
neurons as, 39–40
radial basis, 40–41
reinforcement, 135, 185
sample, 120
unit step, 129
Function estimators, 13–14, 302–8
Fuzziness, 263–98
Bayesian polemics against, 264, 289–93
as description of event ambiguity, 265–68
entropy-subsethood theorem, 265, 293–94
fuzzy entropy theorem, 263, 275–78
proof, 276–78
mathematical, 3
multivalued, 3–11
bivalent paradoxes as fuzzy midpoints, 4–5
defined, 4
in the twentieth century, 5–6
probability and, 9–11, 264–65
randomness vs., 9, 264–65
sets-as-points geometric view of, 7–9, 263–65, 269–75, 300, 306
counting with fuzzy sets, 274–75
fundamental proposition of, 271–72
paradox at the midpoint, 272–74
subsethood, 9–11, 279–81, 287, 288
subsethood theorem, 263, 278–93
algebraic derivation of, 280–82
corollaries to, 263, 291–92
degrees of subsethood, 279–80
fuzzy Bayes theorem and, 290–93
geometric proof of, 263, 282–87
universe as fuzzy set, 268–69
within-set, 276–77
Fuzzy association, 302–4
Fuzzy associative memories (FAMs), 299–338. *See also* Adaptive fuzzy associative memories (AFAMs); Binary input-output fuzzy associative memories (BIOFAMs); Truck backer-upper control systems
bank, 301
cells, 28–29, 328

correlation-minimum inference procedure, 321
decompositional inference, 322–26
adaptive, 326–27
defined, 300
fuzzy and neural function estimators, 302–8
fuzzy Hebb, 308–27
autoassociative matrix, 309
correlation-minimum bidirectional FAM theorem, 310–11, 312, 315
correlation-product bidirectional FAM theorem, 311–13, 315
recalled outputs and "defuzzification," 314–16
as mappings, 306–7
rules (associations), 26–28, 300–301, 341, 380
for inverted pendulum, 318–20
processing of, 29–30
product-space clustering to generate, 28–29
storage of, 29–30, 314
superimposition of, 313–14
virtual-representation scheme for, 31–32
software, 418–24
surface, 320
system architecture, 316–17
Fuzzy Cartesian product, 328
Fuzzy centroid, 315–16, 344, 385
computation of, 386–90
defuzzification, 315–16, 343, 344
Fuzzy cognitive maps (FCMs), 152–58, 306
causal feedback loops in, 152
combining, 155–58
reasoning with, 153–55
of South African politics, 152–55
strengths and weaknesses of, 155
as temporal associative memories (TAMs), 153–54
Fuzzy (continuous) logics, 6
Fuzzy controllers, 379–81. *See also* Target-tracking control systems; Truck backer-upper control systems
Fuzzy entropy, 7, 263, 275–78

Fuzzy function estimators, 302–8
Fuzzy Hamming distance, 274, 275
Fuzzy Hebb procedure, 305
Fuzzy inference, 29
Fuzzy logical inferences per second
    (FLIPS), 18
Fuzzy machine intelligence, 2–3
Fuzzy membership functions, 341
Fuzzy power set, 7, 8, 47, 279, 380
Fuzzy representation of structured
    knowledge, 304–6
Fuzzy sets, 47, 264
    counting with, 274–75
    defined, 380
    as functions, 300
    height and normality of, 310
    numerical and multidimensional
        characteristics of, 299–300
    overlap and underlap within, 269, 271,
        276–77
    as points in cubes, 7–9, 269–71
    sets-as-functions definition of, 270
    sets-as-points geometric view of, 7–9,
        263–65, 269–75, 300, 306
Fuzzy-set samples, 302–3
Fuzzy square, completing the, 271–72
Fuzzy systems, 200–201
    adaptive, 18–19, 29–30
    architecture, 30
    as between-cube mappings, 299–301
    defined, 380
    as function estimators, 13–14
    as parallel associators, 29–32
    as principle-based systems, 32–34
    as structured numerical estimators,
        26–28
Fuzzy theory, 264
    "extension principle" of, 31, 300
Fuzzy unit (fit) value, 8, 47, 270, 383
Fuzzy variables, 301
Fuzzy vector-matrix multiplication, 307–8

G

Gaines, B. R., 5, 269, 273, 295
Game theory, minimax theorem of, 199
Gardner, L. A., 186, 189, 213

Gaussian density, 123
Gaussian density function, 128
Gaussian signal functions, 40–41
Gaussian white noise, 118–19, 127–31,
    132
Geman, 253, 256
Generalization, 20–21
Generalized delta rule, 197. See also
    Backpropagation algorithm
Geometric supervised learning algorithm,
    213
Giles, C. L., 239, 256
Gilmore, 242, 256
Global equilibria, 223–24
Global stability, 68, 69, 224. See also
    Adaptive bidirectional associative
    memories (ABAMs); Convergence
    in AVQ models, 233
    of Cohen-Grossberg systems, 100, 238
    for competitive ABAM system, 238
    for competitive learning law, 238
    of feedback neural networks, 62, 232–42
    of gradient systems, 253
    in neural networks, convergence in, 77,
        78
    for nonadaptive BAM, 238
    for signal Hebbian ABAMs, 237–38
    stability-convergence dilemma and,
        235–36
    stochastic, 224
    structural stability differentiated from,
        242
Gluck, M. A., 51–52, 52, 53, 163, 167,
    170
Gödel, 269
Gotoh, K., 152, 170
Gould, S. J., 21, 34
Gradient descent, 203
    conjugate, 204
    estimated, 188
Gradient-descent algorithm, 189
    estimated, 193–94
    ideal, 193
Gradient-descent optimization, stochastic,
    221. See also Supervised learning;
    Unsupervised learning

Gradient systems
  global stability of, 253
  stochastic
    convergence of, 253
    neural networks as, 221–23
"Grandmother" cells, 95
Greville algorithm, 81–82
Grossberg, S., 2, *34*, 68, 91, 94–99, 100,
        101, 102, *104*, 116, 143, 146, 166,
        170, *171*, 182, 187, *214*, 233, 234,
        236, 237, 238, 240, 241, 247, 251,
        253, *255*, *256*
Grossberg model, additive, 96
Guest, C. C., 68, 102, *104*

H

Haines, K., 92, *104*
Hambly, A. R., 118, *171*
Hamming distance, 87–88, 141
  fuzzy, 274, 275
Hart, H. L. A., 33, *34*
Hartley, R., 253, *257*
Hayek, F. A., 24, 33, *35*
HDTV, 364
Hebb, D. O., 111, 115, *171*, 182, *214*, 308,
        335
Hebb fuzzy associative memories (FAMs),
        308–27
  autoassociative matrix, 309
  correlation-minimum bidirectional FAM
        theorem, 310–11, 312, 315
  correlation-product bidirectional FAM
        theorem, 311–13, 315
  recalled outputs and "defuzzification,"
        314–16
  superimposition in, 313–14
Hebbian ABAMs, 234, 236
  differential, 235, 241–42
  signal, 237–40
Hebbian BAMs, 141
Hebbian correlation decoding, 140–45
Hebbian learning
  bipolar (outer-product), 79–81
  differential, 163–65, 167–68, 182
Hebbian learning law
  deterministic-signal, 165

differential, 115, 117
  pulse-coded, 167–68
  random, 137
  random classical, 137
  random-signal, 131
  signal, 118, 134
Hebbian synapses, 134, 138
Hebb matrix, fuzzy, 309
Hecht-Nielsen, R., 92, *104*, 147, *171*
Height of fuzzy sets, 310
Heisenberg uncertainty principle, 269
Heteroassociative network, 46, 60
Hidden-neuron error, 202–3
High-definition television (HDTV), 364
Higher-order signal Hebbian ABAMs,
        239–40
Hinton, G. E., 255, *256*
Hirsch, M. W., 72, *104*, 231, 239, 242, *256*
Histogram, adaptive, 28–29
Hodgkin, A. L., 100, *104*, *256*
Hodgkin-Huxley membrane equation, 100,
        101
Hoff, M. E., Jr., 190–91, *215*
Hopfield, J. J., 7, *35*, 47, 48, *53*, 62, 64,
        69, 72, 92, 93, *104–5*, 234, *256*,
        257
Hopfield circuit, 62, 100, 234
Hopfield model, 92–94
Hornik, K., 199, 201, *214*
Huber, P. J., 211, 212, *214*
Hume, D., 1, 20, 111, 138, 268, *295*
Huxley, A. F., 100, *104*, *256*
Hwang, 253
Hyperbolic-tangent signal functions, 48–49
Hypercube, 46–47, 270, 380
HyperLogic Corporation, 407, 408
Hyperplane, 188
Hyperrectangles, 282–84, 285

I

Ideal gradient-descent algorithm, 193
Idempotency, 83–84
Identity, law of, 265
Identity (unit) operator, 120
Image transform coding. *See* Transform
        image coding

Implication, logical, 281
Independent random vectors, 124
Indexed random variables, 119
Indication, competition as, 146
Indicator function, 6, 125, 184
Inequality, correction, 142
Inference
    adaptive causal, 158–59
    backward-chaining, 25
    correlation-minimum, 312, 321, 343, 344
    correlation-product, 312
    decompositional, 322–27
    forward-chaining, 25
    fuzzy, 29
    parallel associative, 18
Information
    class-membership, 114
    error, 114
    local, 114–15
    pattern-class, 17, 113
    side, 365–66
Inhibition, shunting, 101
Inhibition topology, lateral, 228
Inhibitory ions, 43
Inhibitory synapses, 60
Inner-product assumptions, 222
Input activation, 39
Input associant, 302
Input autocorrelation matrix, 191–92
Input-output functions, 13
Input values, on-center off-surround flow of, 96
Instantaneous error, 191
Instantaneous mean-squared error, 191
Instantaneous squared error, 207
Instantaneous squared-error vector, 204–5
Intelligence
    machine, 2–3, 12–19. *See also* Artificial intelligence (AI)
    neural, 2–3
Intelligent behavior, 19–34
    defined, 19
    expert-system knowledge trees, 24–25
    generalization and creativity, 20–21
    learning or adapting, 22–24
    rule vs. principles in, 24

symbolic vs. numeric processing, 25–26
Interlingua, 24
Inverted pendulum. *See* Pendulum, inverted
"Invisible hand" equilibria, 89–91
Ionic flow, 43
Ionic pair, antagonistic, 43
Ions, neuronal, 56–57
Ito, T., 212, *214*

J

Jain, A. K., 363, *377*
Jaynes, E. T., 48, *53*, 264, 291, 292–93, *295*
Joint neuronal-synaptic dynamical system. *See* Random adaptive bidirectional associative memories (RABAMs)
Jones, R. S., *103*, 222, *255*
Junction, synaptic, 42–43

K

Kac, M., 268, 287, *295*
Kailath, T., 151, *171*
Kak, A. C., 365, *378*
Kalman, R. E., 151, *171*, 190, *214*
Kalman filter, 151, 190, 379
    "extended," 213
    for target-tracking control, 392–96
        control surfaces, 395–96, 397
        performance, 398, 399, 401, 402
        sensitivity and robustness of, 399, 403
Kanizsa, G., 2, *35*
Kanizsa square, 2
Kant, I., 2, *35*, 269, *295*
Karhunen-Loeve transform (KLT), 365
Kelsen, H., 33, *35*
Kinser, J. M., 102, *105*
Kirchhoff's current law, 57
Kirchhoff's voltage law, 57
Kishi, F. H., 81, *105*
Kline, M., 3, *35*, 269, *295*
Klir, G. J., 31, *35*, 269, 275, 277, *295*, 307, 310, *335*
Klopf, A. H., 159–60, 167, *171*, 182, *214*

Knowledge
  as rule trees in expert systems, 24–25
  structured, neural vs. fuzzy
    representation of, 304–6
Knowledge bases, 24–25
Knowledge trees, 25
Kohonen, T., 44, 46, *53*, 60, 81, 85, *105*,
    136, 147, *171*, 185, *214*, *256*
Kohout, L., 278, *294*
Kong, S.-G., 167, *171*, 229, 232, *256*
Kosko, B., 7, 12, 17, 31, *35*, 40, 47, *53*,
    62, 103, *105*, 117, 152, 168, *171*,
    185, *214*, 239, *256*, 264, 274, *295*,
    300, *335*, *336*, 341, *361*, 383, *406*

L

Language, 25
Large numbers, laws of, 127, 155–56
Lateral inhibition topology, 61, 228
Learning, 2. *See also* Adaptation;
        Competitive learning; Supervised
        learning; Unsupervised learning
    algorithms, 113–14. *See also specific
        algorithms*
    anti-Hebbian, 255
    bipolar Hebbian (outer-product) method
        of, 79–81
    Boltzmann machine, 255
    as change, 22–24, 111–15
    as encoding, and quantization, 111–15
    laws of, 115–19
    in neural networks, 112
    pattern, 15
    reinforcement, 181
Learning-automata algorithms, 181
        Learning Machines (Nilsson), 187
        Learning-rate coefficients, 186
Learning rates, 187, 194–95
Learning theorems, 182, 187
Least-mean-square (LMS)
    algorithm, 181, 190–96, 210, 212
    assumption, 193–94
    convergence interval, 196
Least-square estimation, 288–89
Lebesgue measure, 20
Liar paradox, 4
Limb, J. O., 363, *377*

Limit cycle (limit torus), 15, 143
Lindley, D. V., 264, 291–92, 294, *295*
Linear associative memory (LAM), 82–83
    optimal (OLAM), 81–85, 191, 192
        autoassociative filtering, 83–85
Linear competitive learning law, 115, 117,
    146
Linear differential competitive learning,
    226–27
Linear differential competitive learning
    law, 118
Linear discriminant function, 188
Linear filter, 83
Linearly separable pattern sets, 188
Linear signal functions, 41, 49
Linear stochastic approximation, 190–96
Lippman, A. B., 364, *378*
Lisp (computer language), 24
LMS. *See* Least-mean-square (LMS)
Local conditional covariance matrices, 150
Local error minimum, 198
Local information, 114–15
Logic(s)
    bivalent, 325
    fuzzy (continuous), 6
Logical implication, 281
Logical truth, 268–69, 278–79
Logistic signal functions, 39, 48
Long-term memory, 47
Lukasiewicz, J., 6, 269, 277
Lukasiewicz operator, 309
Lyapunov "energy," 72, 78, 87, 88
Lyapunov exponents, 17
Lyapunov functions, 15, 16, 69–73, 243
    ABAM, 245–46
    bounded decreasing, 73
    defined, 69
    energy of physical system measured by,
        72–73
    quadratic form, 69–71
    signal-energy, 73–74
    stability and, 69, 71–72
Lyapunov minima, 76

M

McClelland, J. L., 196, 201, 203–4, *215*
McCulloch, W. S., 63, *105*

McCulloch-Pitts neurons, 44
McEliece, R. J., 91, *105*
McGillem, C. D., 147, 168, 169, *170*, 213
Machine intelligence, 2–3
  dynamical-systems approach to, 12–19
    function estimators, 13–14
    fuzzy systems and applications, 18–19
    trainable dynamical systems, 13, 14–17
MacQueen, J., 136, 149, *171*
Mahowald, M. A., 98, *105*
Mamdani, E. H., 31, *35*, 309, 317, 323, *336*
Martingale, 193
Martingale assumption, 162–63, 193
Mathai, G., 63, *105*
Mathematical expectation, 124
Mathematical fuzziness, 3
Math-model controllers, 379–81. *See also* Kalman filter
Matrices
  autoassociative, 309
  autocorrelation, 191–92
  covariance, 124, 150
  cross covariance, 124
  splitting, 323
Matrix Euclidean norm, 82
Maximum combination technique, 31
Maximum-entropy estimation, 292
Maximum-entropy signal functions, 48
Maximum-membership defuzzification, 315
Max-min composition, 307–8
Max-min composition recall, 312
Max-product composition operator, 308
Max-product composition recall, 312
Maxwell, T., 239, *256*
Maybeck, P. S., 128, 129, 134, *172*, 206, *214*, 244, 251
Mead, C., 2, 14, *35*, 98, *105*
Mean-squared convergence, 127
Mean-squared derivative, 129
Mean-squared error (MSE), 181, 203, 232
  instantaneous, 191
Mean-squared velocities, 247
Measurable space, 122
Measured voltage, 58
Measure space, 122

Measure theory, 119–27
  measurability and sigma-algebras, 119–22
  probability measures and density functions, 122–27
Membership functions, 6, 341
Membrane, cell, 57–58
Memory
  associative. *See* Adaptive bidirectional associative memories (ABAMs); Bidirectional associative memories (BAMs); Fuzzy associative memories (FAMs); Linear associative memory (LAM); Optimal linear associative memory (OLAM); Temporal associative memories (TAMs)
  capacity, 91–92, 112
  content-addressable (CAM), 68, 69, 232
  fading-memory (recency) effect, 80–81
  long-term (LTM), 47
  short-term (STM), 47
Mendel, J. M., 150, *172*, 190, 198, 213, *214*, 392, *406*
Meyer, B. D., 152, *172*
Midpoints, fuzzy, 4–5, 9, 270–74
Mill, J. S., 158, *172*, 269, *295*
Mingolla, E., 98, *104*
Minimax theorem of game theory, 199
Minsky, M. L., 187–88, 197, 201, 203, *214*
Miyake, S., 212, *214*
Model-free estimators, 13, 25–26
Molecular channels, 43
Monotonicity, signal, 40–41
Monro, S., 186, 210, *215*
Monta, P., 364, *378*
Montagu, A., 1
Moore, J. B., 71, *103*
Motion, Brownian, 116, 118–19, 127–31, 146
  centered, 134
Mulligan, J. H., 63, *105*
Mulloney, B., 62, *105*
Multilayer feedforward network architectures, 201–3
Multiplication, fuzzy vector-matrix, 307–8
Multiplicative activation dynamics, 100–103

Multiplicative activation model, 96–97, 99–103
Multiplicative networks, noise-saturation dilemma in, 94
Multivalence, 3–11
Multivaluedness, 269
Murakami, J., 152, *170*
Muth, F., 162, *172*

N

Natural selection, 3, 76–77, 95, 233
Negative reinforcement, 181
Neocognitron network, 212–13
Nestor, 213
NETalk simulation, 197
Netravali, A. N., 363, 364, *377*, *378*
Neural control systems, 339. *See also* Target-tracking control systems; Truck backer-upper control systems
Neural function estimators, 302–8
Neural intelligence, 2–3
Neural networks, 7. *See also* Adaptive vector quantization (AVQ); Random adaptive bidirectional associative memories (RABAMs)
  association reconstruction in, 314
  autoassociative, 46, 60
  bidirectional, 60–61
  black-box characterization of behavior of, 305
  capacity of, 91
  classification of, 222
  as content-addressable memory, 68, 69, 232
  feedback, 7, 13, 221
    fixed-point stability in, 15, 16
    global stability of, 232–42
  feedforward, 221
    multilayer, 201–3
    noise-saturation dilemma in, 94
    supervised, 17, 223
    unsupervised, 91–92
  as function estimators, 13–14
  globally stable, convergence in, 77, 78
  heteroassociative, 46, 60
  learning in, 112

  noise-saturation dilemma in, 94
  as random field, 64
  as stochastic gradient systems, 221–23
  supervised and unsupervised learning in, 113–15
  taxonomy of models of, 17
  as trainable dynamical systems, 13, 14–17
  unidirectional, 60–61
  as vector stochastic process, 64
Neural representation of structured knowledge, 304–6
Neuron(s), 13
  anatomy of, 42–43
  biological, 14
  competitive signals, 43–44
  fan-out (outstar), 182
  fields, 44
  fluctuation rate, 235
  as functions, 39–40
  McCulloch-Pitts, 44
Neuronal activations as short-term memory, 47
Neuronal drive, 160
Neuronal dynamical system, 44–47, 55
  autonomous, 45
  neuronal activations as short-term memory, 47
  signal state spaces as hypercubes, 46–47
  state spaces of, 45–46
  time in, 45
Neuronal dynamics, 55
  additive, 56–59
    additive external input, 58–59
    membrane resting potentials, 57–58
    membrane time constants, 57
    passive membrane decay, 56–57
Neuronal feedback, additive, 59–61
  bidirectional and unidirectional connection topologies, 60–61
  synaptic connection matrices, 59–61
Neuronal reinforcer, 160
Newton's approximation method, 193
Nguyen, D., 339, 345, *361*
Nikias, C. L., 151, *172*
Nilsson, N., 1, 187, *214*
Noise, 229–30, 255, 268

in additive activation models, 245
Cauchy, 247
colored, 131
Grossberg's interpretation of, 99
OLAM filter effects on, 85
threshold nonlinearity suppression of,
    142–43
white, 118–19, 127–31, 132
"Noise notation," 132
Noise RABAM model, 244
Noise-saturation dilemma, 94–99
    network type and, 94
    saturation theorem, 95–99
Noise suppression, 99
Noise-suppression theorem, RABAM, 95,
    247–53
Noncontradiction, law of, 265, 269, 271,
    272
Normal fuzzy sets, 310
Normalization, 80, 281, 284–85
Normalization rule (conservation law), 97
Noumena, 2
Novel behavior, 20
Novelty filter, 85
Novelty vector, 84, 85
Numbers, symbols vs., 24
Numerical estimators, structured, 26–28
Numeric vs. symbolic processing, 25–26

O

Occurrence probability, 183
Odds-form fuzzy Bayes theorem, 291–93
Ohm's law, 56
Olmsted & Watkins, 407, 408
On-center off-surround flow of input
    values, 96
Operant conditioning, 179, 181–82
Operating range of activation, 94
Operator(s)
    Lukasiewicz, 309
    max-product composition, 308
    projection, 83–84
    signum, 168
    unit (identity), 120
Optical systems, 148

Optimal linear associative memory
    (OLAM), 81–85, 191, 192
    autoassociative filtering, 83–85
Organization of Behavior (Hebb), 115
Orthogonality, 84, 143, 283–84, 286, 288
Outer-product (bipolar Hebbian) learning
    method, 79–81
Outer-product encoding, 86–91
Outer-product laws, 79–80
Outliers, statistical, 211
Output associant, 302
Outputs, summing, 29
Output signal, 39
Outstar (fan-out) neuron, 182
Outstar learning theorems, 182
Overlap, 7, 269, 271, 276–77, 288, 292,
    293–94
OWL demonstration software, 408–18
    adaptive-resonance-theory (ART)
        program, 409–11
    bidirectional-associative-memory
        (BAM) program, 411–13
    BKP program, 413–14
    CL program, 414–16
    RABAM program, 416–18

P

Papert, S., 187–88, 197, 201, 203, *214*
Paradoxes, bivalent, 4–5, 273, 274
Parallel associative inference, 18
Parallel associators, 29–32
Parallel Distributed Processing (PDP)
    (Rumelhart et al.), 196–97
Parameter change, learning and adaptation
    as, 22
Parity function, 201–2
Parker, D. B., 51–52, *52, 53*, 163, 167,
    170, 198, *214*
Parker, T. S., 71, *105*, 230–31, *256*
Passive-decay rate, 56
Passive membrane decay, 56–57
Path enumeration, 25
Path-weighted signals, 59
Pattern-class centroids, 228, 231–32
Pattern-class information, 17, 113
Pattern learning, 15

Pattern recognition (recall), 15, 87, 122, 233
  nearest-neighbor, 116–17
  stochastic, 183–85
  supervised vs. unsupervised, 125, 180, 184–85
Pattern-recognition theory, 149
Patterns, 183
Pattern sets, 188
Pavlov, I. P., 181–82, *214*
Peano, G., 24
Pearl, J., 25, *35*, 152, *172*
Pendulum, inverted, 26–27, 317–22, 333–35
  FAM rules for, 318–20
  input-output product space of, 28
  software instructions, 421–24
  software problems, 337
Perceptron, 181, 187–90, 212
Perceptrons (Minsky & Papert), 187–88
Perkel, D. H., 62, *105*
Permeability, 56
Pineda, F. J., 210, *214*
Pitts, W., 63, *105*
Polynomial estimation, 199
Polyspectra, 151
Popat, A., 364, *378*
Positive reinforcement, 181
Positivity assumptions, 233
Posner, E. C., 91, *105*
Potential difference, 58
Power set, 120
  fuzzy, 7, 8, 47, 279, 380
Power spectral density, 130
Prade, H., 278, *295*, 300, *335*
Pratt, W. K., 112, 125, 150, *172*
Pre-attentive processing, 2–3
Predictive image coding, 363
Principal component axes, 195
Principles, rules vs., 24, 33–34
Probability, 268
  class, 125, 183–84
  fuzziness and, 9–11, 264–65
  occurrence, 183
  reduced to subsethood, 287
  of success, 11
  uniform sampling, 112–13

Probability density functions, 23, 123, 125
Probability function, cumulative, 122–27
Probability measures, 120, 122–27
Probability spaces, 119–31
Processing
  attentive, 3
  pre-attentive, 2–3
  symbolic vs. numeric, 25–26
Product space, 119–20
Product-space clustering, 28–29, 327–35, 364, 368–69, 402
Projection operators, 83–84
Projections, backward and forward, 60
Prolog (computer language), 24
Propositional rules, 24, 200
Prototypes, learned, 112
Pseudo-inverse matrix, 81–82
Psychology, folk, 3
Pulse-coded differential competitive learning (DCL), 167–68
Pulse-coded differential Hebbian learning, 163–65, 167–68
Pulse-coded signal functions, 50–52, 117–18, 163–64
Punctuated equilibrium, 21
Punishment, 181
Pythagorean theorem, 84, 85, 284

Q

Quantization, 111–15, 273–74. *See also* Adaptive vector quantization (AVQ)
Quine, W. V. O., 7, *35*, 266, 269, 273, *295*

R

Radial basis functions, 40–41
Raghuveer, M. R., 151, *172*
Random adaptive bidirectional associative memories (RABAMs), 223, 226, 243–46
  additive, 252–53
  annealing, 253–55
  annealing theorem, 254–55

asymptotic unbiasedness of equilibria, 251
diffusion, 244
noise, 244
noise-suppression theorem, 95, 247–53
    noise-saturation dilemma and, 247–48
    proof of, 249–51
    unbiasedness corollary of, 251, 252
second-order behavior, 247–48
software problems, 258–59
squared velocities, 251
structural stability, 224, 243–46
Random classical differential Hebbian law, 137
Random differential competitive learning law, 137
Random differential Hebbian learning law, 137
Random fields, 119
Random linear competitive learning law, 146
Randomness, 119–31
    ambiguity vs., 265–68
    defined, 119
    fuzziness vs., 9, 264–65
    physical interpretation of, 287
    properties of, 119
    uncertainty as, 264, 266–67
    wide-sense stationary (WSS), 128–29, 191
Random sequence, 119
Random-signal Hebbian learning law, 131
Random supervised competitive learning law, 135
Random variables. *See* Variables
Random vector, 119, 124, 150
Ratio-polynomial signal functions, 49–50
Reasoning with sets, 302
Recall. *See also* Pattern recognition (recall)
    accuracy, 92–93
    asynchronous, 89, 90
    max-min composition, 312
    max-product composition, 312
Recency (fading-memory) effect, 80–81, 138

Receptive field, 41
Recognition-without-definition property, 13–14
Recurrent backpropagation, 210
"Reflectance pattern," 95
Reifsnider, E. S., 51–52, *52*, *53*, 163, 167, *170*
Reilly, D. L., 213, *214*
Reinforcement function, 135, 166, 185
Reinforcement learning, 181
Reinforcer, neuronal, 160
Relative frequency, 11, 288
Representation, virtual, 28, 31–32, 305
Representation theorems, feedforward sigmoidal, 199–201
    fuzzy, 201–2
Rescher, N., 3, 6, *35*, 269, 271, 277, *295*
Resistance, 56
Resonance, 68, 240
Resting potential, cell membrane, 57–58
Resting voltage, 58
Retina, silicon, 98
Reward, 181
Rishel, R. W., 185, *214*
Ritz, S. R., *103*, 222, *255*
Robbins, H., 186, 210, *215*
Robust statistics, 211
Rodemich, E. R., 91, *105*
Rongshu, G., 364, *378*
Rosenblatt, F., 181, 187, *215*
Rosenfeld, A., 365, *378*
Rosser, J. B., 3, *35*
Rounding off, 273–74
Rudin, W., 20, *35*, 84, *105*, 120, 121, 130, *172*, 188, 199, 200, 201, *215*, 232, 250, 252, 253, *256*
Rulebases, 24–25
Rules
    principles vs., 24, 33–34
    propositional, 24
Rule trees, 24–25
Rumelhart, D. E., 136, *172*, 196, 201, 203–4, *215*, 255
Russell, B., 273

S

Sallic, H., 364, *378*
Sample function, 120
Sample space, 120
Samuelson, P. A., 162, *172*, 193, *215*
Saturation theorem, 95–99
Schedules, annealing, 254
Schreiber, W. F., 364, *378*
Scofield, C., 213, *214*
Sebestyen, G. S., 137, *172*
Sejnowski, T. J., 161, *172*, 197, 255, *256*
Selection, natural, 21
Semilinearity, 41
Sensation, Weber law of, 97
Sense modalities, neuronal activations
    identified with, 47
Separation theorem, 188
Sequence, random, 119
Sets. *See also* Fuzzy sets
    Borel, 120
    reasoning with, 302
Sets-as-functions definition of fuzzy sets,
    270
Sets-as-points geometric view of fuzziness,
    7–9, 263–65, 269–75, 300, 306
    counting with fuzzy sets, 274–75
    fundamental proposition of, 271–72
    paradox at the midpoint, 272–74
Shamir, J., 102, *105*
Shen, P., 364, *378*
Shepherd, G. M., 42, *53*
Short-term memory (STM), 47
Shunting activation dynamics, 100–103
Shunting activation models, 97, 98,
    100–103
Shunting inhibition, 101
Shunting networks, noise-saturation
    dilemma in, 94
Side information, 365–66
Sigma-algebra, 119–22, 183, 193
Sigma-count, 274–75
Sigmoidal curve, 39
Signal(s)
    biological, 41–44
    competitive neuronal, 43–44
    Grossberg's interpretation of, 99
    output, 39
    path-weighted, 59
    signed, 43
    suprathreshold, 49
Signal energy, 73
Signal function(s), 48–52
    binary, 43–44
    bipolar, 43–44
    bivalent, 64
    competitive, 226
    exponential-distribution, 49
    Gaussian, 40–41
    hyperbolic-tangent, 48–49
    linear, 41, 49
    logistic, 39, 48
    maximum-entropy, 48
    pulse-coded, 50–52, 117–18, 163–64
    ratio-polynomial, 49–50
    staircase, 40
    threshold, 39–40, 48, 49, 142–43
Signal Hebbian ABAMs, 237–40
Signal Hebbian learning, 138–45
    asymptotic correlation encoding, 138–40
    Hebbian correlation decoding, 140–45
    recency effects and forgetting, 138
Signal Hebbian learning law, 115, 116,
    118, 131, 134, 165
Signal monotonicity, 40–41
Signal-noise decomposition, 140–41
Signal pulse (action), 42
Signal state spaces, 46–47
Signal state vectors, bipolar, 88–89
Signal strength, 196
Signal-to-noise ratio (SNR), 366
Signal vectors, binary, 86–88
Signal velocities, 41
Signed signals, 43
Signum operator, 168
Silicon retina, 98
Silverstein, J. W., *103*, 222, *255*
Simple asynchrony, 64, 93–94
Simulated annealing, 253, 255
Sine function, 219–20
Skinner, B. F., 20, *35*, 179, 181, *215*
Skorokhod, A. V., 118, *172*, 244
Smale, S., 72, *104*, 231, 239, *256*

Smith, C. H., 364, 365, 373, *377*
Software instructions, 407–24
  fuzzy-associative-memory software,
      418–24
    adaptive fuzzy control of inverted
        pendulum, 421–24
    fuzzy target-tracking demonstration,
        419–21
    fuzzy truck backer-upper control
        system, 418–19
  for OWL demonstration software,
      408–18
    adaptive-resonance-theory (ART)
        program, 409–11
    bidirectional-associative-memory
        (BAM) program, 411–13
    BKP program, 413–14
    CL program, 414–16
    RABAM program, 416–18
Sorenson, H. W., 83, *105*, 165, *172*, 213,
    215, 288, *295*
Sorites paradoxes, 5
South African politics, fuzzy cognitive
    map of, 152–55
Sparse coding theorem, 91
Spectral decomposition, 195
Spectral density, power, 130
Splitting matrices, 323
Spurious attractors, 88
Square, fuzzy, 271–72
Squared-error vector, instantaneous, 204–5
Stability, 71, 224. *See also*
        Equilibrium/equilibria; Global
        stability
  asymptotic, 71–72, 238–39
  bidirectional, 68–69, 74–76
  global. *See* Global stability
  Lyapunov functions and, 69, 71–72
  of RABAMs, 224, 243–46
  simple asynchrony and, 94
  structural, 242–43
Stability-convergence dilemma, 224,
    235–36
Stability-plasticity dilemma, 187
Staircase signal function, 40
Stark, H., 129, 162, *172*

State-change patterns, asynchronous,
    64–66
State spaces, 7, 45–47
Statistical correlation, 168–69
Statistics, robust, 211
Steady state, 16, 58
Stearns, S. D., 191, 194, 195, 196, *215*
Stimuli, 19
Stimulus-response pairs, 20–21
Stinchcombe, M., 199, 201, *214*
Stochastic approximation, 114, 161–62,
    181, 193, 198
  backpropagation as, 210–11
  linear, 190–96
  supervised learning as, 185–96
    linear (least-mean-square (LMS)
        algorithm), 190–96, 212
    perceptron, 187–90, 212
Stochastic convergence, 126–27
Stochastic differential equations, 114
Stochastic equilibrium, 133–37, 228–32
Stochastic synapses, 132–33
Stochastic unsupervised learning, 131–37
Strong law of large numbers, 127, 155–56
Structural stability, 242–43
Structured knowledge, neural vs. fuzzy
    representation of, 304–6
Styblinski, M. A., 152, *172*
Subimage classification. *See* Transform
    image coding
Subset asynchrony, 64–66
Subsethood, 9–11, 280–81
  degrees of, 279–80
  elementhood subsumed by, 288
  orthogonality conditions, 288
  probability reduced to, 287
Subsethood measure, 285–87
Subsethood theorem, 263, 278–93
  corollaries to, 263, 291–92
  derivation of
    algebraic, 280–82
    geometric, 263, 282–87
  fuzzy Bayes theorem and, 290–93
Subsets, convex, 188
Success, probability of, 11
Success ratio, 287

Summed absolute error, 211
Superimposition of FAM rules, 313–14
Supersethood, 280–81, 284
Supervised competitive learning (SCL)
    law, 135, 185, 230
Supervised feedforward models, 17
Supervised learning, 23–24, 179–220, 221,
    225
  associative-reward-penalty algorithm,
    212
  backpropagation algorithm, 181, 188,
    196–212, 222, 327
    derivation, 203–10
    feedforward sigmoidal representation
      theorems, 199–201
    history of, 196–99
    multilayer feedforward network
      architectures, 201–3
    robust, 211–12
    as stochastic approximation, 210–11
  backpropagation control systems,
    345–48, 349, 356, 357
  competitive (SCL), 166–67, 222, 228
  geometric algorithm, 213
  neocognitron network, 212–13
  in neural networks, 113–15
  as operant conditioning, 181–82
  in pattern recognition, 125, 184–85
  as stochastic approximation, 185–96
    linear (least-mean-square (LMS)
      algorithm), 190–96, 212
    perceptron, 187–90, 212
  as stochastic pattern learning with
    known class membership, 183–85
  supervised function estimation, 180–81
  unsupervised learning distinguished
    from, 113–14
Supervised reinforcement function, 166
Suprathreshold, 58
Suprathreshold signals, 49
Sutton, R. S., 160, *170*, *172*, *213*
Symbolic vs. numeric processing, 25–26
Symbols, numbers vs., 24
Synapse(s), 14, 42. *See also* Learning
  change in, 112
  drive-reinforcement, 160

  excitatory, 60
  fluctuation rate, 235
  Hebbian, 134, 138
  inhibitory, 60
  local information available to, 114–15
  long-term memory encoding by, 47
  signal velocity computation, 52
  stochastic, 132–33
Synaptic connection matrices, 59–60
Synaptic connection topologies, 221
Synaptic convergence to centroids, 225–32
Synaptic dynamical system, 22
Synaptic efficacies, 60
Synaptic junction, 42–43
Synaptic vectors, 146, 180
  Browning wandering of, 116
  competitive, 231–32
  equilibrium, 229
  equinorm property of, 147–48
Synaptic web, neural network
    representation of, 22
Synchronous state changes, 64
Szu, H., 253, *257*

T

Taber, W. R., 152, *172*, 306, *336*
Tank, D. W., 47, *53*, 234, *256*
Target-tracking control systems, 379–406
  fuzzy controller, 382–92
    adaptive FAM (AFAM), 402–5
    control surfaces, 394–95, 405
    FAM rules, 383–86, 390–91
    fuzzy centroid computation, 386–90
    fuzzy set values, 382–83
    implementation, 390–92
    performance, 397–98, 399, 400
    sensitivity and robustness of,
      399–401, 403, 404
    software instructions, 419–21
  Kalman-filter controller, 392–96
    control surfaces, 395–96, 397
    performance, 398, 399, 401, 402
    sensitivity and robustness of, 399, 403
  real-time target tracking, 381–82
  simulation results, 396–406

Tekolste, R., 68, 102, *104*
Television, high-definition (HDTV), 364
Temporal associative memories (TAMs), 143, 153–54
Thom, 242, 257
Thompson, R. F., 13, *35*, 115, *172*
Three-valued fuzziness, 3
Threshold exponential signal functions, 49
Threshold linear signal functions, 49
Threshold signal functions, 39–40, 48, 49, 142–43
Time in neuronal dynamical systems, 45
Time-varying patterns, 183
Togai, M., 309, 316, *336*
Togai InfraLogic, 407–8
Tom, A., 364, *378*
Trainable dynamical systems, 13, 14–17
Trajectory in state space, 46
Transformation, 19
Transform image coding, 363–78
    adaptive cosine transform coding, 364, 365–66
    adaptive fuzzy associative memory (AFAM) systems for, 366–77
        differential competitive learning (DCL), 368–69, 370–73
        fuzzy set values, 367–68
        product-space clustering to estimate FAM rules, 368–69
        simulation, 373–74
Transversality techniques, 242–43
Tree(s)
    decision, 32–33
    dendritic, 42
    knowledge, 25
    rule, 24–25
Triangular norms (T-norms), 277–78
Truck backer-upper control systems, 339–61
    backpropagation neural, 345–47
        sensitivity and robustness of, 347–48, 349, 356, 357
    fuzzy, 340–45, 346–47
        adaptive, 343, 348–52, 356–60
        sensitivity and robustness of, 347–48, 349, 356, 357

    software instructions, 418–19
    with trailer, 352–60
Truth
    degrees of, 4–5
    factual, 269
    logical, 268–69, 278–79
Tsypkin, Ya. Z., 136, 161, *172*, 179, 185, 198, *215*
Turquette, A. R., 3, *35*

U

Unbiased estimators, 165
Uncertainty
    math-model-controller representation of, 380
    as randomness, 264, 266–67
Underlap, 7, 269, 271, 276–77, 292, 293–94
Unidirectional networks, 60–61
Uniform convergence, 126, 199
Uniform sampling probability, 112–13
Unit (identity) operator, 120
Unit sphere, 148
Unit step function, 129
Universe as fuzzy set, 268–69
Universes of discourse, 300, 301, 323
Unmodelled effects, 115. *See also* Noise
Unsupervised learning, 23–24, 111–77, 221. *See also* Differential competitive learning (DCL); Differential Hebbian learning
    algorithms, 225
    as classical conditioning, 182
    competitive (UCL), 114, 145–51, 175–76, 222, 227. *See also* Target-tracking control systems
        asymptotic centroid estimation, 148–49
        competitive covariance estimation, 149–51
        as correlation detection, 147–48
        as indication, 146
        software problems, 175–76

Unsupervised learning *(cont.)*
  laws of, 115–19
    competitive, 115, 116–17, 238
    differential competitive, 115, 117–18, 137
    differential Hebbian, 115, 117, 137, 167–68
    forgetting law, 138
    linear competitive, 115, 117, 146
    signal Hebbian, 115, 116, 118, 131, 134, 165
    stochastic version, 118, 135–37
  in neural networks, 113–15
  in pattern recognition, 125, 184–85
  probability spaces and random processes, 119–31
    Gaussian white noise as Brownian pseudoderivative process, 127–31
    measure theory, 119–27
  signal Hebbian, 138–45
  stochastic, 131–37
  structural stability of, 242–43
  supervised learning distinguished from, 113–14
Unsupervised networks, 91–92. *See also specific network types*
Upadhaya, B. R., 63, *105*

V

Vagueness, 6
Variables
  fuzzy, 301
  random, 122
    conditionally independent, 125
    convergence of, 126–27
    indexed, 119
    uncorrelated, 124–25
Variance, 124
Variation, 21
Vector(s), 122
  cross-correlation, 191–92
  novelty, 84, 85
  random, 119, 124, 150
Vector-matrix multiplication, fuzzy, 307–8
Velocity(ies)

activation, 41
angular, 317
mean-squared, 247
signal, 41
Velocity-difference property, 167
Venkatesh, S. S., 91, *105*
Virtual representation, 28, 31–32, 305
Visual system, human, 2
Voltage, 58
Voltage drop, 56
Voltage law, Kirchhoff's, 57
Von der Malsburg, C., 116, 136, *172*
von Neumann, J., 3, *34*

W

Wang, Y. F., 63, *105*
Watanabe, H., 309, 316, *336*
Weak law of large numbers, 127
Weber law of sensation, 97
Wee, W. G., 81, *105*
Weighted outer-product law, 80–81
Werbos, P. J., 198, *215*
White, H., 181, 198, 199, 201, 210, 211, *214*, *215*
White noise, 118–19, 127–31, 132
Wide-sense stationary (WSS) random process, 128–29, 191
Widrow, B., 181, 190–91, 194, 195, 196, 198, *215*, 339, 345, *361*
Wiener, N., 12, *35*
Wiener process, 127–31
Williams, R. J., 210, *215*
Williams, W. E., 152, *173*
Wilson, E. O., 22, 24, *35*, 77, *105*
Winston, P. H., 153, *173*
Winter, R., *215*
Wittgenstein, L., 268, *295*
Woods, J. W., 129, 162, *172*

X

XOR software problem, 218–19

Y

Yager t-norm, 297
Yamaguchi, T., 152, *170*
Yamakawa, T., 316, *336*
Yamanaka, Y., 152, *170*
Yao, Y., 15, *35*

Z

Zadeh, L. A., 6, *35*, 269, 270, 274, 277,
    279, 291, *295*, 309, *336*
Zangi, K., 364, *378*
Zero-mean assumption, 247
Zero-mean property, 169
Zeroth-order topology, 44
Zhang, W., 152, *173*
Zipser, D., 136, *172*, 210, *215*

ART.EXE, BAM.EXE, BKP.EXE, CL.EXE and RABAM.EXE by Olmsted & Watkins for NEURAL NETWORKS AND FUZZY SYSTEMS by Bart Kosko

**TRUCK.EXE by HyperLogic Corporation for NEURAL NETWORKS AND FUZZY SYSTEMS
by Bart Kosko**

YOU SHOULD CAREFULLY READ THE FOLLOWING TERMS AND CONDITIONS BEFORE OPENING THIS DISKETTE PACKAGE. OPENING THIS DISKETTE PACKAGE INDICATES YOUR ACCEPTANCE OF THESE TERMS AND CONDITIONS. IF YOU DO NOT AGREE WITH THEM, YOU SHOULD PROMPTLY RETURN THE PACKAGE UNOPENED.

Prentice-Hall, Inc. provides this program and licenses its use. You assume responsibility for the selection of the program to achieve your intended results, and for the installation, use, and results obtained from the program. This license extends only to use of the program in the United States or countries in which the program is marketed by duly authorized distributors.

LICENSE
You may:

a.   use the program;
b.   copy the program into any machine readable form without limit but without redistribution

LIMITED WARRANTY
THE PROGRAM IS PROVIDED "AS IS" WITHOUT WARRANTY OF ANY KIND, EITHER EXPRESSES OR IMPLIED, INCLUDING, BUT NOT LIMITED TO, THE IMPLIED WARRANTIES OF MERCHANTABILITY AND FITNESS FOR A PARTICULAR PURPOSE. THE ENTIRE RISK AS TO THE QUALITY AND PERFORMANCE OF THE PROGRAM IS WITH YOU. SHOULD THE PROGRAM PROVE DEFECTIVE, YOU (AND NOT PRENTICE-HALL, INC. OR ANY AUTHORIZED DISTRIBUTOR) ASSUME THE ENTIRE COST OF ALL NECESSARY SERVICING, REPAIR, OR CORRECTION.

SOME STATES DO NOT ALLOW THE EXCLUSION OF IMPLIED WARRANTIES, SO THE ABOVE EXCLUSION MAY NOT APPLY TO YOU. THIS WARRANTY GIVES YOU SPECIFIC LEGAL RIGHTS AND YOU MAY ALSO HAVE OTHER RIGHTS THAT VARY FROM STATE TO STATE.

Prentice-Hall, Inc. does not warrant that the functions contained in the program will meet your requirements or that the operation of the program will be uninterrupted or error free.

However, Prentice-Hall, Inc., warrants the diskette(s) on which the program is furnished to be free from defects in materials and workmanship under normal use for a period of ninety (90) days from the date of delivery to you as evidenced by a copy of your receipt.

LIMITATIONS OF REMEDIES
Prentice-Hall's entire liability and your exclusive remedy shall be:

1.   the replacement of any diskette not meeting Prentice-Hall's "Limited Warranty" and that is returned to Prentice-Hall, or

2.   if Prentice-Hall is unable to deliver a replacement diskette or cassette that is free of defects in materials or workmanship, you may terminate this Agreement by returning the program.

IN NO EVENT WILL PRENTICE- HALL BE LIABLE TO YOU FOR ANY DAMAGES, INCLUDING ANY LOST PROFITS, LOST SAVINGS, OR OTHER INCIDENTAL OR CONSEQUENTIAL DAMAGES ARISING OUT OF THE USE OR INABILITY TO USE SUCH PROGRAM EVEN IF PRENTICE-HALL OR AN AUTHORIZED DISTRIBUTOR HAS BEEN ADVISED OF THE POSSIBILITY OF SUCH DAMAGES, OR FOR ANY CLAIM BY ANY OTHER PARTY.

SOME STATES DO NOT ALLOW THE LIMITATION OR EXCLUSION OF LIABILITY FOR INCIDENTAL OR CONSEQUENTIAL DAMAGES, SO THE ABOVE LIMITATION OR EXCLUSION MAY NOT APPLY TO YOU.

GENERAL
You may not sublicense, assign, or transfer the license of the program except as expressly provided in this Agreement. Any attempt otherwise to sublicense, assign, or transfer any of the rights, duties, or obligations hereunder is void.

This agreement will be governed by the laws of the State of New York.

Should you have any questions concerning this Agreement, you may contact Prentice-Hall, Inc., by writing to:

> Prentice Hall
> College Division
> Englewood Cliffs, NJ 07632

Should you have any questions concerning technical support you may write to:

> HyperLogic Corporation
> 2411 East Valley Parkway
> Suite 294
> P.O. Box 3751
> Escondido, CA 92025

YOU ACKNOWLEDGE THAT YOU HAVE READ THIS AGREEMENT, UNDERSTAND IT, AND AGREE TO BE BOUND BY ITS TERMS AND CONDITIONS. YOU FURTHER AGREE THAT IT IS THE COMPLETE AND EXCLUSIVE STATEMENT OF THE AGREEMENT BETWEEN US THAT SUPERSEDES ANY PROPOSAL OR PRIOR AGREEMENT, ORAL OR WRITTEN, AND ANY OTHER COMMUNICATIONS BETWEEN US RELATING TO THE SUBJECT MATTER OF THIS AGREEMENT.

Demo Software by Togai InfraLogic, Inc for
Neural Networks and Fuzzy Systems
by Bart Kosko